# Handbook of Facial Growth

**DONALD H. ENLOW, Ph.D.**

*Professor and Chairman, Department of Anatomy,*
*West Virginia University School of Medicine,*
*Morgantown, West Virginia*

*WITH CONTRIBUTIONS BY*
**ROBERT E. MOYERS, D.D.S., Ph.D.** *and* **WILLIAM W. MEROW, D.D.S.**

*ILLUSTRATIONS BY*
**WILLIAM ROGER POSTON, II**
*Supervisor, Biomedical Illustration, West Virginia University Medical Center*

W. B. SAUNDERS COMPANY / Philadelphia / London / Toronto

W. B. Saunders Company:  West Washington Square
Philadelphia, PA 19105

1 St. Anne's Road
Eastbourne, East Sussex BN21 3UN, England

1 Goldthorne Avenue
Toronto, Ontario M8Z 5T9, Canada

**Library of Congress Cataloging in Publication Data**

Enlow, Donald H

Handbook of facial growth.

Bibliography: p.

Includes index.

1. Facial bones – Growth.    2. Face.    I. Moyers, Robert E.,
   joint author.    II. Merow, William W., joint author.
   III. Title. [DNLM: 1. Maxillofacial development.
   WE705 E58ha]

QM535.E48    612'.75    75–293

ISBN 0–7216–3385–4

Handbook of Facial Growth.                                        ISBN 0-7216-3385-4

Last digit is the print number:     9    8    7    6    5    4    3

*To my parents,*
*to Martha,*

*and to three great anatomists and teachers*
*who gave me a philosophy:*

HORACE J. SAWIN

ROBERT S. BENTON

WILFRID T. DEMPSTER

# Preface

This book is a **primer** on the growth of the bones in the face and cranium and the overall process of postnatal facial development. It is not intended as an advanced-level reference work. It does not provide a comprehensive account of the enormous range of subjects in all areas of facial growth. It is not a textbook of cephalometrics. It does not present an exhaustive review and analysis of the research literature. It is not a compendium of statistical "growth data." This book is structured, rather, to provide a basic-level, introductory type of text especially geared for dental and medical students. It deals largely with the fundamental concepts of the facial growth process. Significantly, no such textbook now exists. Yet, in some medical fields and especially in dentistry, "facial growth and development" is among the basic subjects to be regarded as essential.

Some schools offer a separate undergraduate course, or part of a course, dealing with facial growth. Many schools at present do not. One understandable reason is that no single, suitable textbook is at present available. Instructors must pick and choose from an awesome literature to provide reading lists. It is always difficult for a student starting essentially from scratch to appreciate the **basics** this way. Almost all such reading assignments are references from the research literature or chapters from advanced and specialty texts. For the advanced orthodontic student, this is appropriate and proper. For the **beginning** orthodontic student, however, and certainly for all undergraduates, something else is needed.

Many schools provide very little formal instruction on facial growth during the undergraduate period. What is given usually involves only the introductory cephalometrics taught during undergraduate orthodontic and pedodontic courses. If a suitable text were available, however, and if this text were understandable and usable by first-year undergraduates and **beginning** postdoctoral students, the faculty would feel more comfortable in asking their students to become familiar with the basics of the subject.

This text is especially designed for such introductory purposes. While it is hoped that all dental schools will be fortunate enough, someday soon, to have a separate, formal course on Facial Growth and Development (which is obviously appropriate for a dental school curriculum), it is appreciated that there are always scheduling problems. Information dealing with this timely subject must sometimes be incorporated within other, existing courses. With this in mind, the *Handbook of*

*Facial Growth* has been structured to provide the following:

1. It is geared to be understandable by the first-year undergraduate student. The amount of information included and the complexity involved is such that the busy student will not be overwhelmed. The book is also intended as a "primer" for advanced students beginning in specialty fields such as orthodontics, pedodontics, oral surgery, and plastic surgery.

2. Most of the chapters have two parts. Part One is an introductory overview, that is, a digest providing a condensed or abridged version of Part Two. Part Two then presents a more in-depth account and an elaboration of the same overall range of information. More detailed explanations of the underlying theory for the various growth processes are given.

3. If a separate, formal course is provided for either the undergraduate or graduate student, the entire text can be used from front to back. It will accommodate a semester-long lecture or seminar course having one or two meetings a week. If used in more advanced training programs, many instructors will also wish to give supplementary reading assignments in appropriate places so that the student can sample the research literature.

4. If a separate course, or a block included within some other course, is given but time is insufficient for a full-length treatment, the instructor may wish to use only the Part One (abridged version) of each chapter. This would be appropriate for a special "short course" or a two- or three-week period within some other course. The abridged Part One of each chapter can be utilized for postgraduate seminars or continuing education courses in which "facial growth" is a central theme.

5. Instructors in certain specialty undergraduate and graduate courses may wish to utilize particular, selected chapters or parts of chapters because of relevance to parts of their own course content. Oral surgery courses, for example, can refer specifically to sections on the mandibular condyle, the nasal septum, and other such appropriate considerations. Prosthodontic courses can use the parts dealing with bone remodeling processes and the effects of edentulism. Anatomy courses can select the chapter describing facial embryology and some sections covering the postnatal growth process. Periodontic courses may wish to use the parts dealing with the periodontal membrane and other soft tissues. Undergraduate orthodontic and pedodontic courses can use the chapters on cephalometrics, the abridged Part One introductions, and selected topics from the Part Two (more detailed) sections. Radiology courses can utilize the sections on cephalometrics and interpretations of headfilms. The Part One sections can be used in orthodontic and anatomy courses for dental hygiene students. Medical students, residents, and practitioners in neurosurgery, plastic surgery, ENT, and pediatrics will find most of the subject areas useful and relevant.

In any event, the first-year undergraduate will have multiple occasions to use the book, either all at once or part-by-part as the student progresses through the curriculum. Even if only the Part One sections are required at the undergraduate level, many students will find themselves looking ahead to Part Two of each chapter for more information, because the subject is interesting and relevant. It provides a

handbook source, also, whenever the subject of facial growth comes up, which may occur frequently in many courses. The book will continue to be utilized by students in postgraduate specialty programs.

The book is purposely concise, and it is composed largely of explanatory diagrams accompanied by short blocks of descriptive text. It provides a building of concepts one upon another. The idea is to emphasize easily understandable diagrams rather than long pages of burdensome narrative; it is designed to allow one to learn the basics of facial growth in the shortest possible time.

DONALD  H.  ENLOW

# Acknowledgments

The author is greatly indebted to Mr. William Roger Posten, II for his skillful production of many of the new illustrations in this book. We both give our thanks to Mrs. Virginia Swecker for her considerable help with some of the drawings. The new photographs were prepared by Mr. Leland Bowerman and Mr. Lester C. Bond. Several illustrations from the author's previous works have been used, and these were prepared by Mr. William L. Brudon. I am very grateful to my secretarial staff for their tolerant and always diligent efforts: Ms. Mary Martha Kent, Marylynne Ronk, Linda Rogan, MaryBeth Sustarsic, and Joetta Tustin. We all wish to acknowledge the most cordial and helpful assistance, advice, and cooperation of the editors and the staff of Saunders. A series of grants by the National Institute of Dental Research and the National Institute of Child Health and Human Development aided the author in the original, long-term research studies that are the basis for many parts of this book.

# Contents

# Introduction

## THE CHANGING FEATURES OF THE GROWING FACE

The "baby face" has large-appearing eyes, dainty jaws, a smallish pug nose, puffy cheeks with buccal fat pads, a high intellectual-like forehead without coarse eyebrow ridges, a low nasal bridge, a small mouth, velvety skin, and overall wide and short proportions. It is a cute face. It warms the cockles of parental hearts. A parent can worry, though, because the otherwise great little face "has no chin," or the "jaw is much too small," or "the eyes are too far apart." However, these and many of the other features of the baby's face gradually undergo marked changes as the face grows and develops through the years. The chin develops, jaw size catches up, and the eyes appear less wide-set. From the many possible variations that can exist among different individuals, a person's own facial characteristics take on, month by month, definitive adult form. The general features of any fully grown face are quite different from those of the same individual as an infant and young child. Trying to decide which parent the infant "looks like" or which uncle it "takes after" is a fun game but usually a more or less futile one. There is little in the general shape and proportions of the infantile face, at least topographically, to give a hint as to what form it will take in later years.

Outlined below are some major features that contrast the characteristics of the child and adult faces. Study them carefully. Later chapters will explain the actual growth processes that underlie them.

**FIGURE 1–1**

The baby's face appears quite diminutive relative to the much larger, more precocious cranium above and behind it. The respective proportions change significantly, however. The growth of the brain slows considerably after about the third or fourth years of childhood, but the facial bones continue to enlarge markedly for many more years.

The eyes appear large in the young child. As facial growth continues, however, the nasal and jaw regions grow much faster and to a much greater extent than the orbit and its soft tissues. As a result, the eyes of the adult appear smaller in proportion.

The ears of the infant and child appear to be low; in the adult they are much higher with respect to the face. Do the ears actually rise? No; they in fact move downward during continued growth. However, the face enlarges inferiorly even farther, so that the **relative** position of the ears seems to rise.

1

The young child's forehead is upright and bulbous. The forehead of the adult is much more sloping (the amount of slope is sex- and ethnic-related). The forehead region of the child seems very large and high because the face beneath it is still relatively small. The child's forehead continues to enlarge during the early years, but the face enlarges much more, so that the proportionate size of the forehead becomes reduced.

**FIGURE 1–2**

The childhood face appears quite broad. As development continues, **vertical** facial growth bypasses expansion in width to a marked extent, so that a much more narrow facial proportion characterizes the adult. (Even so, some adult faces can still appear rather wide and round. This is because of a wider and rounder type of brain configuration, a common variation. Such faces thus have a more juvenile-like appearance.)

The nasal bridge is quite low in the child. It rises (to a greater or lesser extent in different facial types) to become much more prominent in many adults.

The eyes of the infant seem quite wide-set with a broad-appearing nasal bridge between them. This is because the nasal bridge is so low, and also because much of the width of the bridge has already been attained in the infant. With continued growth, the eyes spread further laterally, but only to a relatively small extent. Actually, the eyes of the adult face are not much farther apart than those in the child. Because of the higher nasal bridge, the increase in the vertical facial dimension, and the widening of the cheekbones, the eyes of the adult thus **appear** much closer together.

**FIGURE 1–3**

The infant and young child have much more of a "pug nose." It protrudes very little and is vertically quite short. The shape and size of the infantile nose, however, give little indication as to what will happen to it during subsequent growth. The lower part of the nose in the adult is proportionately much wider and a great deal more prominent. The extent is ethnic-related.

The whole nasal region of the infant is vertically shallow. The level of the nasal floor lies close to the inferior orbital rim. In the adult, the midface has become greatly expanded, and the nasal floor has descended well below the orbital floor. This change is quite marked because of the enormous enlargement of the nasal chambers. Note the close proximity of the young child's maxillary arch to the orbit, in contrast to their positions in the adult.

The superior and inferior orbital rims of the young child are in an approximately vertical line. Because of the unique human forehead, frontal sinus development, and supraorbital protrusion, however, the upper orbital rim of the adult noticeably overhangs the lower. The orbital opening becomes inclined obliquely forward.

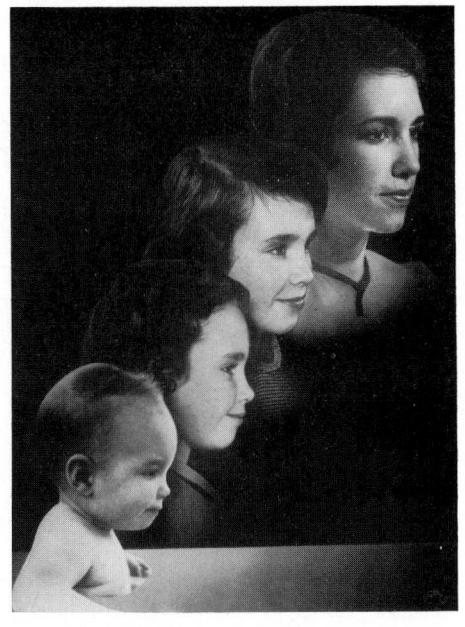

**FIGURE 1–1**

*(Courtesy of William L. Brudon. From Enlow, D. H.: The Human Face. New York, Harper & Row, 1968.)*

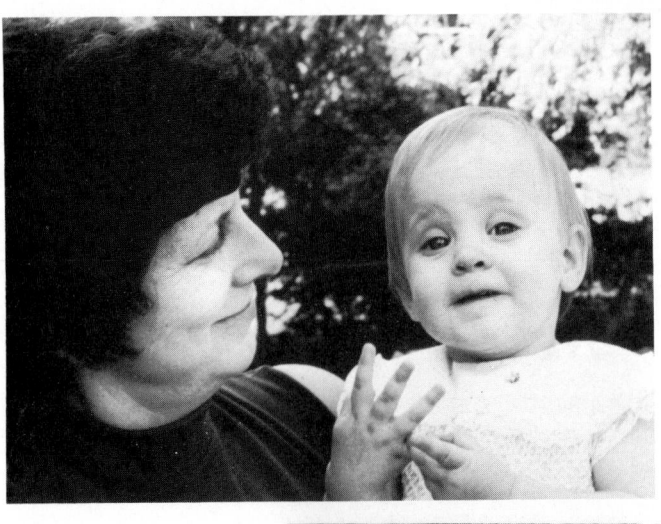

**FIGURE 1–2**

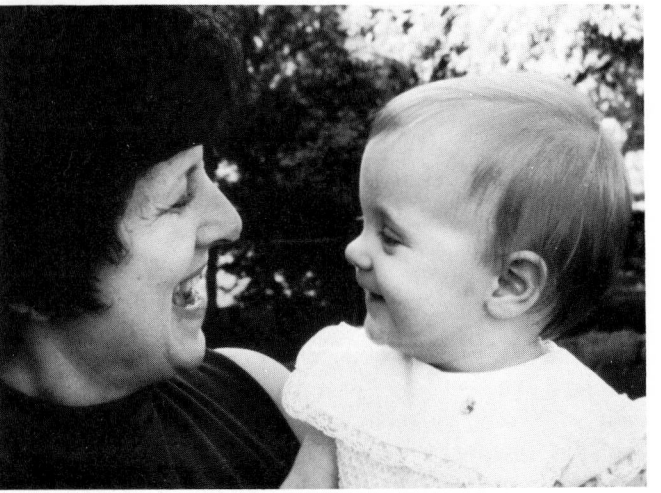

**FIGURE 1–3**

**FIGURE 1–4**

Below the orbit, the nasal chambers in the adult face expand laterally nearly halfway across the orbital floor. In the infant, the breadth of the nasal cavity scarcely exceeds the width of the nasal bridge. During subsequent growth, the inferior portion of the nose expands laterally much more than the superior part.

The tip of the infant's nasal bone protrudes very little beyond the inferior orbital rim. The area **between** the nasal tip and the inferior rim of the orbit (that is, the lateral bony wall of the nose) is characteristically narrow. In the adult, this area becomes markedly expanded. The divergent directions of orbital, nasal, cheekbone, and maxillary arch growth "draw out" the contours among them.

**FIGURE 1–5**

The lateral orbital rim and the cheekbone in the child are more forward-appearing because the whole face is relatively flat and wide. Because of an actual "regressive" mode of growth, however, these facial parts come to lie in a less prominent position in the adult face. In the infant, the protrusive appearance of the cheekbone is augmented by the characteristic infantile buccal fat pad in the hypodermis of the cheek. Adults tending to have a relatively wide, short (thus more childlike) type of face typically show an even greater "cherubic" appearance if they are overweight; the buccal region contains adipose resembling the buccal fat pad of infancy.

While the cheekbone is prominent in early childhood, it is nonetheless quite diminutive and fragile compared with that of the adult. The malar process and the inferior part of the zygoma enlarge considerably during childhood growth, even though they actually grow in a **backward** direction. Because of the differential extents and directions of growth in other parts of the face, these growth increases by the zygoma are often masked. The **protrusive** modes of supraorbital and nasal growth cause the adult forehead and nose to appear progressively more prominent relative to the **retrusively** growing cheekbones and lateral orbital rims. This feature is more noticeable in the male. Compare with Figure 1–7.

The entire face of the adult is much deeper horizontally because of these divergent directions of growth among all the various regional parts. The whole face is **drawn out** in many directions. The adult face has much bolder topographic features, and it is much less "flat."

**FIGURE 1–6**

As the whole face expands, the frontal, maxillary, and ethmoidal sinuses enlarge to occupy spaces not otherwise functionally utilized. Architecturally, the sinuses are leftover "dead" (unused) spaces. They were not created especially to provide "resonance to the voice," nasal drip, or other special functions, although they have become secondarily involved in such roles.

The mandible of the young child appears quite small and "underdeveloped" relative to the upper jaw and the face in general. It is small not only in actual size but also proportionately. The mandible normally lags in early growth, however, and it will usually more or less catch up to the maxilla unless a malocclusion is "programmed" to develop. Because of this, it is sometimes difficult to predict during early childhood possible malocclusions that might or might not become fully expressed during later development. Compare with Figures 1–5, 1–7, and 1–8.

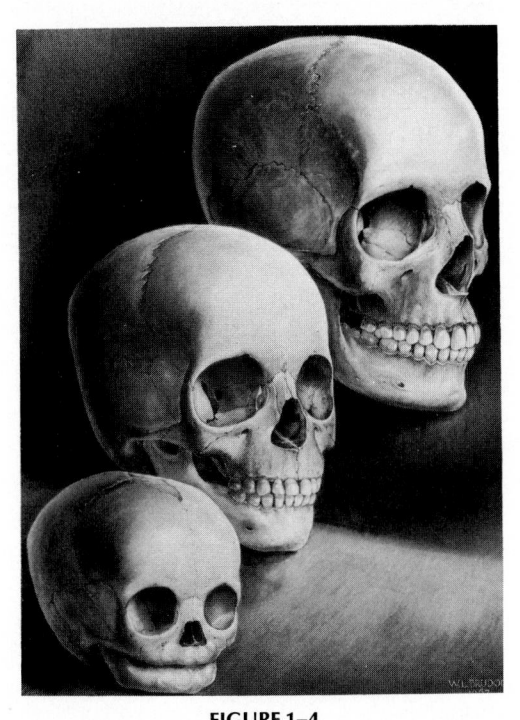

**FIGURE 1–4**

*(Courtesy of William L. Brudon. From Enlow, D. H.:* The Human Face. *New York, Harper & Row, 1968.)*

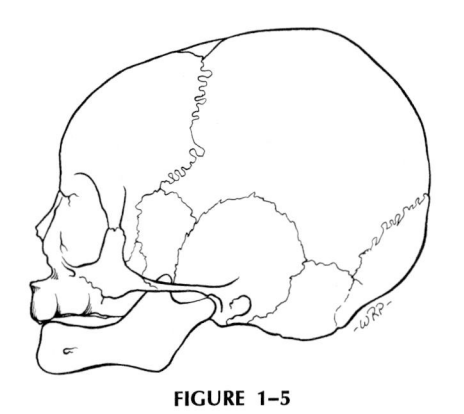

**FIGURE 1–5**

The chin is poorly formed in the infant. Because of remodeling changes that gradually take place, however, the chin becomes more prominent year by year. A "cleft" is sometimes formed in the **fleshy** part of the chin (not usually in the bone itself) when the two sides of the lower jaw fuse during early development. The cleft deepens when the soft tissues of the two sides then continue to expand postnatally. For some reason, this facial feature has become adopted in our society as a symbol of masculinity when it is present in the male. Its presence in the female has no social significance one way or the other.

The young child's mandible appears to be pointed. This is because it is wide, short, and more "V"-shaped. In the adult, the entire lower jaw becomes "squared." With the development of the chin, together with massive growth in the lateral areas of the trihedral eminence, eruption of the permanent dentition, lateral enlargement of each ramus, expansion of the masticatory musculature, and flaring of the gonial regions, the whole lower face takes on a "U"-shaped configuration, resulting in a **considerably** more full appearance. Compare with Figure 1–8.

In the infant and young child, the gonial region lies well inside (medial to) the cheekbone. In the adult, the posteroinferior corner of the mandible extends laterally out to the cheekbone, or nearly so. This gives the posterior part of the jaw a square appearance.

### FIGURES 1–7 AND 1–8

The ramus of the adult mandible is much longer vertically. It is also more upright (this refers to the ramus as a whole and not the misleading "gonial angle" measurement). The elongation of the ramus accommodates the massive vertical expansion of the nasal region and the eruption of the permanent teeth.

The premaxillary region normally protrudes beyond the mandible in the infant and young child, and it lies in line with or forward of the bony tip of the nose. This gives a prominent appearance to the upper jaw and lip. In subsequent facial development, however, the nose becomes much more protrusive, and the tip of the nasal bone comes to lie well ahead of the basal bone of the premaxilla. The lower jaw later catches up and matches most or all of the increased growth of the maxilla.

The forward surface of the bony maxillary arch in the infant has a vertically convex topography. This is in contrast to the characteristically concave contour of this region in the adult. The alveolar bone in this area of the adult face is noticeably more protrusive and proportionately much more massive (in conjunction with the permanent dentition).

The whole face, vertically, is a great deal longer and more sloping as a result of the many changes outlined above.

The quite small mastoid process of the infant develops into the sizeable protuberance of the adult. A bony styloid process is also lacking in the newborn. The ring-shaped bone around the external acoustic meatus faces downward in the infant but is later rotated during growth into a more vertical position.

At birth, the overall length of the cranium is approximately 60 to 65 per cent complete, and it increases rapidly. By five years, it reaches about 90 per cent of its full size. Also, much of the adult width of the cranium is attained by the first or second year.

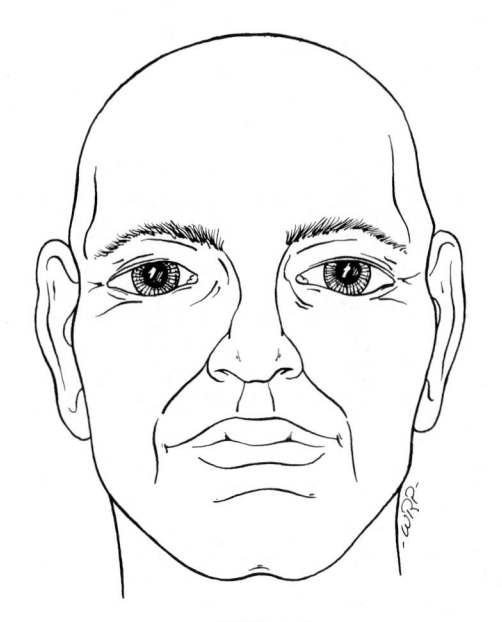

FIGURE 1–6

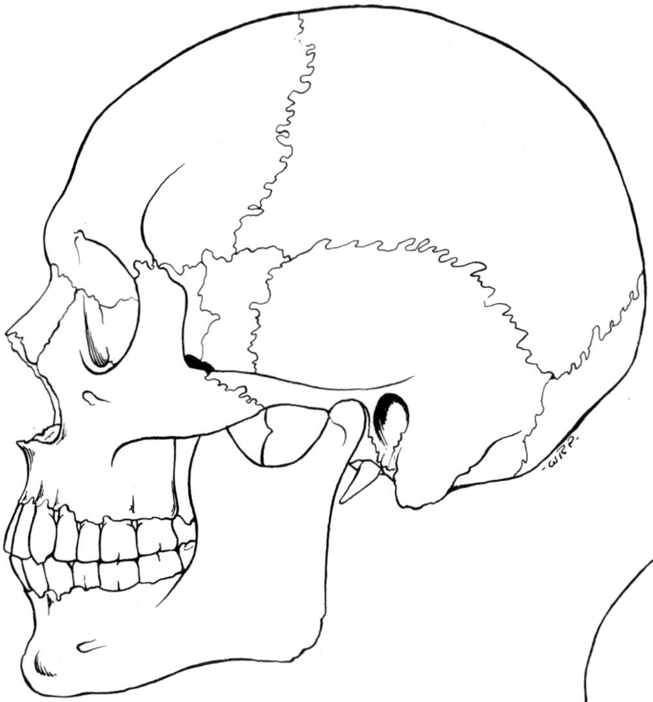

FIGURE 1–7

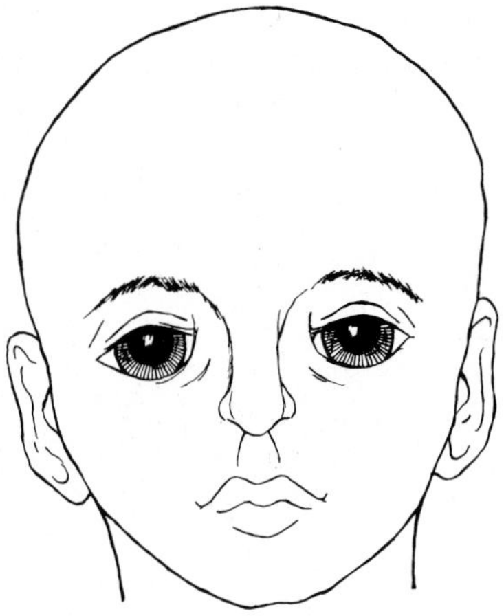

FIGURE 1–8

In the newborn, six fontanelles ("soft spots") are present between the bones of the skull roof. They cover over at different times, but all have been reduced to sutures by the eighteenth month. The sutures of the cranial vault are relatively nonjagged in the baby, and the outer surface of the bone is smooth. A much rougher bone texture characterizes the surface of the adult calvaria, and the suture lines become noticeably much more interlocking. The metopic suture (separating the right and left halves of the frontal bone) usually fuses by the second year.

In the child, the slender neck below a relatively large cranium, particularly in the occipital region, gives a characteristic "boyish" appearance to the whole head. This gradually disappears (to a greater or lesser extent) until about puberty, when the expansion of the neck muscles and other soft tissues causes a proportionate decrease in the prominence of the head relative to neck circumference. This is less noticeable in the female.

**FIGURE 1–9**

The external appearance of the baby's face does not reveal the truly striking enormity of the dental battery developing within it. The teeth are a dominant part of the infant's face as a whole, yet they are not even seen. The parent doesn't usually realize they are already there at all, much less suspect the massiveness of their extent. In this illustration, one is almost overwhelmed by the remarkable extent of **teeth** all over the midfacial region. The average civilian does not appreciate that the mouth of the little child is bounded by a virtual palisade of multitiered primary and permanent teeth in many stages of development. When a crown tip first protrudes through the gingiva as it erupts, the parent naturally believes that the process is just beginning and that the tooth is only a tiny but newsworthy addition to the pink mouth. It is not realized that the whole midface is occupied by a vast magazine of unerupted teeth hidden to the eyes. The thin covering and supporting **bone** of the jaws is a much less commanding feature of the young face.

**FIGURE 1–10**

How does one recognize advancing age in an **adult** face by external appearance? Among soft tissue features are the wrinkles, creases, and permanent lines that develop in characteristic locations, signaling one's years. One such line extends from just above each side of the wings of the nose obliquely outward and downward to the corners of the mouth. It is the line that appears when one smiles, at any age, but it becomes **permanent** and much more **marked** in middle age. Other permanent lines appear beneath the eyes. Lines radiating from the lateral corners of the eyes can sometimes develop. The skin of the forehead develops permanent horizontal creases, and vertical lines may also develop in the area above the bridge of the nose between the eyebrows. Jowl lines frequently occur about halfway along the length of the mandibular corpus, and lines can come down from the corners of the mouth to either side of the chin. The skin, in general, appears much less firmly bound to the hypodermis, and it tends to sag and droop. The "pores" of the skin in the older adult become larger and more noticeable. There are also characteristic skeletal changes. Some of these are described in the later section dealing with edentulism (Chapter 13). The feature we most often associate in one's superficial facial appearance with the onset of middle age, however, is the "smile line" described above.

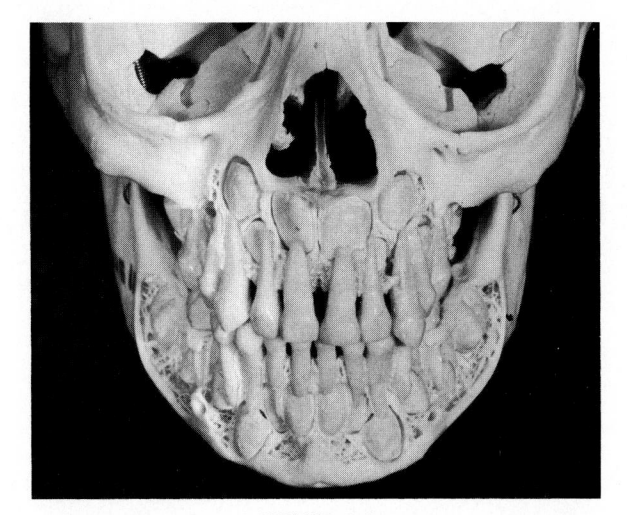

**FIGURE 1–9**

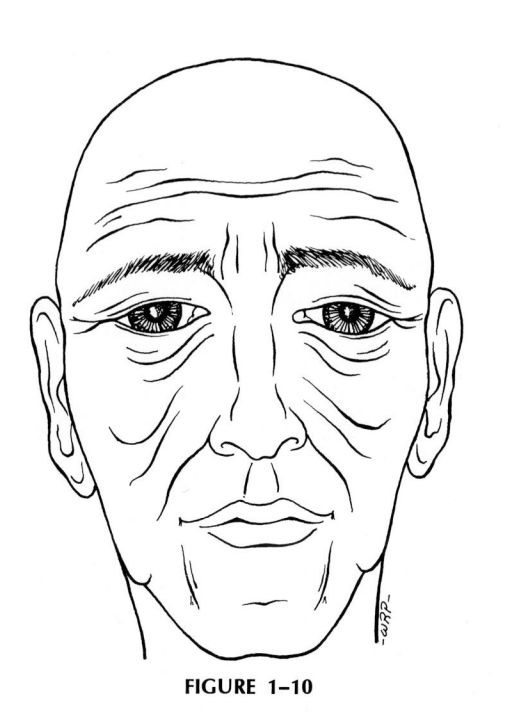

**FIGURE 1–10**

# Chapter 2

# Introductory Concepts of the Growth Process

# PART ONE*

_____

**FIGURE 2–1**

**Concept 1.** Bones grow by adding new bone tissue on one side of a bony cortex and taking it away from the other side. The surface facing toward the direction of progressive growth receives new bone deposits (+). The surface facing away undergoes **resorption** (−). This growth process is termed "drift." It produces a direct growth movement of any given area of a bone.

**FIGURE 2–2**

**Concept 2.** The outside and inside surfaces of a bone are completely blanketed by a mosaic-like pattern of "growth fields." Note that the outside surface, however, is **not** all "depository," as one might presume. About half of the periosteal (external) surface of a whole bone has a characteristic arrangement of **resorptive fields** (darkly stippled areas); a characteristic pattern of **depository fields** covers the remainder (lightly stippled areas). If a given periosteal area has a resorptive type of field, the opposite inside (endosteal) surface of that same area has a depository field. Conversely, if the periosteal field is depository, the endosteal field on the opposite side of the cortex is usually resorptive. These combinations produce the characteristic **growth movements** (that is, drift) of all parts of an entire bone. How can a bone enlarge if half of its **outside** surfaces actually undergo resorption? This is explained in the concepts that follow.

**FIGURE 2–3**

**Concept 3.** Bone produced by the covering membrane ("periosteal bone") comprises about half of all the cortical bone tissue present; bone laid down by the lining membrane ("endosteal bone") makes up the other half. In this diagram, note how the cortex on the right was formed by the periosteum and the cortex on the left by the endosteum as both sides shifted (drifted) in unison to the right.

_____

*Part One is an introductory overview. Part Two provides more in-depth information covering essentially the same range of subjects.

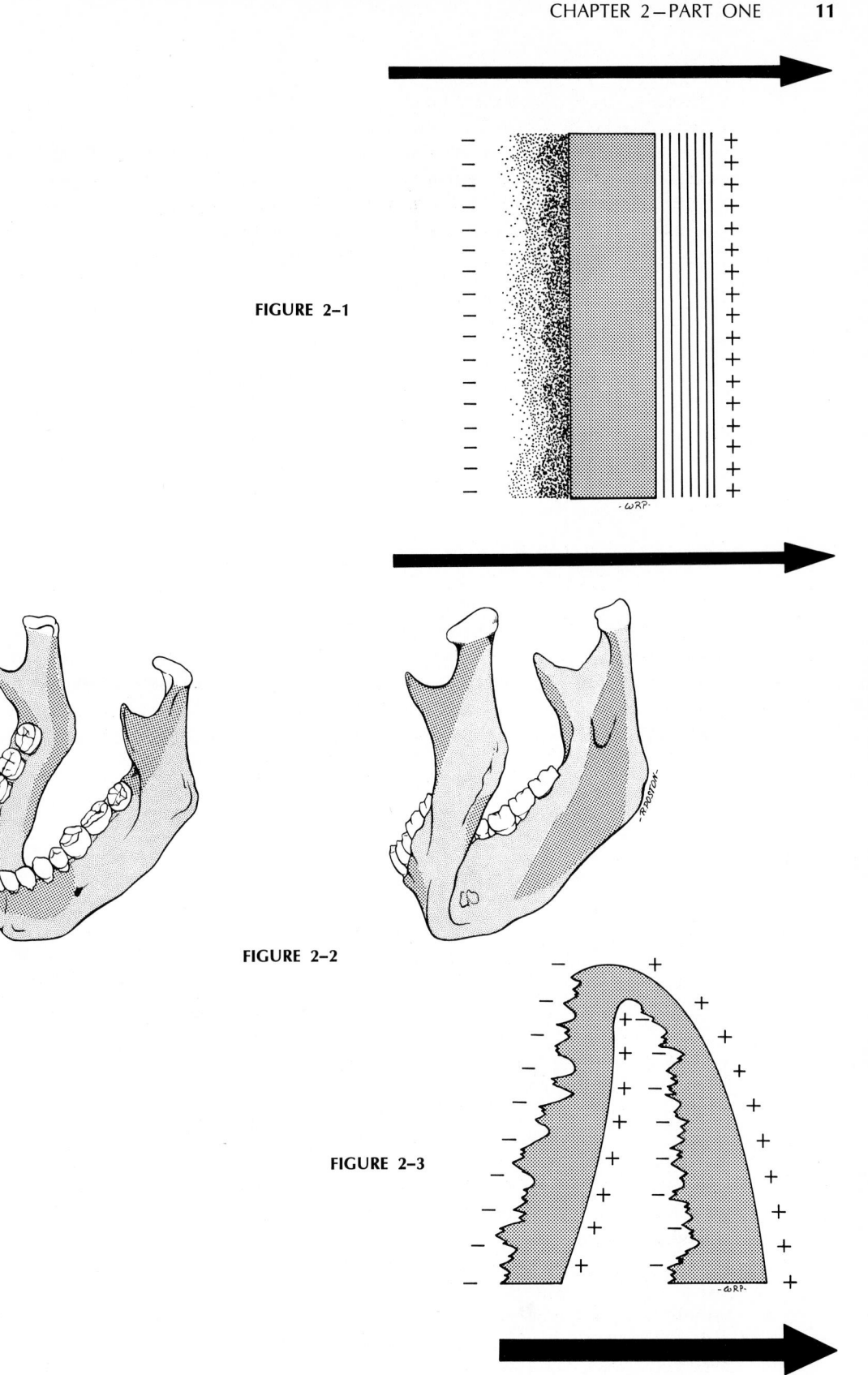

FIGURE 2-1

FIGURE 2-2

FIGURE 2-3

**FIGURE 2–4**

**Concept 4.** The operation of the growth fields covering and lining the surfaces of a bone is actually carried out by the **membranes** and other surrounding tissues rather than by the hard part of the bone. The bone does not "grow itself"; growth is produced by the **soft tissue matrix** that encloses each whole bone. The genetic and functional determinants of bone growth reside in the soft tissues. Growth is not "programmed" within the calcified part of the bone itself.

**FIGURE 2–5**

**Concept 5.** All the various resorptive and depository growth fields throughout a bone do not have the same **rate** of growth activity. Some depository fields grow much more rapidly or to a much greater extent than others. The same is true for resorptive fields. Fields that have some special significance or noteworthy role in the growth process are often termed growth **sites.** The mandibular condyle, for example, is such a growth site. Remember, however, that growth does not occur just at such special growth sites, as is sometimes presumed. The entire bone participates. **All** surfaces are, in fact, sites of growth, whether specially designated or not.

**FIGURE 2–6**

**Concept 6. Remodeling** is a basic part of the growth process. The reason why a bone must remodel during growth is because its regional parts become **moved;** "drift" moves each part from one location to another as the whole bone enlarges. This calls for sequential remodeling changes in the shape and size of each region. The ramus, for example, moves progressively posteriorly by a combination of deposition and resorption. As it does so, the anterior part of the ramus becomes **remodeled** into a new addition for the mandibular corpus. This produces a growth elongation of the corpus. This progressive, sequential movement of component parts as a bone enlarges is termed **relocation.** Relocation is the basis for remodeling. The whole ramus is thus relocated posteriorly, and the posterior part of the lengthening corpus becomes relocated into the area previously occupied by the ramus. Structural remodeling from what **used to be** part of the ramus into what then **becomes** a new part of the corpus takes place. The corpus grows longer as a result.

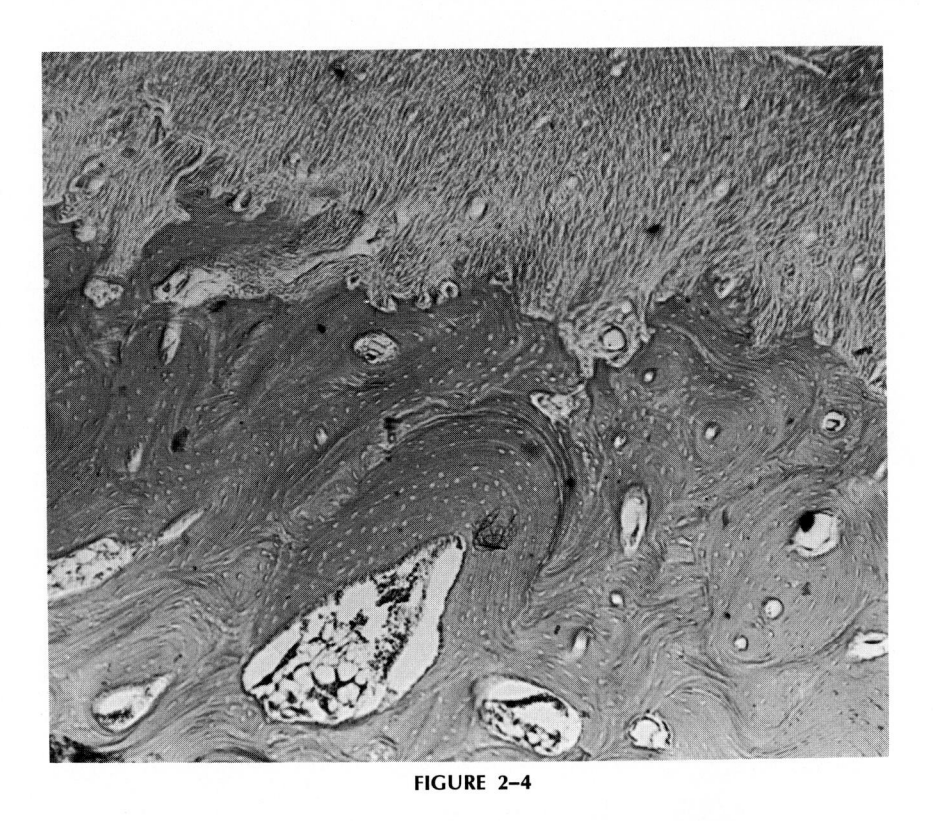

**FIGURE 2-4**

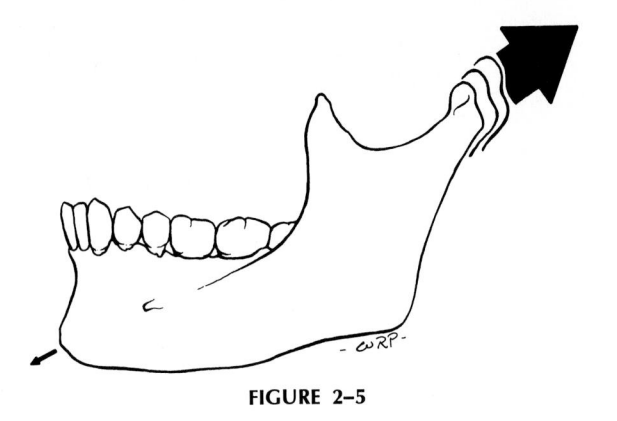

**FIGURE 2-5**

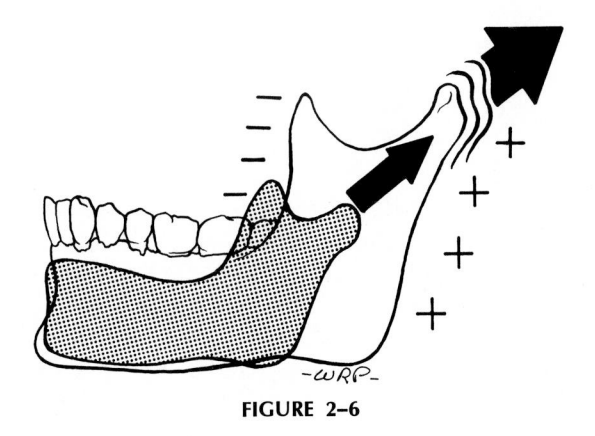

**FIGURE 2-6**

**FIGURE 2–7**

The same deposition and resorption which produce the overall growth enlargement of a whole bone carry out relocation and remodeling at the same time. Growth and remodeling are, in effect, inseparable parts of the same actual process. It can now be understood why about half of any given bone can and must have a resorptive **external** (periosteal) surface as the bone increases in overall size. The reason is that the bone does **not** simply enlarge symmetrically by new bone deposition uniformly over all external surfaces, as shown in the diagram. Rather, each of the regional parts of the bone becomes relocated into a sequentially new position. To do this, some outside surfaces are necessarily resorptive.

**FIGURE 2–8**

In the maxilla, the palate grows downward (that is, becomes **relocated** inferiorly) by periosteal resorption on the nasal side and periosteal deposition on the oral side. This growth and remodeling process serves to enlarge the nasal chambers. What used to be the bony maxillary arch and palate in early childhood are remodeled into what then becomes the nasal chambers of the adult. About half of the palate is thus resorptive and about half depository. The nasal mucosa provides the periosteum on one side, and the oral mucosa provides it on the other side.

**FIGURE 2–9**

**Concept 7.** As a bone enlarges, it is simultaneously carried away from other bones in direct contact with it. This creates the "space" within which bony enlargement takes place. The process is termed **primary displacement** (sometimes also called "translation"). It is a physical movement of a whole bone and occurs while the bone grows and remodels by resorption and deposition. As the bone grows by surface deposition in a given direction, it is simultaneously displaced in the **opposite** direction.

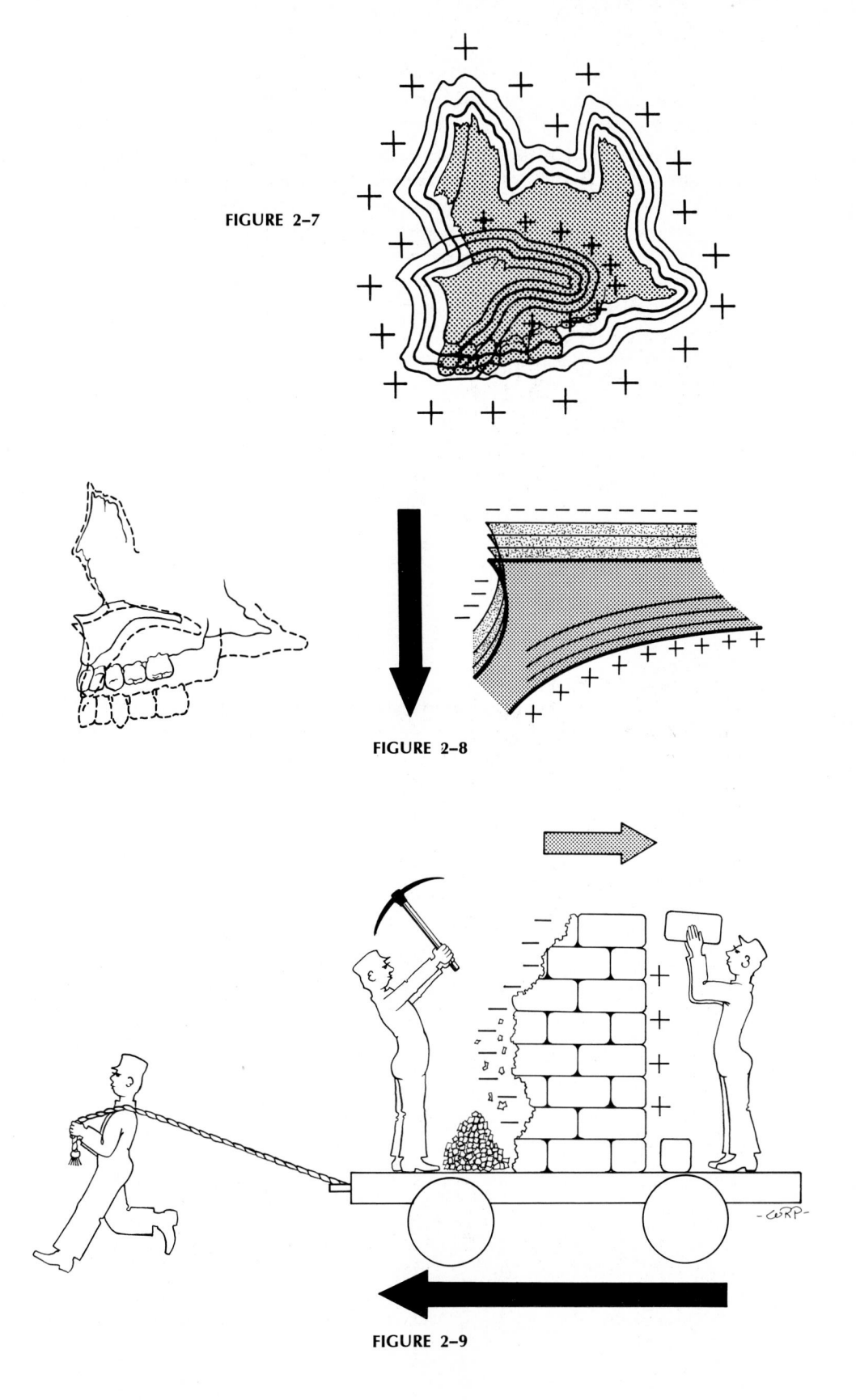

FIGURE 2-7

FIGURE 2-8

FIGURE 2-9

**FIGURES 2–10, 2–11, AND 2–12**

The process of new bone deposition does not cause displacement by **pushing** against the articular contact surface of another bone. Rather, the bone is **carried** away by the expansive force of all the growing soft tissues surrounding it.* As this takes place, new bone is added immediately onto the contact surface, and the two separate bones thereby remain in constant articular junction. The nasomaxillary complex, for example, is in contact with the floor of the cranium (*1*). The whole maxillary region, *in toto,* is **displaced** downward and forward away from the cranium by the expansive growth of the soft tissues in the midfacial region (*2*). This then triggers new bone growth at the various sutural contact surfaces between the nasomaxillary composite and the cranial floor (*3*). Displacement thus proceeds downward and forward as growth by bone deposition simultaneously takes place in an opposite upward and backward direction (that is, **toward** its contact with the cranial floor).

**FIGURE 2–13**

**Concept 8.** A process of **secondary displacement** also occurs during growth. **Primary displacement,** just described, is associated with a bone's **own** enlargement. Secondary displacement, however, is the movement of a whole bone caused by the separate enlargement of **other** bones, which may be nearby or quite distant. For example, increases in size of the bones that compose the middle cranial fossa (in conjunction with growth of the brain) result in a marked displacement movement of the whole maxillary complex anteriorly and inferiorly. This is quite independent of the growth and enlargement of the maxilla itself. The displacement effect is thereby of a secondary type. What happens deep in the cranial base thus affects the placement of the bones in the face. The effects of growth activities in relatively distant locations are passed on, and all such changes must be taken into account when analyzing the growth processes and the facial characteristics of any individual person.

---

*Cartilage may or may not be directly involved as a displacement force in some types of bone-to-bone contacts. This is presently controversial; see Part Two.

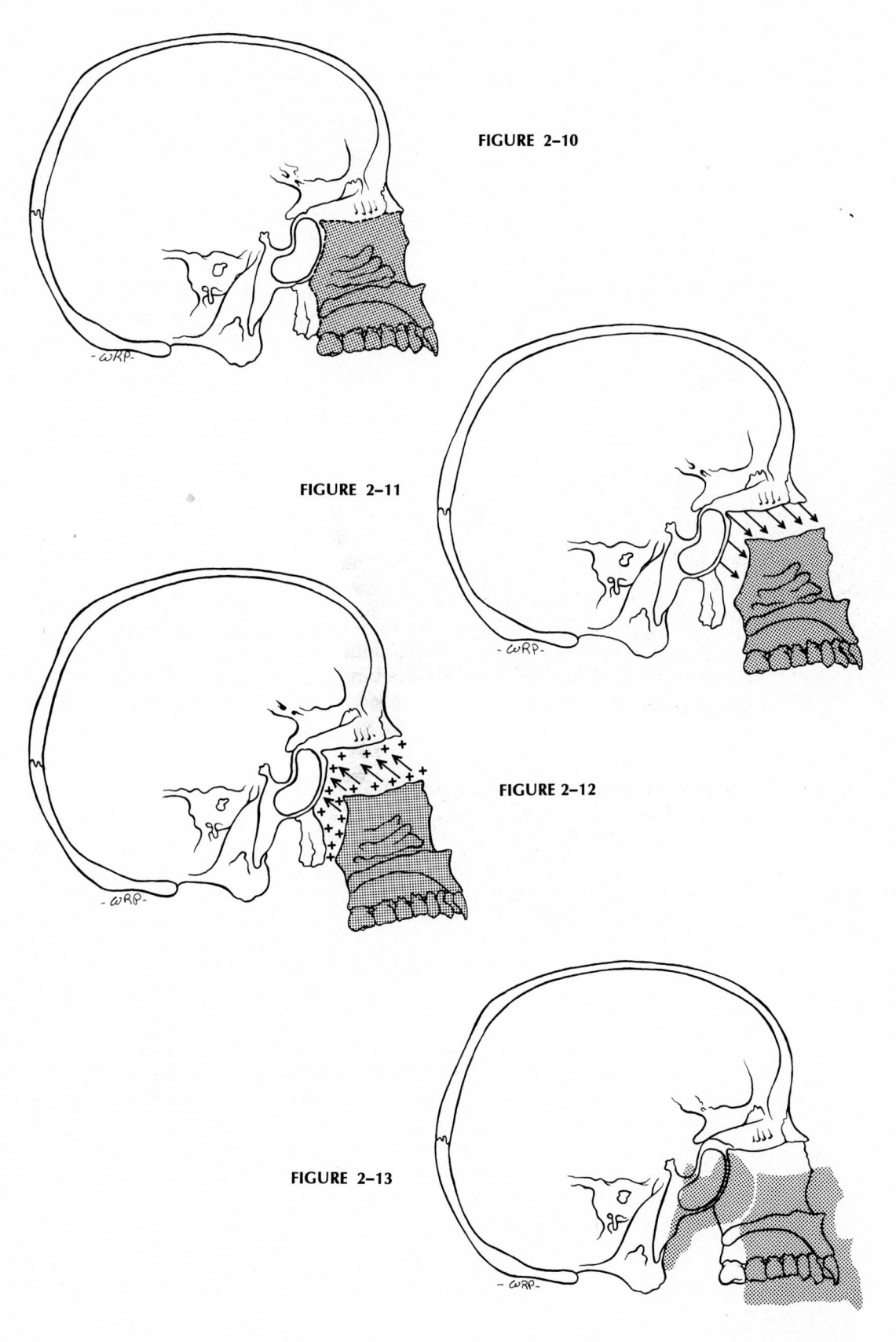

**FIGURE 2–10**

**FIGURE 2–11**

**FIGURE 2–12**

**FIGURE 2–13**

# PART TWO*

_____

## FIGURE 2–14

**Step One** in understanding how the bones of the face and cranium grow is to realize that they **do not** enlarge in the simplified manner shown in this figure. A bone does not increase in size merely by direct, symmetrical, outward expansion of all surfaces and contours, as if magnified by a lens.

## FIGURE 2–15

A bone **does not** grow by generalized, uniform deposition of new bone (+) on all outside surfaces, with corresponding resorption (−) from all inside surfaces. It is not possible for bones having the complex morphology of, say, the mandible or the maxilla to increase in size by such a growth process. Because of the topographically complex nature of each bone's shape, the bone must have a **differential** mode of enlargement, in which some of its parts and areas grow much faster and to a much greater extent than others. Many of the **external** surfaces of most bones are actually **resorptive** in nature. How can a bone increase in size, even though many outside (periosteal) surfaces undergo resorptive removal as the bone grows? Keep this question in mind as the processes of facial growth are explained in the pages that follow.

Two basic kinds of **growth movements** occur during the enlargement of each bone in the facial and cranial skeleton: (1) cortical drift and (2) displacement. **Displacement** is a movement of whole bones away from one another. This creates the space within which growth enlargement of each of the separate bones takes place. **Cortical drift** is the process that produces the enlargement, and it is a direct growth movement produced by deposition of new bone on one side of a cortical plate, with resorption from the opposite side. The process of drift is explained first, then that of displacement.

## FIGURE 2–16

This diagram schematizes the process of cortical drift. The bony cortex **moves** from *A* to *B*. The surface that faces **toward** the direction of movement is depository (+). The opposite surface, facing away from the growth direction, is resorptive (−). If the rates of deposition and resorption are equal, the thickness of the cortex remains constant. If deposition exceeds resorption, cortical thickness increases. It is apparent that the actual bone tissue present in stage *B* is not the same present in *A*, owing to the continuous process of new additions on one side, combined with removal of the older bone from the other side.

---

*The basic concepts of growth briefly summarized in Part One are considered in more detail in Part Two.

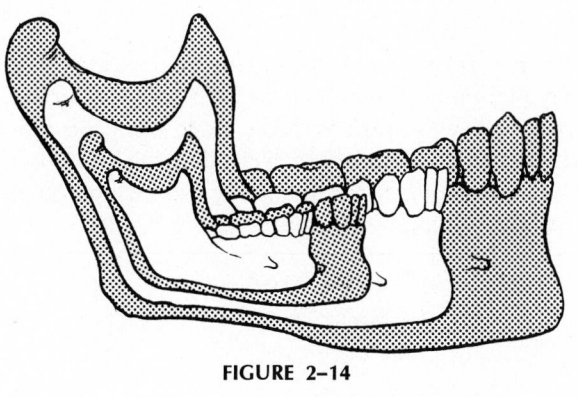

**FIGURE 2–14**

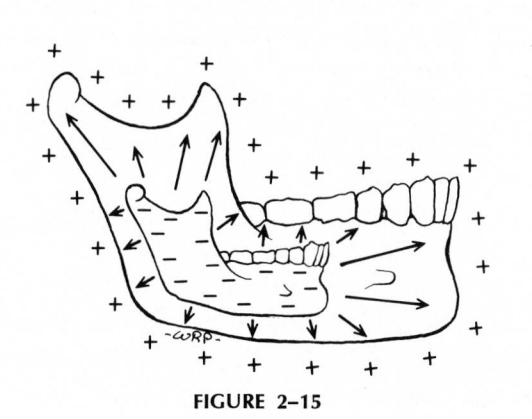

**FIGURE 2–15**

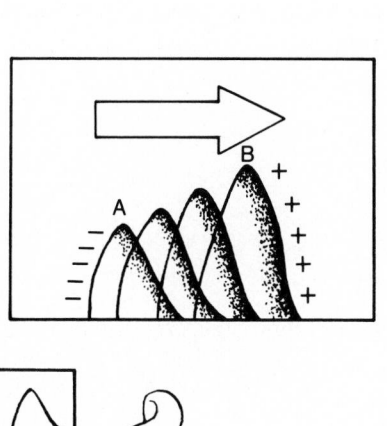

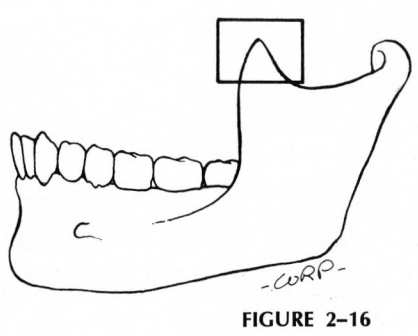

**FIGURE 2–16**

**FIGURE 2–17**

If a metallic marker* is implanted on the depository side of a cortex, it becomes progressively more deeply embedded in the cortex as new bone continues to form on the surface and as resorption takes place from the other side. Eventually, the marker becomes translocated from one side of the cortex to the other. This is not because of its own movement (the marker itself is immobile) but rather because of the "flow" of the drifting bone around it.

**FIGURE 2–18**

Directions of growth sequentially undergo **reversals.** A **reversal line** (small arrow) can be seen in microscopic sections wherever this occurs; it is the interface between the layers of bone that were produced first on one side and then on the other as the **direction** of growth turned about. Why this happens is explained in following paragraphs.

**FIGURE 2–19**

A given bone has **fields** of resorptive (darkly stippled) and depository activity over all its inside and outside cortical surfaces. This is the basis for the **differential growth** process that produces the irregular shape of a bone. The shape is irregular because of its many varied functions (that is, attachments of many different muscles pulling in different directions, articulations with other bones, support of teeth, and so on).

**FIGURE 2–20**

In this diagram, the pattern of growth fields results in a **rotation** of the skeletal part shown.

---

*Metallic implants (tiny pieces of tantalum or some other appropriate metal) are often used as radiographic markers in clinical and experimental work to study bone growth and displacement in head-films. Using the markers as registration points when superimposing serial headfilm tracings, the amount and direction of growth can be readily determined.

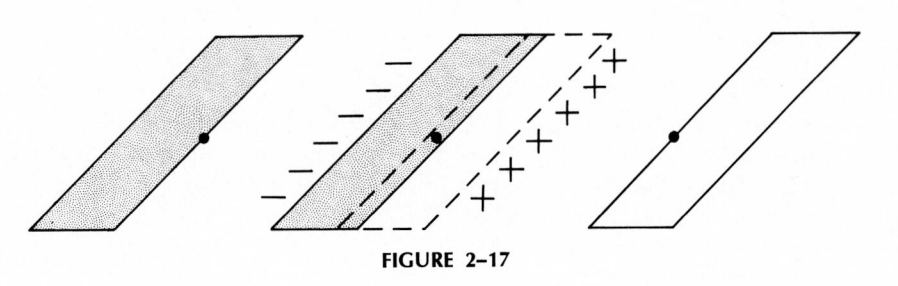

**FIGURE 2-17**

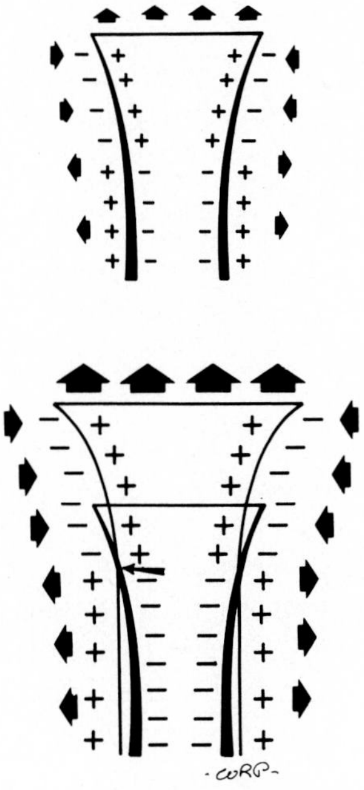

**FIGURE 2-18**

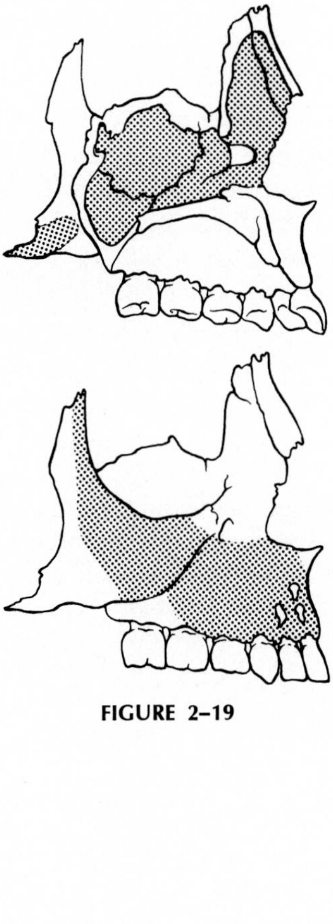

**FIGURE 2-19**

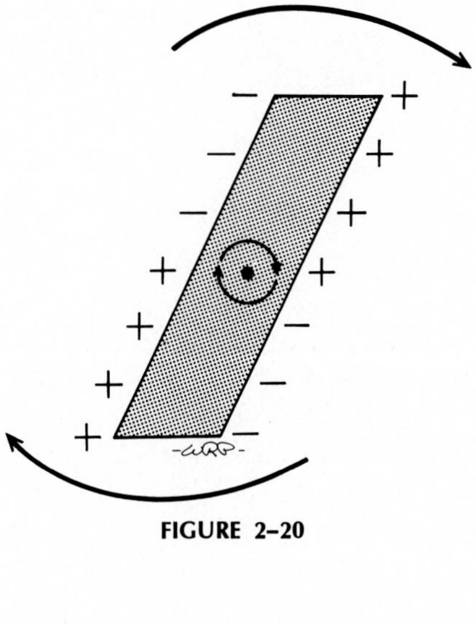

**FIGURE 2-20**

**FIGURE 2–21**

The activities of the growth fields reside in the soft tissue of the **periosteum** and **endosteum,** not in the hard part of the bone itself. The bone does not control and carry out its own growth. The membranes and other soft tissues that **enclose** the bone produce and control the bone's growth. The bone itself is passive. It is a product of the overall growth process, not a pacemaker.

**FIGURES 2–22 AND 2–23**

Layers of bone formed by the covering membrane (periosteal bone tissue) and by the lining membrane (endosteal bone tissue) can occur in the same cortex, as seen by growth stages *1* and *2*. They are separated by a reversal line. Layer *1* was produced during a past growth stage involving an endosteal direction of cortical growth. Layer *2* was then formed following reversal in the course of growth. A given cortex can be composed entirely of either endosteal (*a*) or periosteal (*b*) tissue (Figure 2–23) if reversals are not involved.

**FIGURE 2–24**

In most bones of the face and cranium (and most other bones of the body as well), about half of the total amount of cortical bone tissue is of **endosteal** origin and about half of **periosteal** origin. Approximately half of the **periosteal** surfaces are **resorptive** and about half are **depository** in nature. The same is true for endosteal surfaces. This provides two growth functions: the **enlargement** of any given bone and also the **remodeling** of each bone, a process that accompanies its enlargement.

Four different kinds of remodeling occur in bone tissues. One is **biochemical** remodeling, taking place at the molecular level, that is, the constant deposition and removal of ions to maintain blood calcium levels and carry out other mineral homeostasis functions. Another type of remodeling involves the secondary reconstruction of bone by **Haversian systems** and the rebuilding of cancellous trabeculae. A third kind of remodeling relates to the regeneration and reconstruction of bone during or following **pathology** and **trauma.** The remodeling process that we are dealing with in facial morphogenesis, however, is **growth remodeling.** In order for a bone to grow and enlarge, it must also undergo a simultaneous process of remodeling. The reasons are described below.

**FIGURE 2–25**

The pattern of resorption and deposition seen here produces a combination of growth movements in which part *a* moves to *a'* and *b* moves to *b'*. Surfaces *d* and *g* represent the external (periosteal) sides, and *e* and *f* are internal (endosteal) surfaces. Surface *f* is resorptive, but the **inside** surface at *e* faces the direction of growth; the bone thereby actually enlarges here by an endosteal mode of growth, that is, by the continued adding of new bone on the inside rather than the outside. The bone as a whole thus increases in size, even though about half of its external surfaces are resorptive.

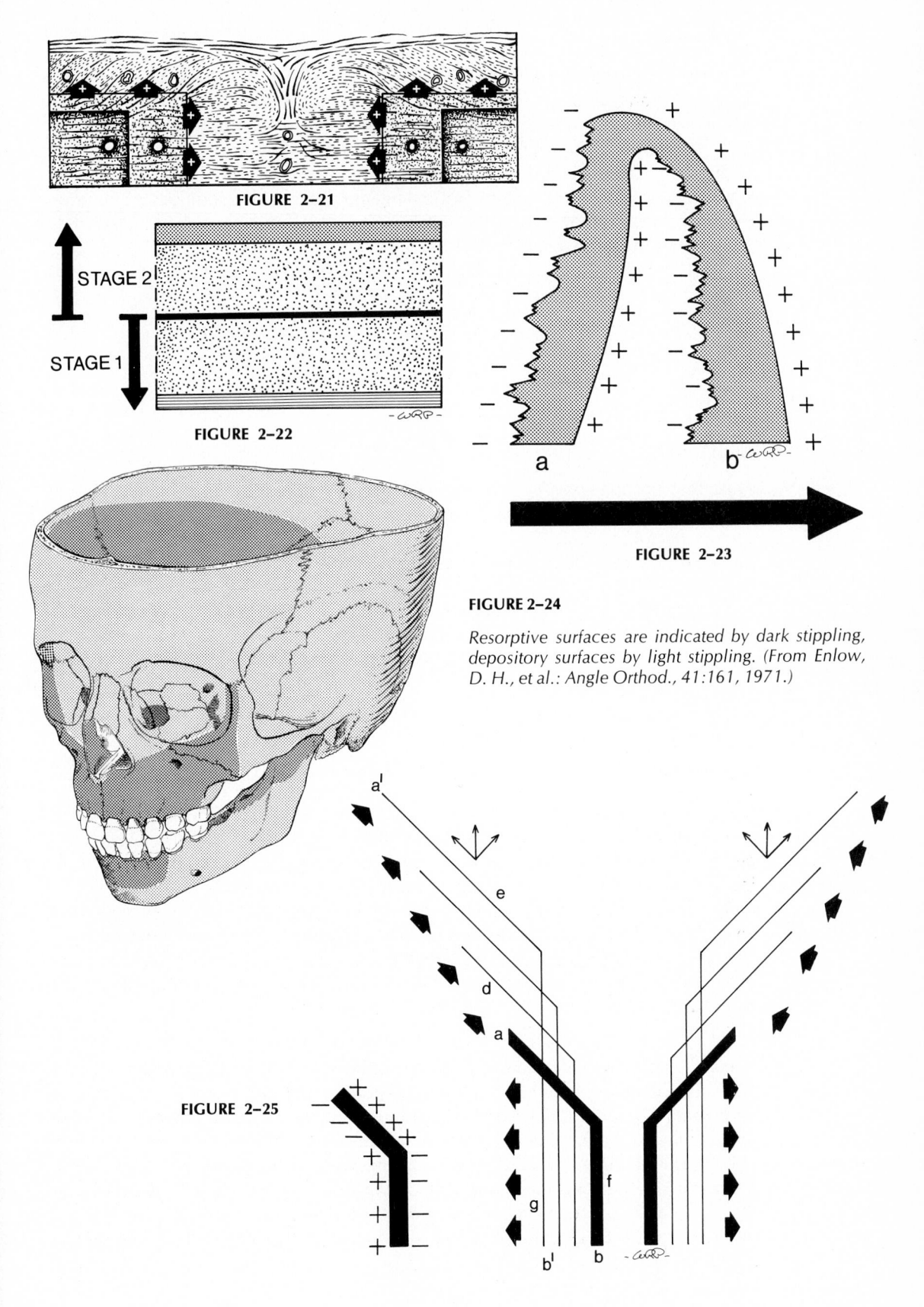

**FIGURE 2–21**

STAGE 2

STAGE 1

**FIGURE 2–22**

a          b

**FIGURE 2–23**

**FIGURE 2–24**

*Resorptive surfaces are indicated by dark stippling, depository surfaces by light stippling. (From Enlow, D. H., et al.: Angle Orthod., 41:161, 1971.)*

**FIGURE 2–25**

a¹

e

d

a

g          f

b¹    b

**FIGURE 2–26**

A useful concept in facial growth is the "V" principle. Many facial and cranial bones, or parts of bones, have a V-shaped configuration. Note that deposition occurs on the **inner** side of the "V"; resorption takes place on the outside surface. The "V" **moves** from position A to B and, at the same time, increases in overall dimensions. The direction of movement is toward the wide end of the "V." Thus, a simultaneous growth movement and enlargement occurs by additions of bone on the inside with removal from the outside. The "V" principle will be referred to many times in later explanations of the facial growth process.

**FIGURE 2–27**

Note that the diameter at A is **reduced** because the broad part of the bone is relocated to position B. This is a remodeling change that converts a **wider** part into a more **narrow** part, as both become sequentially relocated. Periosteal resorption and endosteal deposition of bone tissue carries this out.

**FIGURE 2–28**

If a transverse section of the bone is made at A, it is seen that the periosteal surface is **resorptive;** bone-removing osteoclasts blanket this surface field during the active period of bone growth. The depository endosteal surface is lined with bone-producing osteoblasts. A transverse section at B shows new endosteal bone added onto the inner surface of the cortex. A transverse section made at C shows an endosteal layer that was produced during the inward growth phase. This is covered by a periosteal layer of bone following **outward** reversal, as this part of the bone now increases in diameter. A transverse section at D shows a cortex composed entirely of periosteal bone. The outer surface is depository, and the endosteal surface is resorptive.

If metallic implant markers are nailed into the bone at X, Y, and Z, note that marker X is subsequently released from the bone because that part becomes removed by periosteal resorption. It will lie free in the surrounding soft tissue. Marker Z is also released because of endosteal resorption taking place in this location, and it will lie free in the medullary cavity. Marker Y was originally inserted into the cortex on the periosteal side but becomes translocated over to the endosteal side of the cortex. This is because new cortical bone is added to the left, and older bone is removed from the right, thus changing the relative position of the marker from one side of the cortex to the other.

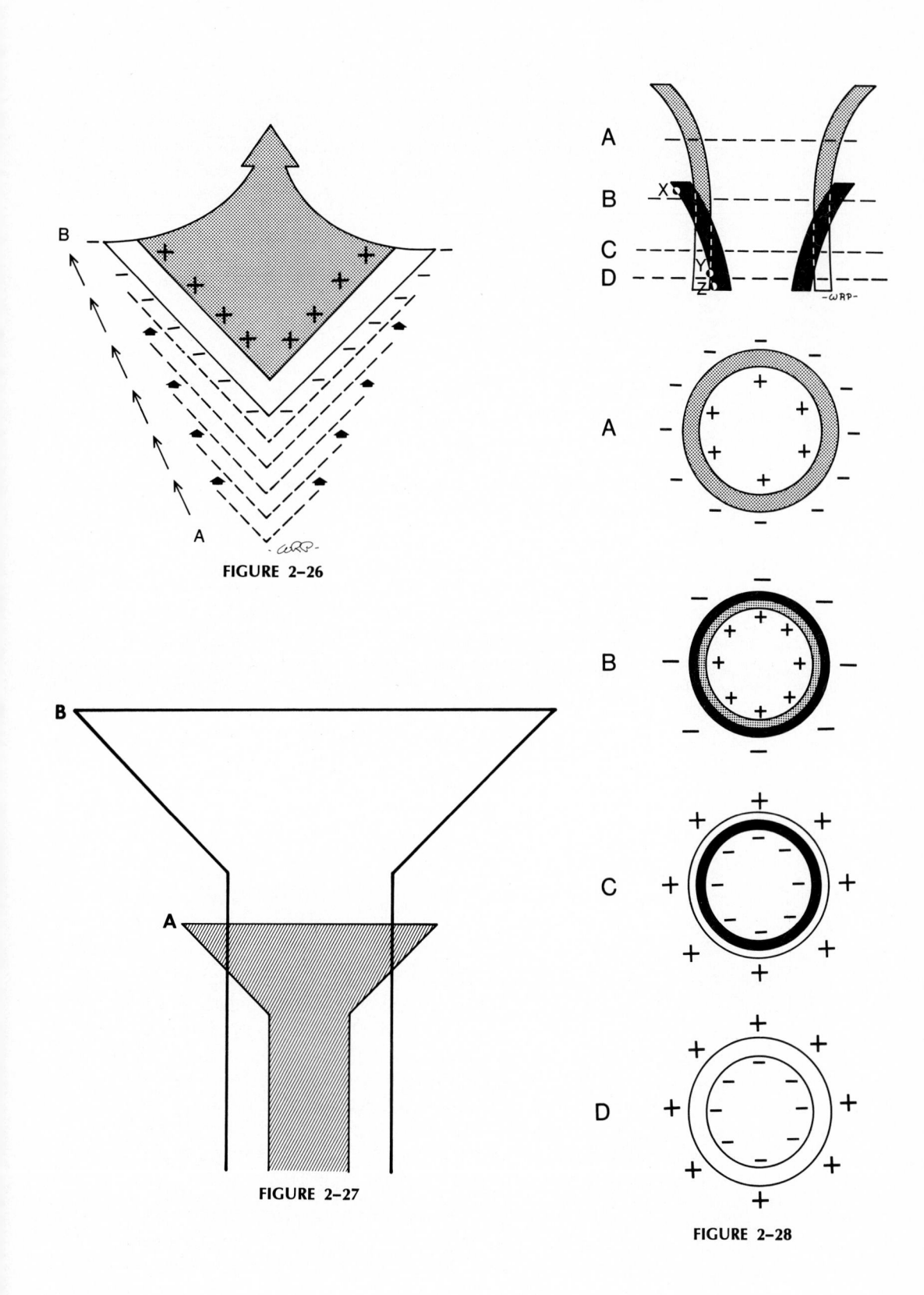

**FIGURE 2-26**

**FIGURE 2-27**

**FIGURE 2-28**

**FIGURE 2–29**

This transverse section through the zygomatic arch demonstrates how a bone grows and remodels laterally as the whole bone simultaneously grows in length. The zygomatic arch is moving laterally and also inferiorly as the entire face widens and expands downward. It does this by progressive deposition on the lateral-facing and downward-facing periosteal and endosteal surfaces, with resorption from the opposite surfaces. The remnants of the old cortical contours can be recognized in microscopic sections.

**FIGURE 2–30**

Why do bones remodel as they grow? The basic, key factor is the process of **relocation.** In this stack of chips, the black chip is at the top end in a. As "growth" continues to take place, the black chip becomes progressively "relocated"—not by its own movement but because new chips have been added on one side and removed from the other. This changes the **relative position** of the black chip within the stack, even though this chip itself does not move. Let the stack of chips represent a whole growing bone having complex topographic shape rather than a perfectly cylindrical form. It is then apparent that the changing relative positions of the black chip would require continuous **remodeling** of the shape and sectional dimensions to conform with each successive position the chip comes to occupy. A **sequence** of continuous remodeling changes is required level by level. Remodeling is a process of reshaping and resizing each level (chip) within a growing bone as it is relocated sequentially into a succession of new levels. This is because additions and/or resorption in the various **other** parts cause changes in the relative positions of all levels up and down the line. Note that the **position** of the condyle in the smallest mandibular stage becomes relocated into the middle of the ramus and then on to the anterior border of the ramus. Continuous remodeling is thus involved as this area changes in relative position.

**FIGURE 2–31**

In the face of a young child, the levels of the maxillary arch and nasal floor lie very close to the inferior orbital rim. The maxillary arch and palate, however, **move** downward. This process involves (in part) an inferior direction of **drift** by the hard palate and the bony maxillary arch. Bone deposition occurs on the downward-facing surfaces, together with resorption from the superior-facing surfaces on the palate. The combination results in a downward **relocation** of the whole palate–maxillary arch composite into progressively lower levels, so that the arch finally comes to lie considerably below the inferior orbital rim. The vertical dimension of the nasal chamber is greatly increased as a result.

About half of the external surfaces involved in this growth and remodeling change are resorptive and half depository. About half of the bone tissue of the palate is thus endosteal and half periosteal. (The cortex on the nasal side of the palate is produced by the endosteum of the medullary cavity.)

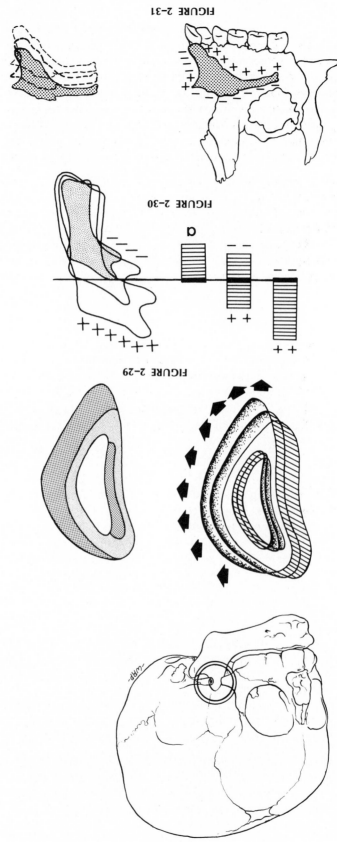

FIGURE 2-31

FIGURE 2-30

FIGURE 2-29

**FIGURE 2-32**

Because of this **relocation** process, the nasal area of the adult occupies an area where the bony maxillary arch **used to be** located during childhood. What was once the bony maxillary arch and palatal region has been converted into the expanded nasal region. This is "growth remodeling"; the basis for it is **relocation.**

As the palate and arch grow downward by constant deposition of new bone on one side and resorption of previously formed bone from the other, the bone tissue that comes to house the teeth at older age periods is not the same actual bone enclosing them during the succession of former growth levels. This is significant because the growth movement and the exchanges of bone involved are **used** by the orthodontist to "work with growth." It is also why orthodontic procedures are often more effective in the growing child than in the adult.

**FIGURE 2-33**

As the mandible grows, the ramus moves in a backward direction by appropriate combinations of resorption and deposition. Approximately half of the external surfaces involved are thus resorptive and half depository; about half of the bone that is produced is thereby endosteal in type and half periosteal. As the ramus is **relocated** posteriorly, the corpus becomes **lengthened by a remodeling conversion** from what was at one time the ramus during a former growth period.

**FIGURE 2-34**

During growth from the fetus to the adult, the "molar" region in the younger mandible, for example, undergoes relocation to occupy the "premolar" region of the larger, older mandible. It is apparent that remodeling is a process of relocation and that the same **deposition and resorption producing growth enlargement also carry out the growth remodeling process.** Growth, indeed, **is** a process of remodeling to provide enlargement. (There is more to growth, however, as will be seen.)

**FIGURE 2-35**

Remodeling maintains the overall morphology of a bone during its growth. Any given bone grows differentially, that is, it increases in some directions much more than others and at varying regional rates. If a bone were to grow uniformly in all directions, relocation would not be involved, and remodeling would not be required as a part of the growth process. Because of the multiple physiologic and mechanical functions of a bone, however, it necessarily has a complex topographic shape. This can only be produced by a differential mode of growth involving remodeling.

The mandible grows differentially in directions that are predominantly posterior and superior. Even though successive remodeling of one part into another constantly takes place as the whole bone enlarges, the form of the bone as a whole is sustained (with some characteristic age changes in shape). It is remarkable that the external morphology of any given bone is relatively constant, even though its substance undergoes massive internal changes and all its parts experience widespread alterations in regional shape and size as they are relocated. This is the special function of "growth remodeling"; it **maintains** the form of a whole bone while providing for its enlargement at the same time.

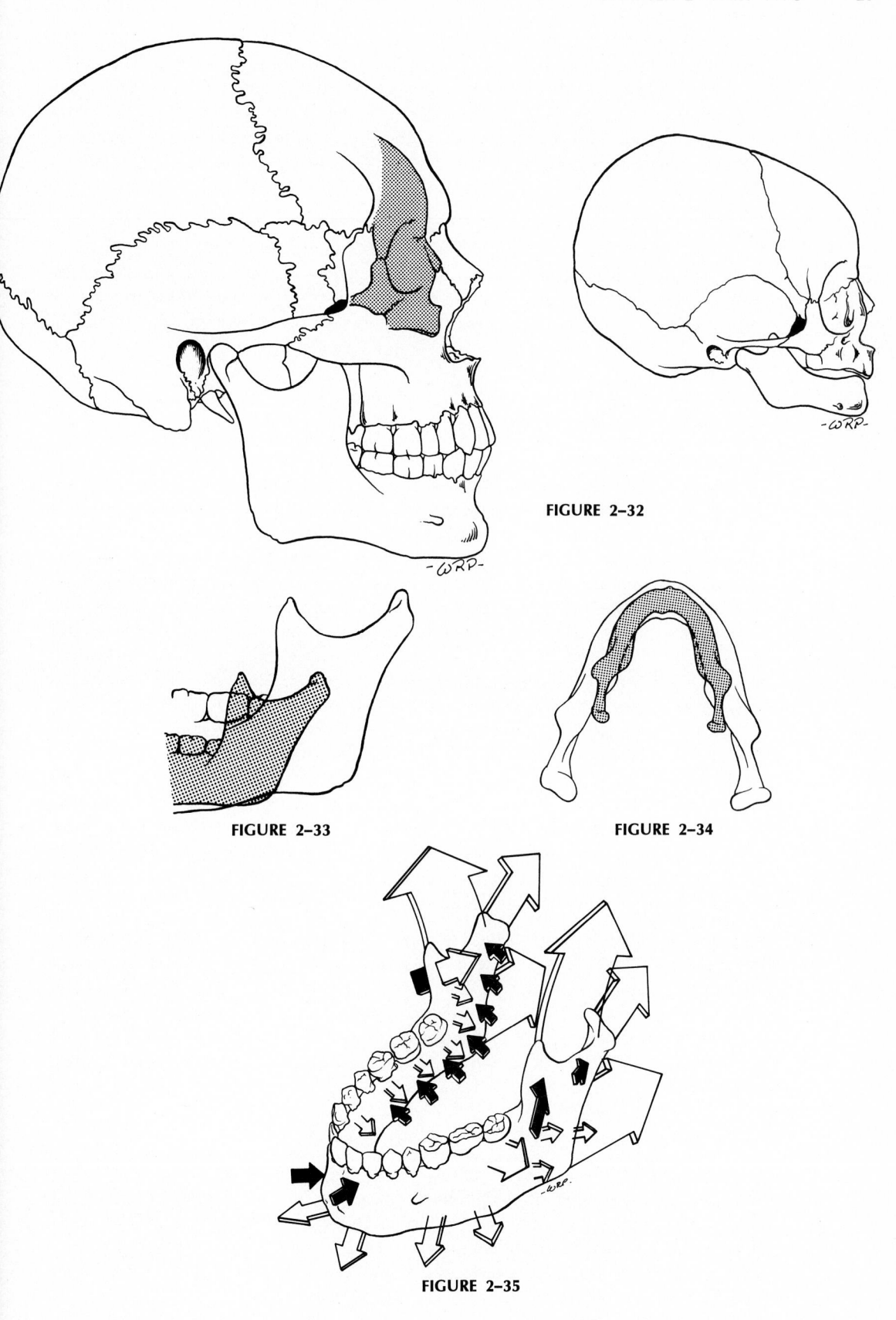

**FIGURE 2–32**

**FIGURE 2–33**

**FIGURE 2–34**

**FIGURE 2–35**

The entire bone takes part in this process of growth enlargement and remodeling. **All** surfaces, inside and out, participate actively at one time or another. The entire periosteal surface and the entire endosteal surface of the cortex (and all trabecular surfaces of the cancellous bone as well) are involved. The bone does not just grow at special sites or "centers." Many divergent directions of growth take place among the many different regions of a bone. It is a three-dimensional growth process.

**FIGURE 2–36**

The resorptive and depository **fields** of growth, mentioned on page 20, blanket all of the outside and inside surfaces. This mosaic pattern is more or less constant for each bone throughout the growth period, unless a major change in the shape of a region becomes involved. As the perimeter of these growth fields enlarges, the parts of the bone associated with them correspondingly increase in size.

**FIGURE 2–37**

During the operation of the relocation process, it is the **growth fields in the soft tissue membranes** that first move and control the relocation movements of the underlying bony parts associated with each field. The growth movement of the bone **follows** the pace-setting movement of the overlying growth field. There is virtually no lag time, however, between the two. Note that resorptive field *a* moves to *a'* and that the corresponding zone of the bone beneath it remodels and moves under its control. The boundaries of this resorptive field are thus changed so that they come to occupy the region that was located just posteriorly. A *reverse line (x)* separates field *a* from the area of the ramus behind it, and it moves to *x'*. The resorptive field in the larger mandible (*a'*) occupies the same **relative** placement as it did when the bone was smaller during former growth stage (*a*). The field, however, (*1*) is now larger and (*2*) has moved to a new location, bringing its part of the bone with it by continuous bone deposition and resorption (that is, remodeling growth). The actual bone tissue present in the ramus of the smaller stage has been replaced by a whole new generation of bone in the location occupied by the ramus of the larger stage following relocation. The **patterns of distribution** of all the various resorptive and despository fields, however, have not changed; the fields have only moved from one position to another as the whole bone enlarges. This requires **reversals** as any one field expands into areas previously occupied by other fields, which in turn have moved on to hold successively new locations.

Keep in mind that the growth field does not reside in the bone itself. The bone is the **product** of the field. The genetic thrust of growth and the cells that carry it out are actually in the enclosing **soft tissues.** The bone, however, contributes feedback information to the soft tissue membranes, so that when the size, shape, biomechanical properties, and so forth of the bone come into equilibrium with functional requirements, the histogenetic activity of the osteogenic membranes then becomes turned off.

**FIGURE 2–38**

The osteocyte itself does not ordinarily contribute to the growth process unless it becomes released with subsequent change into some other cell type. The bone cell cannot divide; it has nowhere to divide to. Its genes are locked away and not available for further growth and differentiation of the bone matrix surrounding it. (Osteolysis can occur, however, a process that involves limited resorption in the immediate area around an osteocyte.)

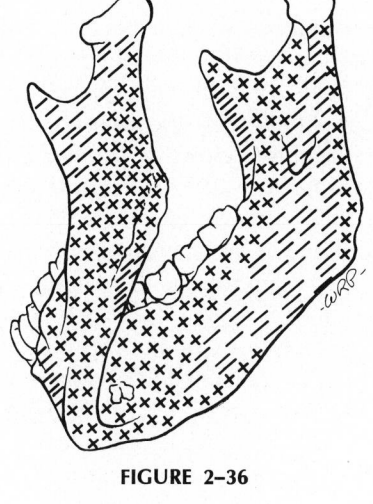

**FIGURE 2–36**

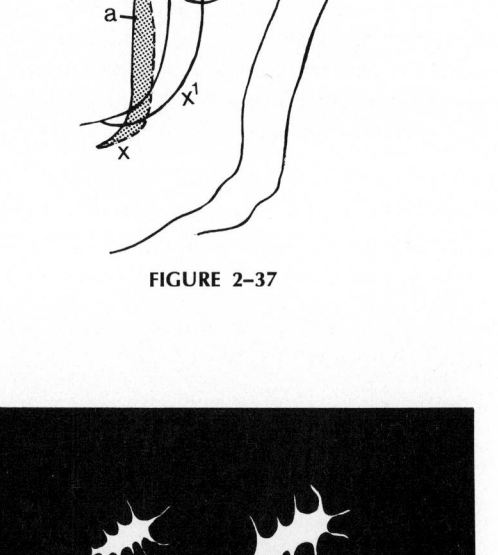

**FIGURE 2–37**

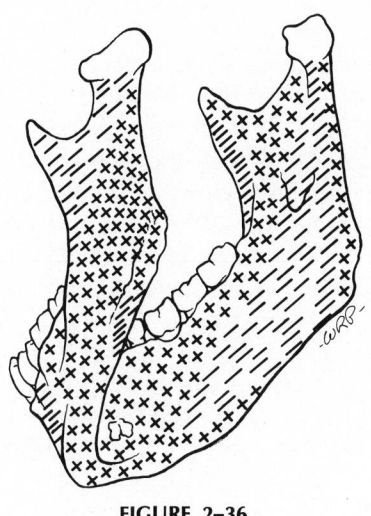

**FIGURE 2–38**

**FIGURE 2–39**

All clinical and basic science workers need to understand the plan of distribution for the major cranial and facial growth fields. The characteristic pattern of resorptive and depository fields for each facial bone is **essential knowledge.** These patterns are the key to the underlying theory used in many clinical procedures. Is one working "with" or "against" growth? In which specific **direction** is a particular region growing? Inward? Outward? Posteriorly? What happens to this region when it becomes relocated? What is the effect on the teeth? What is the relative velocity of growth and remodeling in any given region? Is a bone and periosteal transplant, or the field to which it is moved, unknowingly "programmed" to be **resorptive** when the worker assumes it to be depository? Why do some bone transplants undergo uncontrollable resorption? What determines bone–soft tissue equilibrium for a growth field? What is the prescribed **boundary** of a particular growth field? What happens to the stability of a region if this growth boundary is violated by some clinical procedure? What are the functional factors that **control** the growth in any given region? Will intrinsic functional control processes work against the clinician to cause subsequent loss of treatment results? Will they effectively preclude successful treatment? Can these intrinsic control factors be harnessed in order to control the control process itself?

**Variations** in the shape and size of the face are always the rule. No two faces are quite alike. Morphologic variations, normal and abnormal, are produced by corresponding developmental variations that take place during the growth process. Some can be genetically established by characteristic soft tissue relationships that are hereditary determinants of bone growth (and also cartilage relationships that are genetic determinants, according to some investigators). Other variations are largely determined by functional changes in soft tissue relationships within a given individual during his own development. The results, however, are all based on the following:

a. Fundamental difference in the **pattern** of the fields of resorption and deposition, that is, the number and the configuration of the growth fields in an individual person.

b. The specific placement of the **boundaries** between growth fields, that is, the size of any given growth field.

c. The differential **rates** of deposition and resorption in each field.

d. The **timing** of the growth activities among all the different fields.

**FIGURE 2–40**

Some growth fields have traditionally been singled out for special attention because of their particular role in the growth of a given facial or cranial bone. These special **growth sites** include the sutures of the face and cranium (a, b, c, d, f), the mandibular condyle (e), the maxillary tuberosity (h), the synchondroses of the cranial base (i), and the alveolar bone housing teeth (g). All these growth sites are described in later chapters, but it is to be understood that such special sites **do not** carry out the **entire** growth process for the particular bones associated with them. **All other** inside and outside surfaces of a given bone also actively participate in the overall growth process. The contributions made by these other growth fields are just as basic and essential as the specially designated sites. This important point has not always been understood.

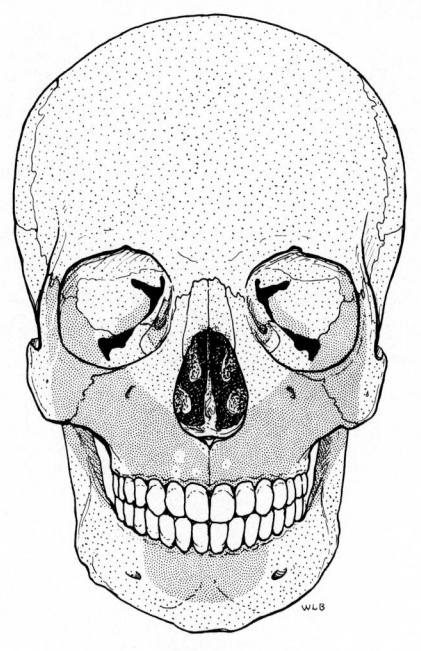

**FIGURE 2–39**

*(After Enlow, D. H.:* The Human Face. *New York, Harper & Row, 1968.)*

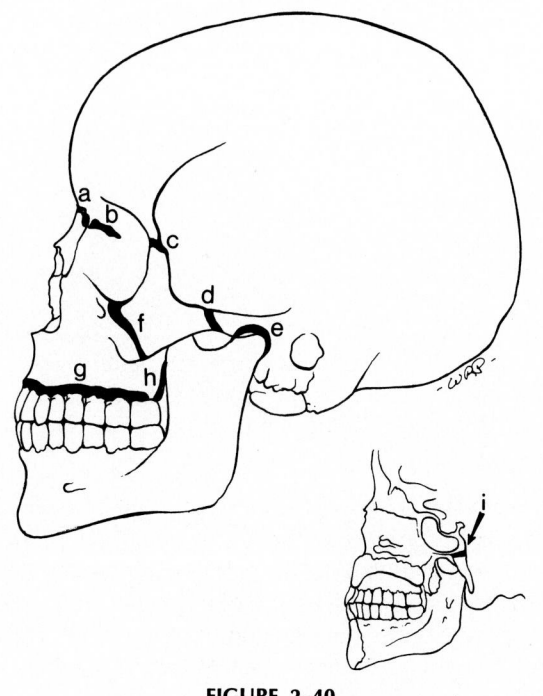

**FIGURE 2–40**

**FIGURE 2–41**

As mentioned, the mandibular condyle is a specially recognized growth site. The condyle and some other special sites are sometimes also termed growth "centers." This label has come into disfavor, however, because it is now believed that such centers do not actually control the growth processes of the bone as a whole. They are not "master centers" that directly regulate the overall growth process of the entire bone and all its regional parts. While these "centers" are important growth sites, they represent only regional fields of growth adapted to the localized morphogenic circumstances in their own particular areas. Such centers, as pointed out above, do not provide for the entire growth of the whole bone, as has sometimes been mistakenly presumed.

**FIGURE 2–42**

Routine headfilms, of course, are **two-dimensional,** and this is a limitation that presents many troublesome problems. Only the anterior and posterior **edges** of the ramus, for example, can be visualized (as at A and B) in lateral cephalograms. Important changes on surfaces in the span **between** these edges (C) cannot be visualized. This is all the more reason for the clinician and researcher to thoroughly understand what happens when such areas grow in a **three-dimensional** manner not representable by the headfilm itself.

**FIGURE 2–43**

In **cortical drift,** the bone moves from A to B by deposition on the side facing the direction of growth movement. The opposite side is resorptive. In the process of **displacement,** however, the whole bone is **carried** from B to A by mechanical force. Drift and displacement are separate processes, but they occur in conjunction with one another.

The growth expansion of a single bone is a process by which the size and shape of the bone develop in response to the composite of all the functional soft tissue relationships associated with that individual bone. The bone does not grow and enlarge in an isolated way, however; its increases in size include articular contacts with **other** bones which are also enlarging at the same time. For this reason, all articular contacts are important—condyles, sutures, and synchondroses—because they are the sites where displacement is involved.

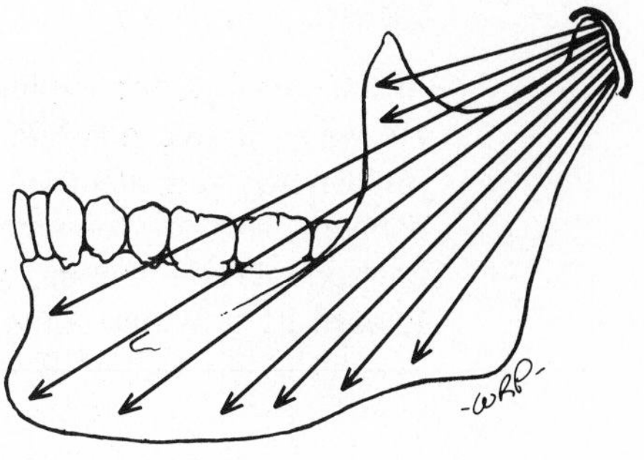

**FIGURE 2–41**

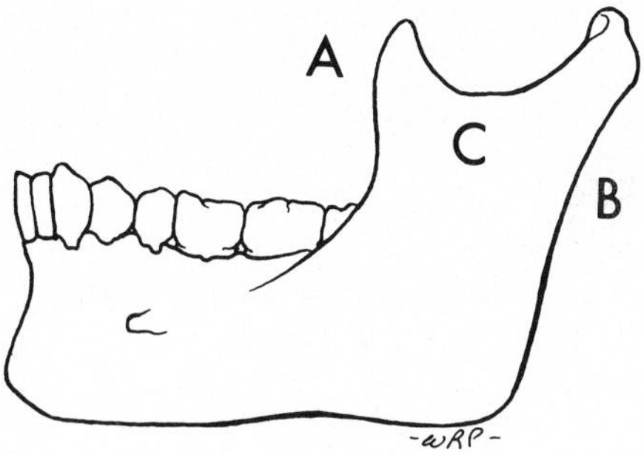

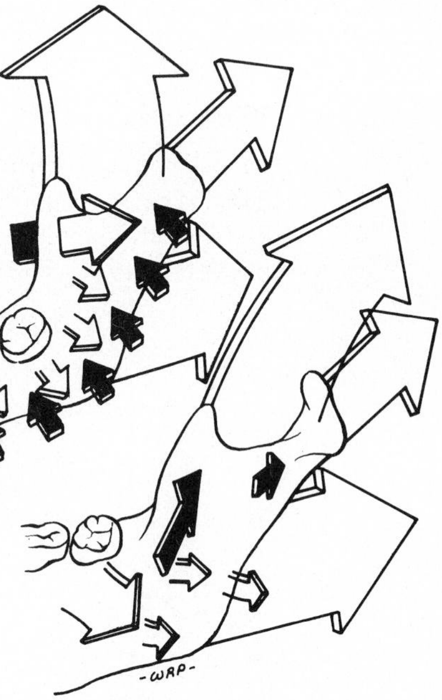

A

C

B

**FIGURE 2–42**

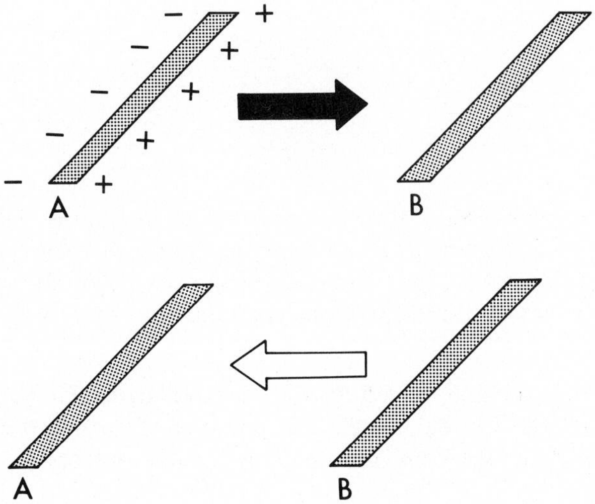

**FIGURE 2–43**

A    B

A    B

**FIGURES 2–44 AND 2–45**

In this analogy, the expansion of a single balloon does not "compete" for space. However, if **two** enlarging balloons are in contact with each other, a displacing **movement** takes place until their positions become adjusted as either one or both expand. This movement proceeds away from the interface between the two balloons. What happens when the mandible, for example, grows in a direction **toward** its articular contact with the cranium? A "displacement" takes place in which the whole mandible moves **away** as it enlarges toward the temporal bone.

**FIGURE 2–46**

Do the balloons **push and shove** each other apart as they expand? Or are the balloons **carried** apart by **other** mechanical forces, and, as they become separated, does growth enlargement occur simultaneously to constantly maintain the contact between them?

**FIGURE 2–47**

This has been, and still is, one of the great historical controversies in craniofacial biology. The mandible grows by deposition and resorption in the manner shown here. The predominant vectors (direction and magnitude) of growth are posterior and superior. Thus, the condyle grows directly **toward** its articular contact in the glenoid fossa of the cranial floor.

As this takes place, the whole mandible is moved forward and downward by the same amount that it grows upward and backward. The direction of growth by new bone additions at the condyle and the direction of displacement are opposite to each other.

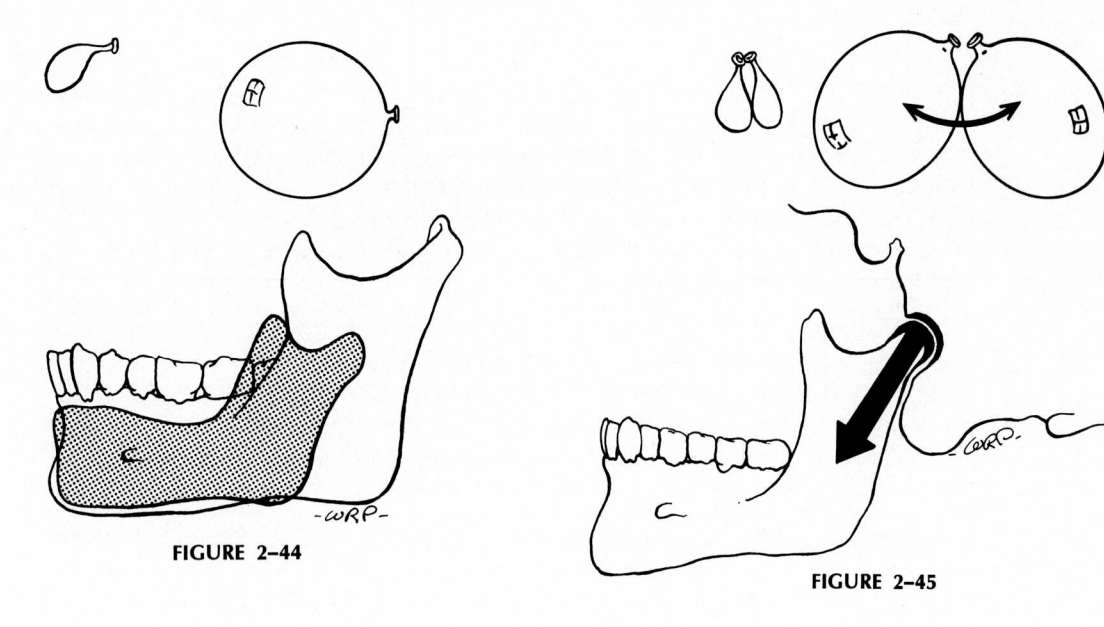

FIGURE 2–44

FIGURE 2–45

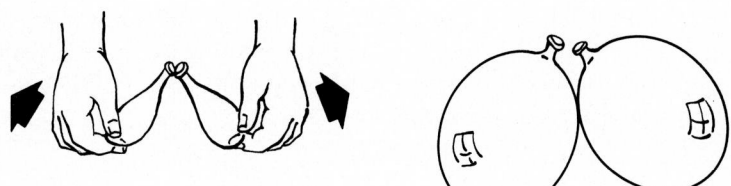

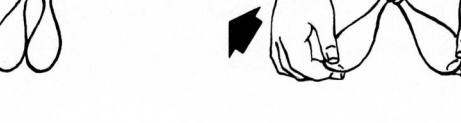

FIGURE 2–46

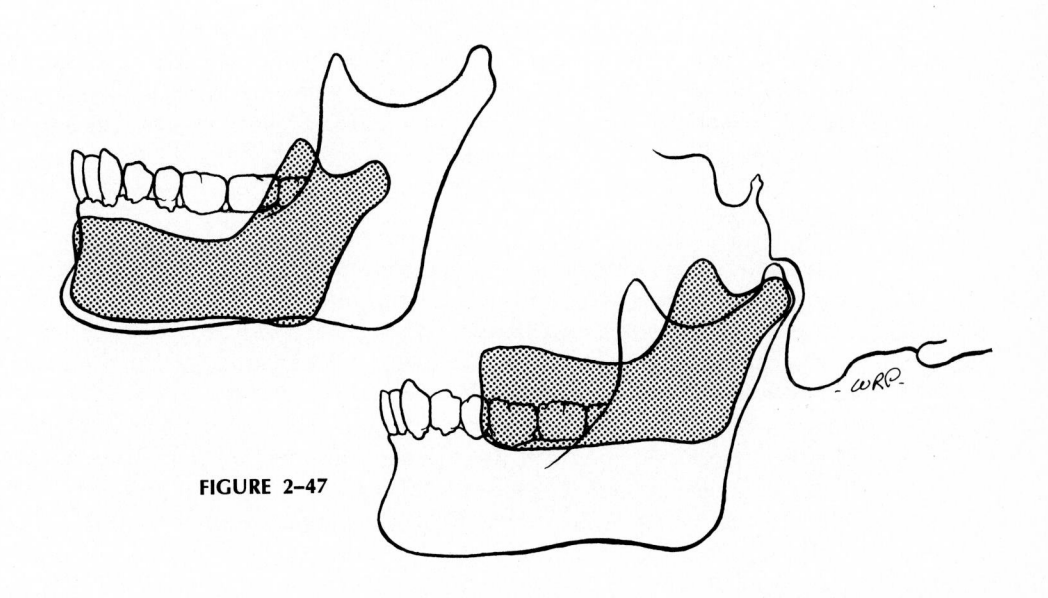

FIGURE 2–47

**FIGURE 2–48**

Is this accomplished by a **shove** against the articular surface caused by the growth of the condyle, or, conversely, by a **carry** of the entire mandible away from the cranial base by other mechanical forces (such as the expansive growth of the contiguous muscle and connective tissue mass)? Simultaneous bone growth occurs at the condyle to maintain constant contact with the temporal bone. As "force *a*" carries the mandible anteriorly and inferiorly, the condyle is triggered to respond by an equal amount of growth at *b*.

Is condylar growth thus the active cause of displacement or the passive (secondary) response to it? Current theory favors the passive **carry** concept rather than an active thrust or push (which was popular for many years); the problem, however, has not yet been resolved to everyone's satisfaction. (See the "functional matrix" concept described on page 80.) The implications are much more than theoretic, as will be seen.

In summary, two basic modes of skeletal movement take part in the growth of the face and cranium. **Drift** involves deposition of bone on the side pointed toward the direction of growth of a given area; resorption usually occurs on the opposite side of that particular bony cortex (or cancellous trabecula). **Displacement** is a separate movement of the **whole bone** by some physical force that carries it, *in toto,* away from its contacts with other bones, which are also growing and increasing in overall size at the same time. This two-phase drift-displacement process takes place virtually simultaneously. The displacement movement, however, is presently believed by many researchers to be the primary change, with the rate and direction of bone growth representing a secondary response. Two kinds of displacment occur, **primary** and **secondary.**

**FIGURE 2–49**

In **primary displacement,** the process of physical carry takes place in conjunction with a bone's **own** enlargement. Two principal growth vectors in the maxilla, for example, are posterior and superior. As this occurs, the whole bone is displaced in opposite anterior and inferior directions. Primary displacement produces the "space" within which the bone continues to grow. The amount of this primary displacement exactly equals the amount of new bone deposition that takes place. The respective directions are always opposite in the primary type of displacement. Because primary displacement takes place at interfaces with other, contiguous skeletal elements, **joint contacts** are important sites involved in this kind of remodeling change.

**FIGURE 2–50**

In **secondary displacement,** the movement of the bone is not directly related to its own enlargement. For example, the anterior direction of growth by the middle cranial fossa displaces the maxilla anteriorly and inferiorly. Maxillary growth and enlargement itself, however, are not involved in this particular kind of displacement movement. Thus, as any bone grows, remodels, and becomes displaced in conjunction with its own growth process, it is also displaced, in addition, by the growth of **other** bones and their soft tissues. This can have a "domino" effect. That is, growth changes can be passed on from region to region to produce a secondary effect in areas quite distant. Such changes are cumulative.

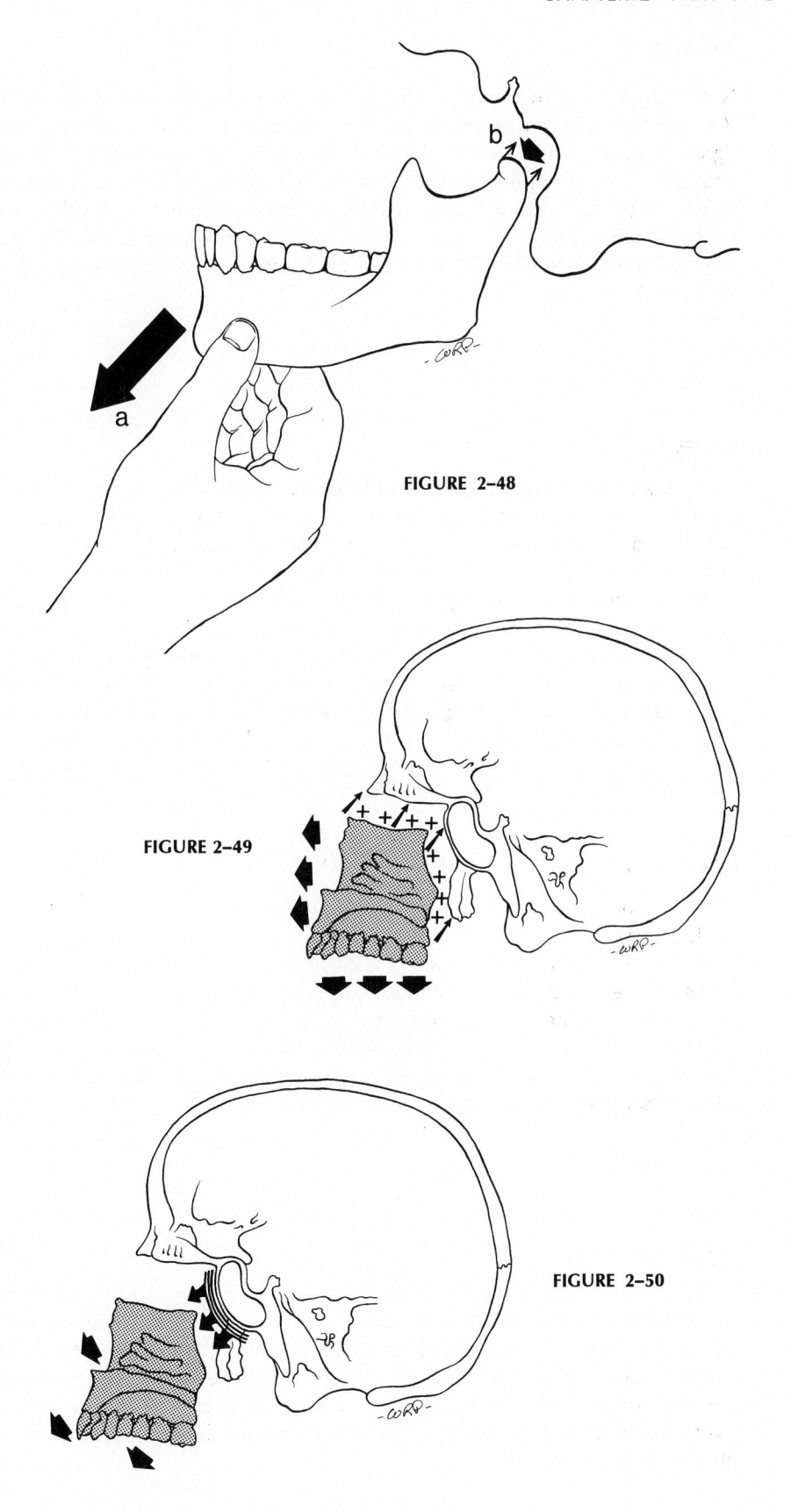

**FIGURE 2–48**

**FIGURE 2–49**

**FIGURE 2–50**

**FIGURE 2–51**

Note that much of the anterior part of the midfacial region is **resorptive** in nature. Yet, the face grows **forward.** How can this be? The face does not, simply, just "grow" directly anteriorly. The forward movement is a **composite** result of growth changes (a) by resorption and deposition that cause the maxilla to **enlarge backward** and (b) by primary and secondary displacement movements that cause it to be **carried forward.** The resorptive nature of the anteriorly facing surface of the premaxilla is concerned with its downward, not forward, growth (as explained in Chapter 3).

**FIGURE 2–52**

To illustrate the composite nature of these different growth processes, the growth of the arm is used for an analogy. The tip of the finger moves away from the shoulder as the whole arm increases in length. Most of this growth movement of the finger, of course, is not a consequence of growth at the fingertip itself. The aggregate summation of linear growth increases by all the bones in the arm at each particular interface between the phalanges, carpals, metacarpals, radius, ulna, and humerus is involved. The contribution by the tip of the terminal phalanx is only a relatively small part of the total. It is the secondary displacement effect produced by all the **other** bones in the arm that causes most of the growth movement of the finger tip.

**FIGURE 2–53**

Similarly, the greater part of the growth movement of the tip of the premaxilla is produced by the growth expansion of all the bones behind and above it and by growth in other parts of the maxilla. The premaxillary tip itself contributes only a very small part of its own forward growth movement. The enlargement of the maxilla proper and the frontal, ethmoid, occipital, sphenoid, and temporal bones provides an aggregate expansion, the sum of which is the basis for most of the total forward movement of the premaxilla. (It contributes somewhat more to its own downward movement, however, as seen in Chapter 3.)

**FIGURE 2–54**

The factor of secondary displacement is a fundamental part of the overall process of craniofacial enlargement. Growth effects of skeletal parts far removed are passed on, bone by bone, to become expressed on the resultant topography of the face. Cranial floor–facial growth imbalances often contribute materially to misalignments and improper positionings of the facial bones. Secondary displacement is one of several basic factors involved in the developmental basis for malocclusions and other types of facial dysplasias.

**FIGURE 2–51**

*(From Enlow, D. H., and R. E. Moyers: J.A.D.A., 82:763, 1971.)*

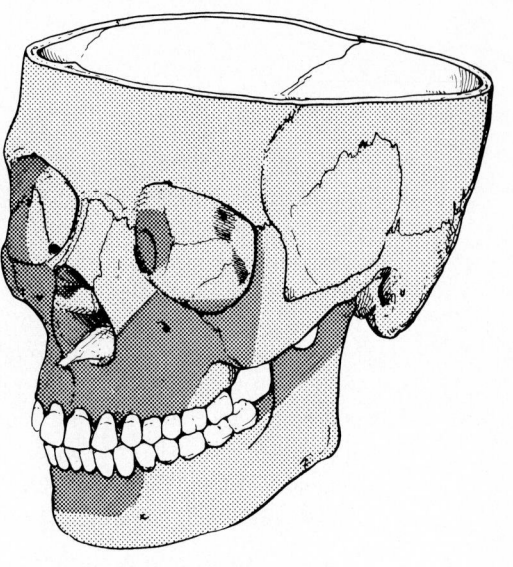

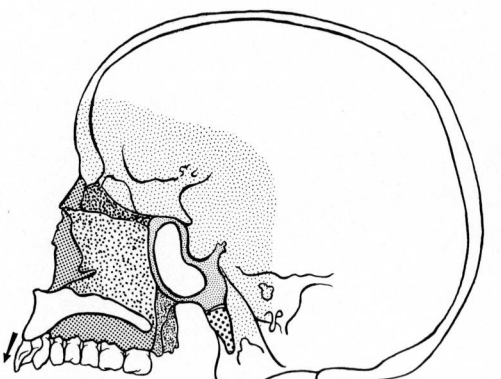

FIGURE 2–53

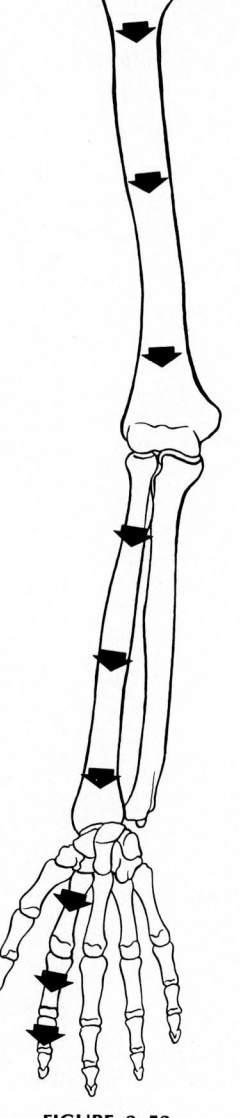

FIGURE 2–54

FIGURE 2–52

**FIGURE 2–55**

Both primary and secondary displacement as well as drift are involved in the growth movements of all bones. A great many different combinations of all three processes are found throughout the craniofacial complex. For example, bones *X* and *Y* are in contact (as by a suture, condyle, or synchondrosis). A growth increment by bone deposition at *a* produces a similar end-effect as deposition at *b*, with accompanying primary displacement of the whole bone to the right. Or increment *c* is added at the contact interface, with accompanying primary displacement of the whole bone to the right. Resorption at *d* occurs, however, producing an end-result equivalent to the above two examples. Or secondary displacement of segment *Y* is caused by separate segment *X* owing to growth addition *e*. Primary displacement accompanies growth at *f*. With resorption at *g*, it is thus seen that this combination also produces end-results similar to all the above examples.

The analysis of composite growth changes is always difficult in headfilm analyses because, as just seen, the **same** growth results can be theoretically attained by many different combinations of drift and displacement. The purpose in Chapter 3 is to analyze just which of the many hypothetical combinations actually take place in each of the many regions of the face and cranium.

**FIGURE 2–56**

The conventional method used to show facial "growth" is to superimpose serial headfilm tracings (i.e., tracings of the same individual at different ages) on the cranial base. Sella is usually used as a registration point for the superimposition. Tracings are ordinarily used instead of the headfilms themselves because superimposed X-ray films pass insufficient light.

Superimposing on the cranial base demonstrates the "downward and forward" (one of the most frequently used cliches in facial biology) expansion of the whole face relative to the cranial base. Great caution must be exercised, however, to be certain one understands possible misrepresentations of just what this really shows.

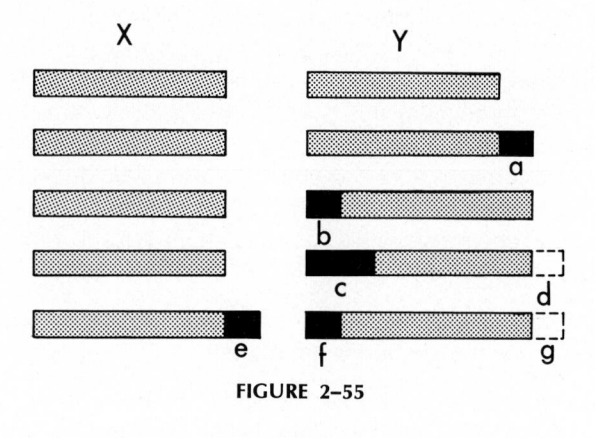

**FIGURE 2–55**

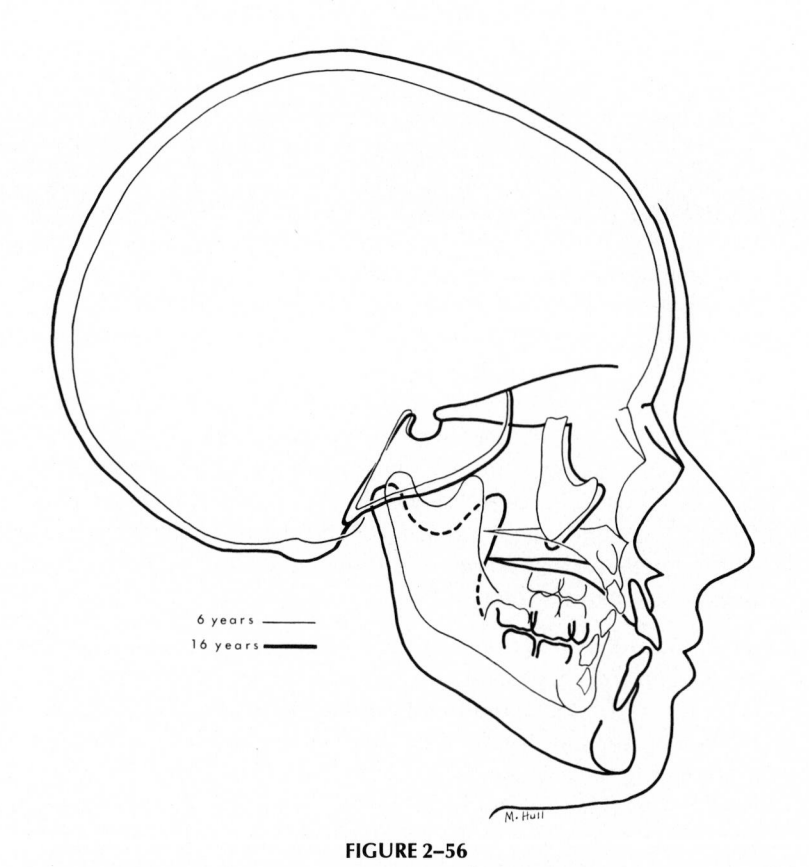

**FIGURE 2–56**

*(From Enlow, D. H.:* The Human Face. *New York, Harper & Row, 1968.)*

**FIGURES 2–57, 2–58, AND 2–59**

First, this method is appropriate and valid because we all naturally tend to visualize facial enlargement in **relation** to the cranium (and brain) behind and above it. That is, the characteristically small face and the earlier-developing, larger brain in early childhood **change** progressively in respective proportions. The face grows and develops rapidly throughout the childhood period; the size and shape of the brain and the cranial vault also change, but much less noticeably. The parent **sees** the structural transformations of the child's face, month by month, as it "catches up" with the earlier-maturing brain and braincase. Superimposing on the cranial base (sella, etc.) thereby represents what one actually visualizes by direct observation as the face progressively enlarges.

Superimposing headfilm tracings "on the cranial base" is **not** valid, however, **if** the following incorrect assumptions are made:

a. The incorrect assumption that the cranial base is truly stable and unchanging. It is not. This notion has often mistakenly been made. The floor of the cranium continues to grow and undergo remodeling changes throughout the childhood period (although this is much more marked in some regions than others at different age levels). Properly taken into account, however, this is not necessarily a factor since the purpose, really, is only to show facial growth changes **relative** to the cranial base, whether or not it is actually stable.

b. The incorrect assumption that "fixed points" indeed actually exist, that is, anatomic landmarks that do not move or remodel. All surfaces, inside and out, undergo continued, sequential growth movements and remodeling changes during morphogenesis (with the exception of no size changes by the ear ossicles). While the **relative** position of some landmarks can remain relatively constant, the structures themselves actually experience significant growth movements and remodeling changes along with everything else (Figure 2–58). Sella *(a)* has often been presumed to be a true "fixed" point or one which represents the "zero growth point" in the head. Of course, it is not. Sella changes during continued growth. This, however, does not invalidate the use of sella to represent a registration point on the cranial base **if** these various considerations are properly taken into account. **Nasion** *(b)* is another such landmark. So many marked growth and remodeling variations are associated with this point relating to age, sex, ethnic, and individual differences, however, that the use of nasion as a cephalometric landmark requires great caution. (**Note:** there are other basic reasons why the use of points such as nasion and sella are misleading if improperly used, as explained in Chapters 4 and 5.)

c. The incorrect assumption that the traditional "forward and downward" picture of facial enlargement, seen when serial tracings are superimposed on the cranial base, represents the **actual mode of facial growth.** Many workers believe, quite incorrectly, that this is how the face really grows, that is, that the facial profile of the younger stage expands straight to the profile of the older stage by direct growth from one to the other. This has been one of our most common misconceptions and one of the most difficult to overcome. The face does not merely "grow" in the manner represented by such an overlay. Growth is a multifactorial, cumulative composite of changes in the many regions of the head, the summation of which produces the "forward and downward" expansion seen in the overlay.

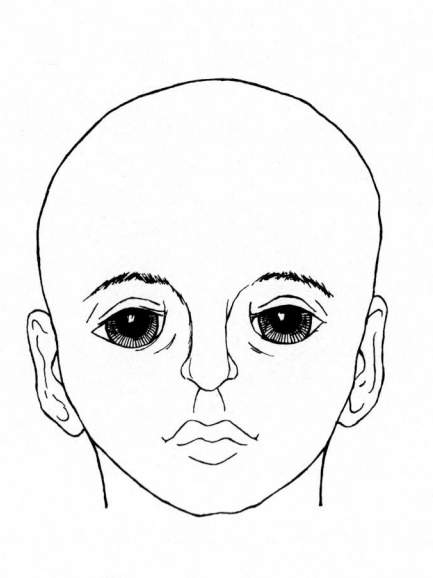

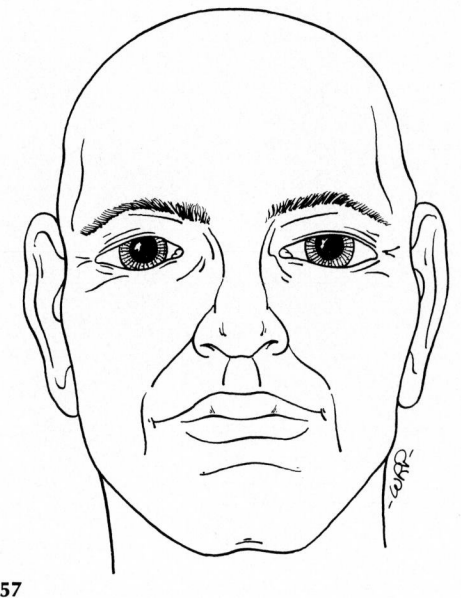

**FIGURE 2–57**

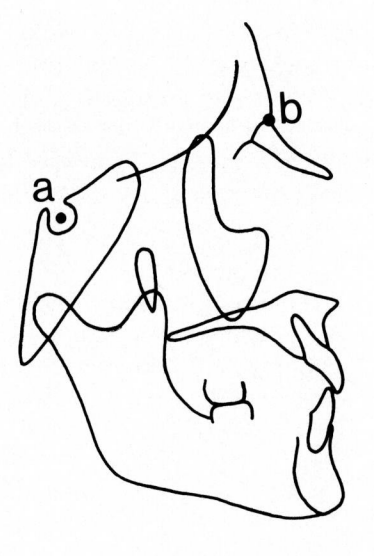

**FIGURE 2–58**

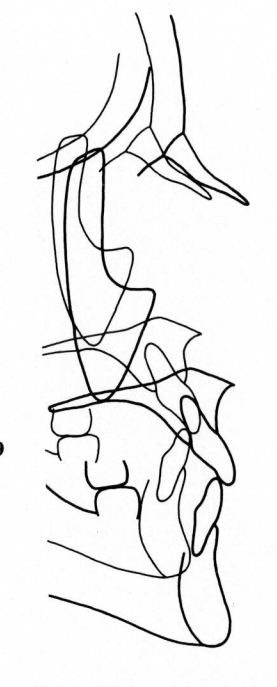

**FIGURE 2–59**

**FIGURE 2–60**

Superimposing headfilm tracings on the cranial base shows the **combined** results of (a) deposition and resorption (drift) and (b) primary and secondary displacement **relative** to a common reference plane (such as sella-nasion of the cranial base). The superimposing procedure, however, does not provide an accurate representation for either (a) drift or (b) displacement in most facial regions. Note that the two placements of the mandible in the preceding Figure 2–56, for example, do not properly represent either its growth by deposition and resorption (a) or its displacement (b), as shown in Figure 2–60. The overlay positions for the mandible in Figure 2–56 (and other facial bones as well) simply indicate their successive **locations** at the two age levels relative to the cranial base, not their actual modes of growth.

One basic problem always encountered with routine methods of superimposing headfilm tracings on the cranial base is that the **separate** effects of (a) growth by deposition and resorption and (b) displacement are **not distinguishable.** This is an important consideration. The purpose of the next chapter is to demonstrate these separate effects and to explain how the process of craniofacial growth is really carried out.

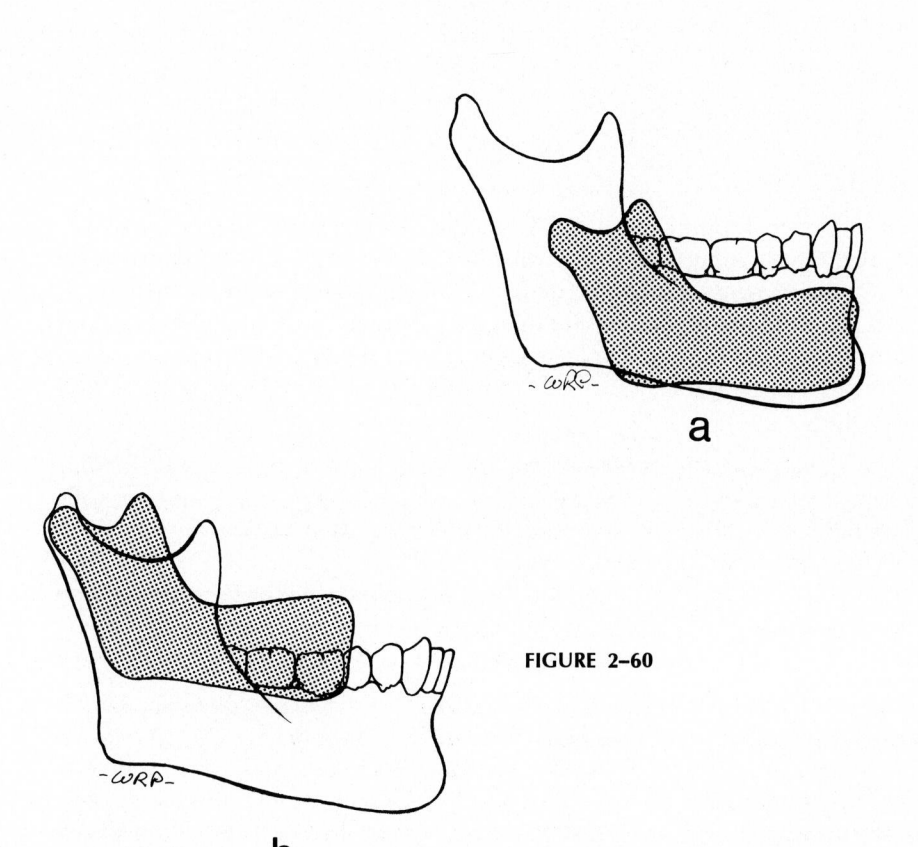

a

b

FIGURE 2-60

# Chapter 3

# The Facial Growth Process

# PART ONE

In the pages that follow, the overall sequence of facial and cranial growth is outlined. Part One is a digest version to provide a general, less detailed understanding of the growth process as a whole. In Part Two, the sequence of growth changes is repeated but with more in-depth information about the underlying theory of the growth mechanisms involved. Before beginning with the regional descriptions of the growth process, the rationale behind these processes is briefly explained below; read carefully.

1. The multiple growth processes in all the various parts of the face and cranium are described separately as individual "regions" or "stages." The sequence will begin arbitrarily with the maxillary arch. Changes are then shown for the mandible followed by growth changes in parts of the cranium and then of the other regions, one by one. Keep in mind that these regional growth processes all take place simultaneously even though they are presented here as a sequence of separate stages.

2. Growth increases are shown in such a way that the same craniofacial form and pattern are maintained throughout; that is, the proportions, shape, relative sizes, and angles are not altered as each separate region enlarges. Thus, the geometric form of the whole face for the first and last stages is exactly the same; only the overall **size** has been changed. Each sequential stage incorporates all the stages that precede it. The final stage is a cumulative composite.

3. Facial and cranial enlargement, in which form and proportions remain constant, constitutes "balanced" growth. However, a perfectly balanced mode of growth in **all** the parts of the face and cranium **never** occurs in real life. Because imbalances always occur during the actual developmental processes, **changes** in facial shape and form always take place as the face grows into adulthood. That is, imbalances in the growth process lead to corresponding imbalances in structure. Most of these "imbalances" are perfectly normal and are a regular part of the developmental and maturation process. This is why the face of a child undergoes sequential alterations in profile and in facial proportions as growth progresses. The mandible of the very young child, for example, is characteristically small in relation to his maxilla but later "catches up" to provide anatomic balance. The forehead is bulbous in the young child but becomes sloping as the frontal sinuses develop. The nasal region is shallow early in postnatal life but later becomes markedly expanded

relative to other cranial and facial regions. Many more such progressively imbalanced changes occur. Thus, **imbalanced** growth is **always** involved in the development of any individual's face. This is also why no two faces are exactly alike; the extent, locations, and patterns of growth changes are highly variable and individualized.

4. The reason why the facial growth descriptions that follow are presented first as a "balanced" series is twofold. They will show just what constitutes the concept of "growth balance" itself and how to understand what this actually means. Second, in order to be able to recognize and explain facial **imbalances,** it is necessary to know what constitutes deviations from the balanced mode of development, that is, exactly **where** disproportions develop to cause a given facial pattern, and **how much** is involved in terms of dimensional and angular departures from balanced growth. Only by understanding the balanced process can one accurately identify, measure, and, importantly, account for the imbalanced.

5. No face has yet been encountered, as mentioned previously, that has a perfect anatomic and geometric "balance" among all of its regions and parts, although a functional equilibrium usually exists. **Variation in form and balance** is the name of our professional game. By analyzing the growth process and the results of facial growth in any given person, we can identify **where** "imbalances" occur, and we can determine just how these developmental variations have caused a given facial pattern. Any face is a composite of a great many regional imbalances, some slight but others sometimes marked. The face of each of us is the aggregate sum of all the many balanced and imbalanced craniofacial parts combined into a composite whole. Regional imbalances often tend to compensate for each other to provide functional equilibrium. The process of **compensation** is a feature of the developmental process; it provides for a certain latitude of imbalance in some areas in order to offset the effects of disproportions in other regions.

6. Because variations in regional cranial and facial balance exist as a **normal** developmental process, many different kinds and categories of facial form and pattern occur. This underlies the characteristic differences associated with age, sex, ethnic group, and individualized features of the face. Some variations, however, exceed the limits of what can be regarded as "normal." Because we can account for the growth of a balanced face, we can also explain many (but not all) developmental and structural factors that relate to the abnormal face. This is the special subject of a later chapter.

7. The regional descriptions of the growth process outlined below are not randomly presented. Rather, a system is used that, in fact, is the same developmental plan utilized in the growth process itself. This is the **counterpart principle** of craniofacial growth. It states, simply, that the growth of any given facial or cranial part relates specifically to **other** structural and geometric "counterparts" in the face and cranium. For example, the maxillary arch is a counterpart of the mandibular arch. These are **regional** relationships throughout the whole face and cranium. If each regional part and its particular counterpart enlarge to the same extent, balanced growth between them is the result. This is the key to what determines the presence or lack of balance in any region. Imbalances are produced by differences in respective amounts or directions of growth between parts and counterparts. Many part-counterpart combinations exist throughout the skull, and these provide a meaningful and easy way to evaluate the growth of the face and the morphologic relationships among all its structural components.

**FIGURES 3–1 AND 3–2**

To illustrate the counterpart principle, an expandable photographic tripod is used here as an analogy. The tripod has a series of telescoping segments in each leg; the length of each segment matches the length of its "counterpart" segments in the other two legs. If all the segments are extended to exactly the same length, the tripod retains geometric balance and overall symmetry. If, however, any one segment is not extended equal to the others, that leg as a whole is either shorter or longer, although the remainder of all the segments in that leg match their respective counterparts. One can thus identify **which** particular segment is different and determine the extent of imbalance. Segment $x$, for example, is short relative to $y$, thus causing a retrusion of $z$.

**FIGURE 3–3**

Many other hypothetical combinations exist. For example, segments $a$, $b$, and $c$ are short with respect to their segment-counterparts in the other legs. Overall symmetry is balanced, nonetheless, because all of these regional imbalances offset each other, and the total length of each leg is therefore the same.

The "test" for a part-counterpart relationship in the face and cranium is not difficult. The question is simply asked: "If a given growth increment is added to a specific bone, **where** must an equivalent increment be added to **other** bones if the same form and balance are to be retained?" The answer to this question then identifies which other specific bones or parts of bones are involved as counterparts. This counterpart concept will be used repeatedly in the present chapter as well as in following chapters dealing with facial variations and abnormalities.

8. The growth process for each region is presented as two separate parts. First, the changes produced by **deposition and resorption** are described and shown by **fine arrows** in the illustrations. Second, the changes produced by **displacement** are described and are represented by **coarse arrows.** These two processes, it is understood, take place at the same time, but they must be described separately because their effects are quite different. Then the question asked is, "Where do counterpart changes also occur if the same pattern is to be maintained?" This identifies the **next** anatomic region, which is then described in turn.

**FIGURES 3–4 AND 3–5**

**Regional Change 1.** Note that two reference lines are used, a horizontal and a vertical,* so that directions and amounts of growth changes can be visualized. The bony maxillary arch lengthens horizontally in a **posterior** direction (this always comes as a surprise to new people in the business, and to some older pros as well). This is schematized by showing a posterior movement of **PTM** (pterygomaxillary fissure). Note its new location behind the vertical reference line.

---

*This vertical line is not arbitrary; it is the *PM* boundary, which is one of the most basic and important natural anatomic planes in the head (see Chapter 4). The horizontal line is the "functional occlusal plane."

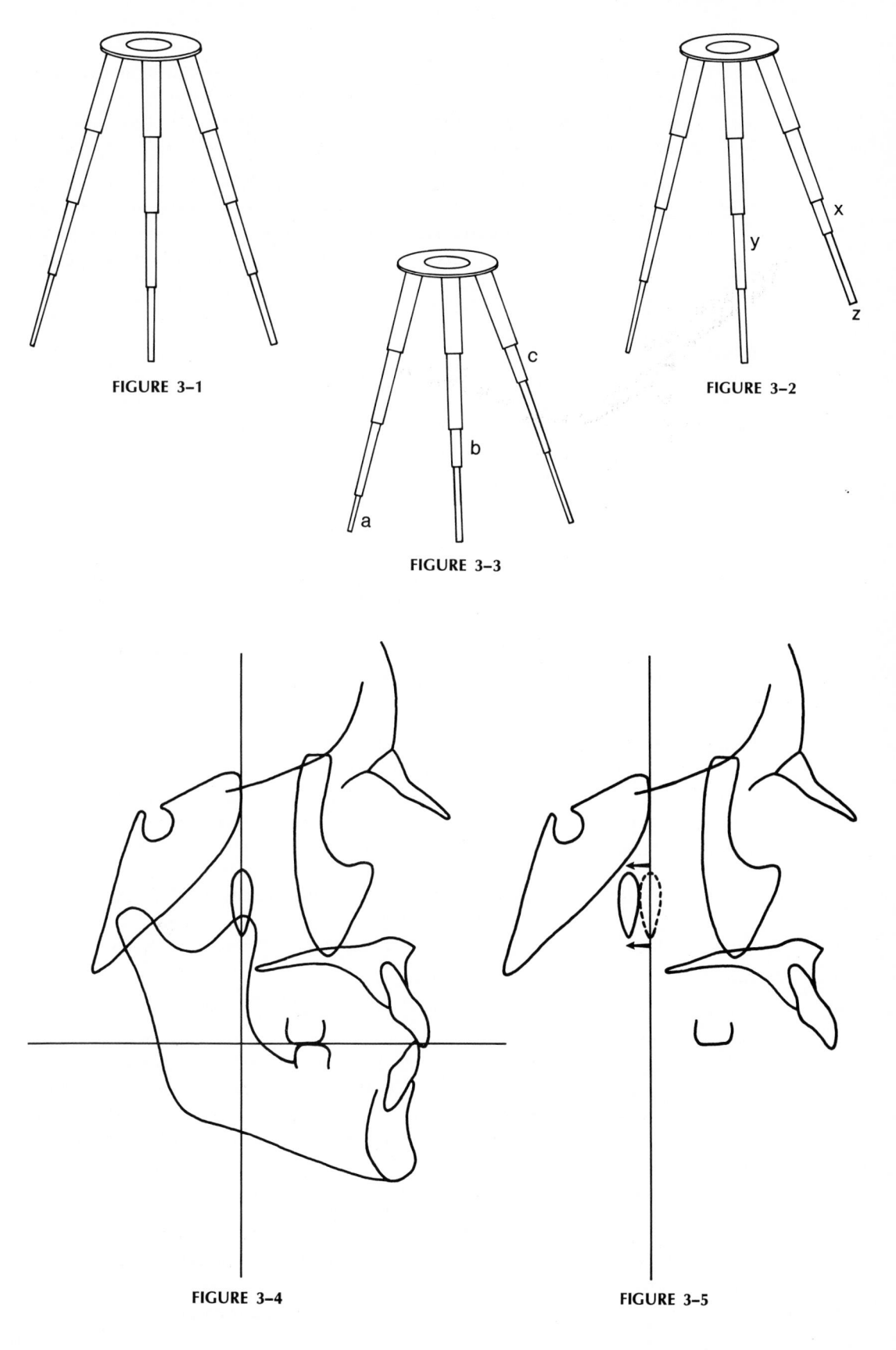

**FIGURE 3-1**

**FIGURE 3-3**

**FIGURE 3-2**

**FIGURE 3-4**

**FIGURE 3-5**

**FIGURE 3–6**

**PTM** is the routine radiographic landmark used to identify the maxillary tuberosity, and it appears on headfilms as an "inverted teardrop" produced by the gap between the pterygoid plates and the maxilla (coarse arrow). The "point" used for headfilm analysis is shown by the fine arrow.

The overall length of the maxillary arch has increased by the same amount that **PTM** moves posteriorly. Bone has been deposited on the posterior-facing cortical surface of the maxillary tuberosity. Resorption occurs on the opposite side of the same cortical plate, which is the inside surface of the maxilla within the maxillary sinus.

**FIGURE 3–7**

**Regional Change 2.** The above stage is the first of the two-part growth process described for each region; that is, growth by deposition and resorption. The second part involves **displacement,** described in the present stage. As the maxillary tuberosity grows and lengthens posteriorly, the whole maxilla is simultaneously **carried** anteriorly. The amount of this forward displacement exactly equals the amount of posterior lengthening. Note that **PTM** is "returned" to the vertical reference line. Of course, it never actually departed from this line because backward growth (Stage 1) and forward displacement (Stage 2) occur at the same time. This is a primary type of displacement because it occurs in conjunction with the bone's own enlargement; that is, as the bone is displaced, it undergoes growth in order to keep pace with the amount of displacement. A protrusion of the forward part of the arch now occurs; not because of direct growth in the forward part itself but rather growth in the **posterior** region of the maxilla as the whole bone is simultaneously displaced anteriorly.

**FIGURE 3–8**

**Regional Change 3.** The question is now asked: "When the elongation of the maxilla in Stage 1 is made, **where** must equivalent changes **also** be made if structural balance is maintained?" In other words, what are the **counterparts** to the bony maxillary arch? Several are involved, including the upper part of the nasomaxillary complex, the anterior cranial fossa, the palate, and the corpus of the mandible. The mandible is described in this stage. The mandible is not to be regarded as a single functional element: it has two major parts, the **corpus** (body) and the **ramus.** These two parts must be considered separately because each has its own separate counterpart relationships with other, different regions in the craniofacial complex.

**FIGURE 3–9**

The bony mandibular arch relates specifically to the bony maxillary arch; that is, the body of the mandible is the structural counterpart to the body of the maxilla. The mandibular corpus now lengthens to match the growth of the maxilla, and it does this by a remodeling conversion from the ramus. The anterior part of the ramus grows posteriorly, a relocation process that produces a corresponding elongation of the corpus. What was ramus has now been remodeled into a new addition onto the corpus. The mandibular arch lengthens by an amount that equals the growth of the maxillary arch (Stage 1), and both elongate in a posterior direction. However, note that the two arches are still offset; the maxilla is in a protrusive position even though upper and lower arch lengths are the same, as seen in Figure 3–10. A "Class II" type of relationship between the maxillary and mandibular molars exists. The proper "Class I" position is seen in Stage I; the mandibular posterior tooth shown in the diagram should normally be about one-half cusp-width ahead of its maxillary antagonist, as seen in Figure 3–4.

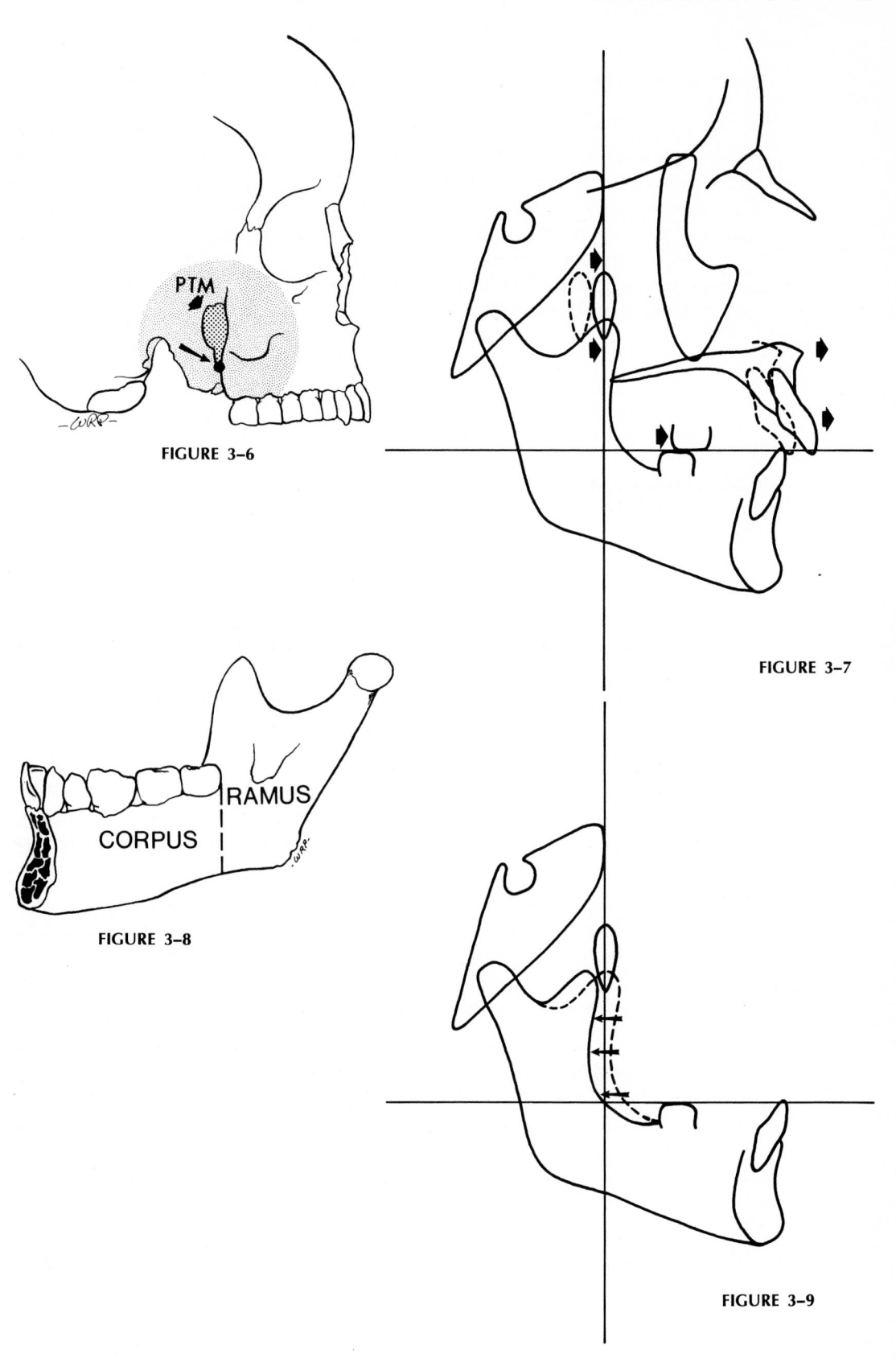

FIGURE 3–6

FIGURE 3–7

RAMUS

CORPUS

FIGURE 3–8

FIGURE 3–9

PTM

**FIGURE 3–10**

**Regional Change 4.** The second of the two growth processes (that is, first, growth by deposition and resorption, and second, displacement) is now described. Remember that these two changes actually occur at the same time. The whole mandible is **displaced** anteriorly, just as the maxilla also becomes carried anteriorly while it simultaneously grows posteriorly. To do this, the condyle and the posterior part of the ramus grow posteriorly. This returns the horizontal dimension of the ramus to the *same breadth* present in Stages 1 and 2 above; the amount of anterior ramus resorption is equaled by the amount of posterior ramus addition. The purpose is not to increase the size of the ramus itself but to relocate it posteriorly in order to provide lengthening for the **corpus.**

**FIGURE 3–11**

**Regional Change 5.** The whole mandible, now, is displaced anteriorly by the same amount that the ramus has relocated posteriorly. This is the primary type of displacement because it occurs in conjunction with the bone's own enlargement. As the bone becomes displaced, it simultaneously grows to keep pace with the amount of displacement. Note the following:

a. The corpus of the mandible grows in a **posterior** direction, just as the maxilla also grows posteriorly (Stage 1). It does this by remodeling from what **was** ramus into what then becomes a posterior addition to the mandibular arch. In this respect, mandibular arch elongation differs from maxillary arch elongation because the maxillary tuberosity is a free surface, unlike the posterior end of the mandibular corpus.

b. The whole ramus has moved posteriorly. However, the only actual change in horizontal dimension involves the mandibular corpus, which becomes longer. The horizontal dimension of the ramus remains constant during **this** particular remodeling stage (the widening of the ramus itself is part of another stage).

c. The anterior **displacement** of the whole mandible equals the amount of anterior maxillary displacement. This places the mandibular arch in proper position relative to the maxillary arch just above it. The arch lengths as well as the positions of the maxilla and mandible are now in balance, and a "Class I" positioning of the teeth has been "returned."

d. Note, however, that the obliquely upward and backward direction of ramus growth must also lengthen its **vertical** dimension in order to provide for horizontal enlargement. This separates the occlusion (contact between the upper and lower teeth) because the mandibular arch is displaced inferiorly as well as anteriorly.

e. In both the maxilla and mandible, the type of displacement is *primary* because it takes place in conjunction with each bone's own enlargement.

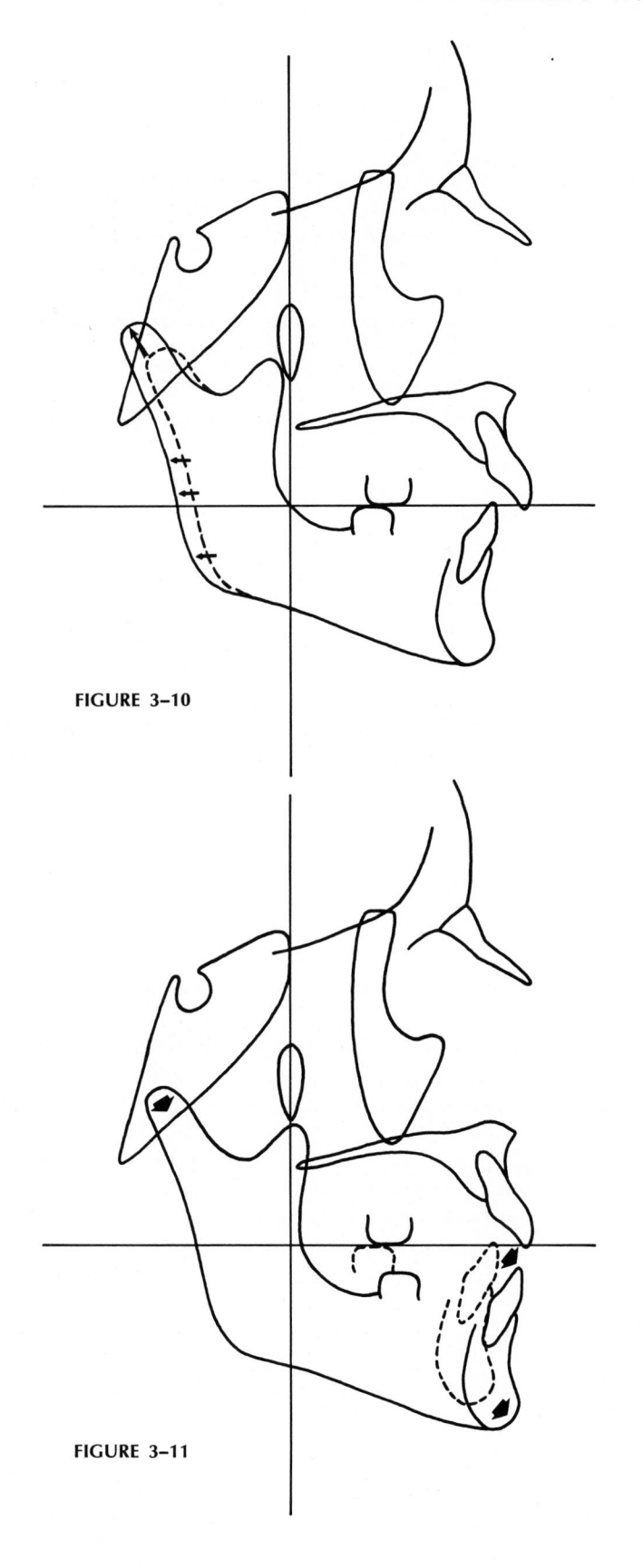

FIGURE 3–10

FIGURE 3–11

**FIGURE 3–12**

*f.* In summary, thus far: the increment of backward growth at the maxillary tuberosity (Stage 1); the amount of forward displacement by the whole maxilla (Stage 2); the extent of remodeling on the anterior part of the ramus and the amount of corpus lengthening (Stage 3); the increment of backward growth by the posterior part of the ramus (Stage 4); and the amount of forward displacement of the whole mandible (Stage 5) are all **precisely equal** in this "balanced" sequence of growth. What happens if they are not all exactly equal (as usually happens), or if differentials in timing occur, is described later.

**FIGURE 3–13**

**Regional Change 6.** While all of the growth and remodeling changes described in the preceding stages have been taking place, the dimensions of the middle cranial fossa have also been increasing at the same time. This is done by resorption of the endocranial side and deposition of bone on the ectocranial side of the cranial floor. The spheno-occipital synchondrosis (a major cartilaginous growth site in the cranium) provides endochondral bone growth in the midline part of the cranial floor.* The total growth expansion of the middle fossa now projects it anteriorly beyond the vertical reference line.

**FIGURE 3–14**

**Regional Change 7.** All cranial and facial parts lying anterior to the middle cranial fossa (in front of the vertical reference line) become **displaced** in a forward direction as a result. The whole vertical reference line moves anteriorly to the same extent that the middle cranial fossa expands in a forward direction. This is because the line represents the anterior boundary between the enlarging middle cranial fossa and the cranial and facial parts in front of it. The maxillary tuberosity remains in a constant position on the vertical reference line as this line moves forward. The forehead, anterior cranial fossa, cheekbone, palate, and maxillary arch all undergo protrusive displacement in an anterior direction. This is a **secondary** type of displacement because the actual enlargement of these various parts is not directly involved. They are simply moved anteriorly because the middle cranial fossa behind them expands in this direction. The floor of the fossa, however, does not **push** the anterior cranial fossa and the nasomaxillary complex forward. Rather, they are **carried** forward as the interface between the frontal and temporal lobes of the cerebrum becomes "separated" by their respective growth increases.

---

*Note the change in the position of the sella turcica. This is a highly variable structure, however, and other patterns of remodeling movements are also common. See Part Two of this chapter.

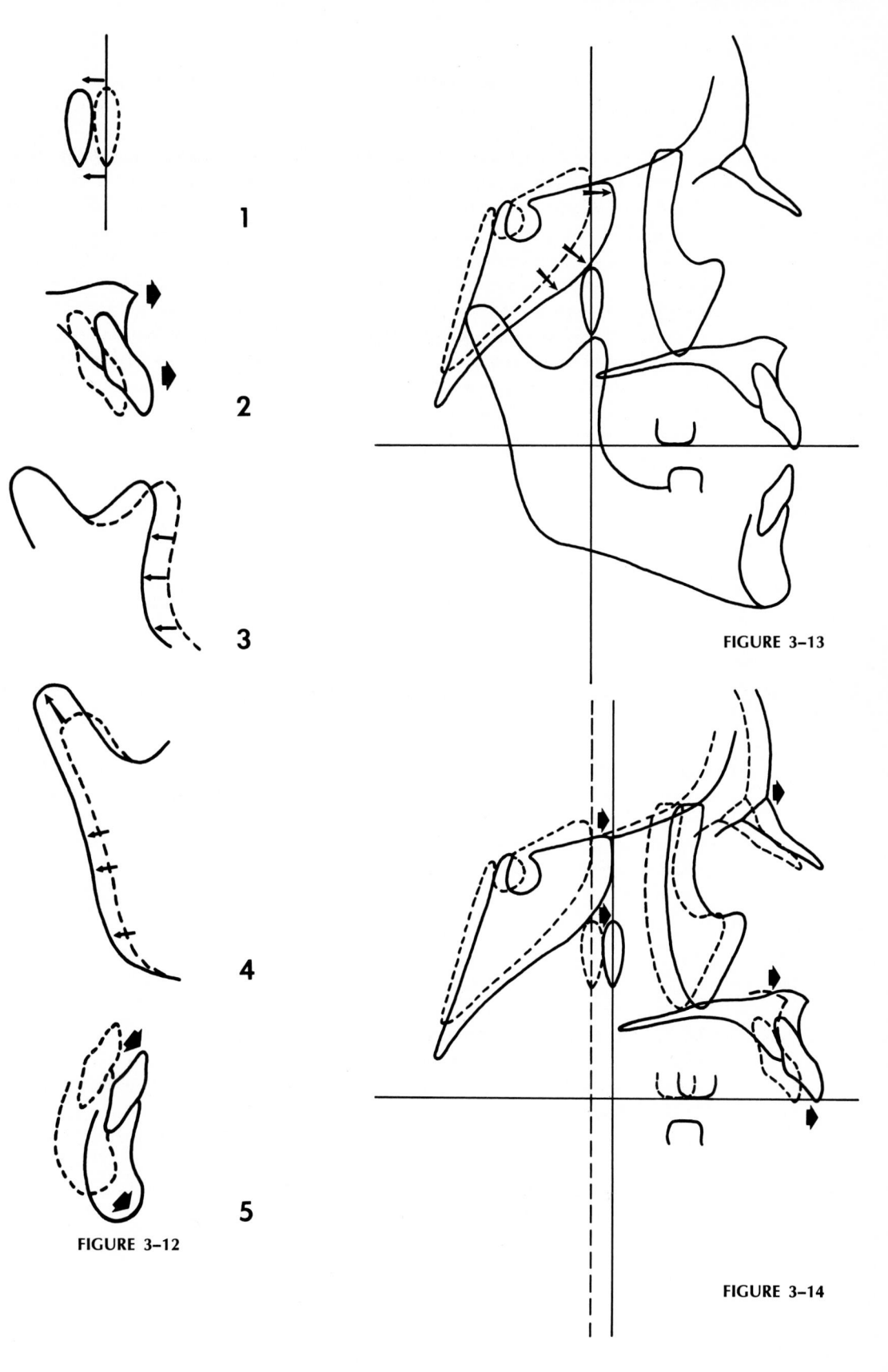

1

2

3

4

5

FIGURE 3–12

FIGURE 3–13

FIGURE 3–14

**FIGURES 3–15 AND 3–16**

**Regional Change 8.**  The expansion of the middle cranial fossa, just described, also has a displacement effect on the mandible. It, too, is a secondary type of displacement. The extent of the displacement effect, however, is much less than that on the maxilla. This is because the greater part of middle cranial fossa growth occurs in front of the condyle and **between** the condyle and the maxillary tuberosity. The spheno-occipital synchondrosis also lies between the condyle and the anterior boundary of the middle cranial fossa. Thus, the amount of maxillary protrusive displacement far exceeds the amount of mandibular protrusive displacement caused by middle fossa enlargement. The result is an offset horizontal placement between the upper and lower arches. The upper incisors show an "overjet," and the molars are in a "Class II" position, even though the mandibular and maxillary arch lengths themselves are matched in respective dimensions. **Sella-nasion** (a much used cephalometric plane) should not be used to represent the "upper face" dimension in comparisons with the entire mandibular dimension, ramus and corpus, as is often done. The comparison is invalid because dissimilar effective spans are being compared and because sella-nasion itself does not represent any anatomically meaningful dimension, either for the cranial base or for the upper face.

**Regional Change 9.**  The question is now asked: "When this change in the middle cranial fossa takes place, **where** must an equivalent change **also** occur if balance is to be maintained?" This identifies the "counterpart" of the middle fossa and shows where facial growth must take place to match it.

Just as the lengthening of the middle cranial fossa places the maxillary arch in a progressively more anterior position, the horizontal growth of the **ramus** places the mandibular arch in a like manner. What the middle cranial fossa does for the maxillary body, in effect, the ramus does for the mandibular body. **The ramus is the specific structural counterpart of the middle cranial fossa.** Both are also counterparts of the **pharyngeal space.** The skeletal function of the ramus is to bridge the pharyngeal space and the span of the middle cranial fossa in order to place the mandibular arch in proper anatomic position with the maxilla.

**FIGURE 3–17**

The horizontal extent of middle cranial fossa elongation is **matched** by the corresponding extent of horizontal increase by the ramus. The horizontal (not oblique) dimension of the ramus now equals the horizontal (not oblique) dimension of the middle cranial fossa. The effective span of the latter, as it relates to the ramus, is the straight-line distance from the cranial floor–condyle articulation to the vertical reference line. Recall that the ramus was previously involved in remodeling changes associated with corpus elongation (Stage 4), but the actual breadth of the ramus was not increased during that particular stage.

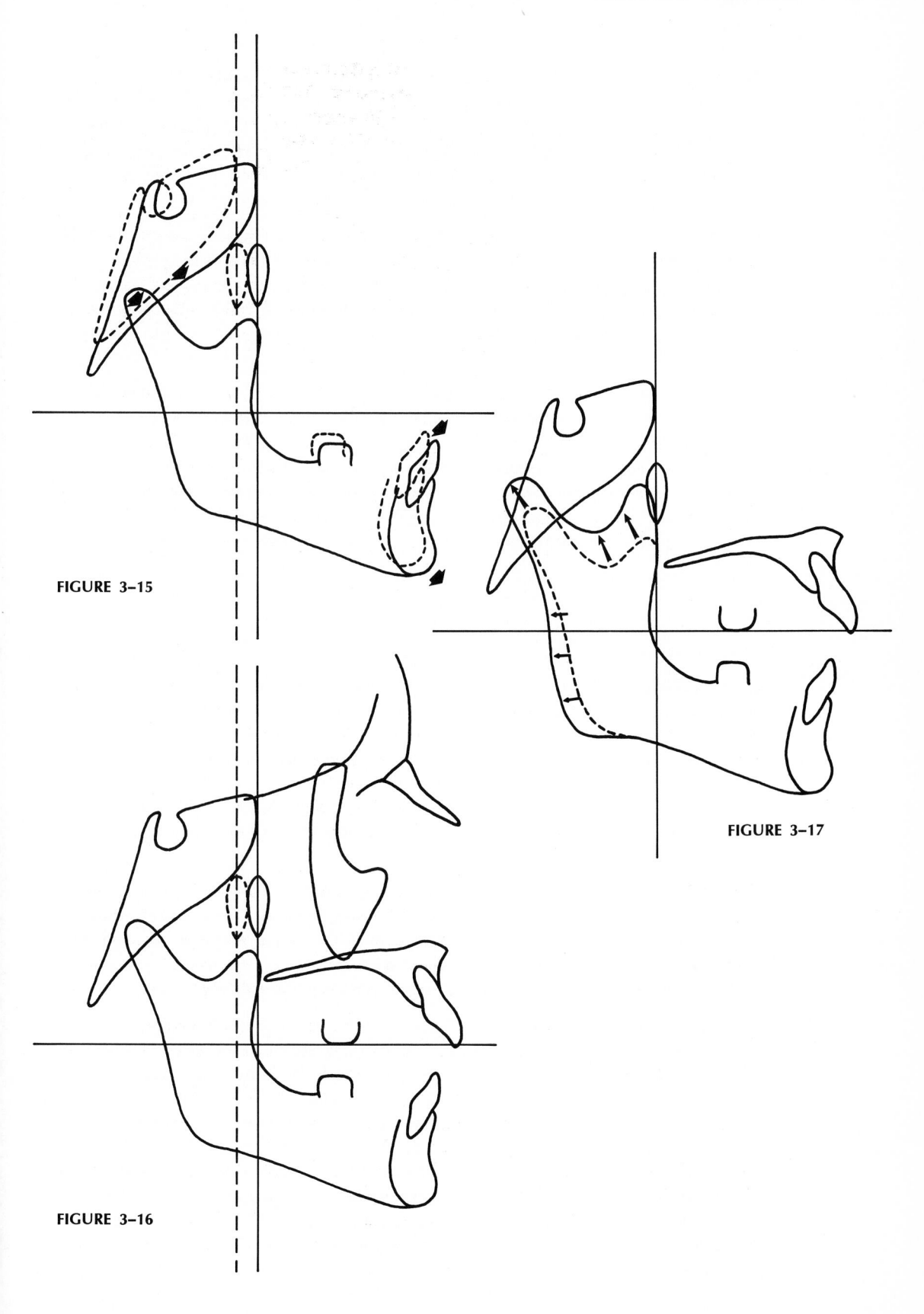

FIGURE 3–15

FIGURE 3–16

FIGURE 3–17

**FIGURE 3–18**

**Regional Change 10.** The entire mandible is displaced anteriorly at the same time that it grows posteriorly. The amount of this anterior displacement equals (a) the extent of posterior ramus and condylar growth (Stage 9); (b) the amount of middle cranial fossa enlargement anterior to the mandibular condyle (Stage 6); (c), the extent of anterior movement of the vertical reference line; and (d) the extent of resultant anterior maxillary displacement (Stage 7).

The oblique manner of condylar growth necessarily produces an upward and backward projection of the condyle with a corresponding downward as well as forward direction of mandibular displacement. The ramus thus becomes vertically as well as horizontally enlarged. This results in a **further** descent of the mandibular arch and separation of the occlusion (it was also previously lowered during Stages 5 and 8).

Note that the protrusion of the maxilla during Stage 7 has now been matched by an equivalent amount of mandibular protrusion. The molars have once again been "returned" to Class I positions, and the upper incisor has no overjet. Note also that the anterior border of the ramus lies ahead of the vertical reference line. The "real" junction between the ramus and corpus, however, is the lingual tuberosity housing the last molar, not the "anterior border." The lingual tuberosity lies on the vertical reference line behind the anterior border, which overlaps this tuberosity (not shown in the figure; observe this on a dry mandible).

**FIGURE 3–19**

**Regional Change 11.** The **floor** of the anterior cranial fossa and the forehead grow by deposition on the ectocranial side with resorption from the endocranial side. This displaces the nasal bones anteriorly. The posterior-anterior length of the anterior cranial fossa is now in balance with the extent of horizontal lengthening by its structural counterpart, the maxillary arch (Stage 1). Because these two regions have undergone equivalent growth increments, the profile retains its originally balanced form. (Actually, age differentials, both in timing and amount, always occur, but our present purpose is to describe perfectly "balanced" growth.)

The enlarging brain displaces the bones of the **calvaria** (domed skull roof) outward. Each bone enlarges by sutural growth. As the brain expands, the sutures respond by depositing new bone at the contact edges of such bones as the frontal, parietal, and temporal. This expands the perimeter of each. At the same time, bone is laid down on **both** the ectocranial and endocranial sides to increase the thickness.

The upper part of the face, which is the ethmomaxillary (nasal) region, also undergoes equivalent growth increments. This facial area increases horizontally to an extent that matches (if the same balance is retained) the expansion of the anterior cranial fossa above and the maxillary arch and palate below it. These areas are all counterparts to each other. The growth process involves direct bone deposition on the forward-facing cortical surfaces of the ethmoid, the frontal process of the maxillary, and the nasal bones. Most of the internal surfaces of the nasal chambers are resorptive. In addition, anterior displacement takes place in conjunction with growth at the various maxillary and ethmoidal sutures. The composite of these changes produces an enlargement of the nasal chambers in an anterior (and also lateral) direction.

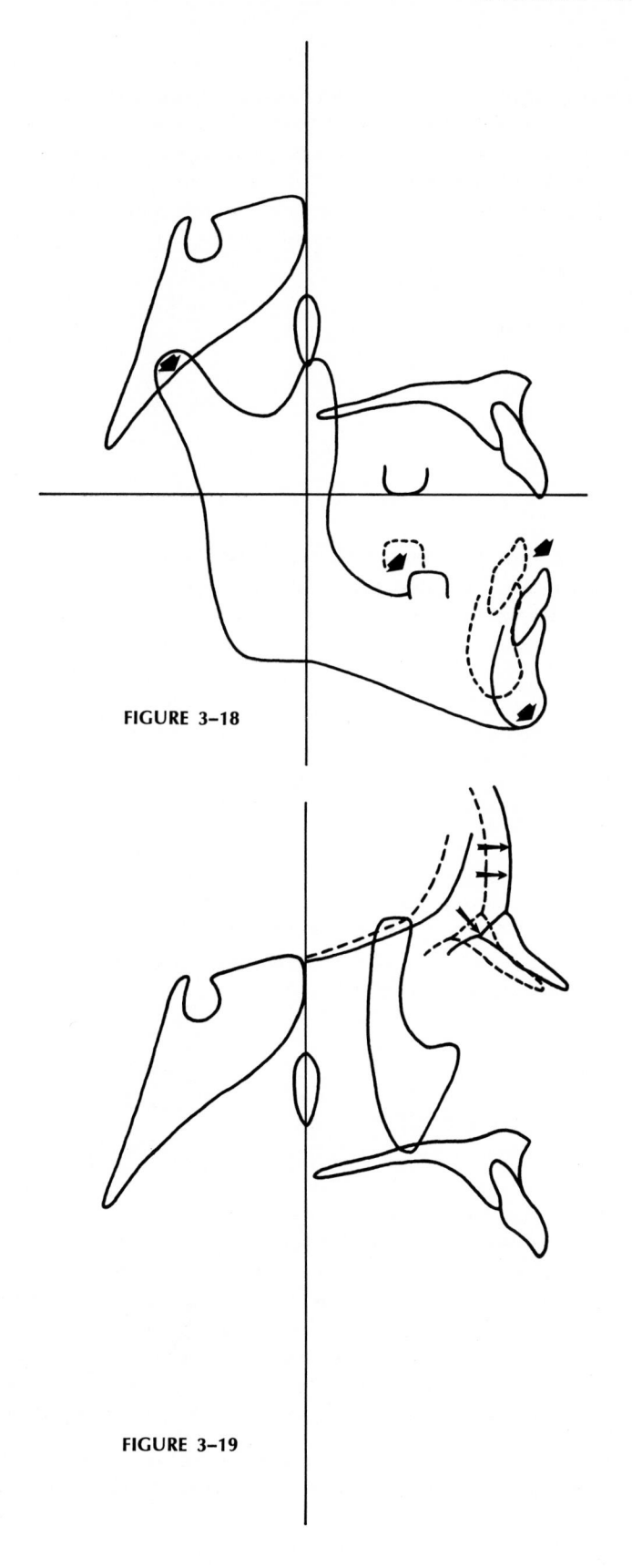

**FIGURE 3-18**

**FIGURE 3-19**

FIGURE 3–20

**Regional Change 12.**   The **vertical** lengthening of the nasomaxillary complex, like its horizontal elongation, is brought about by a composite of (a) growth by deposition and resorption and (b) a primary displacement movement associated directly with its own enlargement. The latter is considered in a later stage. The combination of resorption on the superior (nasal) side of the palate and deposition on the inferior (oral) side produces a downward growth movement of the whole palate from 1 to 2. This **relocates** it inferiorly, a process that provides for the vertical enlargement of the overlying nasal region. The extent of nasal expansion is considerable during the childhood period.

FIGURE 3–21

The anterior part of the bony maxillary arch has a periosteal surface that is **resorptive** (the human, with reduced jaws, is the only species having this), because this area grows **straight downward.** In other animals, the premaxillary region grows forward as well as downward to produce an elongate muzzle.

FIGURE 3–22

As seen in this diagram, the labial (external) side of the premaxillary region faces **away** from the downward direction of growth, and it is thus resorptive. The lingual side faces toward the downward growth directions and is depository. This growth pattern also provides for the remodeling of the alveolar bone as it adapts to the variable positions of the incisors.

FIGURE 3–23

**Regional Change 13.**   Vertical growth by **displacement** is associated with growth at the various **sutures** of the maxilla where it contacts the other separate bones above it. Bone is added at these sutures (only the frontomaxillary and frontonasal are schematically represented here) as the whole maxilla is displaced inferiorly at the same time. The addition of new sutural bone does not "push" the maxilla downward. Rather, the maxilla is **carried** inferiorly by other physical growth forces. This triggers bone deposition in the sutures, and new bone is simultaneously laid down on the sutural edges, keeping the bone-to-bone junction intact.

The increment of bone growth in the suture exactly equals the amount of inferior displacement of the whole maxilla. This is **primary** displacement because it takes place in conjunction with the bone's own enlargement.

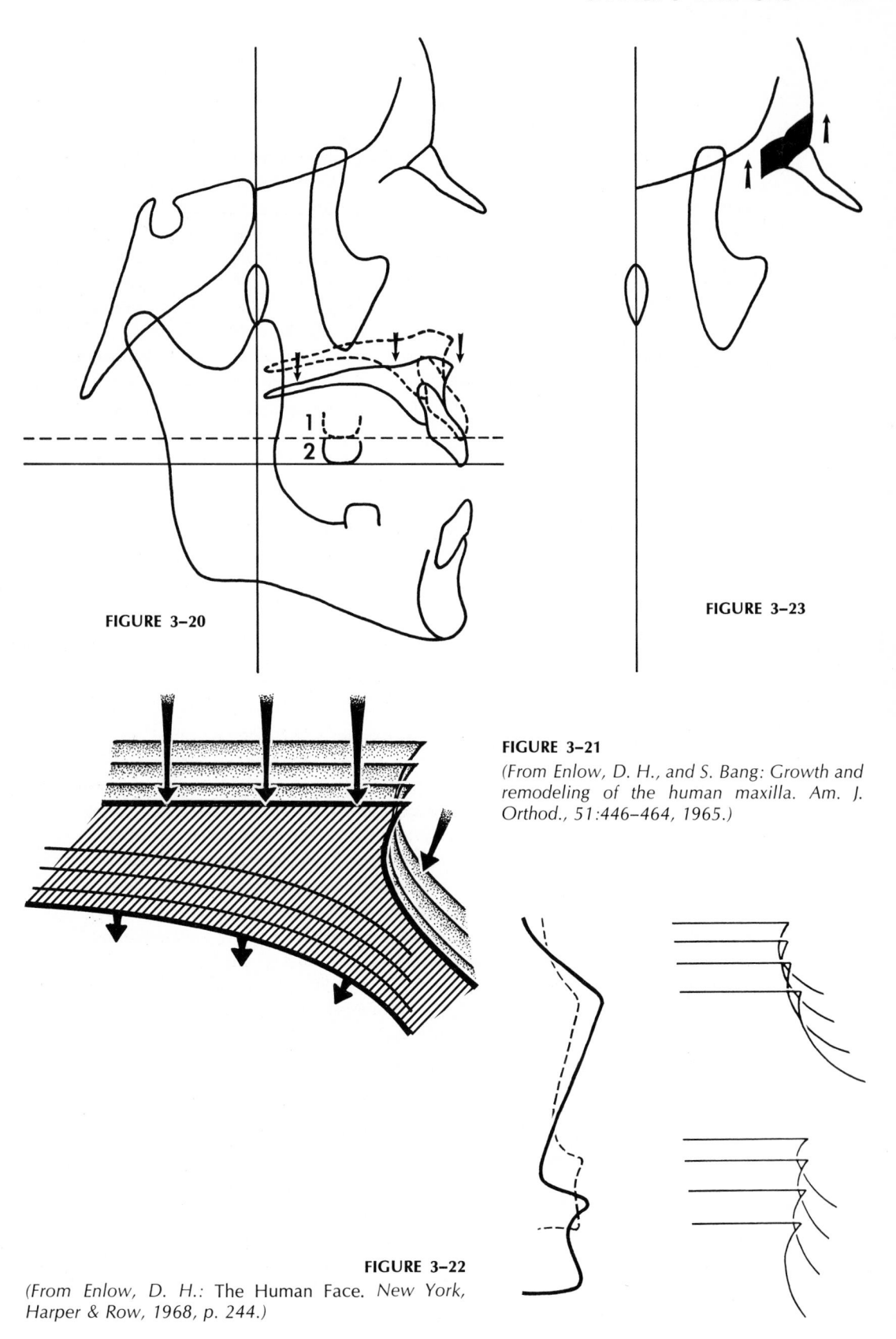

**FIGURE 3–20**

**FIGURE 3–23**

**FIGURE 3–21**
*(From Enlow, D. H., and S. Bang: Growth and remodeling of the human maxilla. Am. J. Orthod., 51:446–464, 1965.)*

**FIGURE 3–22**
*(From Enlow, D. H.: The Human Face. New York, Harper & Row, 1968, p. 244.)*

**FIGURE 3–24**

Of the total extent of downward movement by the palate and maxillary arch, that part from *2* to *3* is produced in association with sutural growth and primary **displacement.** The part from *1* to *2*, which is about half of the total, is direct cortical growth and relocation by resorptive and depository remodeling. Similarly, the movement of the teeth from *2* to *3* is by the downward displacement of the whole maxilla, carrying the dentition passively with it. The movement from *1* to *2* is produced by each tooth's **own** movement as bone is added and resorbed on appropriate lining surfaces in each socket. This is the **vertical drift** of the tooth, a process that is carried out by the **same** deposition and resorption of alveolar bone that produces the familiar "mesial drift" of the dentition (see Chapter 11). Vertical drift takes place **in addition to** eruption, which is a separate growth movement. The vertical drift process is important to the clinician because it provides a great deal of growth movement to "work with" during treatment. Tooth movement from *2* to *3* can also be clinically influenced. This involves the use of "heavy" or orthopedic forces to cause remodeling changes in the size or shape of the whole maxilla or of other separate bones (in contrast to remodeling of the alveolar bone supporting the teeth).

**FIGURE 3–25**

**Regional Change 14.** In three previous stages (5, 8, and 10) it was seen that the mandibular corpus becomes lowered because of the vertical enlargement of both the ramus and middle cranial fossa. Their combined vertical dimensions represent the growth counterpart of the vertical dimension of the nasomaxillary complex and the dentition. In other words, the amount of vertical separation between the upper and lower arches caused by the vertical growth of both the middle cranial fossa and the ramus must be balanced by an equivalent amount of vertical growth in the nasomaxillary complex and the alveolar-dental region of the mandible. The maxillary arch has grown downward to level 3 in Stage 13. Now the mandibular teeth and alveolar bone grow **upward** to attain full occlusion. This is produced by a superior **drift** of each mandibular tooth, together with a corresponding increase in the height of the alveolar bone. The extent of this upward growth movement plus that of the downward growth movement by the maxillary arch equals the combined extent of vertical growth by the ramus and middle cranial fossa **if** the pattern of the face is not changed. Note this factor: the extent of downward drift of the maxillary teeth greatly exceeds the extent of upward drift by the mandibular teeth. Much less growth is thus available to work with in major orthodontic movements of the mandibular as compared to the maxillary teeth.

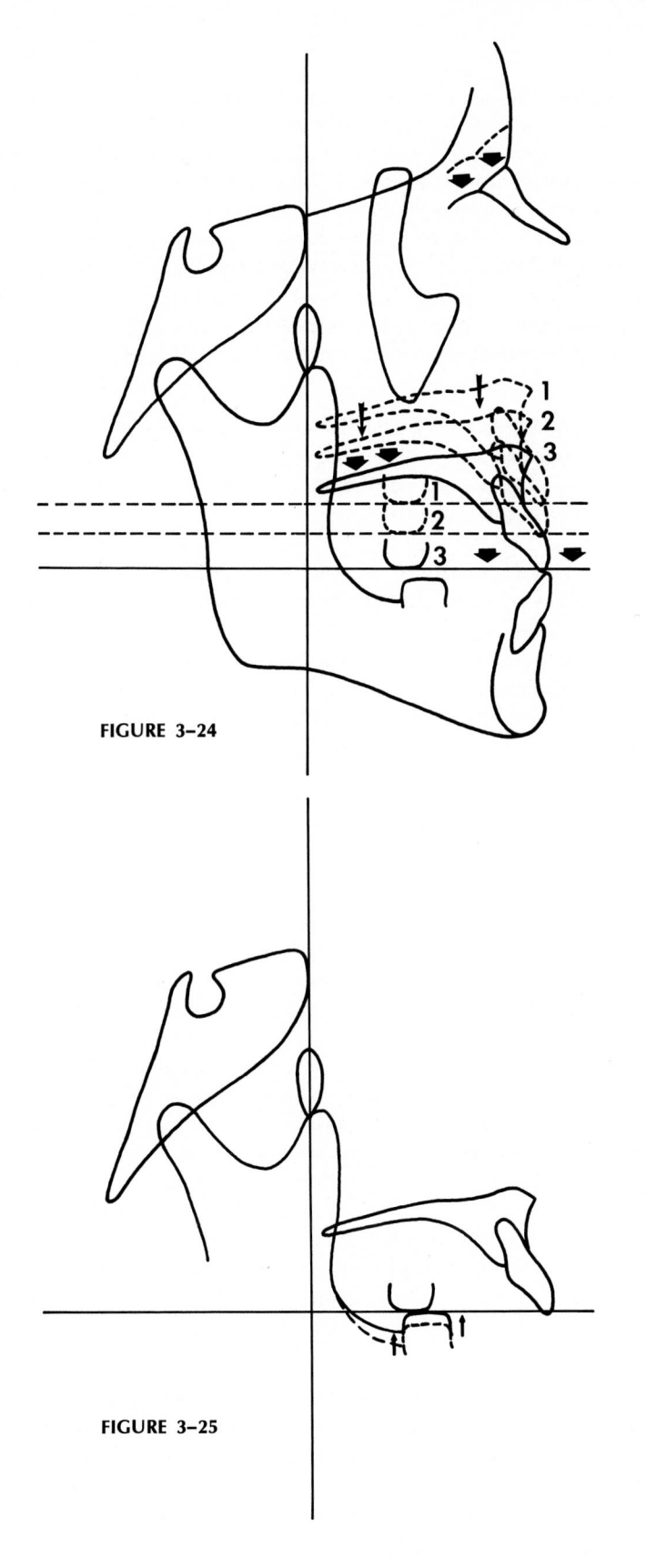

**FIGURE 3–24**

**FIGURE 3–25**

FIGURE 3–26

**Regional Change 15.** While the upward growth movements of the mandibular teeth and alveolar sockets are taking place, remodeling changes also occur in the incisor alveolar region, the chin, and the corpus of the mandible. The lower incisors undergo a lingual tipping (a "retroclination") so that the uppers overlap the lowers to give a proper amount of **overbite.** This involves a posterior rotation movement of the mandibular incisors as they simultaneously drift superiorly. The movement of the teeth is accompanied by **resorption** on the outside surface of the alveolar region just above the chin (and deposition on the lingual side). The alveolar bone thus moves backward as the incisors undergo lingual drift. This does not occur to the same extent in individuals having an "end-to-end" incisor relationship.

Bone is progressively added to the external surface of the chin itself, as well as along the underside and other external surfaces of the corpus. This is slow growth that proceeds gradually throughout childhood. At birth, the mental protuberance is small and inconspicuous. Many anxious parents naturally worry about the chinless appearance of their little child. However, the whole mandible usually tends to lag in differential growth timing and will later catch up to the maxilla in the normal face. The chin takes on more noticeable form year by year. The combination of new bone growth on the chin itself and the posterior direction of bone growth in the alveolar region just above it gradually causes the chin to become more prominent. The whole mandible, meanwhile, is also becoming displaced anteriorly in conjunction with continued growth at the condyle and overall mandibular lengthening.

FIGURE 3–27

**Regional Change 16.** The forward part of the zygoma and the malar region of the maxilla grow in conjunction with the contiguous maxillary complex, and their respective modes of growth are similar. Just as the maxilla lengthens horizontally by posterior growth, the malar area also grows posteriorly by continued deposition of new bone on its posterior side and resorption from its anterior side. The front surface of the whole cheekbone area is thus actually resorptive. This remodeling process keeps its position in proper relationship to the lengthening maxillary arch as a whole. They **both** grow backward, thereby maintaining the proper anatomic positions between them. The amount of deposition on the posterior side, however, exceeds resorption on the anterior surface, so that the whole malar protuberance becomes larger. Another way of understanding the rationale for the growth of the zygomatic process of the maxilla is to compare it with the coronoid process of the mandible. Just as the coronoid process grows backward by anterior resorption and posterior deposition to keep pace with the overall posterior elongation of the whole bone, the zygomatic process similarly grows posteriorly by anterior resorption and posterior deposition.

Note that the vertical length of the lateral orbital rim increases by sutural deposits at the frontozygomatic suture. The zygomatic arch also enlarges considerably by bone deposition along its inferior edge. The arch grows laterally (not seen, of course, in lateral headfilms) by bone deposition on the lateral surface, together with resorption from the medial side within the temporal fossa.

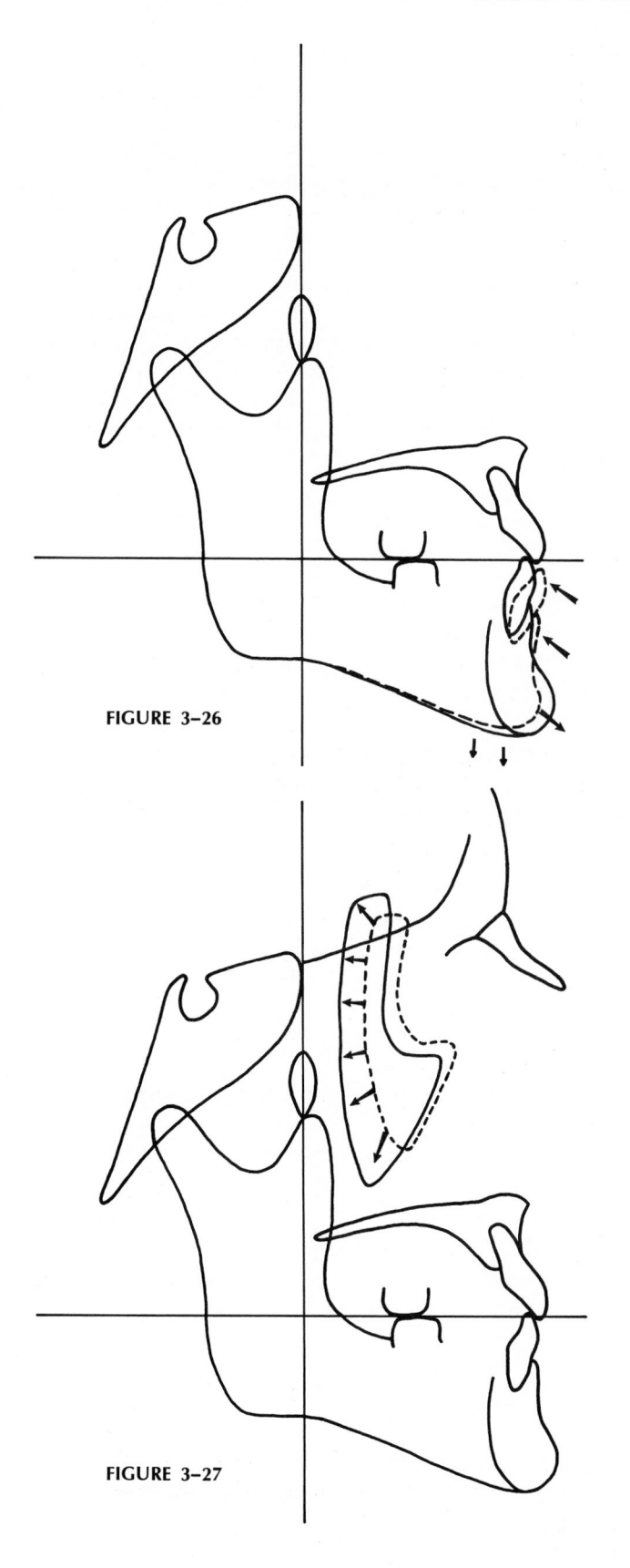

**FIGURE 3–26**

**FIGURE 3–27**

FIGURE 3–28

**Regional Change 17.** Just as the whole maxillary complex is displaced anteriorly and inferiorly as it simultaneously enlarges in overall size, the zygoma is also moved anteriorly and inferiorly by primary **displacement** as it enlarges.The zygoma thereby proportionately matches the maxilla in (1) the directions and amount of horizontal and vertical growth and (2) the directions and amount of primary displacement.

This completes the introductory survey of the regional growth changes taking place in the cranial base and face. The final result is a craniofacial composite that has the same form and pattern present when the first stage was begun. Only the overall size has been altered. All the growth changes among the specific parts and counterparts have been purposefully balanced to give an understanding of the meaning of "balanced growth" and to provide a basis for analyzing **imbalanced** growth changes in a later chapter.

FIGURE 3–29

Here, the first and last stages are superimposed with sella as a registration point. When the sequence of changes described in Stages 1 through 17 are considered, it is apparent that the face does not simply grow directly from one profile to the other. Rather, all the regional changes just outlined are involved. The overlay seen here is the traditional way of representing the results of the overall process of facial enlargement. This overlay does **not,** however, represent the actual growth processes themselves: that is, the changes produced (1) by resorption and deposition and (2) by primary and secondary displacement. Many workers have not always appreciated this basic and important fact. The overlay shows the cumulative summation of **both** (1) and (2).

For quick review before going on to Part 2, all 17 stages are briefly summarized below:

FIGURE 3–30

*Stage 1.* The bony maxillary arch lengthens by posterior growth at the maxillary tuberosity.

FIGURE 3–31

*Stage 2.* The whole maxilla is displaced anteriorly by the same amount it grows posteriorly. These two processes take place simultaneously.

FIGURE 3–32

*Stage 3.* The bony mandibular arch lengthens by remodeling conversion from the anterior part of the ramus. It now matches the maxilla in length, but the two arches are still offset.

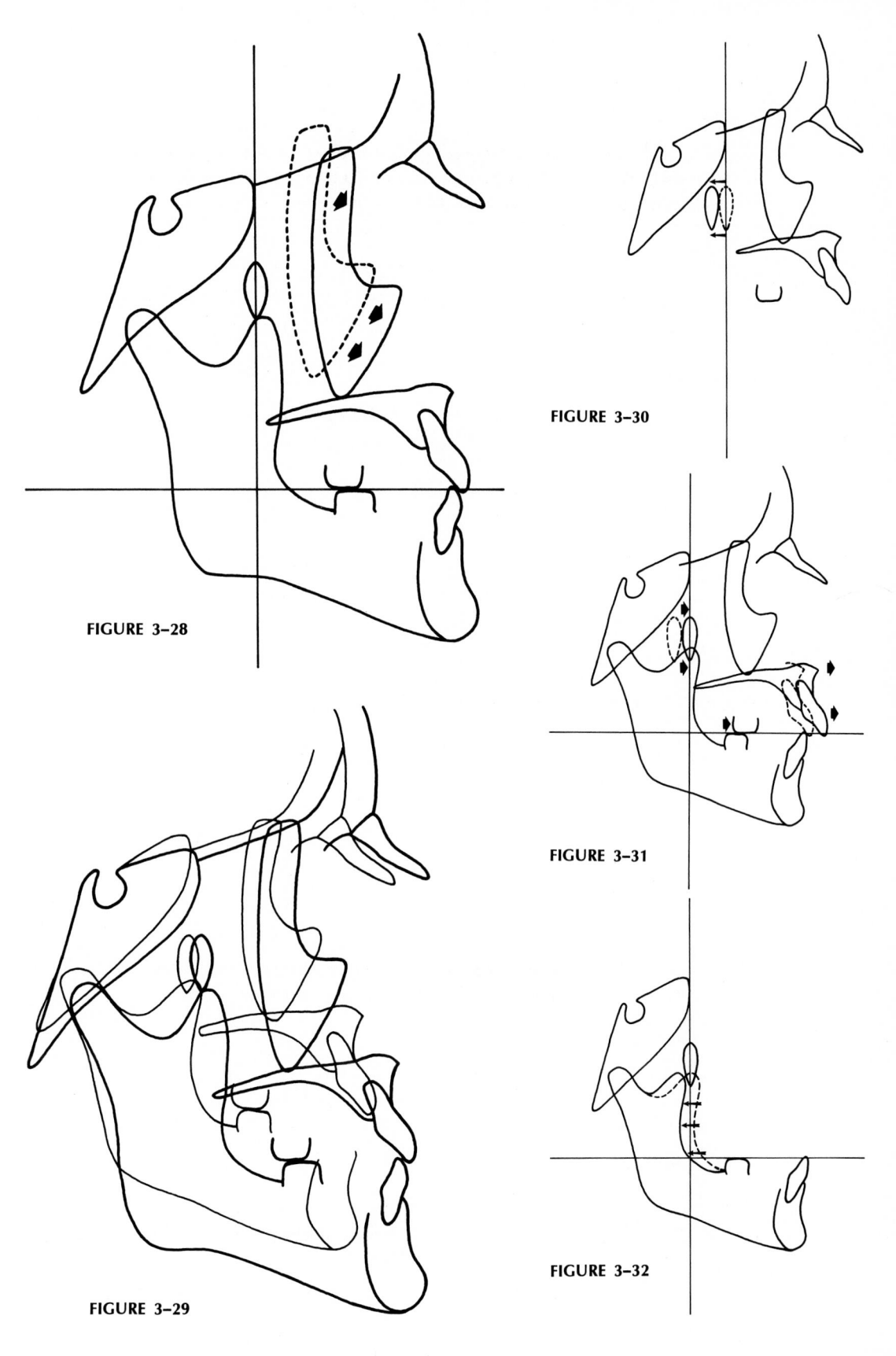

**FIGURE 3–28**

**FIGURE 3–29**

**FIGURE 3–30**

**FIGURE 3–31**

**FIGURE 3–32**

**FIGURE 3–33**

*Stage 4.* The whole ramus grows posteriorly to allow for corpus lengthening. Bone growth occurs at the mandibular condyle and along the posterior part of the ramus to the same extent that the anterior part has undergone resorption.

**FIGURE 3–34**

*Stage 5.* The whole mandible is displaced anteriorly (and inferiorly) by the same amount that the maxilla was displaced in Stage 2. This properly places the mandibular arch relative to the maxillary arch, although the occlusion is now separated because of the vertical growth of the ramus.

**FIGURE 3–35**

*Stage 6.* The middle cranial fossa enlarges by endocranial resorption and ectocranial deposition, as well as by growth at the spheno-occipital synchondrosis and cranial floor sutures.

**FIGURE 3–36**

*Stage 7.* The whole maxillary region and the anterior cranial fossa are displaced anteriorly because of middle cranial fossa expansion (Stage 6).

**FIGURE 3–37**

*Stage 8.* Middle cranial fossa enlargement also causes a forward and downward displacement of the mandible, but to a much lesser extent than that of the maxilla.

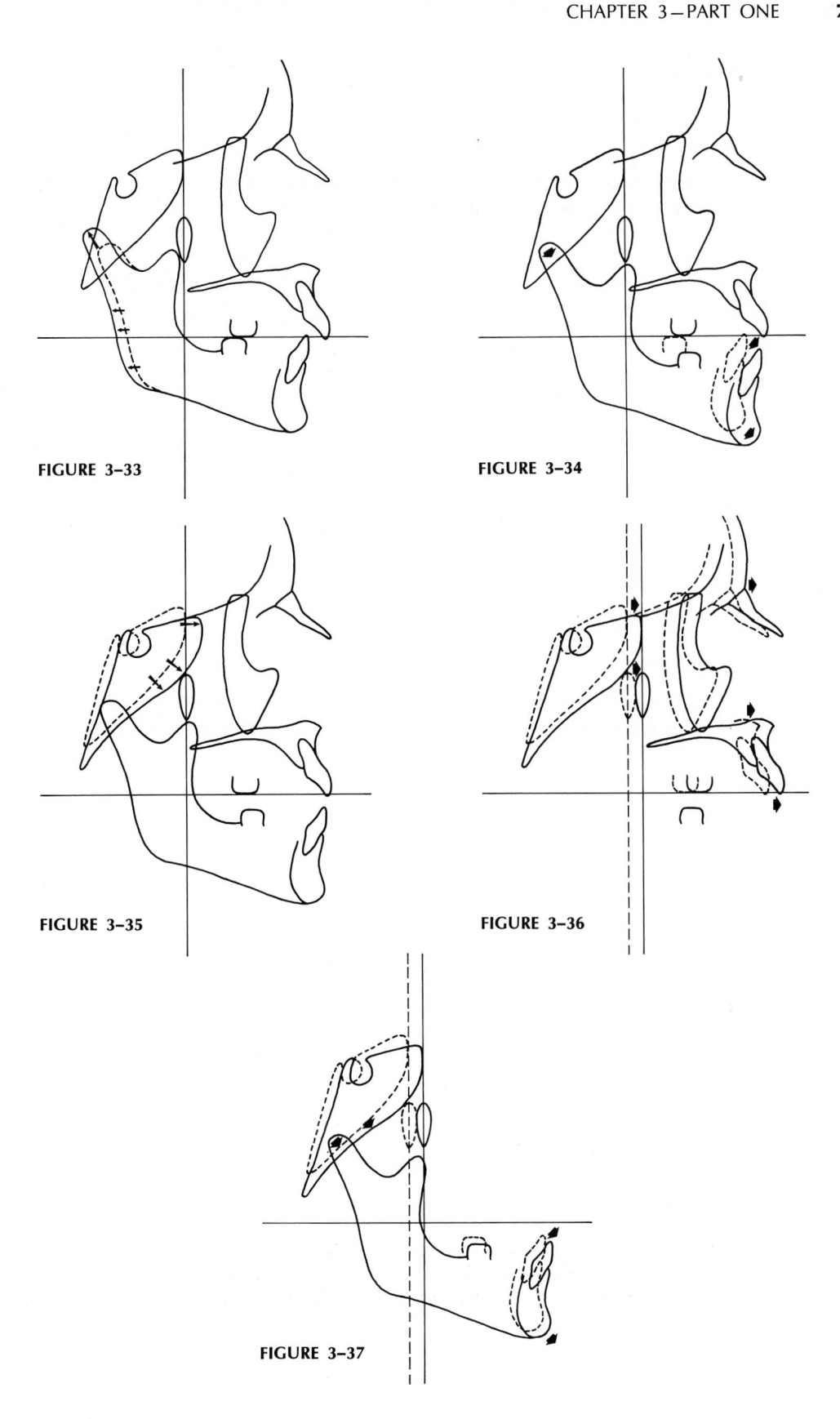

FIGURE 3-33

FIGURE 3-34

FIGURE 3-35

FIGURE 3-36

FIGURE 3-37

**FIGURE 3–38**

*Stage 9.*   The horizontal dimension of the ramus enlarges to match the amount of horizontal enlargement by the middle cranial fossa.

**FIGURE 3–39**

*Stage 10.*   The whole mandible is simultaneously displaced anteriorly (and inferiorly) as the ramus increases in size (Stage 9).

**FIGURE 3–40**

*Stage 11.*   The anterior cranial fossa increases horizontally. This has been matched by the amount of horizontal elongation of the maxilla (Stage 1).

**FIGURE 3–41**

*Stage 12.*   The maxillary arch and palate grow downward by resorption on the nasal side and by deposition on the oral side. The teeth actively drift inferiorly at the same time, by remodeling growth within the alveolar sockets, from *1* to *2*.

**FIGURE 3–42**

*Stage 13.*   The whole nasomaxillary complex is simultaneously displaced inferiorly. This is associated with bone growth at sutures (but not actually caused by sutural bone growth). The teeth are carried passively downward from *2* to *3*.

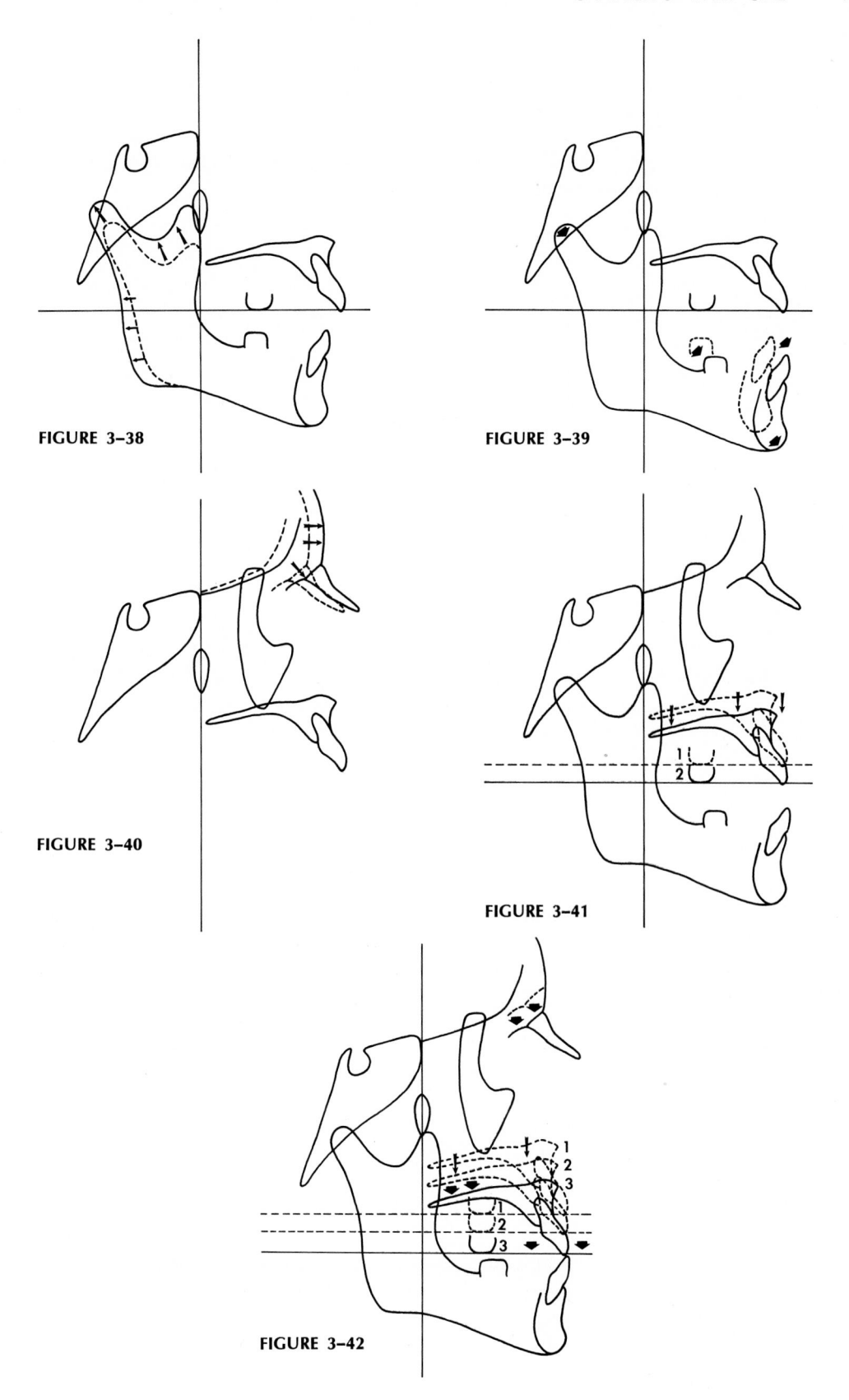

**FIGURE 3–38**

**FIGURE 3–39**

**FIGURE 3–40**

**FIGURE 3–41**

**FIGURE 3–42**

**FIGURE 3–43**

*Stage 14.* The alveolar bone of the mandible grows superiorly. The teeth drift superiorly to maintain occlusal contact.

**FIGURE 3–44**

*Stage 15.* The mandibular incisors drift lingually, and the alveolar bone moves backward by resorption on the labial side. Bone is added onto the chin and around the outside surfaces of the corpus.

**FIGURE 3–45**

*Stage 16.* The malar protuberance grows posteriorly in proportion to the amount of posterior maxillary growth. It also increases vertically to match vertical maxillary growth.

**FIGURE 3–46**

*Stage 17.* The malar region is displaced anteriorly and inferiorly to match the corresponding extent of primary maxillary displacement.

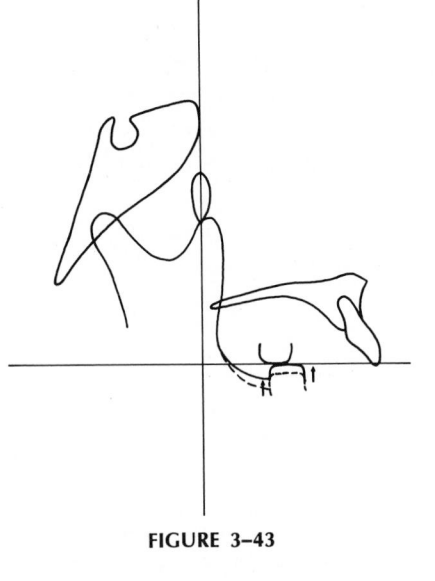

**FIGURE 3–43**

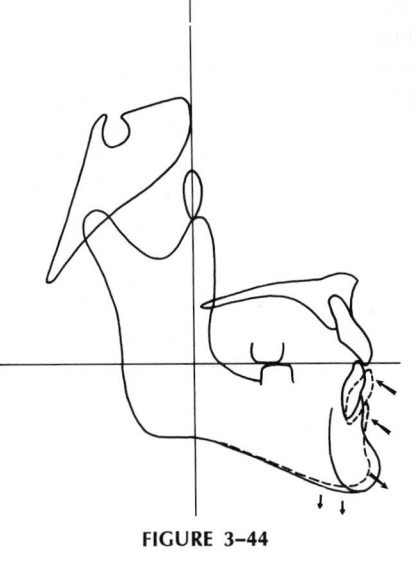

**FIGURE 3–44**

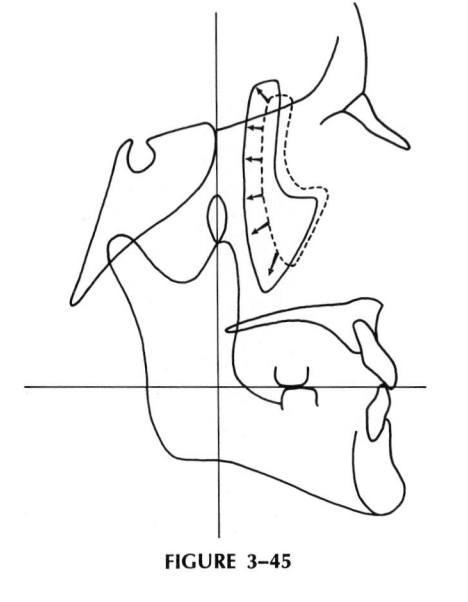

**FIGURE 3–45**

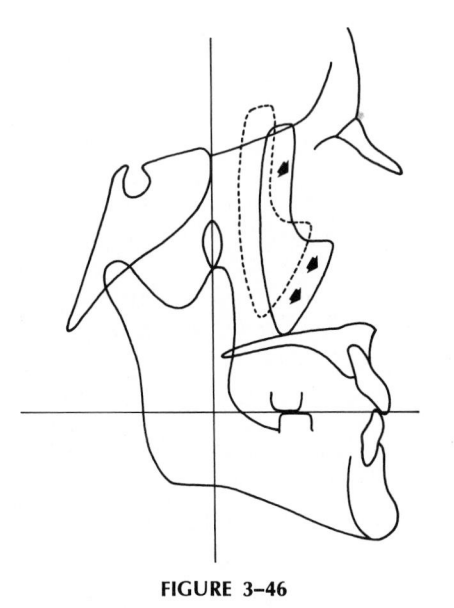

**FIGURE 3–46**

# PART TWO

The regional stages of growth described in Part One are now repeated, stage by stage, in order to explain more fully the actual growth processes involved. Three-dimensional representations of whole bones are used to complement the more simplified two-dimensional descriptions previously given.

### FIGURE 3–47

*Stage 1.* It will be recalled that the horizontal lengthening of the bony maxillary arch is produced by growth at the maxillary tuberosity. This is represented by a backward movement of **PTM** from the vertical reference line.

### FIGURE 3–48

The area shown here is the specific growth field that carries out the change just described. It is a **depository** field in which the backward-facing periosteal surface of the tuberosity receives continued deposits of new bone as long as growth in this part of the face continues. Because the posterior surface of the maxillary tuberosity points toward the direction of arch elongation, this is the particular surface that is depository. The arch also widens, and the lateral surface is, similarly, depository. The endosteal side of the cortex within the interior of the tuberosity is resorptive. The cortex thus drifts progressively posteriorly and also, to a lesser extent, in a lateral direction. Deep to the tuberosity is the maxillary sinus. It increases in size as a result of the same process. In the newborn, this sinus is quite small, but it becomes greatly expanded as growth continues and eventually occupies the greater part of the large suborbital compartment.

In Chapter 11 the various kinds of bone tissues are described as they relate to the many different circumstances of regional growth. Included are those types that characteristically grow rapidly. The bone tissue of the maxillary tuberosity is one of these, because this area provides much of the postnatal lengthening of the arch and nearly all of it after about 5 or 6 years of age. This requires a fairly rapid and sizable amount of continued new bone formation. The maxillary tuberosity is a major "site" of maxillary growth. It does not, however, provide for the growth of the whole maxilla but relates only to that part associated with the posterior part of the lengthening arch. Many other basic and important sites of growth also exist throughout the various parts of the maxilla.

### FIGURES 3–49 AND 3–50

*Stage 2.* The whole maxilla undergoes a simultaneous process of **primary displacement** in an anterior direction as it grows and lengthens posteriorly. The nature of the force that produces this anterior movement has, historically, been a subject of great controversy. One early theory (long since dropped) suggested that additions

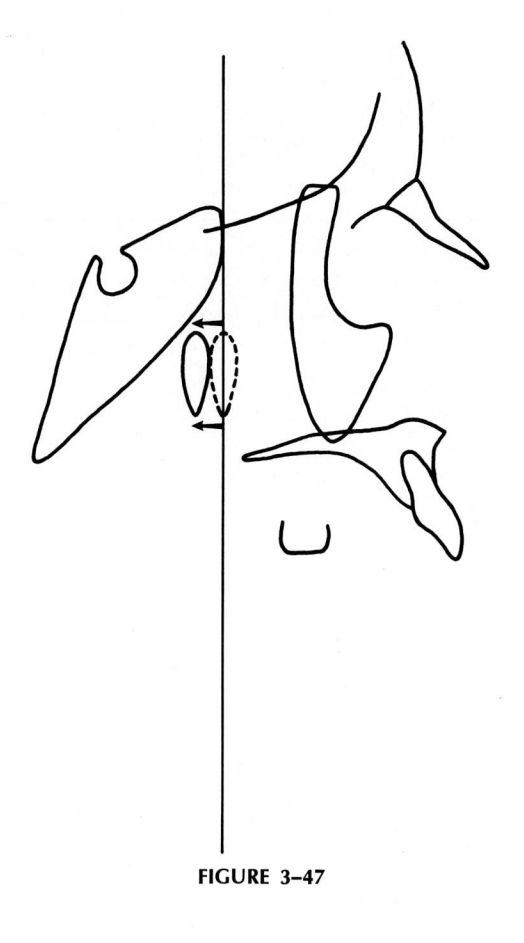

**FIGURE 3–47**

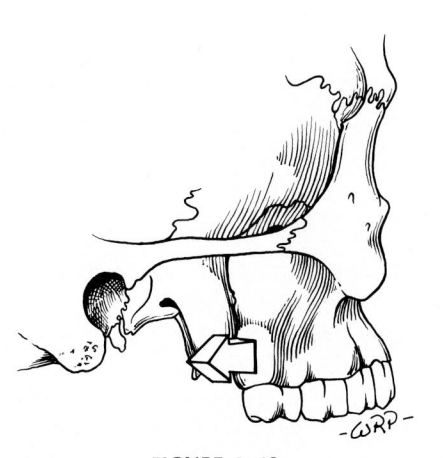

**FIGURE 3–48**

of new bone on the posterior surface of the elongating maxillary tuberosity "push" the maxilla against the adjacent muscle-supported pterygoid plates. This presumably would cause a resultant shove of the entire maxilla anteriorly because of its own posterior bone growth activity. The idea was abandoned, however, when it was realized that the bone is pressure-sensitive and does not have the physiologic capacity to push itself away from the other bones by its own growth.

Another theory held that bone growth within the various maxillary **sutures** produces a pushing apart of the bones, with a resultant thrust of the whole maxilla anteriorly (and inferiorly as well). Although this explanation is still heard, it has been largely rejected for the reason just mentioned: bone tissue is not capable of growth in a field that requires the levels of compression needed to produce a "pushing" type of displacement. The sutural connective tissue is not adapted to a pressure-related growth process (in contrast to cartilaginous types of bone-to-bone contacts, which are much more compression-tolerant). The suture is essentially a **tension-adapted** tissue. Its collagenous fiber construction is a design for traction resistance across the connective tissue bridge between separate bones. The presence of any unusual pressure on a suture triggers bone resorption, not deposition. The sutural membrane cannot withstand any undue amount of compression because pressure affects its vascular and cellular components. It is believed that the stimulus for sutural bone growth is the tension produced by the **displacement** of that bone. The deposition of new bone is the response to displacement rather than the force that causes it.

Thus, as the entire maxilla is carried forward (and downward) by displacement, tension is produced within the sutural membranes. This, in turn, presumably triggers the sutural membranes to form the new bone tissue that enlarges the overall size of the whole bone and sustains constant bone-to-bone sutural contact. Although the "sutural push theory" is not tenable, some students of the problem are now looking anew at growth mechanisms within sutures, but not in the old conceptual way. The same problem exists for the periodontal membrane and the eruption and drifting process of teeth. We still have much to learn about the growth behavior of sutures and other soft tissues associated with bone, and we may yet find that sutures, the periosteum, and the periodontal membrane participate in the displacement process in a way that is presently unknown or only speculative (such as the possible presence of actively contractile myofibroblasts that would cause teeth to erupt and drift, and sutures to slide over one another). See Chapter 11 for a discussion of the sutural growth process.

**FIGURE 3–51**

Another explanation for maxillary displacement is the now famous "nasal septum" theory. This was developed largely by Scott, and the premise for the idea is quite reasonable. It developed from the criticisms of the "sutural theory" described above. The nasal septum hypothesis was soon adopted by many investigators around the world and became more or less the standard explanation replacing the sutural theory. **Cartilage** is specifically adapted to certain pressure-related growth sites, because it is a special tissue uniquely structured to provide the capacity for growth in a field of compression (see Chapter 11). Cartilage is present in the epiphyseal plates of long bones, in the synchondroses of the cranial base, and in

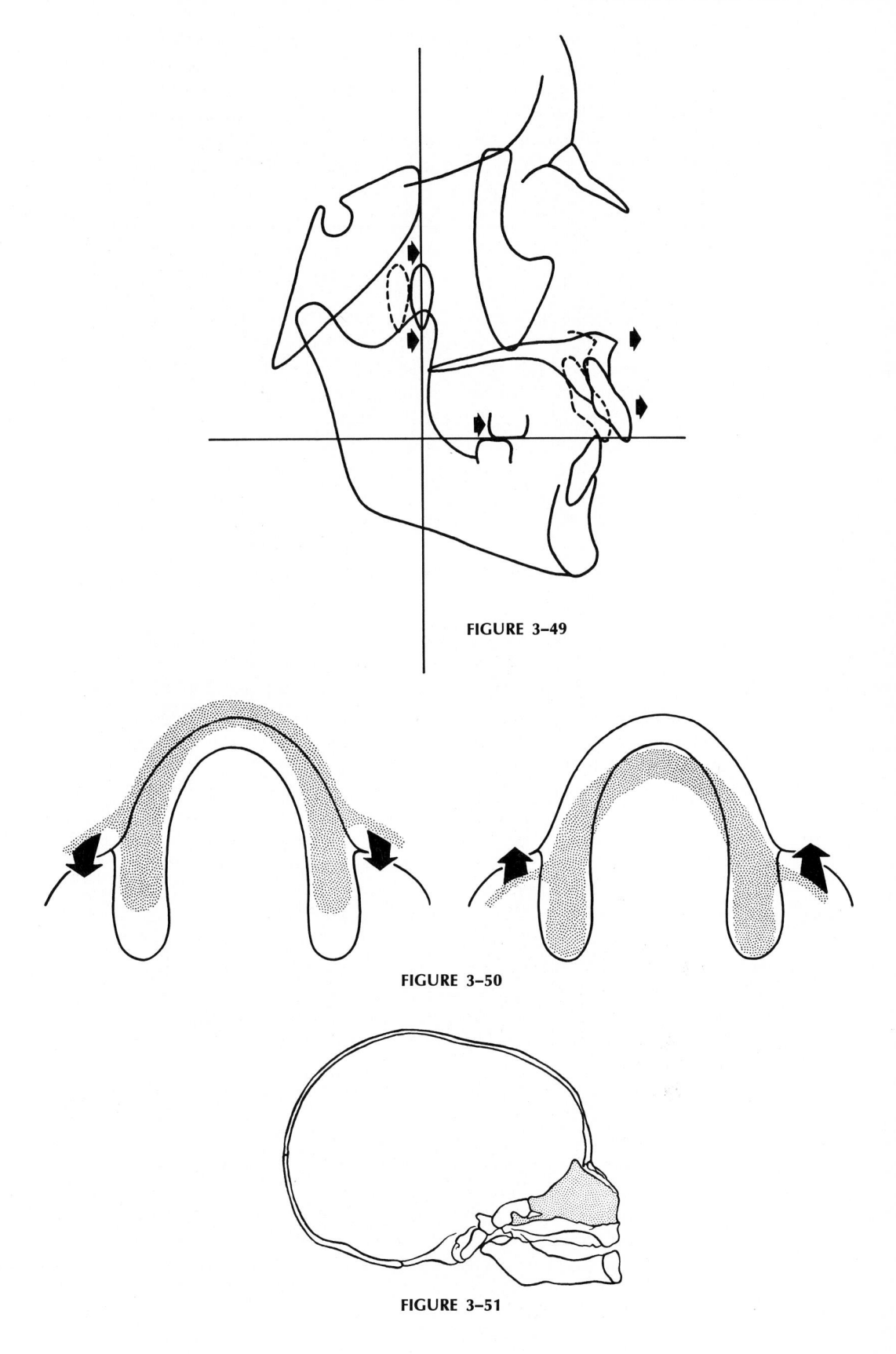

FIGURE 3–49

FIGURE 3–50

FIGURE 3–51

the mandibular condyle, where it provides linear growth by endochondral proliferation. Whereas the cartilaginous nasal septum itself contributes only a small amount of actual endochondral growth, the basis for the theory is that the pressure-accommodating **expansion** of the cartilage in the nasal septum provides a source for the physical force that displaces the maxilla anteriorly and inferiorly. This sets up fields of tension in all the maxillary sutures. The bones then secondarily, but almost simultaneously, enlarge at their sutures in response to the tension created by the displacement process.

As with any important explanatory theory, the nasal septum concept has subsequently received a great deal of laboratory study to test its validity. Ingenious experiments, such as those carried out by Sarnat, suggest that the nasal septum does indeed appear to be a factor contributing to the primary displacement of the nasomaxillary complex. Some other experimental studies, however, have not conclusively shown any special role played by the nasal septum in the displacement process. It is held by Latham that the nasal septum, together with the septopremaxillary ligament, between the septum and the premaxilla, is an operative force **early** in postnatal as well as prenatal growth, but that it is replaced by other displacement forces as the face outgrows the capacity to be moved by the septum. Others suggest that the septum functions essentially to support the roof of the nasal chamber and does not actively participate in the displacement movements of the nasal floor.

There are two basic reasons why this problem is still unresolved and why such divergent opinions are held. The first is that the cause of maxillary displacement is probably multifactorial in nature. If the nasal septum is indeed involved, other factors probably contribute as well, and it is difficult to separate their respective effects in controlled laboratory experiments. The second reason is that experimental studies usually involve surgical removal of parts (such as the septum) to test the nature of their functional roles in growth. It is always difficult in such studies to account for the variables introduced by the experimental procedure itself, such as the destruction of tissues, blood vessels, and nerves that can also play a role in the growth process. Critics of these studies point out that the experimental removal of a given part does not necessarily demonstrate what the actual role of that part is when present *in situ*. It merely shows how the growth process functions in the absence of that part rather than in its presence.

At present, the nasal septum theory is still accepted as a reasonable explanation by a number of investigators, although it is universally realized that much more needs to be understood. Many teachers use the septum as a "symbol" for the force that causes displacement but make it clear that other factors are involved and that there is much to be explained.

A notable advance was made with the development of the **functional matrix** concept, largely by Moss. It deals with the determinants of bone and cartilage growth in general. The functional matrix concept states, in brief, that any given bone grows in response to functional relationships established by the sum of all the soft tissues operating in association with that bone. This means that the bone itself does not regulate the rate and directions of its own growth; the functional soft tissue matrix is the actual governing determinant of the skeletal growth process. The course and extent of bone growth are secondarily dependent upon the growth of

pace-making soft tissues. Of course, the bone and any cartilage present are also involved in the operation of the functional matrix, because they participate in giving essential feedback information to the soft tissues. This causes the soft tissues to inhibit or accelerate the rate and amount of subsequent bone growth activity, depending on the status of functional and mechanical equilibrium between the bone and its soft tissue matrix. The genetic determinants of the growth process reside wholly in the soft tissues and not in the hard part of the bone itself.

The functional matrix concept is basic to an understanding of the fundamental nature of a bone's role in the overall process of growth control. This concept has had great impact in the field of facial biology.

The functional matrix concept also comes into play as a source for the mechanical force that carries out the process of displacement. According to this now popular explanation, the facial bones grow in a subordinate growth control relationship with all the surrounding soft tissues. As the tissues continue to grow, the bones become passively (that is, not of their own doing) **carried** along (displaced) with the soft tissues. Thus, for the nasomaxillary complex, the expansion of the facial muscles, the subcutaneous and submucosal connective tissues, the oral and nasal epithelia lining the spaces, the vessels and nerves, and so on all combine to move the facial bones passively along with them as they grow. This continuously positions each bone and all its parts in correct anatomic positions to carry out its functions, because the functional factors are the very agents that **cause** the bone to develop into its definitive shape and size and to occupy the location it does.

How does the functional matrix explanation relate to the nasal septum theory? This is still highly controversial. The problem, at this writing, is that nobody really understands just how much genetic influence resides within cartilage, and especially the different structural and functional varieties of cartilage. The "forward and downward" displacement of the maxilla could be accounted for solely by the functional matrix mechanism, without direct and purposeful participation by the expanding nasal septum. However, there is still disagreement as to the genetic role of the cartilage in the septum. Does it have an intrinsic genetic capacity to determine the course of its own growth expansion and, with it, resultant displacement of the midface? Or is the growth of the cartilage itself, like that of the bone, controlled by the capsular soft tissue matrix? Further discussion is deferred until the mandibular condyle and the synchondroses are considered in later sections of this chapter.

The functional matrix concept, in general, is established and valid, and it is basic in helping us understand the complex interrelationships that operate during facial growth. It is to be realized, however, that this principle is not intended to explain **how** the growth control mechanism actually functions. This concept describes essentially **what** happens during growth; it does not account for the regulatory processes at the cellular and molecular levels that carry it out.

In summary, Stage 2 of the growth sequence involves a forward **displacement** of the entire maxilla. The amount of this movement equals the increment of new bone growth on the posteriorly facing surface of the maxillary tuberosity. The **timing** of these two growth changes is such that the addition of the new bone coincides with the process of displacement or lags imperceptibly behind it, since the movement of displacement must occur in order to provide the available space for growth expansion. The force that causes the displacement movement, according to current theory, is the "functional matrix" (although some have not entirely abandoned the nasal septum).

**FIGURE 3–52**

*Stages 3 and 4.*   These stages involve the lengthening of the mandibular corpus to an extent that matches its counterpart, the bony maxillary arch. In this growth process, one notable structural difference between the mandible and maxilla exists: the mandible has a **ramus.** The maxillary tuberosity if a **free** skeletal surface. Posterior to it is the oropharyngeal space and the separate (in childhood) pterygoid plates. This maxillary surface grows directly posteriorly. The posterior growth of the mandibular bony arch, however, must proceed into a region already occupied by the ramus. This requires a **remodeling conversion** from ramus to mandibular corpus. That is, the whole ramus becomes relocated posteriorly, and the former anterior part of the ramus is structurally altered into an addition for the corpus, which thereby becomes lengthened by this remodeling process.

**FIGURE 3–53**

The growth movement of the ramus in a backward direction has usually been pictured as essentially a two-dimensional process. This is not merely an incomplete explanation; it is inaccurate as well.

**FIGURE 3–54**

The problem is that some of the key anatomic parts that participate in the relocation and remodeling processes of the ramus and corpus cannot be seen or represented in conventional two-dimensional headfilms and tracings. Among these is the **lingual tuberosity.** This is an important structure because it is the direct anatomic equivalent of the maxillary tuberosity. Just as the maxillary tuberosity is a major site of growth for the upper bony arch, so is the lingual tuberosity a major site of growth for the mandible. Yet this structure is not even included in the basic vocabulary of cephalometrics. The reason, simply, is that it is not recognizable in the headfilm. This presents a severe handicap, because the lingual tuberosity is not only a major growth and remodeling site but also the effective boundary between the two basic parts of the mandible: the ramus and the corpus. The inaccessibility of the lingual tuberosity for routine cephalometric study is a great loss. Nonetheless, the changes this important structure undergoes during growth **must** be understood, all the more so since it cannot be visualized, at least directly, in headfilms.

**FIGURE 3–55**

The lingual tuberosity grows posteriorly by deposits on its posterior-facing surface, just as the maxillary tuberosity undergoes comparable growth additions. Ideally, the maxillary tuberosity exactly overlies the lingual tuberosity (that is, both are aligned on the vertical reference line). Moreover, the lingual and maxillary tuberosities ideally have equal rates and amounts of respective growth. While these growth changes often do not match one another, however, such a "balanced" growth process is described here (the consequences of variations are explained later).

Note that the lingual tuberosity protrudes noticeably in a lingual (medial) direction and that it lies well toward the midline from the ramus. The prominence of the tuberosity is increased by the presence of a large resorptive field just below it. This resorptive field produces a sizable depression, the **lingual fossa.** The combination of resorption in the fossa and deposition on the medial-facing surface of the tuberosity itself greatly accentuates the contours of both regions.

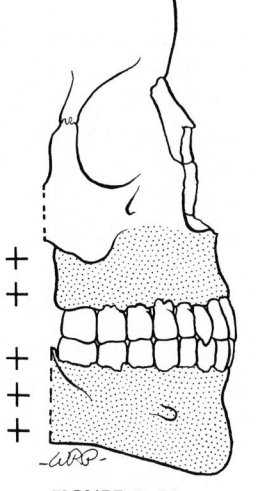

**FIGURE 3–52**

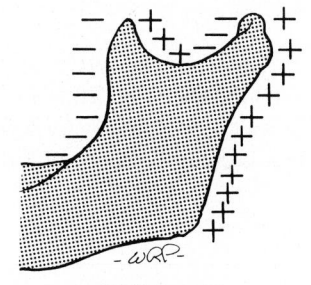

**FIGURE 3–53**

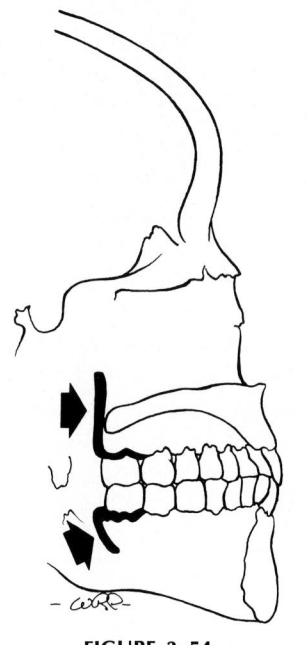

**FIGURE 3–54**

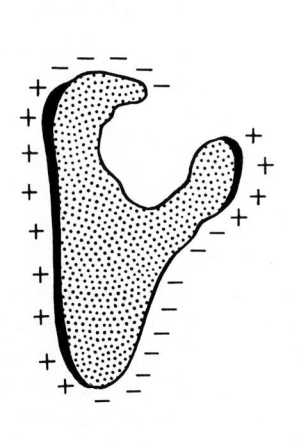

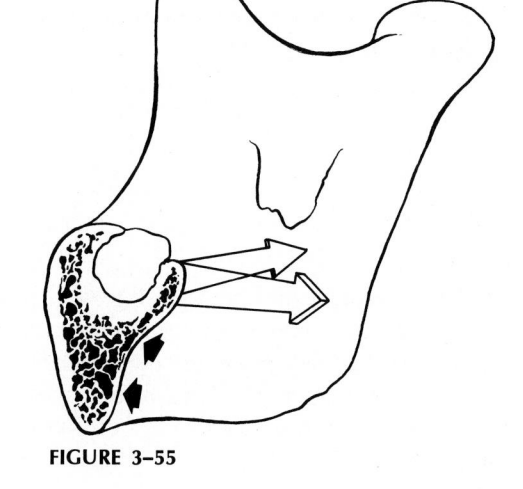

**FIGURE 3–55**

**FIGURE 3–56**

The tuberosity grows in an almost directly posterior direction, with only a relatively slight lateral shift. The latter is because bicondylar width does not increase nearly as much as mandibular length during the childhood period, since most of the lateral growth of the cranial base is complete by about the second and third years.

**FIGURE 3–57**

The posterior growth of the tuberosity is accomplished by continued new deposits of bone on its posterior-facing exposure. As this takes place, that part of the **ramus** just behind the tuberosity grows **medially.** This area of the ramus is coming into line with the axis of the arch in order to join it and thus become a part of the corpus, thereby lengthening it. As pointed out above, the whole ramus lies well lateral to the dental arch.

Deposition on the lingual surface of the ramus just behind the tuberosity produces the medial direction of drift that shifts this part of the ramus into alignment with the axis of the corpus. Keep in mind that the whole ramus is also becoming relocated in a posterior direction at the same time. What has happened, in summary, is that bony **arch length** has been increased and the corpus has been lengthened by (1) deposits on the posterior surface of the lingual tuberosity and the contiguous lingual side of the ramus, and (2) a resultant lingual shift of this part of the ramus to become added to the corpus.

**FIGURE 3–58**

The presence of resorption on the anterior border of the ramus is usually described as "making room for the last molar." It is doing much more than this! The resorptive nature of this region is directly involved in the whole process of progressive relocation of the entire ramus in a posterior direction; this movement continues from the tiny mandible of the fetus to the attainment of full adult mandibular size. The overall extent of ramus movement amounts to several inches, not merely the width of a molar.

**FIGURE 3–59**

Another key point is that the traditional description of posterior ramus movement implies a **straight-line** backward growth process in a two-dimensional plane, as represented by a and b. This is not the case at all. Such a picture of ramus growth shows, simply, resorption on the anterior edge and deposition on the posterior edge; this is incomplete. Growth actually takes place as indicated by c. In d, the growth direction thus follows the x arrows rather than the straight-line axis shown by the y arrows.

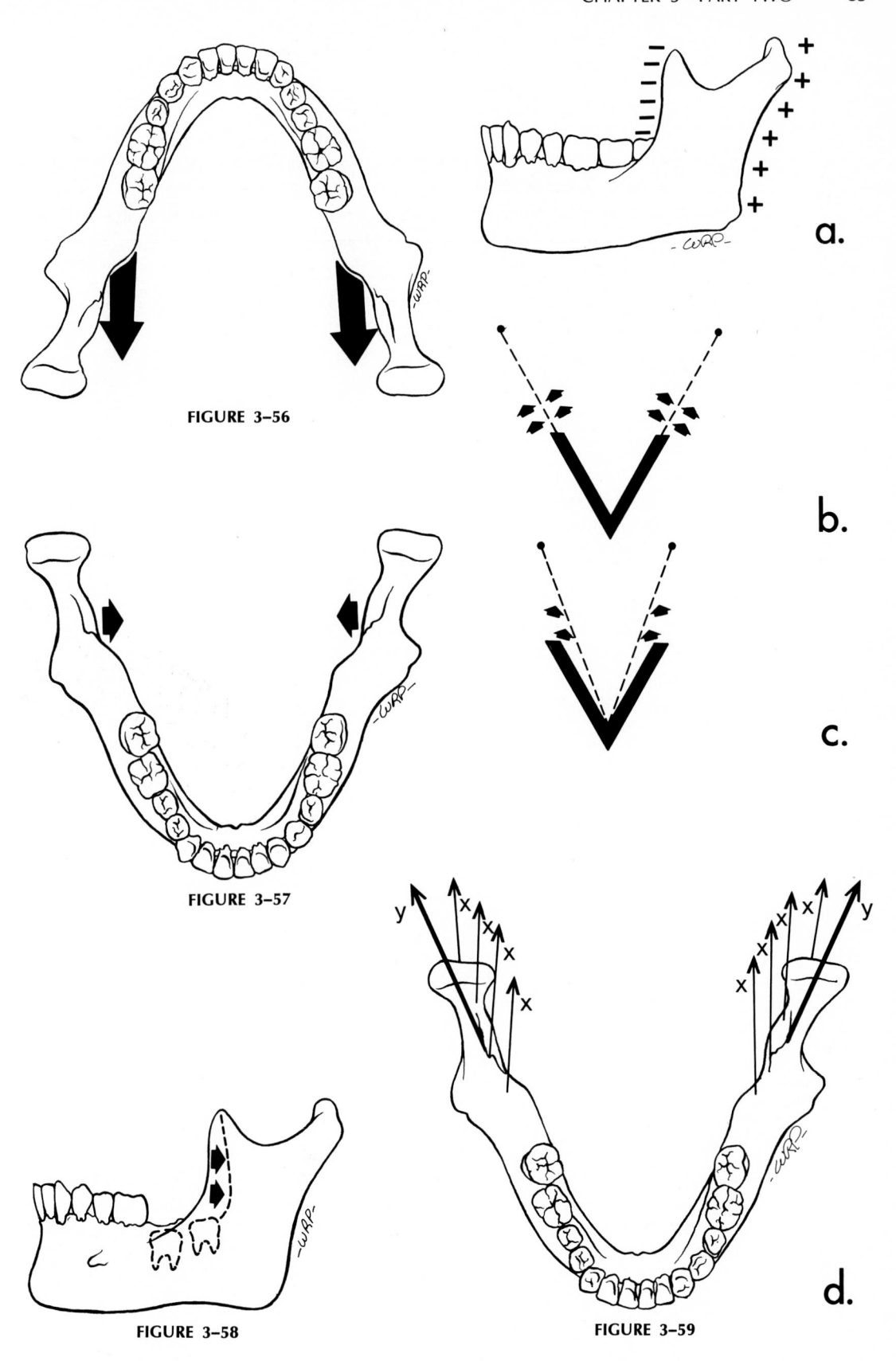

FIGURE 3–56

FIGURE 3–57

FIGURE 3–58

FIGURE 3–59

a.

b.

c.

d.

**FIGURE 3–60**

Growth and remodeling activity does not occur **only** on the anterior and posterior margins. Surfaces **between** the anterior and posterior borders, shown by *a, b,* and *c* in this diagram, also directly participate, because the alignment of the ramus is such that these surfaces become directly involved.

**FIGURE 3–61**

If growth were to occur simply as schematized in diagram *1,* resorption on one end and deposition on the other (as by activity on only the anterior and posterior margins of the ramus) would move it along the axis of the arrow with no need for activity in the area between the ends. However, the various parts of the ramus are, in fact, oriented so that the span between also necessarily comes into play, as shown by diagram *2.* Several different regional directions of alignments are involved in the various parts of the ramus, as described next.

**FIGURE 3–62**

The **coronoid process** has a propeller-like twist, so that its lingual side faces three general directions all at once: posteriorly, superiorly, and medially.

**FIGURE 3–63**

When bone is added onto the lingual side of the coronoid process, its growth thereby proceeds **superiorly,** and this part of the ramus thereby becomes increased in vertical dimension. Notice that each coronoid process lengthens vertically, even though additions are made on the medial (lingual) surfaces of the right and left coronoid processes. This is an example of the enlarging V principle, with the V oriented vertically (see page 24).

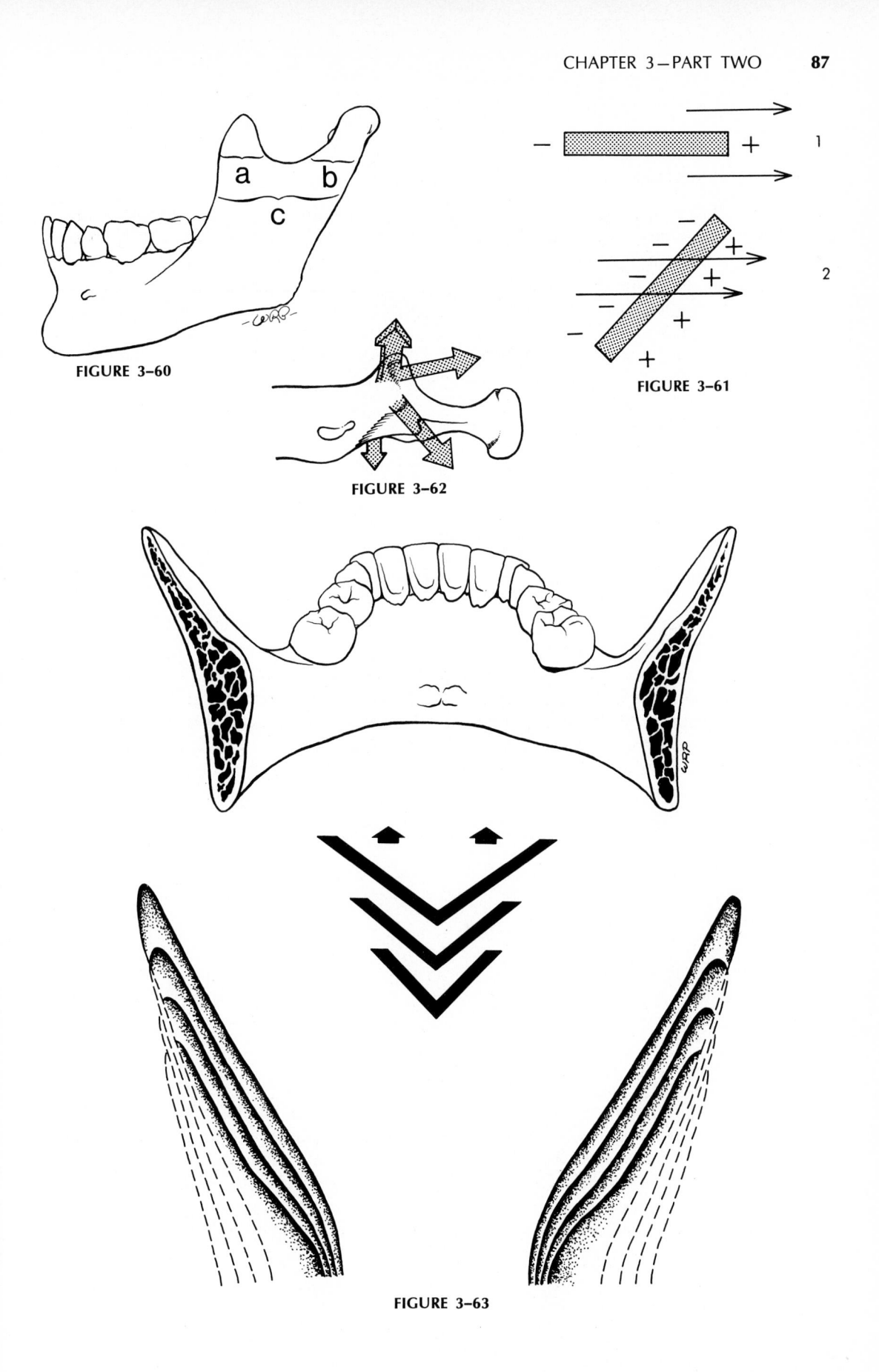

FIGURE 3-60

FIGURE 3-61

FIGURE 3-62

FIGURE 3-63

**FIGURE 3–64**

These **same** deposits of bone on the lingual side also bring about a **posterior** direction of growth movement, because this surface also faces posteriorly. This produces a backward movement of the two coronoid processes, even though deposits are added on the inside (lingual) surface. This is also an example of the expanding V principle, with the V oriented horizontally. Notice, further, that this enables the whole posterior part of the mandible to **widen** (although not very much except during the period of fetal and early childhood cranial base growth in width), even though deposition occurs on the inside of the V.

**FIGURE 3–65**

These **same** deposits of bone on the lingual side also function to carry the base of the coronoid process in a **medial** direction in order to add this part to the lengthening corpus, which lies well medial to the coronoid process. This, again, is an example of the V principle, because a wider part undergoes relocation into a more narrow part as the whole V moves toward its wide end. Thus, the area occupied by the anterior part of the ramus in mandible *1* becomes relocated and remodeled into the posterior part of the corpus in mandible *2*.

**FIGURE 3–66**

In all the above relationships, the **buccal** side of the coronoid process has a **resorptive** type of periosteal surface. This surface faces away from the combined superior, posterior, and medial directions of growth. The remainder of most of the superior part of the ramus, including the whole area just below the mandibular (sigmoid) notch and the **superior** (not lateral or medial) portion of the condylar neck, grows superiorly by deposition on the lingual side and resorption from the buccal side.

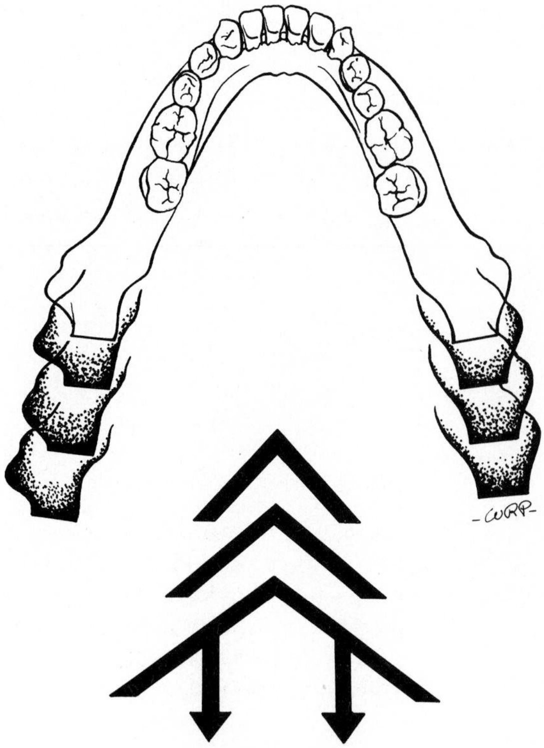

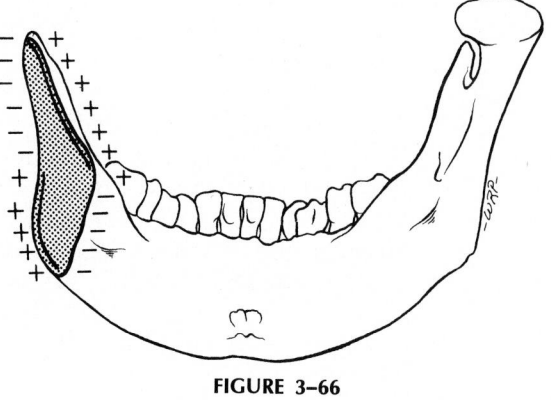

**FIGURE 3–64**

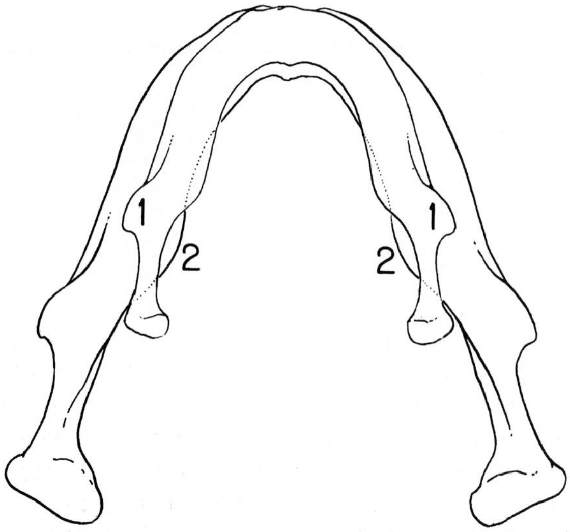

**FIGURE 3–65**

*(Adapted from Enlow, D. H., and D. B. Harris: A study of the post-natal growth of the human mandible. Am. J. Orthod., 50:25–50, 1964.)*

**FIGURE 3–66**

**FIGURE 3–67**

The lower part of the ramus below the coronoid process also has a twisted contour. Its buccal side faces posteriorly toward the direction of backward growth and thus, characteristically, has a depository type of surface. The opposite lingual side, facing away from the direction of growth, is resorptive.

**FIGURE 3–68**

A single field of surface resorption is present on the inferior edge of the mandible at the ramus-corpus junction. This forms the **antegonial notch** by remodeling from the ramus just behind it as the ramus relocates posteriorly. The size of the notch can be increased whenever a downward rotation of the corpus relative to the ramus takes place (see page 132). Other kinds of important mandibular rotations can also involve a sizable resorptive field on the ventral edge of the ramus, as explained later.

**FIGURE 3–69**

The posterior edge of the ramus is a major growth site. The condyle has an obliquely upward and backward growth direction; the angle of growth involved (that is, how much upward and how much backward) is variable and depends on whether an individual is a "horizontal or vertical grower" with respect to the mandible. However, the growth of the posterior border of the ramus necessarily keeps pace with any given amount of condylar growth. Although correlated, these two regional growth sites are essentially separate. Together they represent the most active areas during mandibular growth in distance moved and amount of bone deposited. Because of the relatively rapid rate of ramus growth, the bone tissue in the posterior part of the ramus is characteristically one of the more fast-growing types (Chapter 11).

The gonial region is anatomically variable and therefore much variation is involved in its pattern of growth. Depending on the presence of inwardly or outwardly directed gonial flares, the buccal side can be either depository or resorptive, with the lingual side having the converse type of growth. However, many different histogenetic combinations can be encountered because of the topographic complexity and variability of this region.

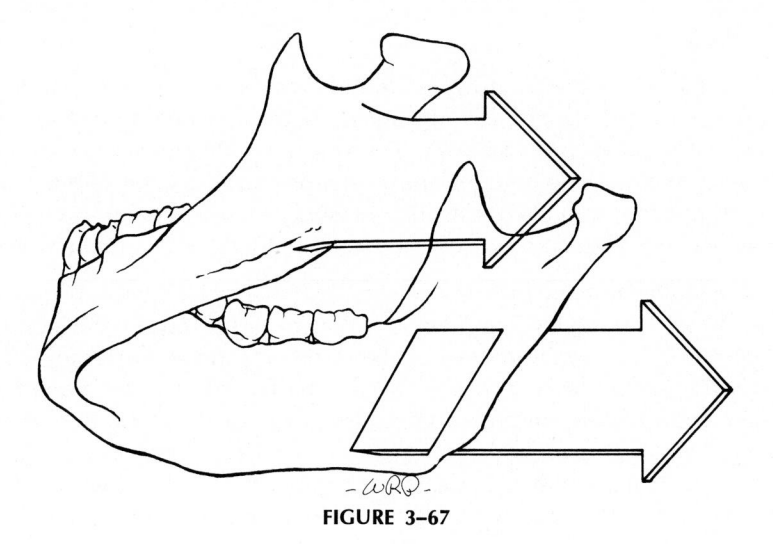

**FIGURE 3–67**

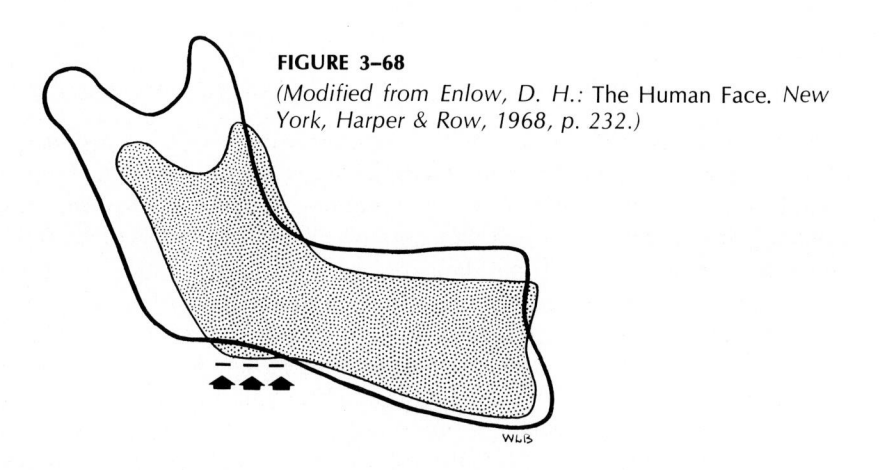

**FIGURE 3–68**
*(Modified from Enlow, D. H.:* The Human Face. *New York, Harper & Row, 1968, p. 232.)*

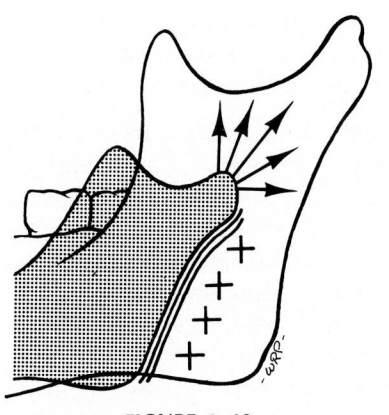

**FIGURE 3–69**

**FIGURE 3–70**

While the whole ramus grows posteriorly and superiorly, the mandibular foramen likewise drifts backward and upward by deposition on the anterior and resorption from the posterior part of its rim. The foramen maintains a constant position about midway between the anterior and posterior borders of the ramus. Even when the ramus undergoes marked alterations associated with edentulism (during which it may become quite narrow), this foramen usually sustains a midway location.

The **mandibular condyle** is an anatomic part of special interest because it is a major site of growth, having considerable clinical significance. The condyle **does not** "regulate" the overall growth of the entire mandible, including all of its many regional parts, however. It is not a "master center" of growth, and all the other growth fields are not dependent upon it for direct control. The condyle is a major field of growth, nonetheless, and it is an important one. It is a **regional** field of growth that provides an adaptation for its own localized growth circumstances, just as all the other regional fields accommodate their own particular (but different) localized growth circumstances. The growth of the mandible as a whole is a result of all the different regional forces and functional agents of growth control acting upon it to produce the topographically complex shape that is the aggregate expression of all these localized factors.

**FIGURE 3–71**

The condylar growth mechanism itself is a clear-cut process. **Cartilage** is present because variable levels of **compression** occur at its articular contact with the temporal bone of the cranium. An endochondral growth mechanism is required, because the condyle grows in a direction toward its articulation in the face of direct pressure. An intramembranous type of growth could not operate, because the periosteal mode of osteogenesis is not pressure-adapted. Endochondral growth occurs only at the articular contact part of the condyle, since this is where pressure exists at levels beyond the tolerance of the bone's soft tissue membrane. The endochondral bone tissue *(b)* formed in association with the condylar cartilage *(a)* is laid down only in the medullary portion of the condyle. The enclosing bony cortices *(c)* are produced by periosteal-endosteal osteogenic activity; these membranes are not subject to the compressive forces of articulation but rather are essentially tension-related, because of muscle and connective tissue attachments. The real functional significance of the condylar cartilage thus involves an adaptation for regional compression, and this regional, endochondral, bone-forming mechanism develops as a specific response to this particular **local** circumstance. The cartilage itself does not contain genetic programming that directly determines and governs the course of growth in the other areas of the mandible.

The condylar cartilage is a **secondary** type of cartilage, which means that it does not develop by differentiation from the established **primary** cartilages of the skull (that is, the cartilages of the pharyngeal arches, such as Meckel's cartilage, and the definitive cartilages of the cranial base). Phylogenetically, the original cartilage and bone that provided for mandibular articulation (the separate articular bone attached by a suture to the dentary bone) become converted to an ear ossicle (the malleus) in mammals. Thus, a "secondary" cartilage was developed on the dentary bone to provide for lower jaw articulation with the cranium. It is believed that the unique connective tissue covering (capsule) of the condylar cartilage is actually an original periosteum. Its undifferentiated connective tissue stem cells, however, develop into

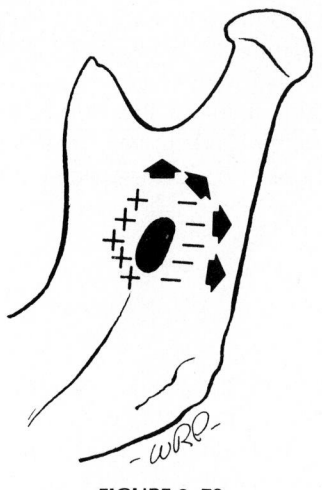

**FIGURE 3–70**

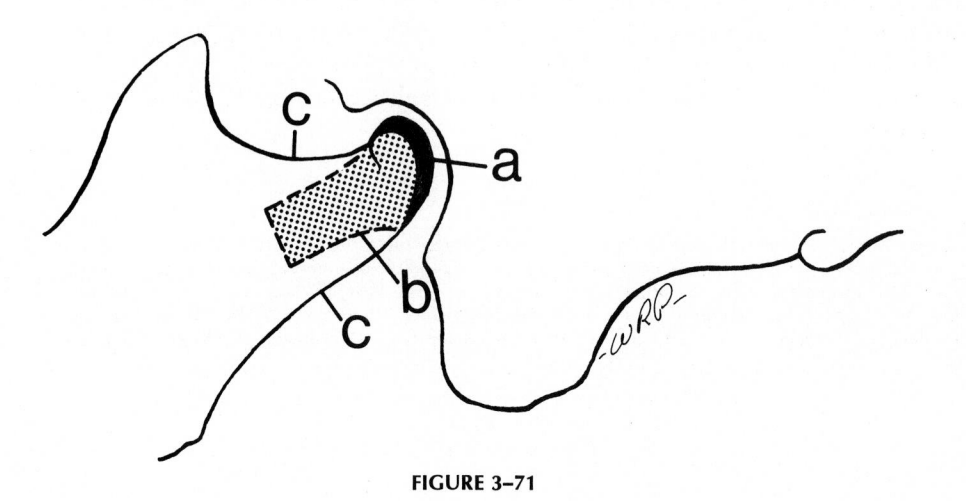

**FIGURE 3–71**

chondroblasts rather than osteoblasts because of the compressive forces acting on this membrane. An adventitious type of "secondary" cartilage develops, rather than bone, because of the functional and developmental conditions imposed upon this part of the mandible. It is thus not an "endochondral" bone in the sense that, phylogenetically, the bones of the cranial base are endochondral in type. The mandible is essentially a membrane bone in which one part (that is, what has become the condyle in mammals) develops in response to a phylogenetically altered developmental situation. This involves the ectopic presence of pressure that, in turn, causes localized ischemia and anoxia, factors that are known to induce chondrogenesis from the pool of undifferentiated connective tissue cells rather than osteogenesis.

The condylar cartilage differs in histologic organization from most other growth cartilages involved in endochondral bone formation. It is not directly comparable with an epiphyseal plate. While different authors have somewhat different working meanings for the terms "primary" and "secondary" cartilages*, it is now generally recognized that the secondary cartilage of the condyle is not the pacemaker for the growth of the mandible. Its contribution is to provide **regional adaptive growth** (that is, "secondary" growth, another definition for the term). It maintains the condylar region in proper anatomic relationship with the temporal bone as the whole mandible is simultaneously being carried downward and forward (see Stage 5). Thus, the condyle is not a "primary" center of growth. It is now believed that the condyle does **not** establish the rate or the amount of overall mandibular growth. The condyle does, however, have a special multidirectional capacity for growth and remodeling in selective response to varied mandibular displacement movements and **rotations** (see page 132). The special structure of the condyle provides for this, unlike the committed unidirectional linear growth of epiphyseal plates produced by the characteristic linearly oriented direction of chondrocyte proliferation.

---

*One common definition is that secondary growth cartilages are those, simply, that have a type of structure that puts them in a separate category from the typical epiphyseal growth plates of long bones. Articular cartilages are another example of the secondary type. As shown by Moss, the condylar cartilage is comparable, both in structure and growth behavior, with an articular rather than an epiphyseal plate cartilage. It is not actually "articular," however, because of its special fibrous covering.

**FIGURE 3–72**

A unique **capsular layer** of poorly vascularized connective tissue covers the articular surface of the condyle (a). This membrane is highly cellular early in development but becomes densely fibrous with age. Just deep to it is a special layer of "precartilage" cells (b). This is the predominant site for cellular proliferation, and it is responsible for the feeding process providing cartilage for endochondral replacement by bone as the deeper layers advance.

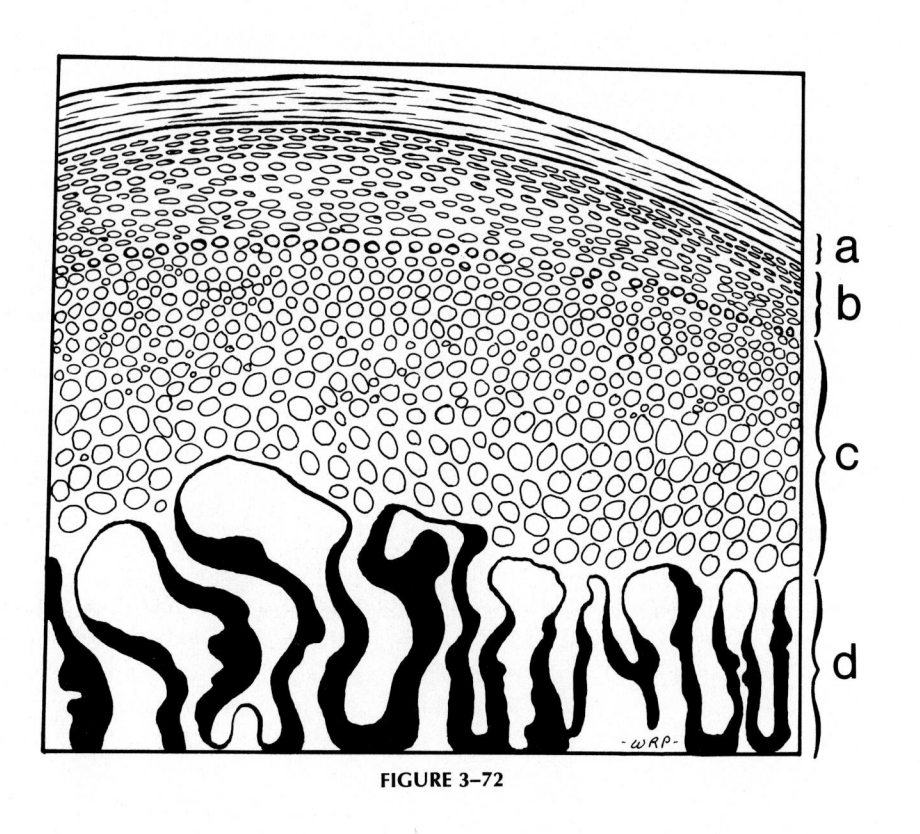

**FIGURE 3–72**

**FIGURE 3–73**

The proliferative process produces the "upward and backward" growth movement of the condyle. The condylar cartilage **moves** by prechondroblast cell divisions on the articular side with an equal amount of cartilage removed from the opposite (internal) side. The removal phase involves replacement with endochondral bone. A trail of continually forming new endochondral bone thus follows the moving cartilage, as schematized by layer d in Figure 3–72.

The precartilage cells are closely packed, and very little intercellular matrix is present. This is due to their rapid proliferative activity. A relatively thin transitional zone of immature hyaline cartilage then occurs deep to the proliferative layer with a somewhat increased amount of matrix. This zone does not appear to contribute materially to the cell division process. The deeper cells become transformed into the next layer as it moves up. This layer is composed of densely packed chondrocytes that are undergoing hypertrophy (c). The matrix is also noticeably scant.

The small amount of matrix in the deepest part of the hypertrophied zone becomes calcified, and a zone of resorption and bone deposition follows (d). Unlike the arrangement in typical **primary** growth cartilage, these various zones **do not have linear columns of daughter cells.** This is a notable histologic difference between primary and secondary types of growth cartilages. As pointed out by Koski, the arrangement of the daughter cells in the condylar cartilage thus does not reflect the direction in which the condyle is growing.

**FIGURE 3–74**

While all this is going on, the periosteum and endosteum are active in producing the **cortical** bone that encloses the medullary core of endochondral bone tissue (which takes the brunt of whatever compressive forces are acting on the condyle). This cortical layer of intramembranous bone continues down onto the condylar neck. The anterior margin of the condylar neck is depository. This surface is part of the sigmoid notch, the entire edge of which grows superiorly and thus receives new bone.

The posterior edge of the condylar neck, which grades onto the posterior border of the ramus, is also depository and grows posteriorly.

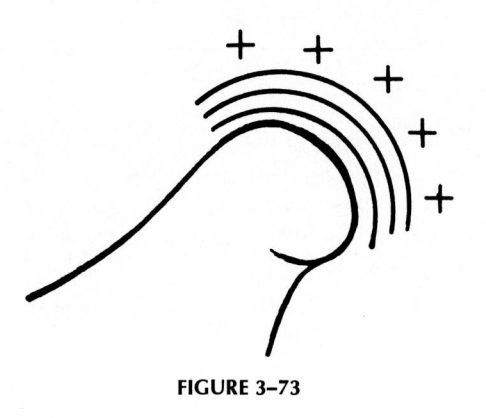

**FIGURE 3–73**

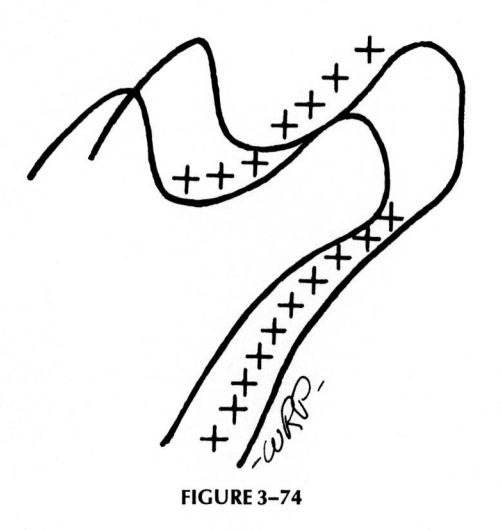

**FIGURE 3–74**

**FIGURES 3-75 AND 3-76**

The lingual and buccal sides of the neck, however, characteristically have **resorptive** surfaces (darkly stippled). This is because the condyle is quite broad and the neck is narrow. The neck is progressively relocated into areas previously held by the much wider condyle, and it is sequentially derived from the condyle as the condyle **moves** in a superoposterior course. What used to be condyle in turn becomes the neck as one is remodeled into the other. This is done by periosteal resorption combined with endosteal deposition.

**FIGURE 3-77**

Explained another way, the **endosteal** surface of the neck actually faces the growth direction; the periosteal side points away from the course of growth. This is another example of the V principle, with the V-shaped cone of the condylar neck growing toward its wide end.

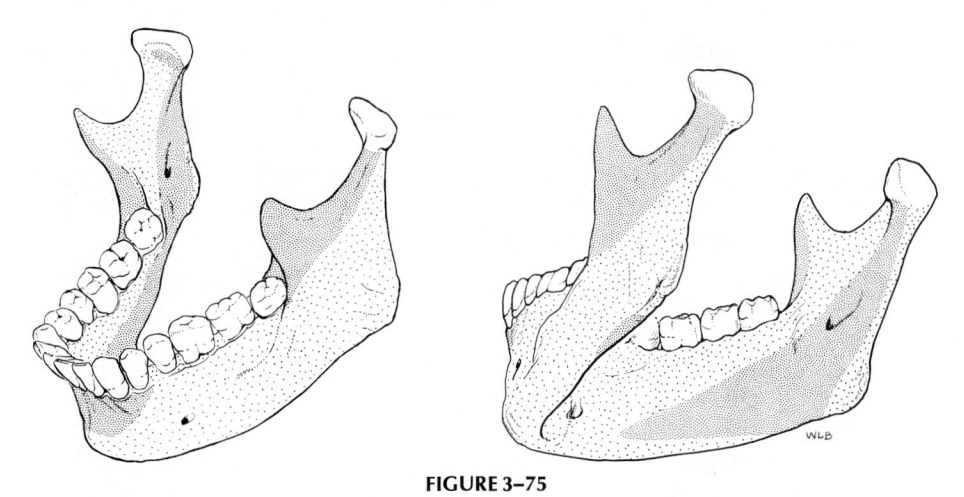

**FIGURE 3–75**

*(From Enlow, D. H.:* The Human Face. *New York, Harper & Row, 1968, p. 135.)*

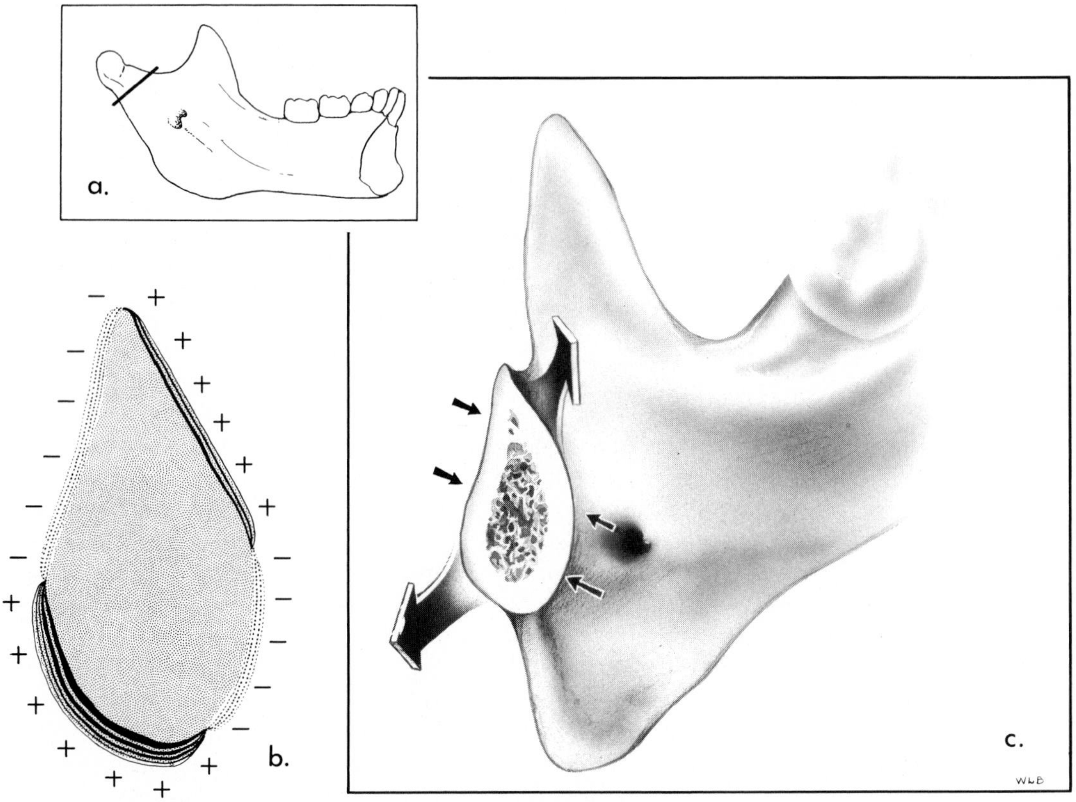

**FIGURE 3–76**

*(Adapted from Enlow, D. H., and D. B. Harris: A study of the postnatal growth of the human mandible. Am. J. Orthod., 50:25–50, 1964.)*

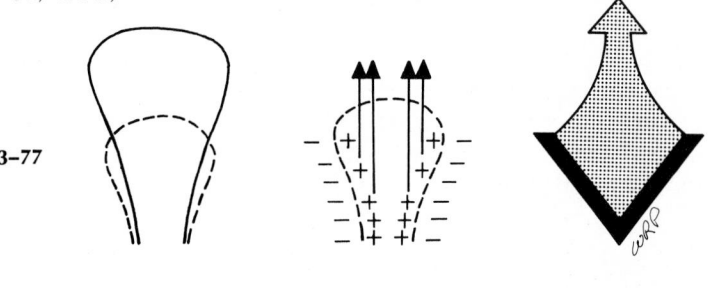

**FIGURE 3–77**

**FIGURES 3–78 AND 3–79**

*Stage 5.* What is the physical force that produces the forward and downward primary **displacement** of the mandible? For many years it was presumed that growth of the **condylar cartilage,** because it is known that cartilage is a special pressure-adapted type of tissue, creates a "thrust" of the mandible against its articular bearing surface in the glenoid fossa. The proliferation of the cartilage **toward** its contact thereby presumably **pushes** the whole mandible away from it.

We are presently in a period of conceptual transition. Some students of facial biology still accept this explanation. However, a growing number of investigators feel that this is either an incomplete or an incorrect answer. The reasons follow.

A great question was created when it was pointed out that mandibles totally lacking condyles exist in nature. Yet, their morphology is more or less normal in all other respects; only the condyle and a part of the condylar neck are congenitally missing. Moreover, these **bilaterally** condyle-lacking mandibles occupy a normal **position;** the bony arch is properly placed for occlusion, and the mandible functions (albeit with distress) in movements even though it lacks an articulation. These revealing observations suggested two conclusions. First, the condyles may not play the role of a "master center" regulating the growth processes in the other parts of the mandible. Second, the whole mandible can become **displaced** anteriorly and inferiorly into its functional position without a "push" against the cranial base. Many experimental studies have subsequently been carried out with similar results, although investigators are still arguing about the proper way to interpret their meaning.

**FIGURES 3–80 AND 3–81**

These observations led to a consideration of the "functional matrix" by students of facial biology. The idea is essentially that the mandible is **carried** forward and downward, just as the maxilla is presumably carried in conjunction with the growth expansion of the soft tissue matrix associated with it. It is a passive type of carrying in which mandibular growth itself does not directly participate; mandibular enlargement is an effect rather than a cause of the displacement movement. Thus, as the mandible is displaced away from its cranial base articular contact, the condyle **secondarily** (but almost simultaneously) grows toward it, thereby closing the potential space without an actual gap being created (unless the condyle does not develop at all, as mentioned previously). There is still, however, actual pressure being exerted on the articular surface; it is presumably a relief of the **amount** of pressure that results from the enlarging soft tissue mass which favors and stimulates condylar growth.

The clinical implications are apparent. Just how involved is the condyle as a factor in facial abnormalities? What happens to the mandible if the condyle is injured during the childhood period? To the orthodontist, a key question is the effect of controlled pressures and tensions acting on the condyle. Which type of force should be used? How can overall mandibular length be increased or decreased for Class II and III individuals by physiologic or mechanical intervention of the condylar growth mechanism?

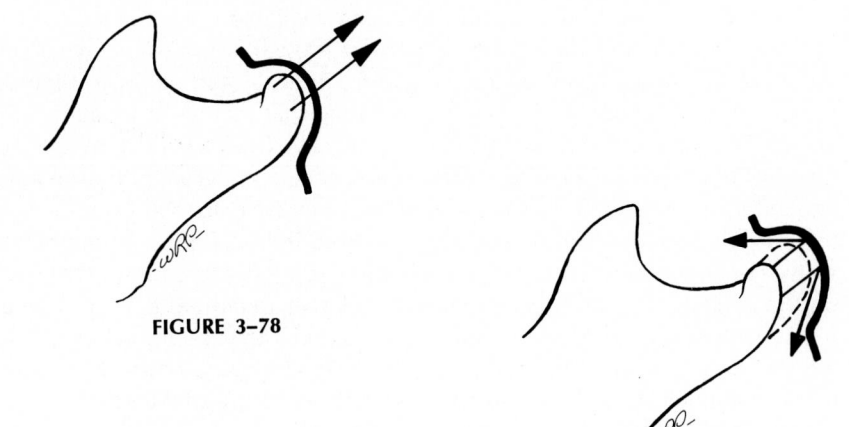

**FIGURE 3–78**

**FIGURE 3–79**

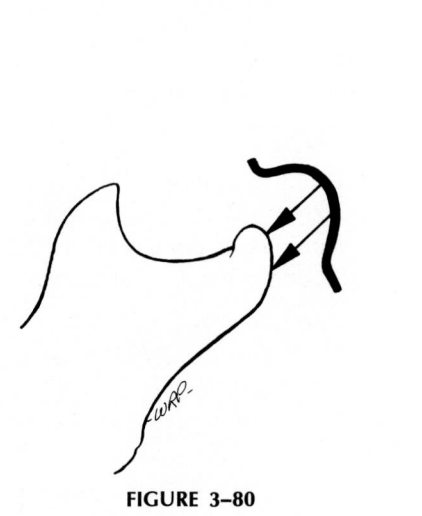

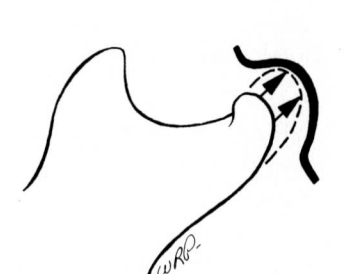

**FIGURE 3–81**

**FIGURE 3–80**

The current thinking is that the condylar cartilage **does** have a measure of intrinsic genetic programming. This, however, appears to be restricted to a capacity for continued cellular proliferation. That is, the cartilage cells are coded and geared to divide and continue to divide, but extracondylar factors are needed to sustain this activity. The **rate** and **directions** of condylar growth are presumably subject to the influence of extracondylar agents, including intrinsic and extrinsic biomechanical forces and physiologic inductors. It is believed that increased amounts of pressure on the cartilage serve to inhibit the rate of cell division and growth. Decreased amounts of pressure appear to stimulate and accelerate growth. Presumably, forces applied to the mandible in such a way that they increase the level of pressure on the condyle would result in a shorter mandible **if** this is done during the period of active condylar growth. Similarly, a **release** of some of the compressive force acting on the condylar cartilage would produce a larger mandible if done during the active growth period. These conclusions are based largely on animal experiments and are not, at least at present, usable for everyday clinical practice. Moreover, recent research studies (as by McNamara) show that the nature of the condylar stimulus is more complex than simple forces acting directly on the condyle; rather, nerve-muscle–connective tissue pathways are involved, and changes utilize a **composite** of such tissue responses and chain feedbacks with the condyle as well as the other parts of the mandible which also participate. Sensory nerve input from the periodontal membranes and from the soft tissue matrix throughout the face pick up stimuli that are passed on via motor nerves to muscles which, in turn, affect the course of growth and remodeling by the condyle and all other areas of the growing mandible.

**FIGURE 3–82**

*Stage 6.* It is often assumed that the face is more or less independent of the cranial base; that facial growth processes and the topographic features of the face are unrelated to the size, shape, and growth of the floor of the cranium. This is not the case at all. What happens in the cranial base very much affects the structure, dimensions, angles, and placement of the various facial parts. The reason is that the cranium is the **template** upon which the face develops. The growth of the middle cranial fossa is described below, that of the anterior cranial fossa later. How differences in the structure of the cranium as a whole affect facial pattern is explained in other chapters.

**FIGURE 3–83**

As each temporal lobe of the cerebrum grows, the middle cranial fossa (which houses the bottom part of the temporal lobe) correspondingly expands by a like amount. The bony surface of the whole cranial **floor** is predominantly resorptive (darkly stippled) in nature. This is in contrast to the endocranial surface of the **calvaria** (skull roof), which is predominantly depository (lightly stippled). The reason for this major difference is that the inside (meningeal surface) of the skull roof is not compartmentalized into a series of confined pockets. The cranial floor, in contrast, has the **endocranial fossae** and other depressions, such as the sella turcica and the olfactory fossae. Why this calls for a difference in the mode of growth is explained below.

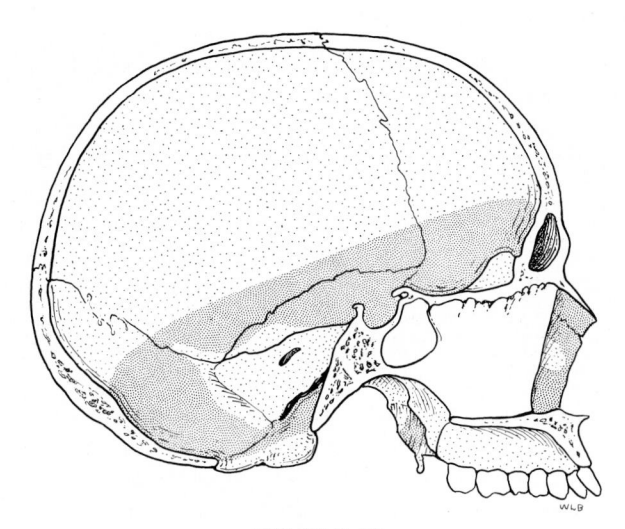

**FIGURE 3–82**

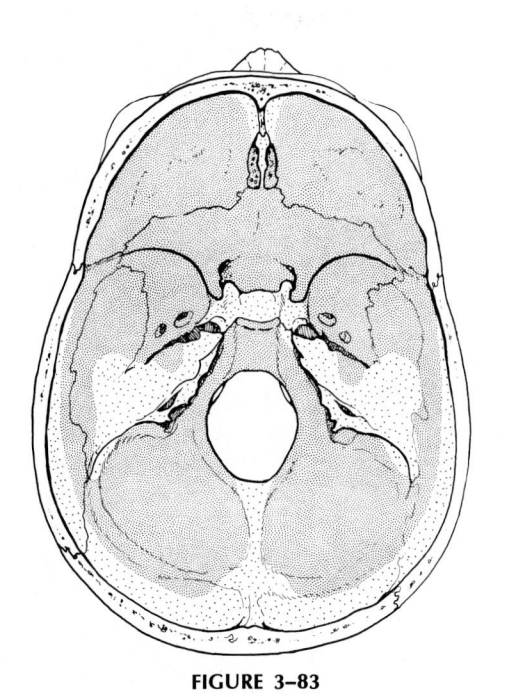

**FIGURE 3–83**
*(From Enlow, D. H.:* The Human Face. *New York, Harper & Row, 1968, p. 197.)*

**FIGURES 3–84, 3–85, 3–86, AND 3–87**

As the brain expands *(a)*, the separate bones of the calvaria are correspondingly displaced in outward directions *(b)*. This is believed to be a passive movement on the part of the bones themselves. In Figure 3–85, the primary displacement causes tension in the sutural membranes, which, according to present theory, respond immediately by depositing new bone on the sutural edges *(a)*. Each separate bone (the frontal, parietal, and so forth) thereby enlarges in circumference. At the same time, the whole bone receives a small amount of new deposition on the flat surfaces of **both** the ectocranial and endocranial sides *(b)*. The endosteal surfaces of the inner and outer cortical tables are resorptive. This increases the thickness of the bone and expands the medullary space between the inner and outer tables. The deposition of bone on the ectocranial surface, however, is **not** the growth change that causes the entire bone to move outward and to increase in its circumference. This is brought about by brain expansion and sutural bone growth, respectively. The arc of curvature of the whole bone decreases, and the bone becomes flatter (Fig. 3–87). While **remodeling** is not extensive in any of these "flat" bones because of their relatively simple contours and morphology, reversals can occur in areas adjacent to the sutures. Here, either outside or inside surface resorption can take place, depending on the local nature of the curvature. This functions to progressively reduce the curvature. The outward movement of each bone in Figure 3–87 is caused by the expansion of the brain. As this happens, each bone also simultaneously changes curvature by the process shown in Figure 3–86.

**FIGURE 3–88**

The cranial floor requires an entirely different mode of growth because of its topographic complexity and the tight curvatures of its fossae. The endocranial side (in contact with the dura which functions as a periosteum) is characteristically **resorptive** in most areas. The reason for this is that the sutures cannot provide for the entire process of growth enlargement. For example, in diagram *a* sutures are located at *1* and *2*, and these produce growth in the direction of the arrows. However, the two sutures present cannot produce the growth for the **other** directions also needed to accommodate brain expansion, as shown in diagram *b*. Growth is accomplished by direct cortical drift, involving deposition on the outside with resorption from the inside; it is the key remodeling process that provides for the direct enlargement of the various endocranial fossae in **conjunction** with sutural (and also synchondrosis) growth.

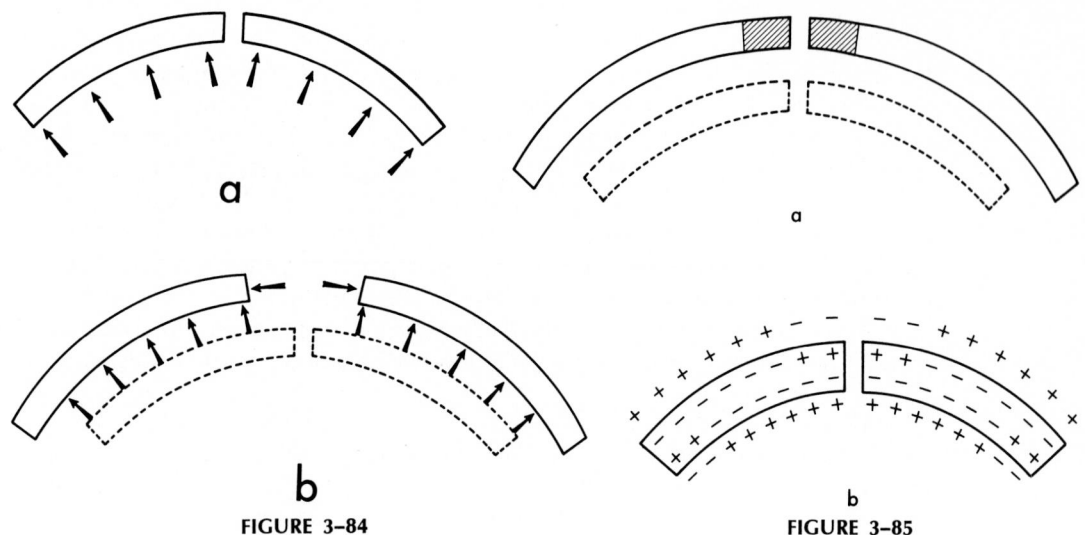

a

b

**FIGURE 3-84**

a

b

**FIGURE 3-85**

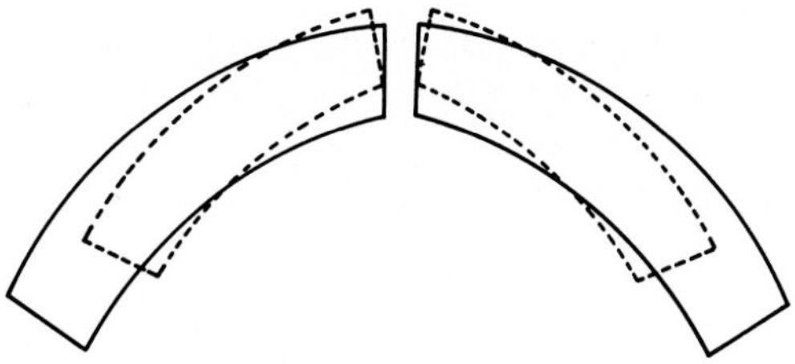

**FIGURE 3-86**

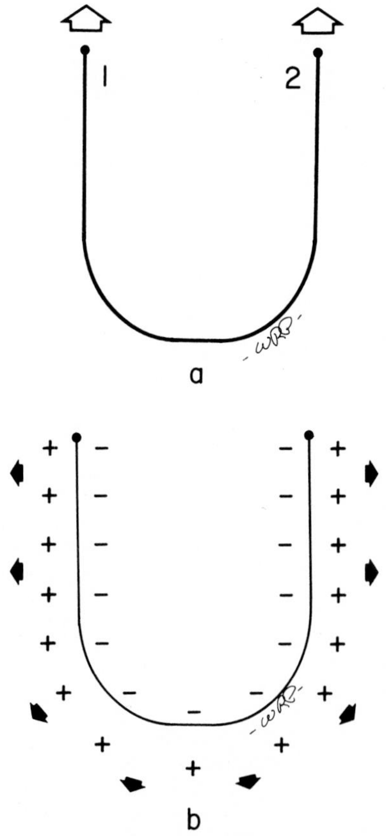

a

b

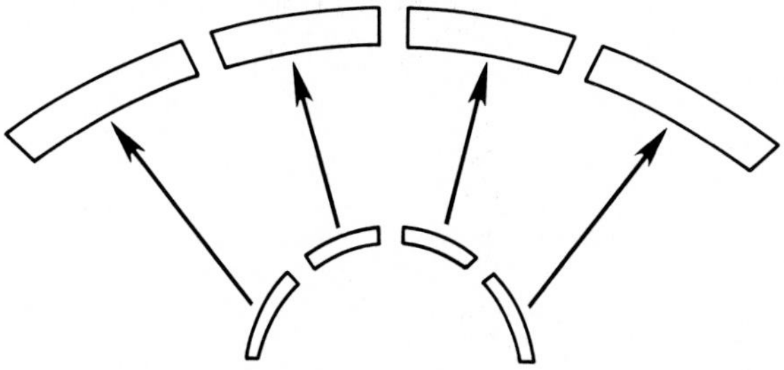

**FIGURE 3-87**

**FIGURE 3-88**

**FIGURE 3–89**

The various endocranial compartments are separated from one another by elevated bony partitions. The middle and posterior fossae are divided by the petrous elevation; the olfactory fossae are separated by the crista galli; the right and left middle fossae are separated by the longitudinal midline sphenoidal elevation just below the sella turcica; and the right and left anterior and posterior cranial fossae are divided by a longitudinal midline bony ridge. All these elevated partitions, unlike most of the remainder of the cranial floor, are **depository** in nature *(a)*. The developmental basis for the depository nature of these partitions is schematized in diagram *b*. The reason, simply, is that, as the fossae expand outward by resorption, the partitions between them must enlarge inward, in proportion, by deposition.*

The **midventral segment** of the cranial floor grows much more slowly than the floor of the laterally located fossae. This accommodates the slower growth of the medulla, pons, hypothalamus, optic chiasma, and so forth, in contrast to the massive, rapid expansion of the hemispheres. Because the floor of the cranium enlarges by remodeling growth in addition to sutural growth, these differential extents and rates of expansion can be carried out. A markedly decreasing gradient of sutural growth occurs as the midline is approached, but direct remodeling growth occurs to provide for the varying extents of expansion required among the different midline parts themselves and between the midline parts and the much faster growing lateral regions.

**FIGURE 3–90**

Unlike the skull roof, the floor of the cranium provides for the passageway of cranial nerves and the major blood vessels. Because the expansion of the hemispheres would cause marked displacement movements of the bones in the cranial floor if only a sutural growth mechanism were operative (as in the skull roof), the process of **remodeling** growth in the cranial base provides for the stability of these nerve and vascular passageways. That is, they do not become disproportionately separated because of the massive expansion of the hemispheres of the brain, as would happen if the bones enlarged primarily at the sutures. The foramen enclosing each nerve or blood vessel also undergoes its own drift process (+ and −) to constantly maintain proper position. The foramen at *a* moves to *b* by deposition and resorption, keeping pace with the corresponding movement of the nerve or vessel it houses. This drift movement is much less than the remodeling movement of the lateral walls of the fossa *(c)*.

---

*The activity of the bone lining the sella, however, is quite variable and can be either depository or resorptive in different areas. Several reasons apparently contribute to this, including the varying degrees of cranial base flexure and the variable amounts of downward and forward displacement of the midventral segments of the whole cranial base by the different shapes and proportionate sizes of the cerebral lobes. The sella turcica, however, must remain in contact with the hypophysis and also adjust to the variable size of the gland itself. If the pituitary fossa is carried downward by whole cranial base displacement farther than necessary, the floor of the sella will correspondingly rise by surface deposition to maintain contact with the pituitary. Or the floor may be partly or entirely resorptive in other individuals to adjust to the balance between cranial base displacement and hypophyseal contact. A common combination is a resorptive posterior lining wall of the hypophyseal fossa and a depository surface on the sphenoidal part of the clivus (as in Figure 3–96). This causes a backward flare of the dorsum sellae to accommodate a pituitary gland that is being displaced to a lesser extent than the sphenoidal body below it. The jugum sphenoidale, like the floor of the sella turcica, shows variations for the same reasons cited earlier. Its dorsal surface may be resorptive in some individuals but depository in others. Melsen reports that it is always depository, but this is not in accord with the author's findings.

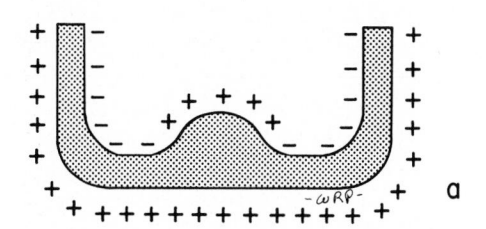

a

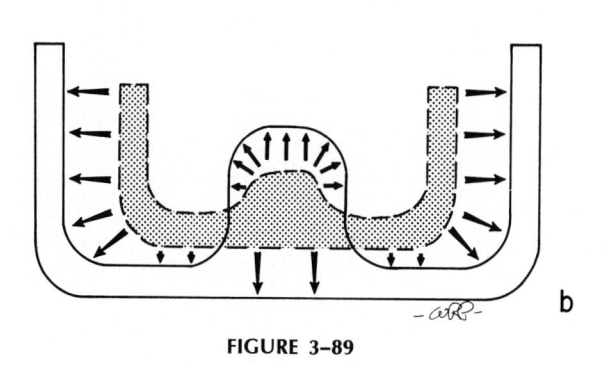

b

**FIGURE 3-89**

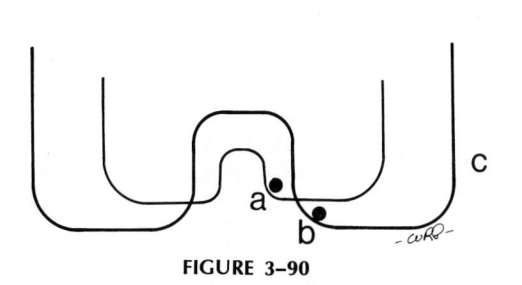

c

**FIGURE 3-90**

**FIGURE 3–91**

The differential remodeling process maintains the proportionate placement of the spinal cord, even though the floor of the posterior cranial fossa, which rims the cord, expands to a considerably greater extent than the foramen magnum. Note the much larger growth increments of the hemispheres and the squama of the occipital bone, in contrast to the smaller growth increments of the spinal cord and the foramen magnum. Differential remodeling, not merely sutural growth, provides for this.

**FIGURE 3–92**

The midline part of the cranial base is characterized by the presence of **synchondroses.** They are "left over" from the primary cartilages of the early cartilaginous cranial base after the endochondral ossification centers appear during fetal development. A number of synchondroses are operative during the fetal and early postnatal period (see Chapter 10). During the childhood period of development, however, it is the spheno-occipital synchondrosis (arrow) that is the principle "growth cartilage" of the cranial base. Like all "growth cartilages" associated directly with bone development, the spheno-occipital synchondrosis provides a pressure-adapted bone growth mechanism. This is in contrast to the tension-adapted sutural growth process. Compression is involved in the cranial base, unlike the calvaria, presumably because it supports the weight of the brain and the face, which bear down on the fulcrum-like synchondrosis in the midline part of the cranial floor. The spheno-occipital synchondrosis is retained throughout the childhood growth period as long as the brain and cranial base continue to grow and expand. It ceases to be active at about 12 to 15 years of age, and the sphenoid and occipital segments then become fused in this midline area before about 20 years of age.

The presence of the spheno-occipital synchondrosis provides for the elongation of the **midline** portion of the cranial base by its pressure-adapted mechanism of endochondral ossification (see Chapter 11). The floor of the cranium also has sutures in the lateral areas, but (1) the force of the compression is accommodated by the synchondrosis, not the sutures, and (2) the expansion of the laterally located **hemispheres** produces tension in these lateral sutural areas, unlike the slower growing midline part of the cranial base not related directly to the hemispheres.

**FIGURE 3–93**

The structure of the synchondrosis is similar to the basic plan for all primary types of growth cartilages, in contrast to the secondary variety which is basically different (see page 94 describing the secondary condylar cartilage). Like the epiphyseal plate of long bones, the synchondrosis has a series of "zones," including the familiar reserve, cell division, hypertrophy, and calcified zones. Like the epiphyseal plate but unlike the condylar cartilage, the chondrocytes in the cell division zone are aligned in distinctive columns that point toward the line of growth. Unlike the epiphyeseal plate, the synchondrosis has **two** major directions of linear growth. Structurally, the synchondrosis is essentially two epiphyseal plates back-to-back separated by a common zone of reserve cartilage. Functionally, however, the behavior of the synchondrosis and the nature of its participation in the overall growth process are less clear.

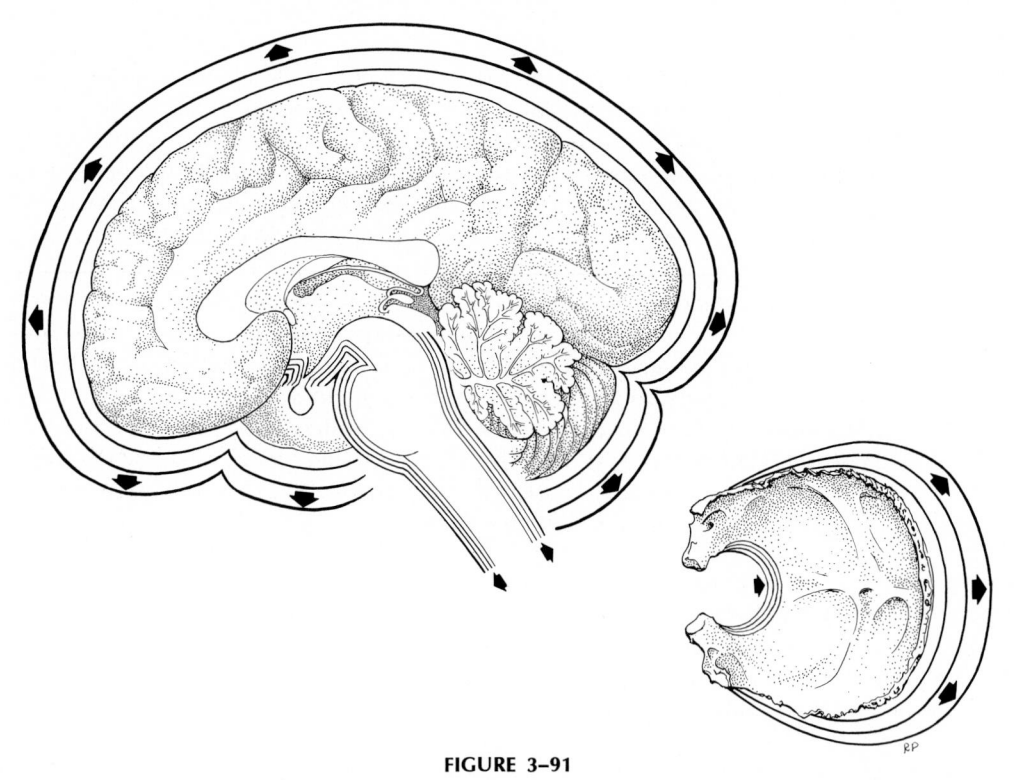

**FIGURE 3–91**

*(Modified from Enlow, D. H.: The Human Face. New York, Harper & Row, 1968, p. 202.)*

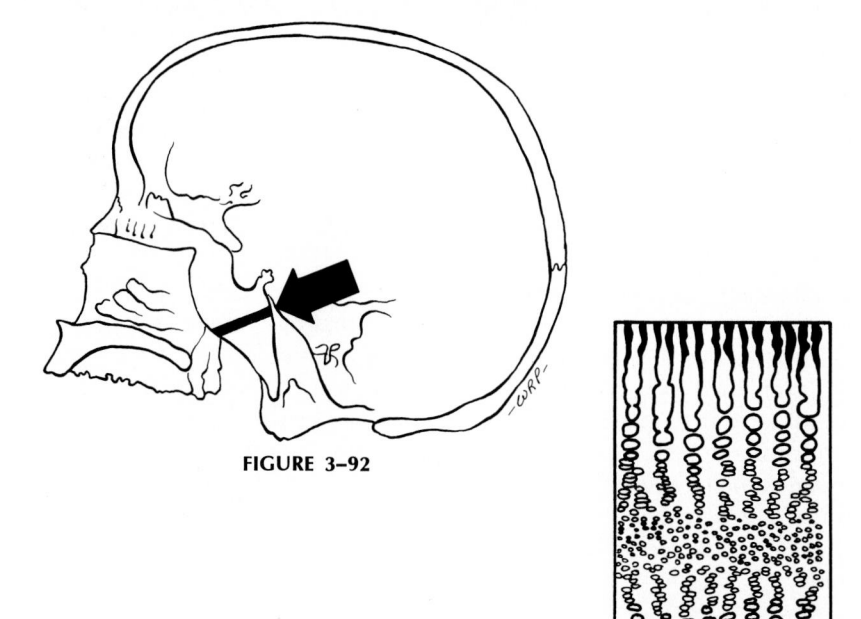

**FIGURE 3–92**

**FIGURE 3–93**

**FIGURE 3–94**

As with growth in all regions of the skull, endochondral bone growth by the spheno-occipital synchondrosis involves primary **displacement.** Thus, the sphenoid and the occipital bones **move apart** by this primary displacement process. At the same time, new endochondral bone is laid down in the medullary regions of each bone, and cortical bone tissue is formed by the periosteum and/or the endosteum around this core of endochondral bone tissue. Each whole bone (the sphenoid and the occipital) thereby becomes lengthened. Both bones also increase in girth by periosteal and endosteal activity. The interior of the sphenoid bone becomes hollowed to form the sizeable **sphenoidal sinus.** This sinus is just behind and in direct line with the bony nasal septum of the nasomaxillary complex. As the midface is displaced forward and downward, the sphenoid retains contact with it, and the sphenoidal sinus is "drawn out" by the enlargement of this part of the sphenoidal body. Sphenoidal sinus expansion does not "push" the maxilla, however. The sinus is secondarily formed as the body of the sphenoid bone expands, keeping constant relationship with the midface.

Two key questions exist with regard to the lengthening of the cranial base at a synchondrosis and the process of displacement that accompanies the elongation of each new bone. First, do the synchondroses **cause** displacement by the process of growth expansion, or is their endochondral growth a **response** to displacement caused by other forces (such as, perhaps, brain expansion)? Second, does the cartilage have an intrinsic genetic program that actually regulates the rate, amount, and direction of growth by the cranial base? Or is the cartilage more or less dependent on some **other** pacemaker for growth control?

Traditionally, the cranial base cartilages (and the whole cranial base in general) have been regarded as essentially autonomous growth units that develop in conjunction with the brain but independent of it. Cranial base growth has been presumed to be controlled by a genetic code residing within the cartilage cells of the synchondrosis. In other words, the shape, size, and characteristics of the cranial base have evolved in direct phylogenetic association with the brain it supports (that is, a "phylogenetic type" of functional matrix), but the cranial base itself has developed a genetic capacity for its own growth that is at least partially separate and independent of the brain and which may function without it during ontogenetic growth (as can happen in agenesis of the brain).

Experimental studies, such as those by Koski, show that the independent proliferative capacity of a synchondrosis does not approach that of epiphyseal plates in long bones. This suggests that, whatever capacity the cranial base (not calvaria) does have for continued growth, extrinsic control factors are also required. Just what these factors actually are and to what extent they are involved is not presently understood, since the cranial base **can** develop to a greater or lesser extent, even though the brain is malformed or absent. This is a major problem of our time and one that has more than academic significance to students of facial biology. In contrast, the calvaria appears to be largely dependent on its surrounding matrix for growth control.

**FIGURE 3–95**

As previously pointed out, the contribution of the synchondrosis relates to the midventral axis of the cranium and not the entire cranial floor. Note that the overall enlargement of the midline part of the cranial base (a) is much less than the expansion of the more laterally located middle (and posterior) cranial fossae (b). This is

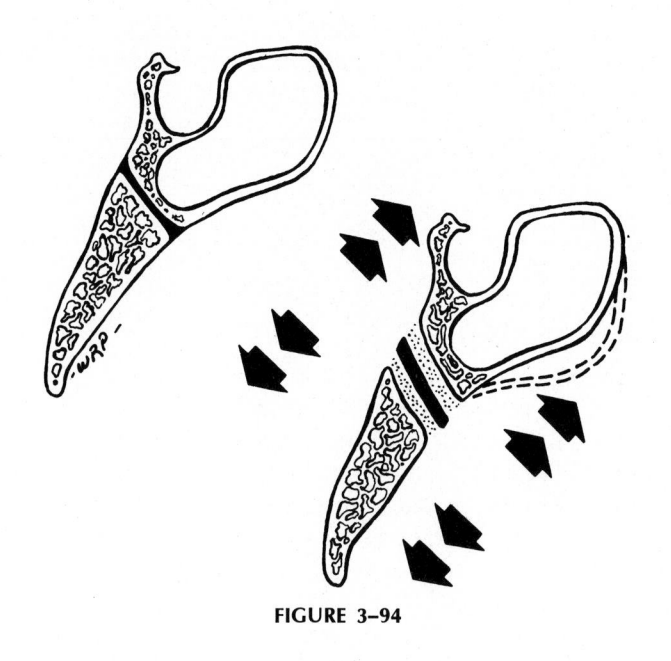

**FIGURE 3–94**

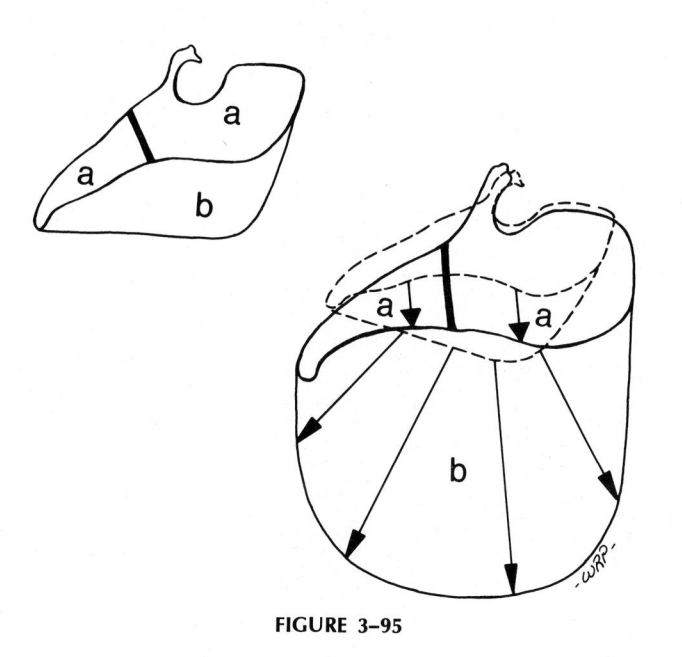

**FIGURE 3–95**

because the lateral fossae house the various lobes of the hemispheres, which enlarge considerably more than the medulla, pituitary gland, hypothalamus, and so forth. Endocranial resorption occurs on both the endocranial surface of the clivus and, laterally, the floor of the middle cranial fossa. For the clivus, this produces an anteroinferior drift movement **in addition to** to the linear growth by the synchondrosis.* For the middle (and also posterior) cranial fossa, it produces a massive enlargement in conjunction with the sutural growth also taking place. The clivus also lengthens by bone deposition on the ectocranial side of the occipital bone at the lip of the foramen magnum. For more detailed descriptions of the regional growth process in each separate bone in the cranium, see Enlow, *The Human Face,* 1968.

**FIGURES 3–96 AND 3–97**

*Stages 7 and 8.*    The expansion of the middle cranial fossa has a major **secondary** displacement effect on the anterior cranial floor, the nasomaxillary complex, and the mandible. Because the posterior boundary of the facial complex exactly coincides with the boundary between the anterior and middle cranial fossae, the horizontal enlargement of the middle fossa produces a like amount of forward displacement for both the anterior cranial fossa and the nasomaxillary complex. The amount of horizontal displacement for the mandible, however, is much less because most of the enlargement of the middle cranial fossa takes place **anterior** to the mandibular condyle.

The enlarging middle cranial fossa does not in itself **push** the mandible, anterior fossa, and maxillary complex forward. Visualize the enlarging temporal and frontal lobes of the cerebrum as two expanding rubber balloons in contact. They are each displaced **away** from one another, although the net effect is a forward direction. The temporal and frontal ''balloons'' have fibrous attachments to the middle and anterior cranial fossae, respectively. As **both** balloons expand, these two fossae are thus **pulled away** from each other. This sets up tension fields in the various frontal, temporal, sphenoidal, and ethmoidal sutures, which presumably trigger sutural bone growth (in addition to direct cortical growth by resorption and deposition). Both fossae are thus enlarged, and the nasomaxillary complex is carried along anteriorly with the anterior cranial fossa to which it is attached. At about 5 or 6 years of age, frontal lobe growth and anterior cranial fossa expansion are largely complete. The temporal lobe and middle fossa, however, continue to enlarge for several years. Expansion of the temporal lobe displaces the frontal lobe forward, and this in turn causes tension in the suture systems between these two areas. The anterior fossa and the maxillary complex are carried anteriorly by the frontal lobe, which moves forward because of temporal lobe enlargement behind it. (**Note:** This ''tension'' trigger in response to **brain** enlargement is the present-day theory. Time may or may not prove this to be correct, but the rationale is at least reasonable as we now see it.)

---

*The dorsum sellae, however, shows much variation in shape and size. In some individuals it flares markedly in an upward and backward direction, and the sphenoidal part of the clivus may be correspondingly depository rather than resorptive, as shown in Figure 3–96. Melsen claims that this surface is always depository, but the author's findings are in disagreement. Some of the bone tissues in this and other areas of the cranial base described by Melsen as ''appositional mature lamellar bone'' are actually of an endosteal rather than a periosteal type.

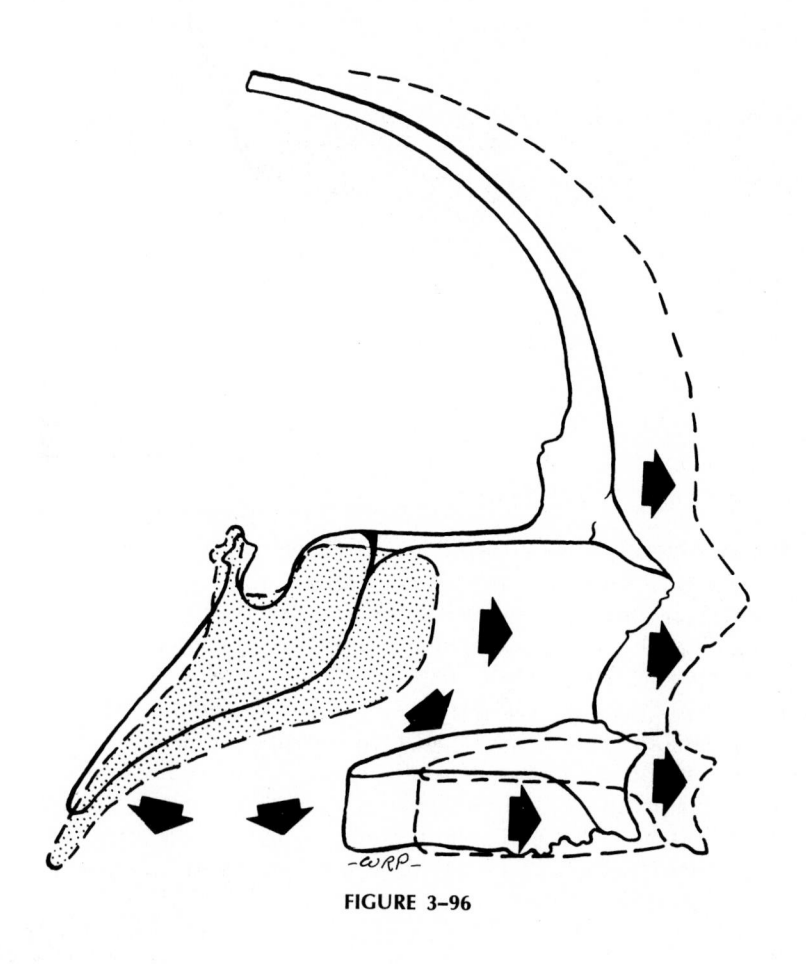

**FIGURE 3–96**

**FIGURE 3–97**

Resorption occurs from the lining side of the forward wall of the middle cranial fossa (*1*), deposition takes place on the orbital side of the sphenoid (*2*), and forward displacement of the anterior cranial fossa occurs as the frontal lobes are displaced anteriorly (*3*). The petrous elevation (*4*) increases by deposition on the endocranial surface, and lengthening of the clivus occurs by growth at the spheno-occipital synchondrosis (*5*). The foramen magnum is progressively lowered by resorption on the endocranial surface and deposition on the ectocranial side. This also contributes to the lengthening of the clivus (*6*).

The vertical enlargement of the middle cranial fossa has a major effect on the vertical placement of both the mandibular and maxillary arches. The effect is a progressive separation of the arches.

**FIGURES 3–98, 3–99, AND 3–100**

*Stages 9 and 10.*    As the horizontal enlargement of the middle cranial fossa advances the nasomaxillary complex by forward displacement, the horizontal span of the pharynx correspondingly increases. The skeletal dimension of the pharynx is established by the size of the middle cranial fossa. The ramus of the mandible bridges the pharynx, and as this space enlarges, the ramus increases to an equivalent extent to maintain the same facial form. The effective **horizontal** dimensions of the ramus and the middle cranial fossa (not their respective oblique dimensions) are direct counterparts to each other. One structural function of the ramus, in spanning the middle cranial fossa, is to provide an intrinsic capacity for whatever adaptation is required to place the corpus in a continuously functional position relative to the maxillary arch. If this is successful, a normal or Class I occlusion is achieved in a given individual. If less than adequate, the greater or lesser degree of failure of its adaptive or compensatory function contributes, in part, to the basis for a malocclusion.

In Stages 3 and 4, the ramus becomes relocated posteriorly by its own growth movement. The purpose is to horizontally lengthen the **corpus** and to displace it anteriorly. While this is occurring, the middle cranial fossa is also enlarging, however, so that now the ramus correspondingly increases in breadth to equal it. This is done by the **same** process that carried out Stages 3 and 4, except that the amount of posterior ramus growth now **exceeds** the amount of anterior resorption. The horizontal (PA) breadth of the ramus is thereby increased.

**FIGURE 3-97**

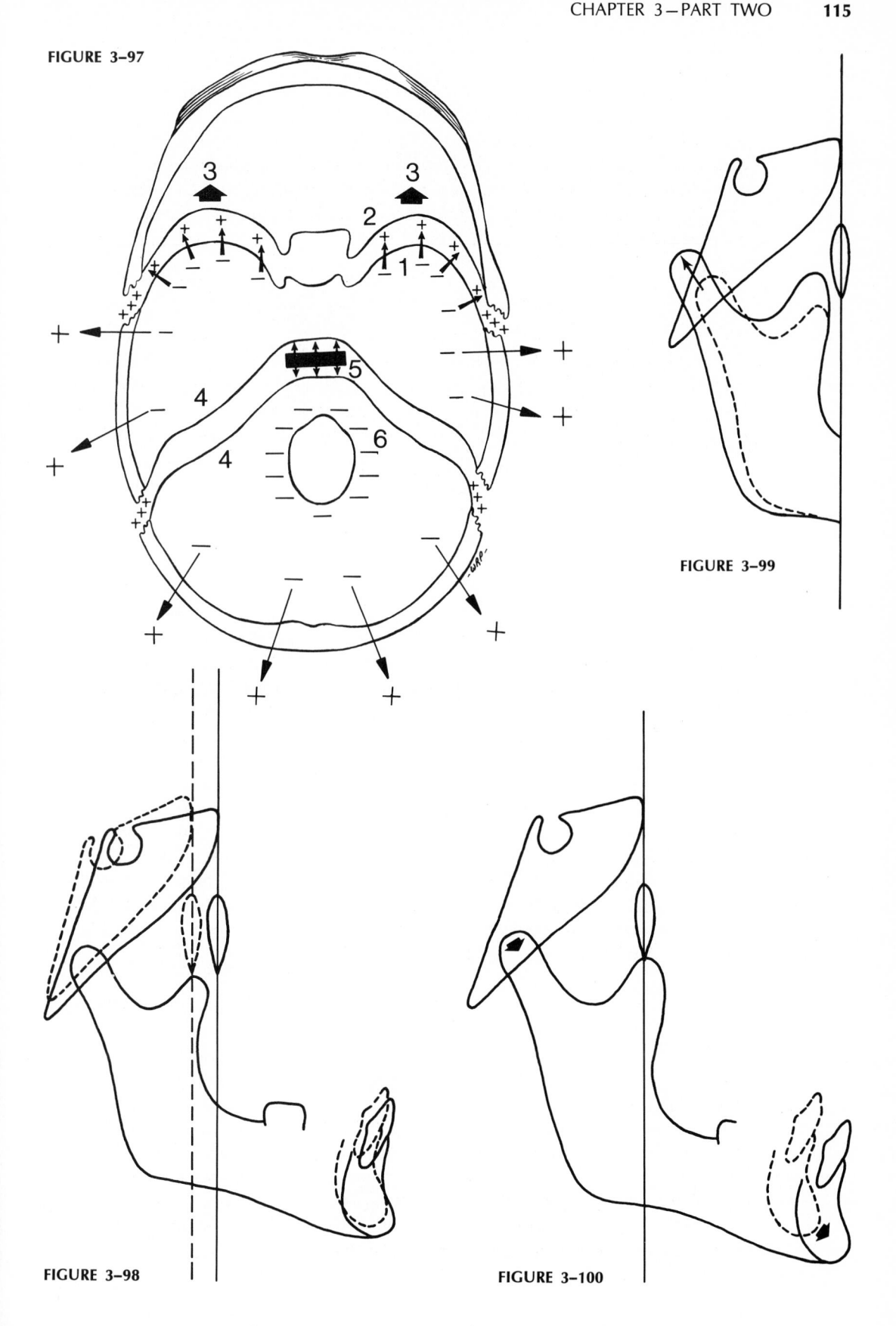

**FIGURE 3-99**

**FIGURE 3-98**

**FIGURE 3-100**

**FIGURE 3–101**

The ramus normally becomes progressively more upright during mandibular development. As long as the ramus is actively growing in a posterior direction, this is accomplished simply by greater amounts of bone addition on the inferior part of the posterior border than on the superior part. A correspondingly greater amount of resorption of the anterior border takes place in the inferior part than in the superior part. A "rotation" of ramus alignment thus occurs. Condylar growth may also become directed in a more vertical course.

**FIGURE 3–102**

The reason the ramus becomes more upright is that it must lengthen vertically to a much greater extent than it broadens horizontally. In this schematic diagram, the pharynx (and middle cranial fossa) enlarges horizontally from a to a'. The ramus enlarges, correspondingly, from b to b' to match it. It also **lengthens vertically,** however. Angle c is thereby reduced to c' in order to accommodate the vertical increase, which allows for the considerable extent of vertical nasomaxillary growth.

**FIGURES 3–103 AND 3–104**

However, vertical lengthening of the ramus continues to take place **after** horizontal ramus growth slows or ceases (when the horizontal growth of the middle cranial fossa begins to stop). This is to match the continued vertical growth of the midface. To achieve this, condylar growth may become more vertically directed, and a different pattern of ramus remodeling can also become operative. The direction of deposition and resorption reverses. A **forward** growth direction can occur on the **anterior** border in the upper part of the coronoid process. Resorption takes place on the upper part of the posterior border. A posterior direction of remodeling takes place in the lower part of the posterior border. The result is a more upright alignment and a longer vertical dimension of the ramus without a material increase in breadth. This remodeling change, when it occurs, appears to be more marked in the later periods of childhood, after the enlargement of the temporal lobes has slowed or largely ceased (with a corresponding decrease in the horizontal enlargement of the ramus) and when the backward relocation of the ramus to provide for corpus lengthening has decreased.

**FIGURE 3–105**

The ramus thus undergoes a remodeling alteration in which its angle becomes changed in order to retain constant positional relationships between the upper and lower arches. This is why the "gonial angle" closes during growth. If mandible a in Figure 3–104 is superimposed over b in the anatomically functional position, it is seen that all the complex remodeling changes in Figure 3–104 serve, simply, to change the ramus angle without increasing its breadth. (See also Stage 15 and pages 94 and 202–207 for further information on mandibular rotations.) This also accommodates the growing muscle sling and muscular adaptations associated with mandibular rotations.

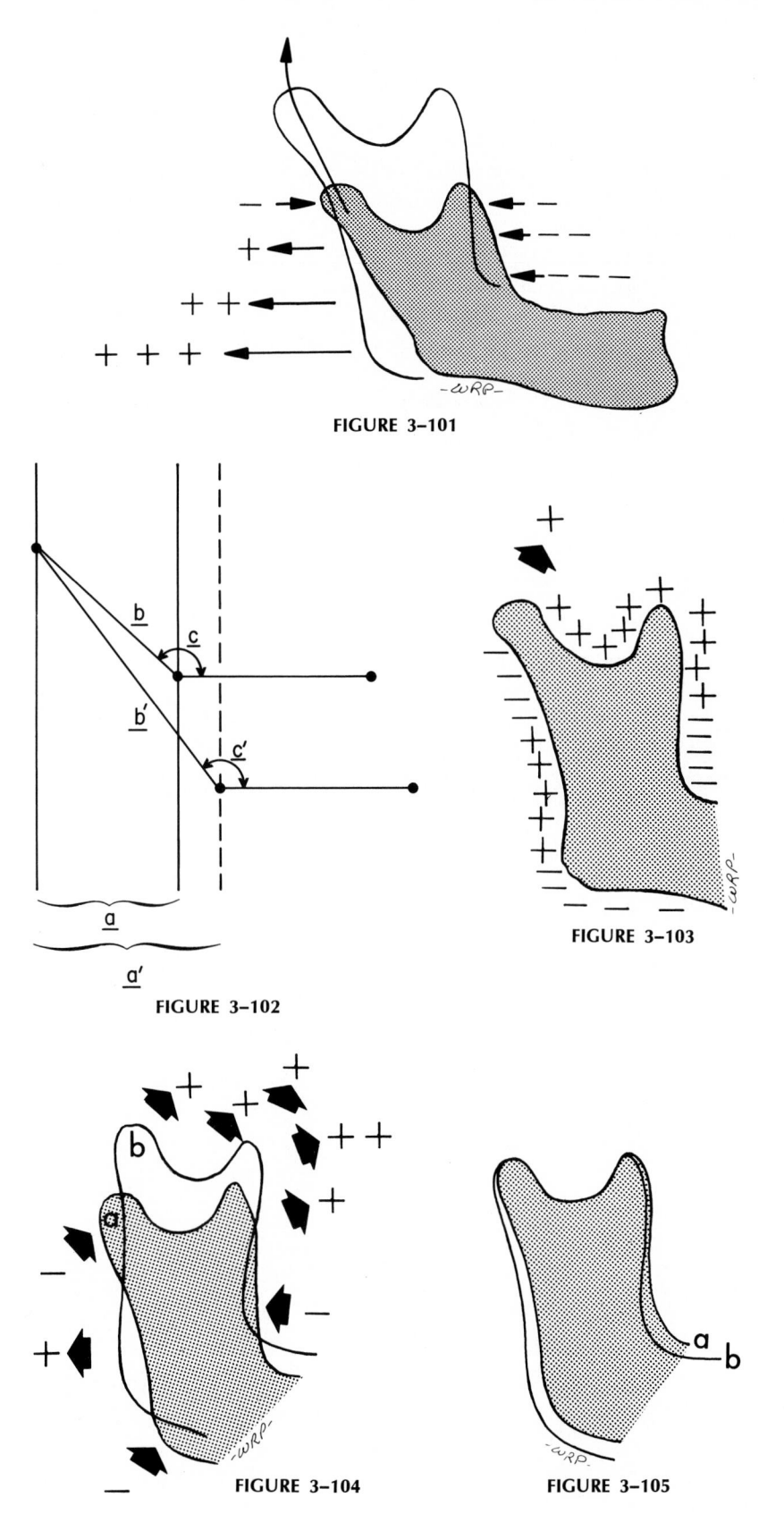

FIGURE 3-101

FIGURE 3-102

FIGURE 3-103

FIGURE 3-104

FIGURE 3-105

**FIGURE 3–106**

*Stage 11.* The anterior cranial fossa enlarges in conjunction with the expansion of the frontal lobes. Wherever sutures are present, they contribute to the increase in the circumference of the bones involved. Thus, the sphenofrontal, frontotemporal, sphenoethmoidal, frontoethmoidal, and frontozygomatic sutures participate in a traction-adapted bone growth response to brain and other soft tissue enlargement. The bones all become **displaced** as a consequence. This is a primary type of displacement, because the enlargement of each bone is involved. Together with this, the bones also grow outward by ectocranial deposition and endocranial resorption. The composite of all these processes produces the growth changes seen in Figure 3–106.

**FIGURE 3–107**

The cranial bones increase in size by sutural bone growth as the forehead becomes displaced anteriorly. The nasomaxillary complex is carried anteriorly as well. Note that the maxillary tuberosity now lies ahead of the vertical reference line. The tuberosity, however, simultaneously grows posteriorly by an equivalent amount (described previously in Stage 1); the floor of the anterior cranial fossa and the bony maxillary arch are counterparts

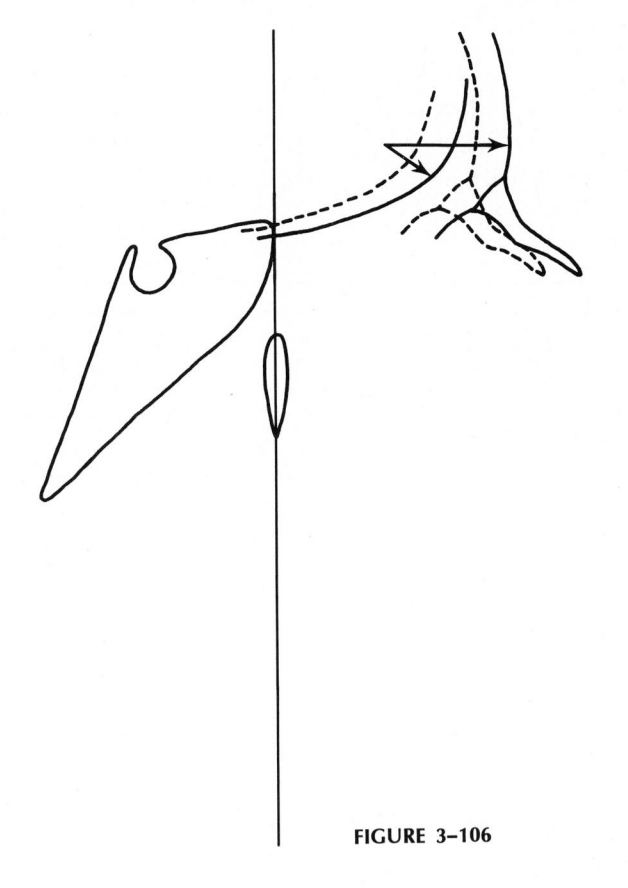

**FIGURE 3–106**

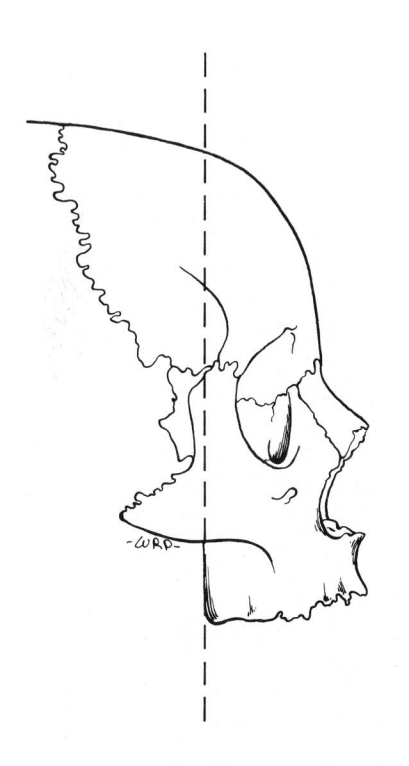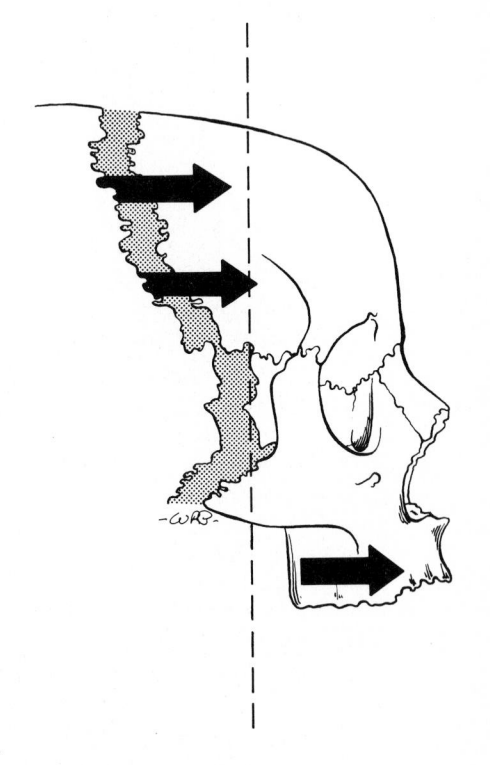

**FIGURE 3–107**

**FIGURE 3–108**

As previously pointed out, sutural growth alone cannot accomplish the extent of cranial fossa expansion required. In addition to bone additions at the various sutures, direct cortical growth also takes place. About midway up the forehead, however, the endocranial side becomes depository rather than resorptive. This reversal line encircles the inner side of the skull and separates the resorptive growth fields of the basicranium from the separate field of the roof (see Figure 3–82).

**FIGURE 3–109**

As long as the frontal lobe of the cerebrum grows, the **inner** table of the forehead correspondingly drifts anteriorly. When frontal lobe enlargement slows and largely ceases sometime before about the sixth or seventh year, the growth of the inner table stops with it. The outer table, however, continues to drift anteriorly. This progressively separates the two tables, and an enlarging frontal sinus results. The size of the sinus, however, and the amount of forehead slope vary considerably according to age, sex, and ethnic characteristics (see Chapter 6). The reason the frontal sinus develops is that the upper part of the nasomaxillary complex continues to grow anteriorly, and the outer table of the forehead moves with it.

**FIGURE 3–110**

Note that the floor of the anterior cranial fossa is also the roof of the underlying orbital cavity. The endocranial side is resorptive, and the orbital side of the very thin bony plate is depository; it moves progressively downward and outward. While this serves to enlarge the bottom part of the cranial fossa, does it also reduce the size of the orbital cavity? The answer is no, because the whole orbit is also becoming **displaced** at the same time in association with growth at the various orbital sutures. This is described later.

*Stage 12.* The vertical lengthening of the nasomaxillary complex involves (a) remodeling growth (deposition and resorption on the various bony cortices) and (b) displacement. The process of remodeling growth is described in this stage, followed by displacement in Stage 13.

**FIGURE 3–111**

With the exception of the uppermost part of the roof in each nasal chamber, the lining surfaces of the bony walls and floor are predominantly resorptive. The mucosal side of each nasal bone is also resorptive. These regional patterns cause a lateral and anterior expansion of the nasal chambers and a downward movement of the palate; the oral side of the bony palate is depository. The nasal side of the cribriform plate (roof of the nasal chamber) is depository, and the cranial side is resorptive. This enlarges the small, paired olfactory fossae and lowers them in conjunction with the downward cortical drift of the entire anterior cranial floor.

The ethmoidal conchae (not diagrammed) have, generally, depository surfaces on their lateral and inferior sides and resorptive surfaces on the superior and medial-facing sides of the thin bony plates. This moves them downward and laterally as the whole nasal region expands in like directions. (The developmentally separate inferior concha, however, can show remodeling variations because it is carried inferiorly, to a greater extent than the others, by maxillary displacement.) The lining cortical surfaces of the maxillary sinuses are all resorptive except the medial nasal wall, which is depository because it moves laterally during nasal expansion.

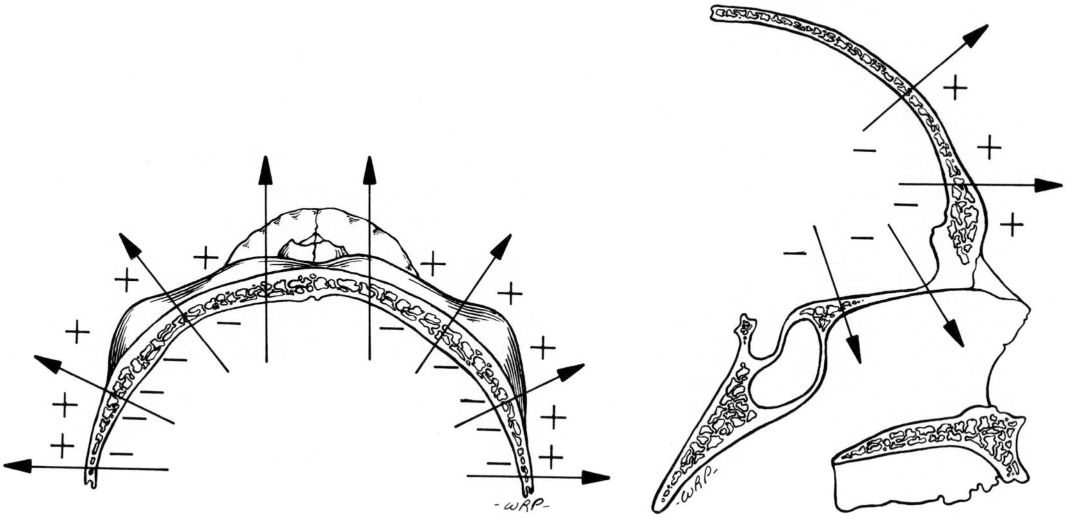

FIGURE 3–108

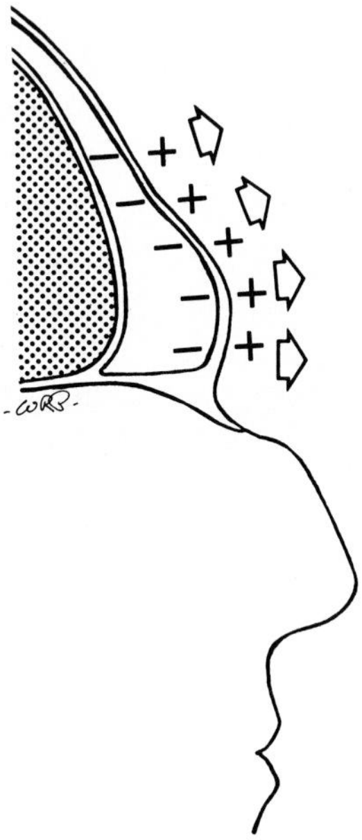

FIGURE 3–109

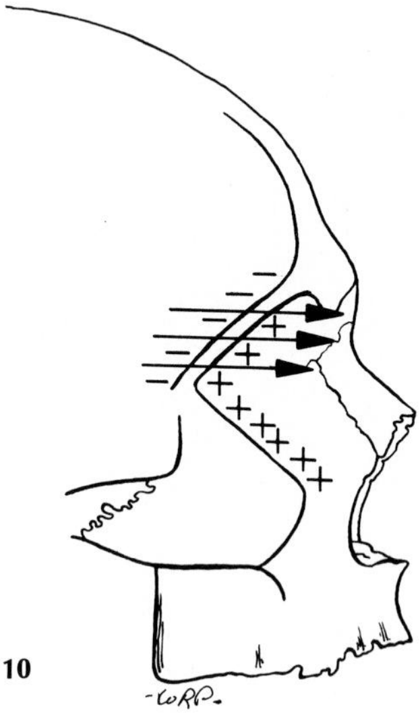

FIGURE 3–110

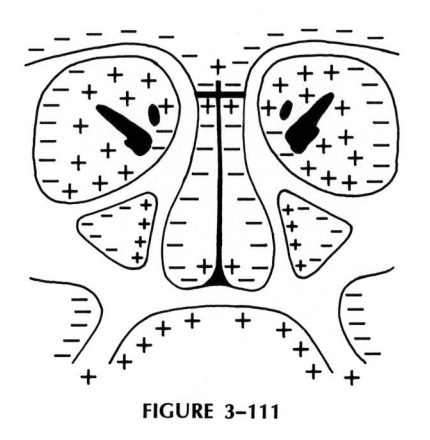

FIGURE 3–111

**FIGURE 3–112**

The bony portion of the internasal septum (the vomer and the perpendicular plate of the ethmoid) lengthens vertically at the various sutural junctions (and to a much lesser extent by endochondral growth where the cartilaginous part contacts the perpendicular plate of the ethmoid). The bony septum drifts laterally also in relation to variable amounts and directions of **septal deviation.** The remodeling patterns involved are individually variable, and the thin plate of bone typically shows alternate fields of deposition and resorption on the right and left sides, which produce a drift and buckling to one side or the other.

**FIGURE 3–113**

Note that the breadth of the nasal bridge does not markedly increase from early childhood to adulthood. The medial wall of each orbit (lateral walls of the nasal chambers between the orbits), however, expands and balloons out considerably in a lateral direction. The ethmoidal sinuses are thereby enlarged greatly.

The **lacrimal** bone is an important participant during orbital growth and remodeling. This diminutive, thin plate of bone plays a key role in providing adjustments for the major differential movements of all the bones surrounding it. The lacrimal is an island of bone with its entire perimeter bounded by sutural contacts separating it from the ethmoid, maxillary, and frontal bones. As these separate bones enlarge or become displaced in many directions and at different rates, the sutural system of the lacrimal provides for the "slippage" of these other bones. It does this by collagenous linkage adjustments within the sutural membrane (see Chapter 11). The lacrimal and its sutures make it possible for the maxilla to "slide" downward along the contact with the medial orbital wall, as the whole maxilla becomes displaced inferiorly.

**FIGURE 3–114**

The lacrimal bone undergoes a remodeling rotation. This is because the more medial **superior** part remains with the lesser-expanding nasal bridge, while the more lateral **inferior** part moves markedly outward to keep pace with the great expansion of the ethmoidal sinuses. This remodeling change is illustrated by *a;* the primary displacement that accompanies it is shown by *b.*

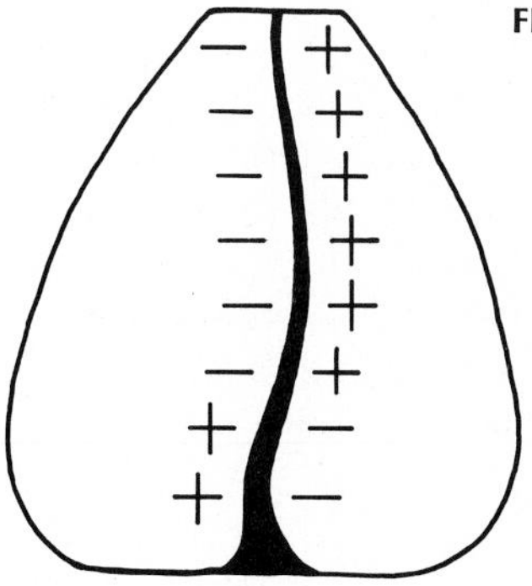

**FIGURE 3–112**

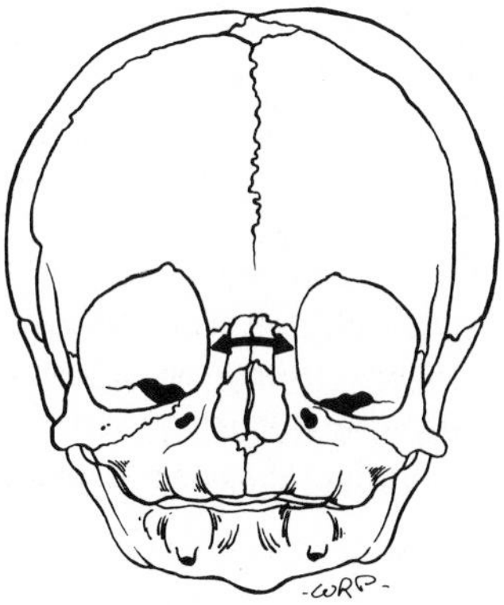

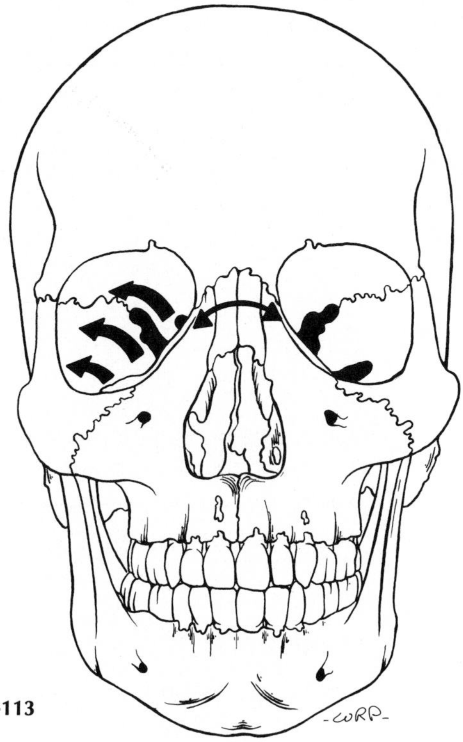

**FIGURE 3–113**

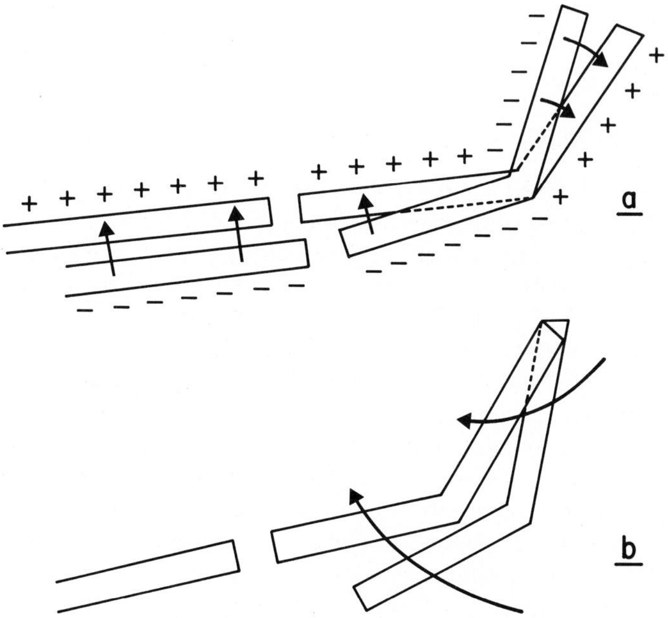

**FIGURE 3–114**

**FIGURE 3-115**

In the growth of the bony maxillary arch, area A is moving in three directions by bone deposition on the external (buccal) surface: it lengthens **posteriorly** by deposition on the posterior-facing maxillary tuberosity; it grows **laterally** by deposits on the buccal surface (this widens the posterior part of the arch); and it grows **downward** by deposition of bone along the alveolar ridges and also on the lateral side, because this outer surface slopes (in the child) so that it faces slightly downward. The endosteal surface is resorptive, and this contributes to maxillary sinus enlargement.

**FIGURE 3-116**

Note that a major change in surface contour occurs along the vertical crest just below the malar protuberance (small arrow). This crest is called the "key ridge." A **reversal** occurs here. Anterior to it the entire external surface of the maxillary arch (the protruding "muzzle" in front of the cheekbone) is **resorptive.** This is because that part of the bony arch in area b is **concave,** and the labial (outside) surface faces upward rather than downward. The resorptive nature of this surface provides an inferior direction of arch growth in conjunction with the downward growth of the palate. This is in contrast to area a, which grows downward by periosteal **deposition.**

**FIGURE 3-117**

Surface a is resorptive; b is depository. A reversal occurs precisely at "A point" (indicated by arrow; this is a much-used cephalometric landmark). Periosteal surface c is resorptive; d is depository; surface e is resorptive; periosteal surface f is depository.

**FIGURES 3-118, 3-119, AND 3-120**

Note that each **tooth** is moved inferiorly by this same remodeling process. Deposition and resorption of bone on the lining surfaces of the alveolar socket (surfaces d and e) produce a **vertical drift** of the tooth. The same alveolar remodeling changes that produce the mesial (or distal) drift of a tooth, as well as any tipping or lateral (buccal) shifts that occur, **also** carry out this vertical drifting process. It is important to understand that the vertical drift of the dentition is **in addition to** the eruption of each tooth. Vertical drift is an important process and should not be referred to simply as "eruption." Eruption is associated with the development of the tooth, and it is a process that brings a tooth into definitive crown height above the bone and gingiva. The vertical drift movement carries the whole tooth **and its socket** inferiorly in conjunction with continued midfacial growth. The purpose is to vertically enlarge the dimension of the nasal chamber by relocating the entire palatal-maxillary arch complex downward. The distance covered by this movement is significant. By harnessing the movement of inferior drift, the orthodontist can guide the teeth into new positions, thereby taking advantage of the growth process.

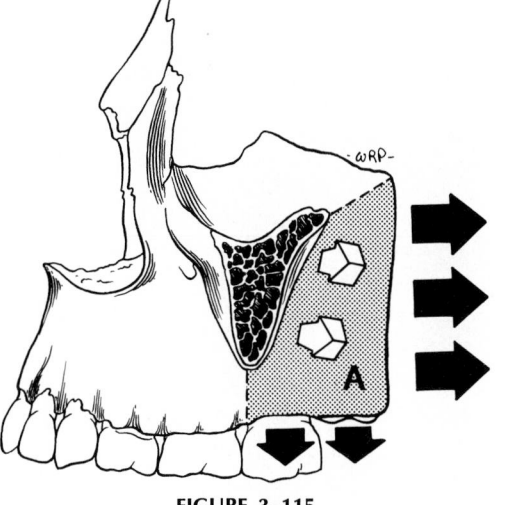

**FIGURE 3–115**

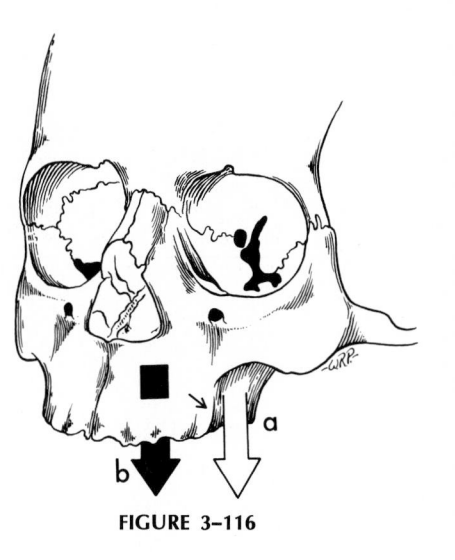

**FIGURE 3–116**

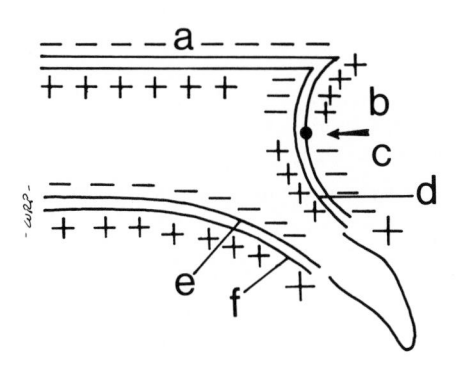

**FIGURE 3–117**

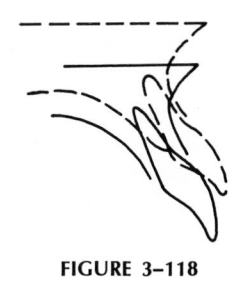

**FIGURE 3–118**

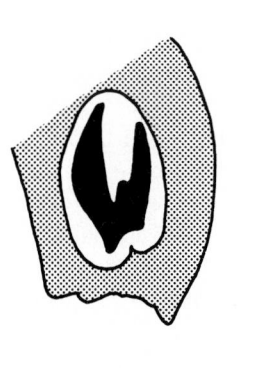

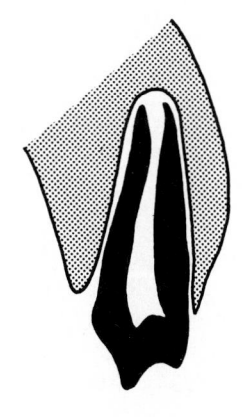

**FIGURE 3–119**

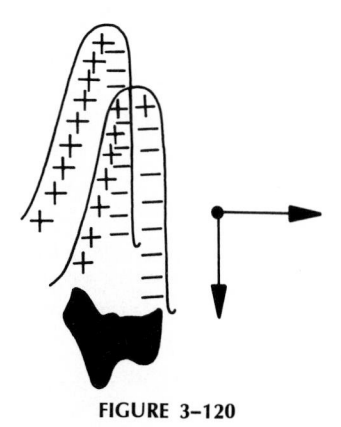

**FIGURE 3–120**

**FIGURE 3–121**

Even though the external (labial) side of the whole anterior part of the maxillary arch (the protruding "muzzle") is resorptive, with bone being added onto the **inside** of the arch, the arch nonetheless increases in width, and the palate becomes wider. This is another example of the V principle.

The rotations, tipping, and inferior drift of the teeth, in combination with the characteristic external resorptive surface of the forward part of the maxilla, sometime result in a localized protrusion of a tooth root tip through the bony cortex. Such penetration results in a "defect" (that is, a tiny hole in the bone) termed a "fenestra."

**FIGURE 3–122**

*Stage 13.*   This stage involves the **primary displacement** of the whole ethmomaxillary complex in an inferior direction. Its displacement movement accompanies simultaneous enlargement (by the process of resorption and deposition) in all areas throughout the entire nasomaxillary region.

**FIGURE 3–123**

New bone is added at the frontomaxillary, zygotemporal, zygosphenoidal, zygomaxillary, ethmomaxillary, ethmofrontal, nasofrontal, frontolacrimal, palatine, and vomerine sutures. These sutural deposits, as explained in Stage 2, are believed to be the **response** to displacement and not the cause of it. That is, the whole region is carried inferiorly (and anteriorly), and this is presumed to result in a tension stimulus that triggers sutural osteogenesis. The bones thereby remain in constant sutural contact. The process of displacement produces the "space" within which the bone enlarges. Sutural bone growth does not **push** the nasomaxillary complex down and away from the cranial floor (although future research will probably show that the suture is more active in this process than presently suspected, but not by the traditional push idea). The displacement of the bones is a nonparticipating "carry" produced by the expanding soft tissue functional matrix (or, according to older theory, the nasal septum). As the bones of the ethmomaxillary region are displaced downward *(a)*, sutural bone growth *(b)* takes place at the same time in response to it, thus enlarging the bones as the soft tissues grow. This places all the bones in new positions in conjunction with the generalized expansion of the soft tissue matrix and maintains continuous sutural contact as the bones become "separated."

It is believed that more vertical displacement (and thus more vertical sutural growth) takes place in the posterior part of the face than in the anterior portion. More direct cortical remodeling growth (deposition and resorption), however, occurs in the anterior part. The latter feature also relates to the downward occlusal plane inclination that is often present (page 218). The balance between the greater or lesser amounts of displacement and remodeling growth in the posterior and anterior parts of the maxilla, however, in general is apparently a response to the clockwise or counterclockwise rotatory displacement caused by the downward and forward growth of the middle cranial fossa. The nasomaxillary complex must correspondingly undergo a compensatory growth rotation in order to retain its proper vertical position relative to the vertical reference (*PM*) line and to the neutral orbital axis (see also Chapter 4).

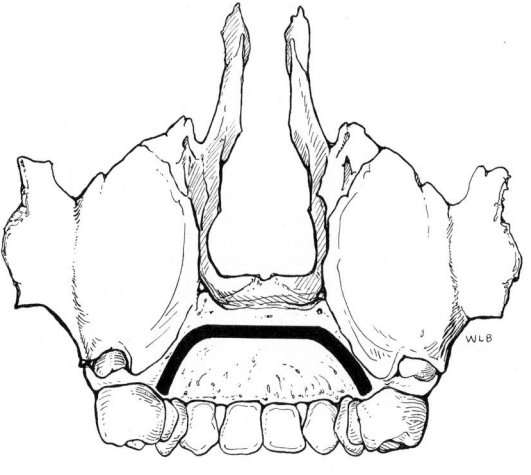

**FIGURE 3–121**

*(From Enlow, D. H., and S. Bang: Growth and remodeling of the human maxilla. Am. J. Orthod., 51:446–464, 1965.)*

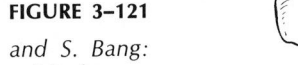

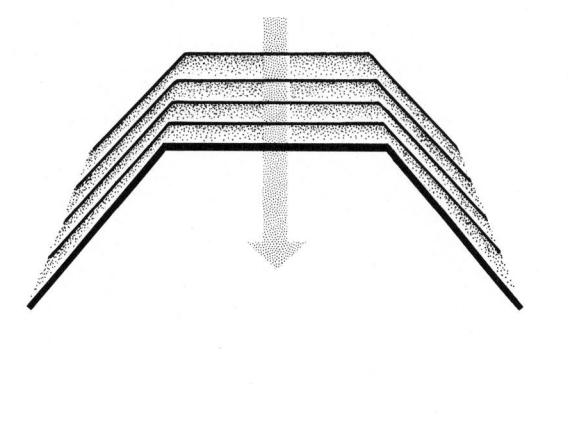

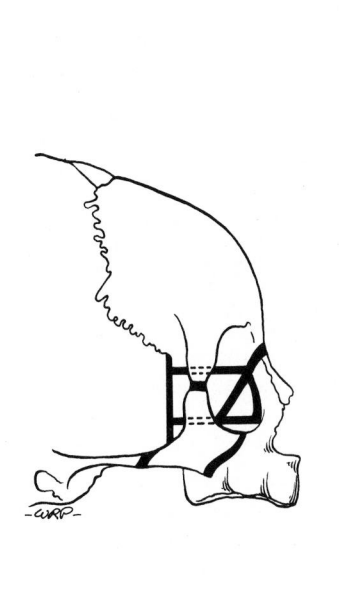

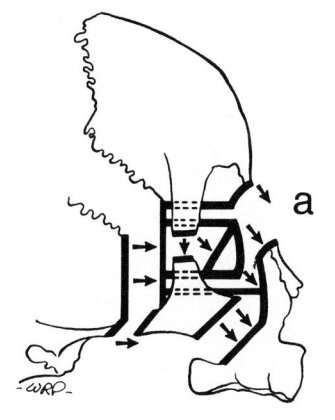

**FIGURE 3–122**

**FIGURE 3–123**

**FIGURE 3–124**

Most sutures in the facial complex do not simply grow in perpendicular directions to the plane of the sutural line. This was pointed out in a previous stage with respect to the lacrimal sutures. Because of the multidirectional mode of primary displacement and the differential extents of growth among the various bones, a slide or slippage of bones **along** the plane of the interface can be involved, as shown by the studies of Latham. As the maxillary complex is displaced downward and forward, or as it grows by deposition and resorption, it undergoes a slide at sutural junctions with the lacrimal, zygomatic, nasal, and ethmoidal bones. This is schematized by a slip of *b* over the sutural front of *a*. The process requires relinkages of the collagenous fiber connections across the suture (see Chapter 11).

**FIGURE 3–125**

The present stage deals with the downward, primary displacement of the nasomaxillary complex. However, it is apparent that the downward **and** forward directions of movement occur at the same time and that they are produced by the same actual displacement process. Moreover, the sutures do not represent special "centers" of growth. A suture is just another regional site of growth adapted to its own localized, specialized circumstances, just as all the other parts of the bone have their own regional growth processes. This point is often misunderstood. It is not possible for a bone to grow **just** at its sutures; nor is it possible for a bone to have "generalized surface growth" without sutural involvement (in areas where sutures are present, of course; nonsuture regions may enlarge by direct remodeling). The old idea that "the sutural growth system closes down at a given age, but the bone continues to enlarge simply by generalized **surface** deposition" is not valid. For example, bone additions on surface *x* enlarge the surface area of the bone, but additions must **also** be made by deposits at sutural surfaces *y* in order to maintain morphologic form. It would not be possible for the bone to enlarge in surface area without corresponding additions at the sutural contacts.

**FIGURE 3–126**

As pointed out in Stage 12, the downward movement of the teeth from *1* to *2* is accomplished by a **vertical drift** of each tooth in its own alveolar socket as the socket itself drifts (remodels) inferiorly by deposition and resorption. The movement of the dentition from *2* to *3,* however, is a passive **carry** of the teeth, the entire bony arch, and all the tooth sockets as the whole maxilla is displaced downward. The teeth are sometimes used as "handles" for orthodontic appliances to retard or accelerate the extent of such displacement of the **whole** maxilla. The movement from *1* to *2,* however, is harnessed and utilized to guide the placement of each **individual** tooth rather than the entire bone. Controlling the tooth movement from *1* to *2* requires alveolar remodeling; controlling the movement from *2* to *3* may involve "orthopedic forces" to alter the form and size of non–tooth-bearing areas of the bone. Both types of movements, however, occur only during the period of active growth and nasal chamber expansion. They are thus not available in treatment procedures involving adult patients.

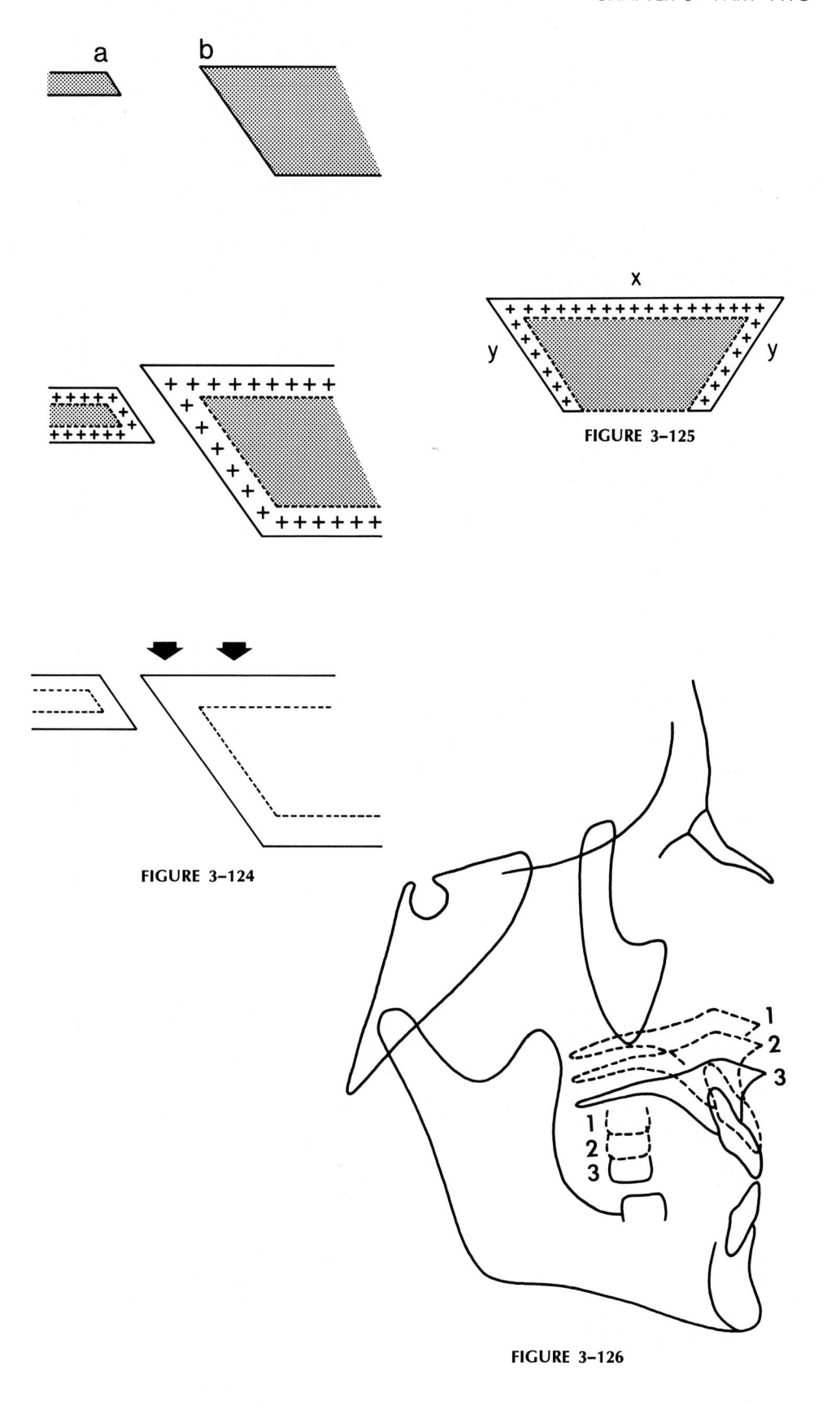

a

b

x

y                y

FIGURE 3–125

FIGURE 3–124

FIGURE 3–126

**FIGURE 3–127**

*Stage 14.* The growth and remodeling changes of both the ramus and middle cranial fossa (Stages 5, 8, and 10) produce a lowering of the mandibular arch. This accommodates the vertical expansion of the nasomaxillary complex. To bring the upper and lower teeth into full occlusion, the mandibular teeth must drift (not simply erupt) vertically. The amount can vary considerably among different individuals having different facial types, and it can also vary between the anterior and posterior parts of the arch. The latter is involved in occlusal plane rotations (see page 218). Significantly, the amount of upward mandibular tooth drift is much less than the downward drift and displacement of the maxillary teeth. This is one of several reasons why orthodontic procedures, at least at present, often attack the maxillary dentition, even though a given malocclusion can be based largely on an improper positioning of the mandible. That is, an "imbalance" is clinically produced in the maxilla to offset the effect of an existing skeletal imbalance in the mandible (or cranial base), because it is the maxilla that can most readily be altered and controlled. (This paradoxical situation may change with future procedural developments when intrinsic control mechanisms of facial growth are better understood.) Because imbalances may still exist, however, long-term stability can be involved, and retention is a problem.

**FIGURES 3–128 AND 3–129**

*Stage 15.* During the descent of the maxillary arch (Stage 13) and the vertical drift of the mandibular teeth (Stage 14), the anterior mandibular teeth simultaneously drift **lingually** and superiorly. This produces a greater or lesser amount of anterior **overbite.** The remodeling process that brings this about involves periosteal resorption on the labial side of the labial bony cortex *(a),* deposition on the alveolar surface of the labial cortex *(b),* resorption on the alveolar surface of the lingual cortex *(c),* and deposition on the lingual side of the lingual cortex *(d).*

**FIGURE 3–130**

At the same time, bone is progressively added onto the surface of the basal bone area. The reversal between these two growth fields usually occurs at the point where the concave surface contour becomes convex. The result of this two-way growth process is a progressively enlarging mental protuberance. Man is one of only two species having a "chin" (the elephant is the other). Whatever its mechanical adaptations, the chin is a phylogenetic result of facial rotation into a vertical position, decreased prognathism (as described in Chapter 4), the marked extent of vertical facial growth, and the development of an overbite (in comparison with an end-to-end type of occlusion).

There is considerable variation in the placement of the reversal line between the resorptive alveolar and the depository chin areas; it may be fairly high or low. Variations also occur in the relative amounts of resorption and deposition. There are, correspondingly, marked variations in the shape and the size of the chin among different individuals. It is one of the most variable areas in the entire mandible.

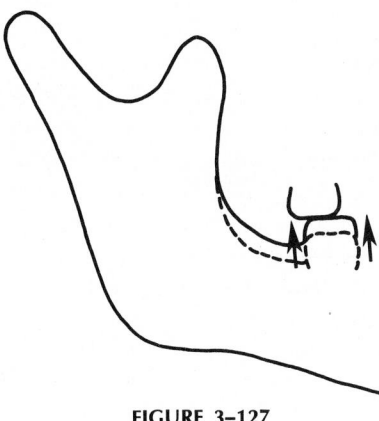

**FIGURE 3–127**

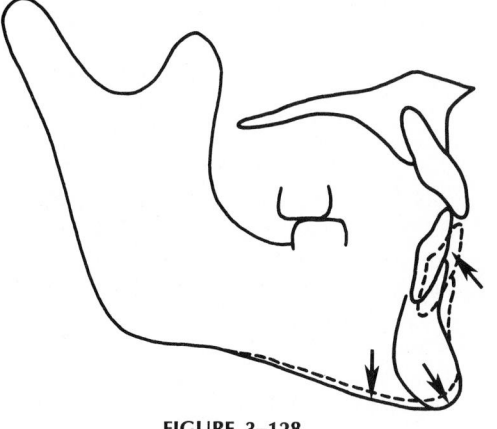

**FIGURE 3–128**

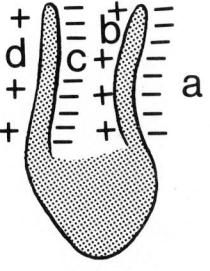

**FIGURE 3–129**

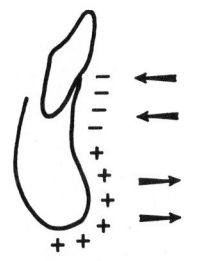

**FIGURE 3–130**

**FIGURE 3–131**

Except for a resorptive zone on the lingual side, the remainder of the perimeter of the mandibular corpus receives progressive deposits of bone. This enlarges the breadth of each side of the corpus; side *b* grows to a slightly greater extent than side *a* because bony arch width increases slightly during postnatal mandibular growth, but not as much as the bony maxillary arch increases in width. The ventral border of the corpus is also depository; this is a slow growth process, however. The amount of upward alveolar growth exceeds the extent of downward enlargement by the "basal bone." (**Note:** Basal bone is a term sometimes used to denote that part of the corpus not involved in "alveolar" movements of the teeth. This area has a higher resistance threshold to extrinsic forces than alveolar bone, which is extremely labile. There is no distinct structural line, however, separating basal from alveolar bone tissue. This is more of a physiologic than an anatomic difference.)

**FIGURE 3–132**

Whenever a change in the angle between the ramus and corpus develops, multiple sites of remodeling can be involved. The trajectory of condylar growth may be a factor, as shown by *a, b,* and *c.*

**FIGURE 3–133**

If condylar growth has slowed or largely ceased, combinations such as resorption at *d* and *e* with deposition at *f* and *g* can produce angular changes of the ramus relative to the corpus by direct remodeling. Such remodeling processes can either close or open the "gonial angle." (See Stage 9 and pages 116 and 202–207.)

**FIGURES 3–134, 3–135, 3–136, AND 3–137**

The corpus may also participate in alterations of the ramus-corpus angle. Increased growth along the ventral part of the mandibular corpus often occurs in edentulism and in some types of malocclusions in which the corpus rotates inferiorly relative to the ramus (that is, the gonial angle is opened). The size of the antegonial notch is determined largely by the nature of the ramus-corpus angle and by the extent of bone deposition on the underside (inferior margin) of the corpus just anterior to the notch. The notch itself is also increased in size due to its resorptive periosteal surface. A mandible characteristically has a less prominent antegonial notch *(b)* if the angle between the ramus and corpus becomes closed, or a much more prominent antegonial notch *(a)* if it becomes opened. The antegonial notch itself is resorptive because it is relocated posteriorly, as the corpus lengthens, into the former gonial region of the ramus.

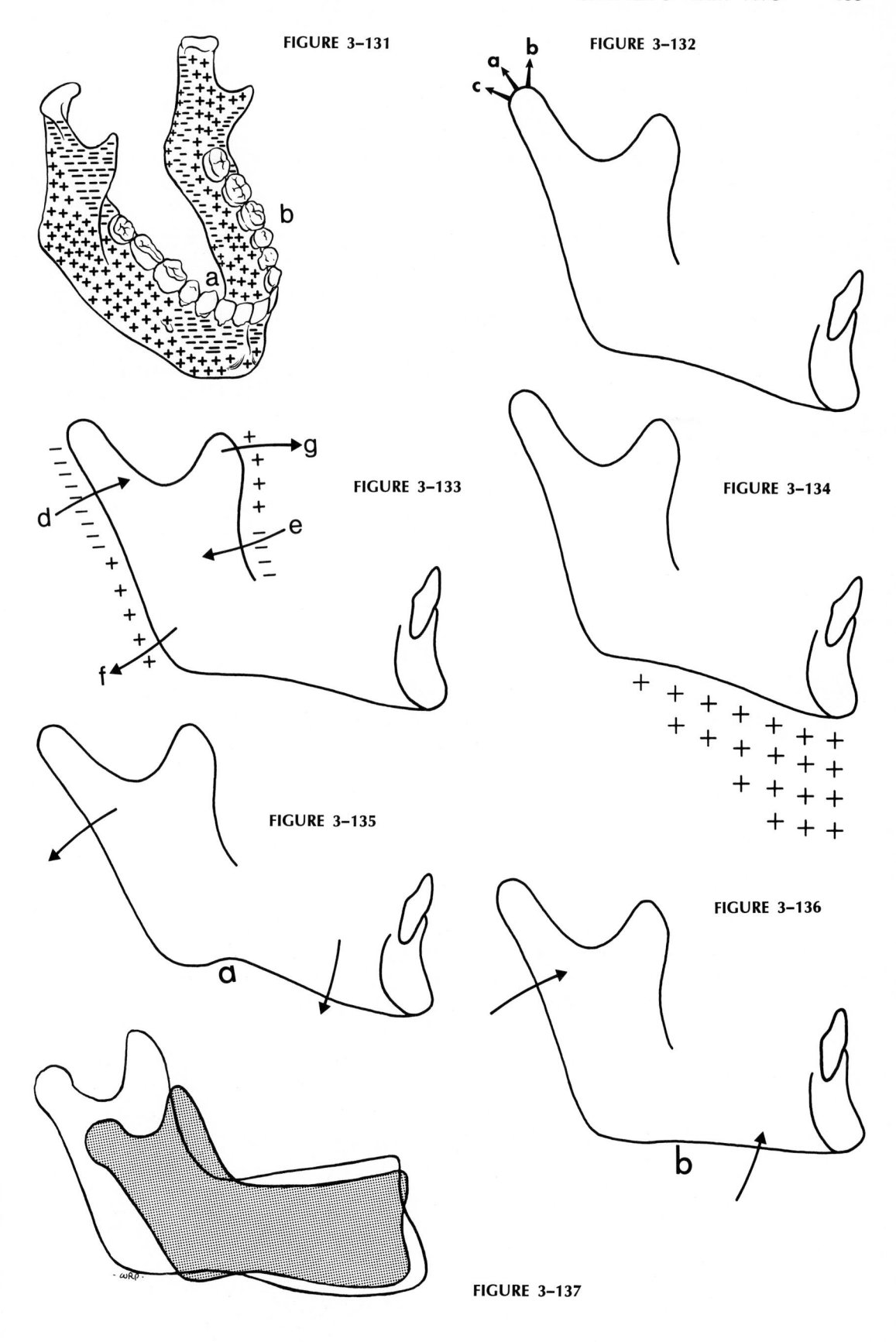

FIGURE 3–131

FIGURE 3–132

FIGURE 3–133

FIGURE 3–134

FIGURE 3–135

FIGURE 3–136

FIGURE 3–137

**FIGURES 3–138, 3–139, AND 3–140**

*Stage 16.*   The growth changes of the malar complex are similar to those of the maxilla itself. This is true for the remodeling process as well as the displacement process.

The posterior side of the malar protuberance is depository. Together with a resorptive anterior surface, the cheekbone relocates **posteriorly** as it enlarges. It would seem untenable that the whole front surface of the cheek area can actually be **resorptive.** However, as the maxillary arch grows posteriorly, the malar region must also move backward at the same time to keep a constant relationship with it. The extent of malar relocation is somewhat less in order to maintain **relative** position along the increasing length of the maxillary arch. The zygomatic process of the maxilla thus behaves in a manner similar to that of the coronoid process of the ramus. Both move posteriorly as the maxillary and mandibular arches also grow posteriorly.

The inferior edge of the zygoma is heavily depository. The anterior part of the zygomatic arch becomes greatly enlarged vertically as the face develops in depth.

**FIGURE 3–141**

The zygomatic arch moves **laterally** by resorption on the medial side within the temporal fossa and by deposition on the lateral side. This enlarges the temporal fossa and keeps the cheekbone proportionately broad in relation to face and jaw size and masticatory musculature. The anterior trough of the temporal fossa moves posteriorly by the V principle.

**FIGURE 3–142**

As the **malar** region grows and becomes relocated posteriorly, the contiguous **nasal** region is enlarging in an opposite, anterior direction. This draws out and greatly expands the contour between them, resulting in a progressively more protrusive-appearing nose and a horizontally deeper face.

The remodeling changes of the **orbit** are complex. This is because many separate bones are involved in the orbit, including the maxilla, ethmoid, lacrimal, frontal, and zygomatic and the greater and lesser wings of the sphenoid, and because many different rates and amounts of remodeling growth and displacement occur among these different bones and their parts.

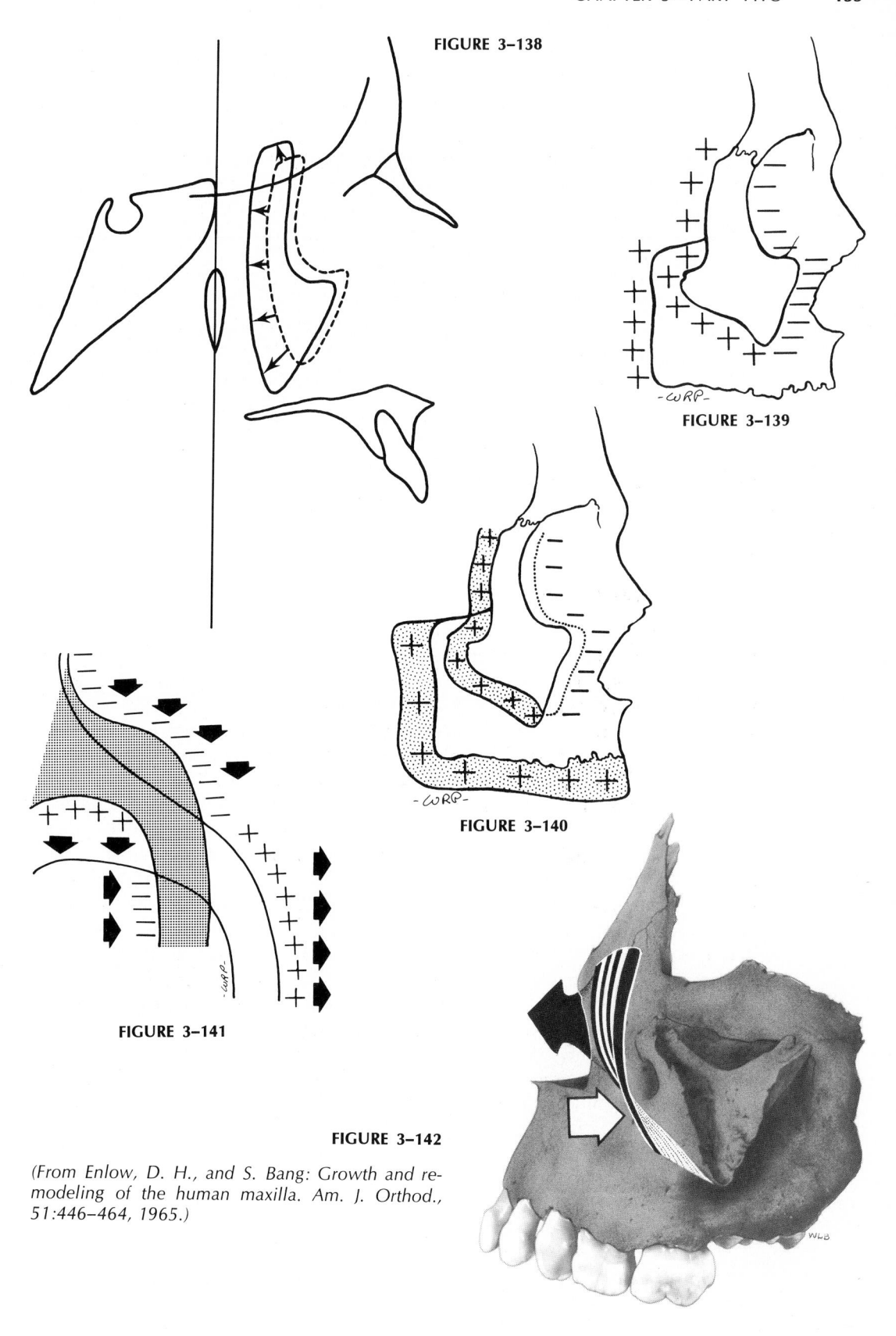

**FIGURE 3–138**

**FIGURE 3–139**

**FIGURE 3–140**

**FIGURE 3–141**

**FIGURE 3–142**

*(From Enlow, D. H., and S. Bang: Growth and remodeling of the human maxilla. Am. J. Orthod., 51:446–464, 1965.)*

**FIGURE 3–143**

The remodeling activities in the medial wall of the orbit, including the lacrimal and ethmoid, were described in Stage 12. In the remainder of the orbit, most of the roof and the floor are depository. The orbital roof is also part of the floor of the anterior cranial fossa. As the frontal lobe of the cerebrum expands downward, the orbital roof also grows inferiorly by resorption on the cranial side and deposition on the orbital side. It would seem that a depository type of orbital roof and orbital floor would decrease the size of the cavity. However, two changes come into play that actually increase it, although the amount is relatively small in the older child. First, the orbit grows by the V principle. The cone-shaped orbital cavity grows and **moves** in a direction toward its wide opening; deposits on the inside thus enlarge rather than reduce the volume. Second, the factor of **displacement** is directly involved. In association with sutural bone growth at the many sutures within and outside the orbit, the orbital floor is displaced in a progressive downward direction along with the rest of the nasomaxillary complex.

**FIGURES 3–144, 3–145, AND 3–146**

Note that the floor of the nasal cavity in the adult is positioned **much lower** than the floor of the orbital cavity. Compare this with the situation in the child. As described earlier, about half the process of palatal descent is produced by downward **displacement** of the whole maxilla (associated with maxillary sutural growth). The greater part of the orbital floor is a part of the maxillary bone. Thus, both the orbital and nasal floors are regional portions of the **same** bone, and the same displacement process that carries the palate downward **also** carries the floor of the orbit down at the same time. The extent of this downward displacement, however, greatly exceeds the smaller amount required for oribtal enlargement; that is, a lesser increase is needed for the eyeball and other orbital soft tissues than for the marked expansion carried out by the nasal chamber. The floor of the orbit offsets this by growing upward. Deposition takes place on the intraorbital side of the orbital floor and resorption on the maxillary sinus side. This maintains the orbital floor in proper position with respect to the eyeball above it. The **downward displacement** movement of the maxilla is thereby compensated for by **upward growth** to an amount that accommodates the relatively small enlargement of the orbital soft tissues. The nasal floor, in contrast, doubles the amount of displacement movement by **additional** downward cortical drift. Thus, the orbital and nasal floors are displaced in the same direction because they are parts of the same bone, but they undergo remodeling growth in opposite directions.

**FIGURE 3-143**

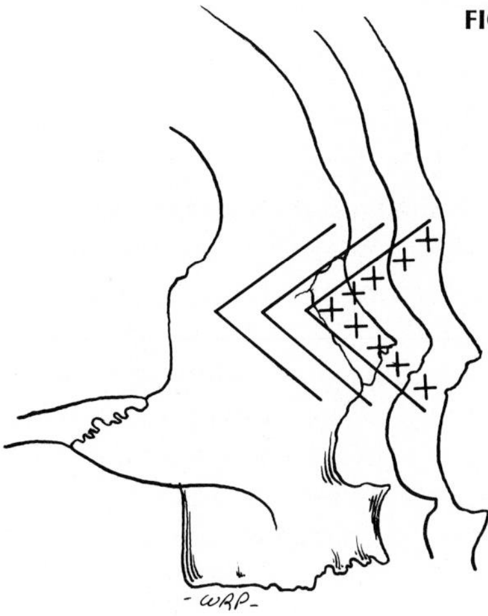

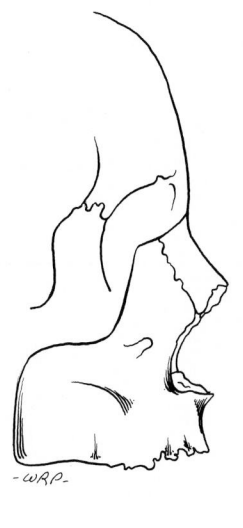

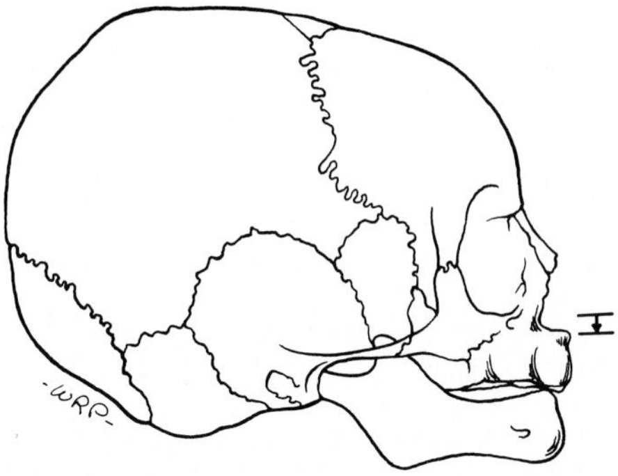

FIGURE 3-145

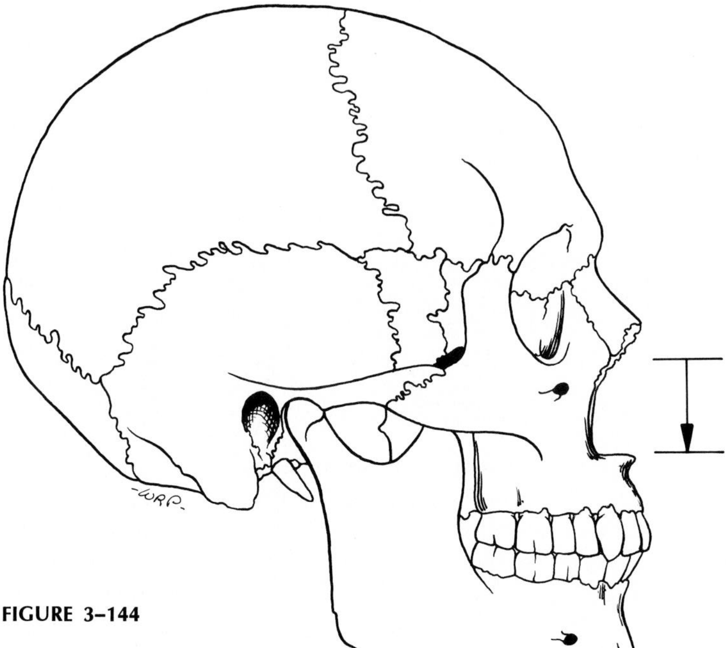

FIGURE 3-144

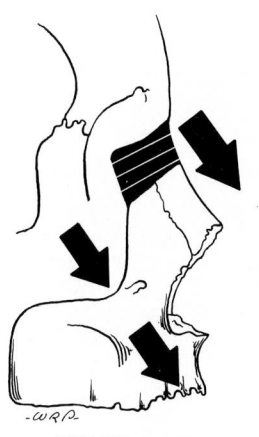

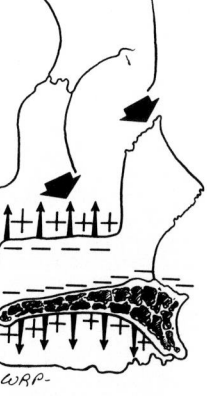

FIGURE 3-146

**FIGURES 3–147 AND 3–148**

The floor of the orbit also grows laterally. It slopes in a lateral manner, and deposits on the floor thus move it in this same direction. The lateral wall of the orbital **rim** grows by resorption on the medial side and by deposition on the lateral side. This intraorbital field of resorption continues directly onto the anterolateral surface of the orbital roof beneath the overhanging supraorbital ridge. This is the only part of the orbital roof and lateral wall that is resorptive, and it provides for the lateral expansion of the domed roof as well as for the forward growth of the supraorbital ridge. The cutaneous side of the supraorbital ridge is depository, and this combination causes the superior orbital rim to become protrusive. An upper orbital rim that extends forward beyond the lower rim is a characteristic of the adult face.

The lateral orbital rim undergoes remodeling growth in a posterior direction at the same time. The lateral growth change increases the side-to-side dimension of each orbit and also contributes to the lateral movement of the whole orbit involved in the small amount of increase in the interorbital dimension. The backward growth change of the lateral orbital rim keeps it in proper location with respect to the backward direction of growth by the zygoma. The forward growth of the superior orbital ridge combined with the backward growth of the lateral rim and cheekbone causes the lateral rim in the human face to slant obliquely upward and forward, in contrast to the case in other mammals. This reflects the forward rotation of the entire upper part of the human face and the backward rotation of the lower part.

Note that the resorptive face of the cheekbone area combined with the depository nature of the whole nasal region of the maxilla greatly expands the surface contour between them and deepens the topography of the face. The medial rim of the orbit is only slightly in front of the lateral rim in the young child. In the adult, the medial rim has grown forward with the anterior-growing nasal wall, and the lateral rim has grown backward with the cheekbone. The medial and lateral rims are thus drawn apart in divergent posterior-anterior directions as the face deepens.

*Stage 17.* The zygoma becomes **displaced** anteriorly and inferiorly in the same directions and amount as the primary displacement of the maxilla. The malar protuberance is a part of the maxillary bone and is carried with it. The separate zygomatic bone is displaced inferiorly in association with bone growth at the frontozygomatic suture, and anteriorly in relation to growth at the zygotemporal suture. The force that causes it is the same as for the maxilla: the functional matrix or, according to older theory, the nasal septum. The growth changes of the malar process are similar to those of the mandibular coronoid process. Both grow backward, along with the backward elongation of each whole bone, by anterior resorption and posterior deposition. Both become displaced anteriorly along with each whole bone.

**Note this important feature of facial growth.** In many of the growth and remodeling processes described throughout this chapter, one major difference exists between the female and the male. In the female, skeletal growth changes in the face slow and cease shortly after puberty. In the male, however, topographic and dimensional changes continue through the late adolescent period. The facial similarities that exist between the sexes during childhood, therefore, are altered markedly in the teens.

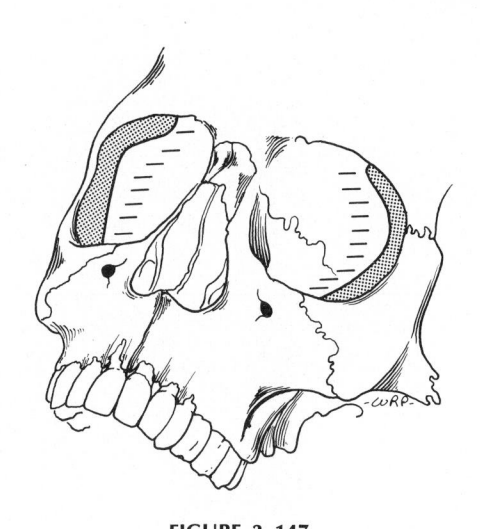

FIGURE 3–147

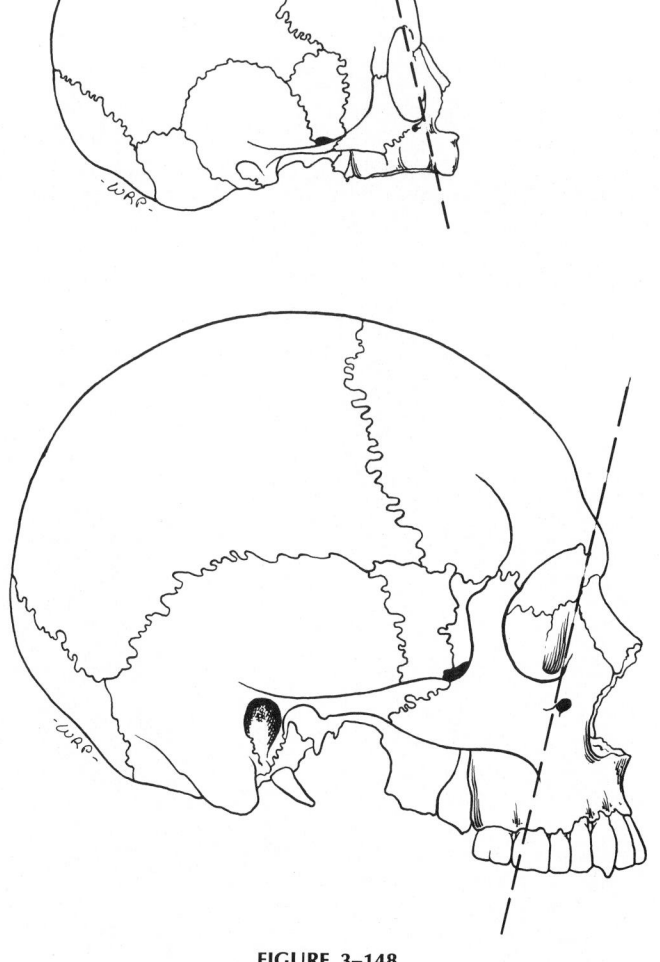

FIGURE 3–148

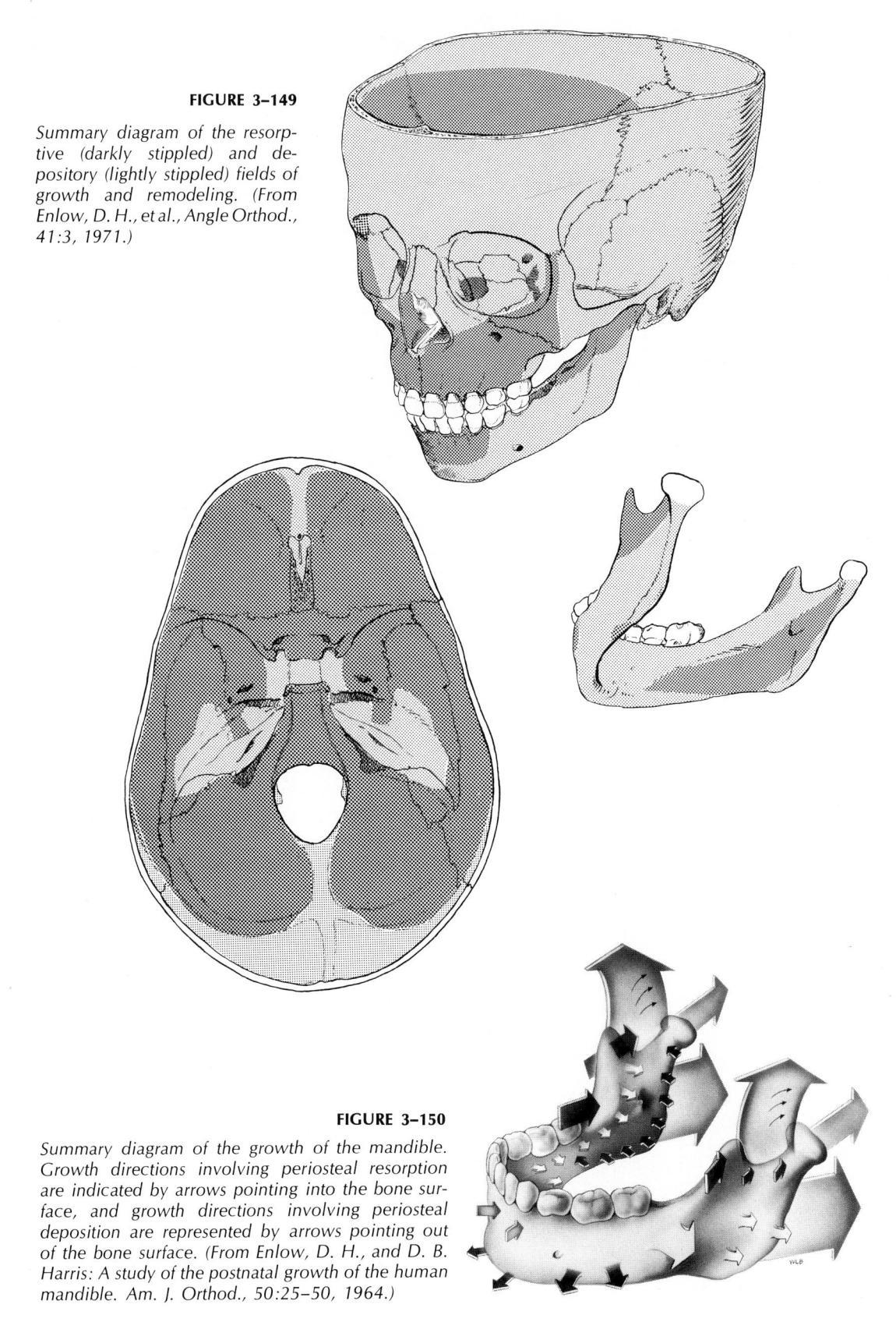

**FIGURE 3–149**

Summary diagram of the resorptive (darkly stippled) and depository (lightly stippled) fields of growth and remodeling. (From Enlow, D. H., et al., Angle Orthod., 41:3, 1971.)

**FIGURE 3–150**

Summary diagram of the growth of the mandible. Growth directions involving periosteal resorption are indicated by arrows pointing into the bone surface, and growth directions involving periosteal deposition are represented by arrows pointing out of the bone surface. (From Enlow, D. H., and D. B. Harris: A study of the postnatal growth of the human mandible. Am. J. Orthod., 50:25–50, 1964.)

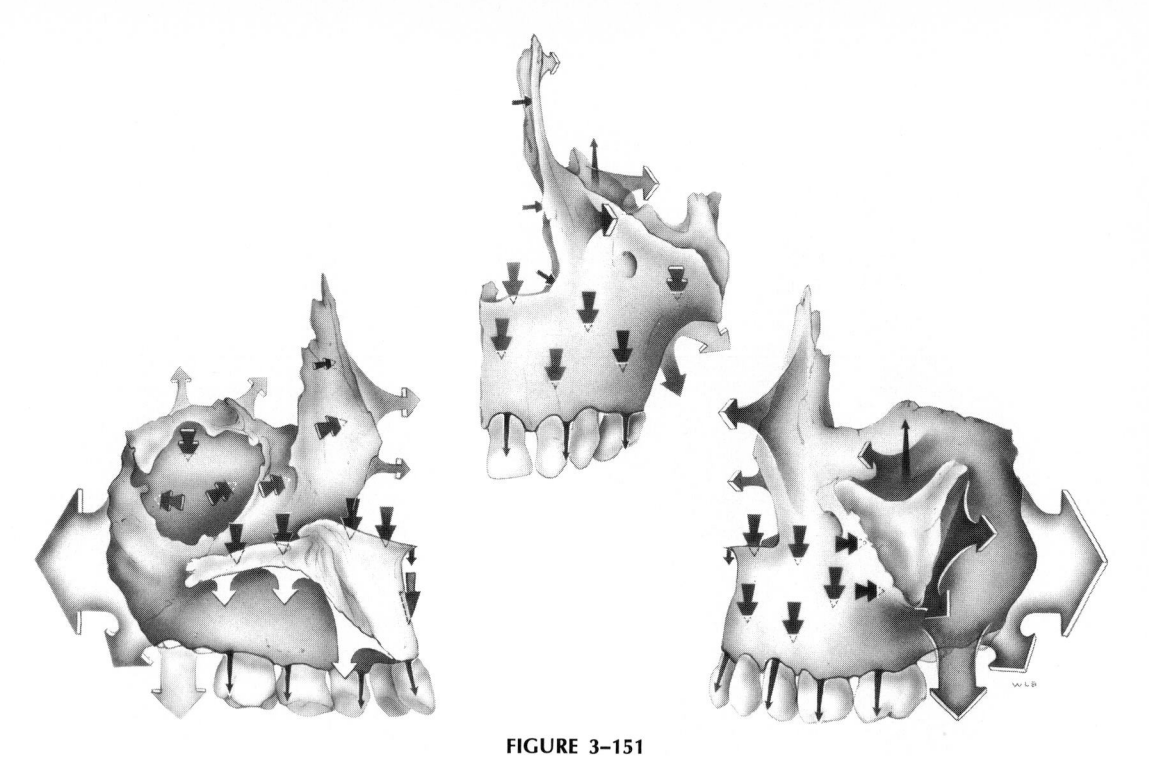

**FIGURE 3–151**

*Summary diagram of maxillary growth. Growth directions involving surface resorption are represented by arrows entering the bone surface. Directions of growth involving surface deposition are shown by arrows emerging from the bone surface. (From Enlow, D. H.:* The Human Face. *New York, Harper & Row, 1968, p. 164.)*

**FIGURE 3–152**

*Transverse sections made at a, b, c, d, and e are characterized by the growth and remodeling patterns shown. (Adapted from Enlow, D. H., and D. B. Harris: A study of the postnatal growth of the human mandible. Am J. Orthod., 50:25–50, 1964.)*

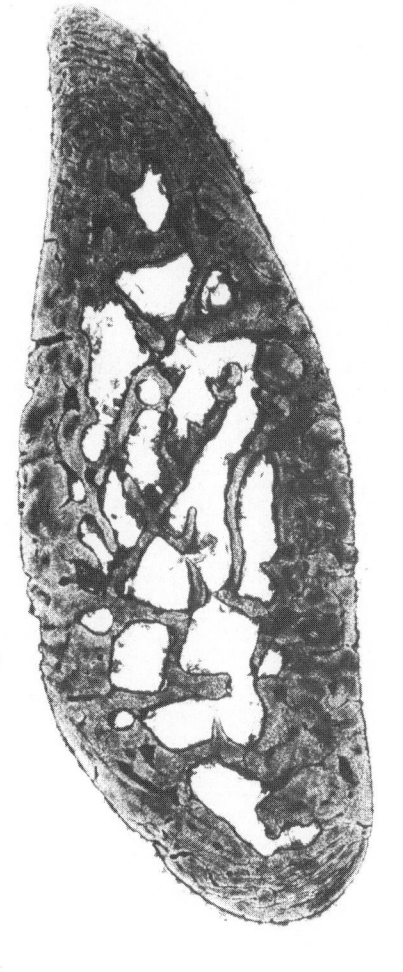

**FIGURE 3–153**

*This is a transverse section through the neck of the mandibular condyle. The superior margin, which is part of the mandibular (sigmoid) notch, is depository. The lower margin, which is the posterior border of the ramus, is also depository. Both the buccal and the lingual surfaces are resorptive. (From Enlow, D. H., and D. B. Harris: A study of the postnatal growth of the human mandible. Am. J. Orthod., 50:25–50, 1964.)*

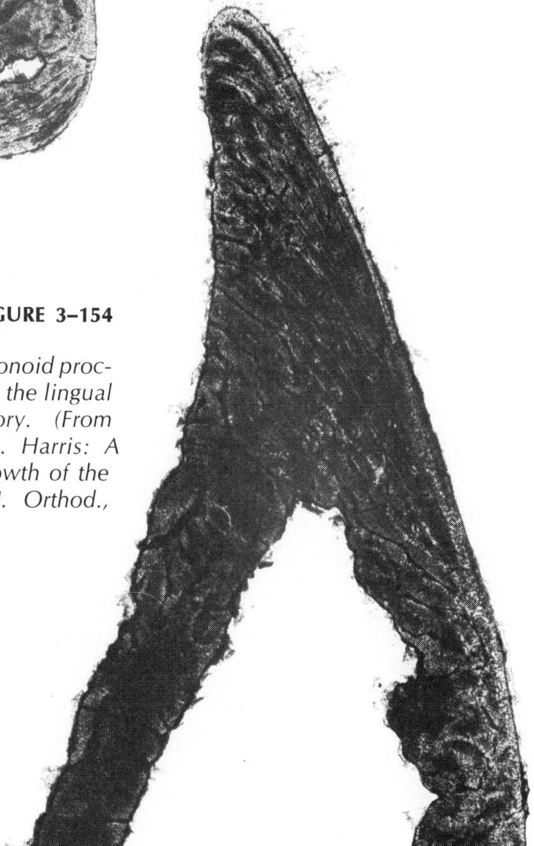

**FIGURE 3–154**

*The buccal side of the coronoid process (left) is resorptive, and the lingual side (right) is depository. (From Enlow, D. H., and D. B. Harris: A study of the postnatal growth of the human mandible. Am. J. Orthod., 50:25–50, 1964.)*

**FIGURE 3–155**

*This is a transverse section through the posterior part of the mandibular corpus. The lower two-thirds of the lingual surface (on the right) is resorptive and forms the lingual fossa. The upper third, which is the lingual tuberosity, is depository. Except for a tiny patch of surface resorption near the top, the entire buccal side (left) is depository. Note the reversal between the depository buccal and the resorptive lingual sides on the inferior margin of the corpus. (From Enlow, D. H., and D. B. Harris: A study of the postnatal growth of the human mandible. Am. J. Orthod., 50:25–50, 1964.)*

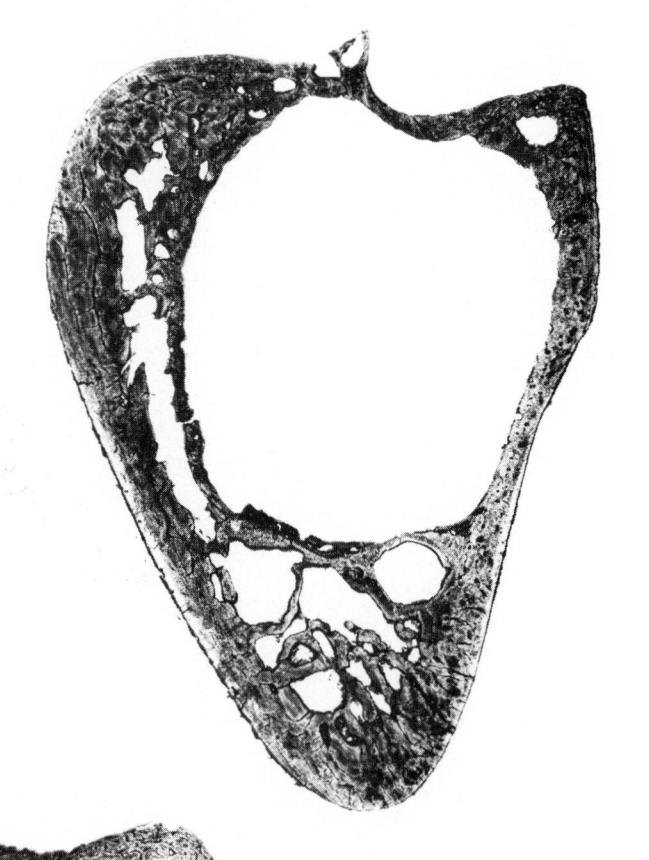

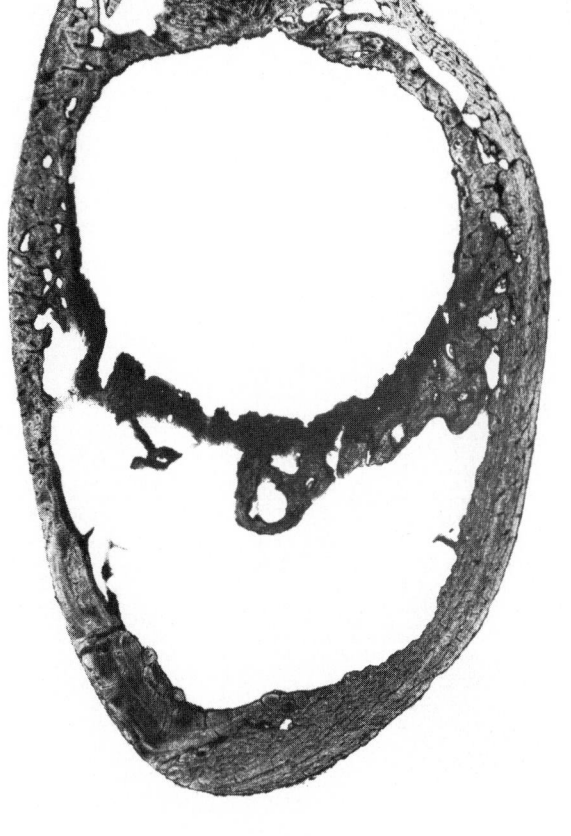

**FIGURE 3–156**

*This transverse section of the mandible was taken from the level of the second molar. On the lingual side (left), the entire surface is resorptive except for a depository zone near the superior border. The entire buccal side (right) is depository. (From Enlow, D. H., and D. B. Harris: A study of the postnatal growth of the human mandible. Am. J. Orthod., 50:25–50, 1964.)*

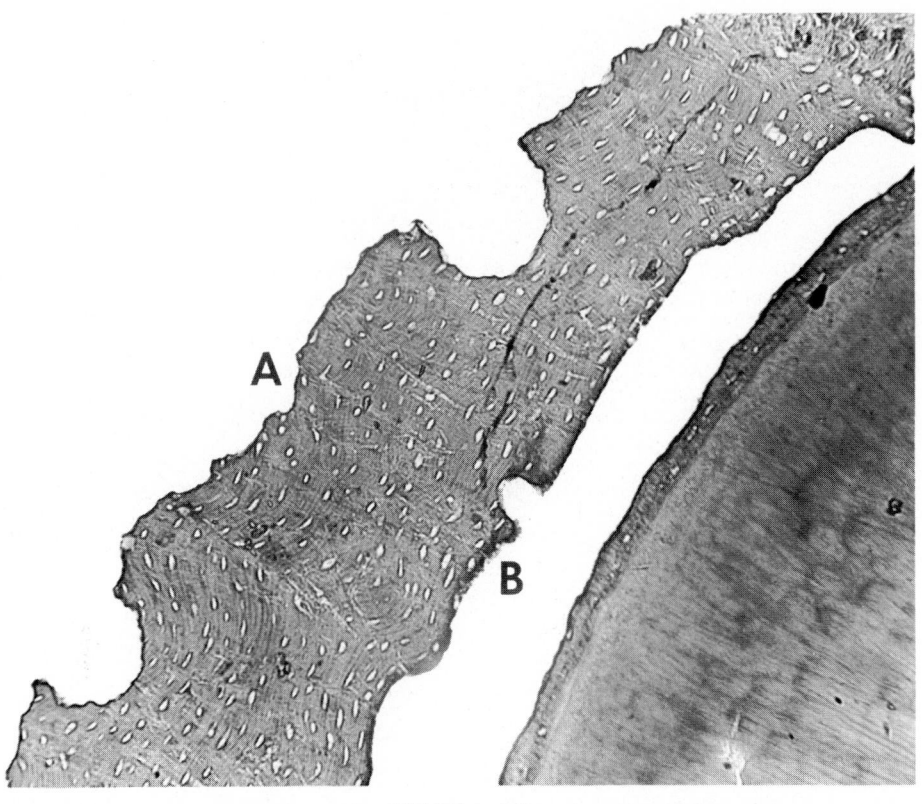

**FIGURE 3–157**

*The resorptive nature of the external (periosteal) surface of the human premaxillary cortex is shown (A). The opposite alveolar lining surface is depository (B). The bone of the cortex was laid down by the perio-dontal membrane on the one side and resorbed by the periosteum from the other. This combination* **moves** *this part of the bony arch, and also the tooth, in a downward direction (see* **vertical tooth drift** *on page 124). The bone tissue on the outer (labial) surface, therefore, was actually formed by the periodontal membrane on the inside and subsequently translocated to the periosteal side of the moving cortical plate. (From Enlow, D. H.:* The Human Face. *New York, Harper & Row, 1968, p. 152.)*

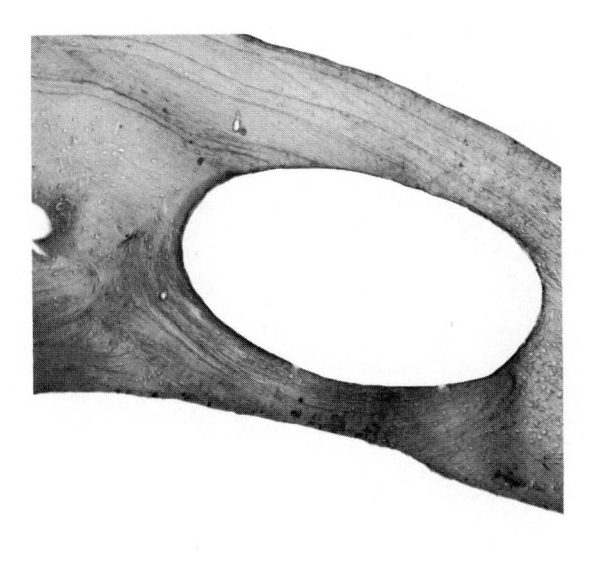

**FIGURE 3–158**

*The frontal process of the maxilla has a depository outer surface (top) and a resorptive mucosal surface within the nasal chamber (bottom). Note the corresponding drift of the cancellous space in a like direction by resorption on the upper side and deposition on the lower side. (From Enlow, D. H.:* The Human Face. *New York, Harper & Row, 1968, p. 156.)*

**FIGURE 3–159**

*The maxillary portion of the orbital floor is composed of a single, thin plate of periosteal lamellar bone. The upper (orbital) surface is depository, and the opposite side (bottom), which lines the maxillary sinus, is resorptive. (From Enlow, D. H., and S. Bang: Growth and remodeling of the human maxilla. Am. J. Orthod., 51:446–464, 1965.)*

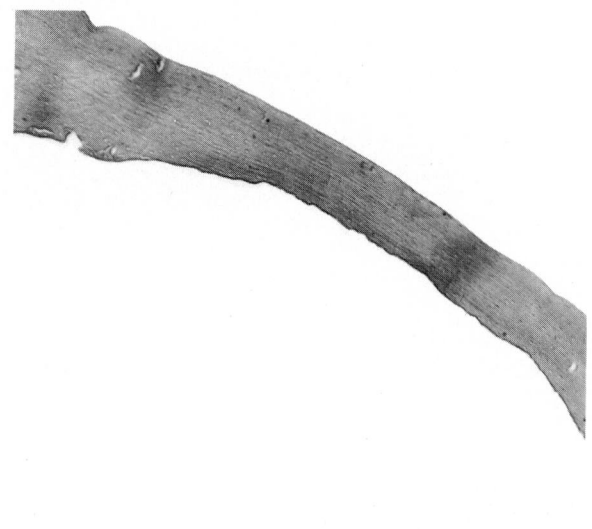

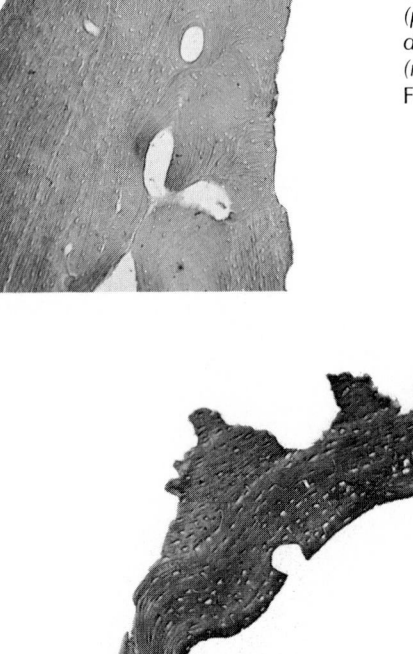

**FIGURE 3–160**

*This is a transverse section through the lateral nasal wall (part of the maxillary bone). The external side (left) is depository, and the opposite side within the nasal chamber (right) is resorptive. (From Enlow, D. H.:* The Human Face. *New York, Harper & Row, 1968, p. 156.)*

**FIGURE 3–161**

*Lacrimal bone. Surface A is the external side of the cortex that forms a portion of the medial lining wall of the orbital cavity. In the cephalic part of the lacrimal, the surface is characteristically resorptive, as seen in this section. The contralateral side within the nasal chamber is depository (B). A similar combination is seen within the lacrimal groove (C and D). (From Enlow, D. H.:* The Human Face. *New York, Harper & Row, 1968, p. 162.)*

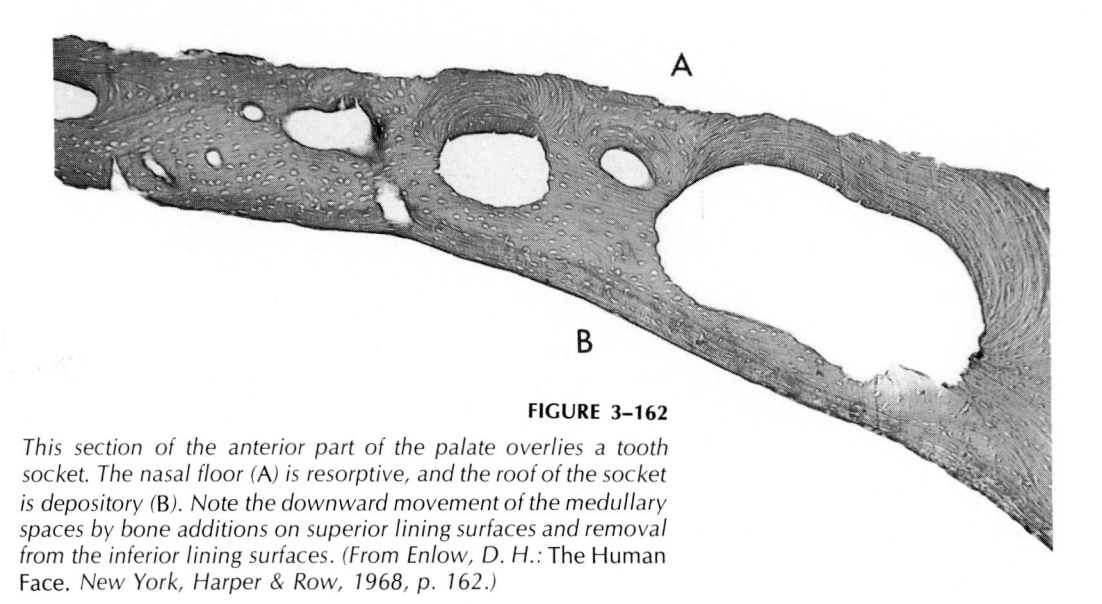

**FIGURE 3–162**

*This section of the anterior part of the palate overlies a tooth socket. The nasal floor (A) is resorptive, and the roof of the socket is depository (B). Note the downward movement of the medullary spaces by bone additions on superior lining surfaces and removal from the inferior lining surfaces. (From Enlow, D. H.:* The Human Face. *New York, Harper & Row, 1968, p. 162.)*

# The Plan of the Human Face

## PART ONE

The human face is certainly different from that of other mammals. The long, narrow, functional muzzle that slopes gracefully onto the streamlined cranium of a typical mammal is in marked contrast to the muzzleless, broad, vertical, flattened human face enveloped by an enormous balloon-shaped cranium with a bulbous forehead overhanging tiny, retrusive jaws, a small mouth, a chin, and a curious vestige of a narrow fleshy snout protruding in front of an owl-eyed face showing changing expressions. While somehow beautiful to our eyes, this has to be an "odd" design by ordinary mammalian standards. What are the factors that underlie the functional, developmental, and phylogenetic reasons for this specialized facial configuration? There have been many theories over the years, but, regrettably, we may never know for sure what the **primary** factors were that initiated the long evolutionary chain of interrelated adaptations throughout the whole body that relate to the many design features of our facial heritage. We can, however, partially explain the anatomic, developmental, and functional meaning of each factor in this series of mutually dependent changes. And we can propose some pretty reasonable phylogenetic explanations as well. More than merely being interesting, this helps all of us to better understand the basic plan of facial construction. We can thereby evaluate more meaningfully the various facial dimensions, planes, angles, and so on that are so important to the clinician and to basic science researchers.

**FIGURE 4–1**

**Concept 1.**   Man is one of the few truly bipedal mammals that exists. Our upright posture involves a great many anatomic and functional adaptations throughout every part of the body, and no one of these would work without all the others. We have "feet," and the human foot stands by itself, as it were, as a unique anatomic feature of man. The designs of the toes, foot bones, arch of the foot, ankle, leg bones, pelvis, and vertebral column all interrelate in the anatomic composite that provides upright body stance. The head is in a balanced position on an upright spine. The arms and hands have become freed. The manipulation of food and other objects and defense, offense, and so forth utilize primarily the hands rather than the shortened jaws.

**FIGURE 4–2**

**Concept 2.**   The enormous enlargement and the resultant configuration of the brain has caused a "flexure" (bending) of the human cranial base. This relates to two key features. First, the spinal cord is aligned vertically, a change that permits upright, bipedal body stance with free arms and hands. Second, the orbits have undergone a rotation in conjunction with frontal lobe expansion. This aligns them so that they point in the forward direction of upright body movement. The body has become vertical, but the neutral visual axis is thereby still horizontal, as in other mammals. (Note: The muzzle of a typical animal points obliquely downward in the "neutral" position, not straight forward. This positions the orbital axis approximately parallel with the ground.) The cranial base of the typical mammal is flat, in contrast to the human cranium, and the spinal cord passes into a horizontally directed vertebral column.

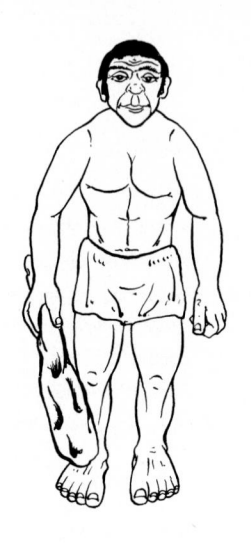

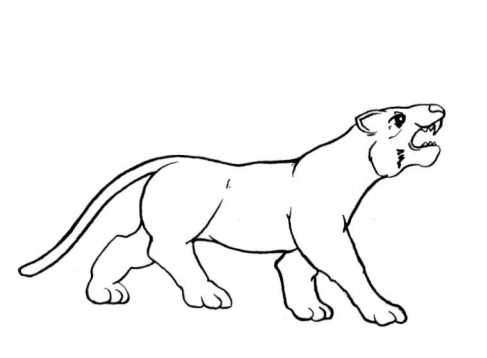

**FIGURE 4–1**

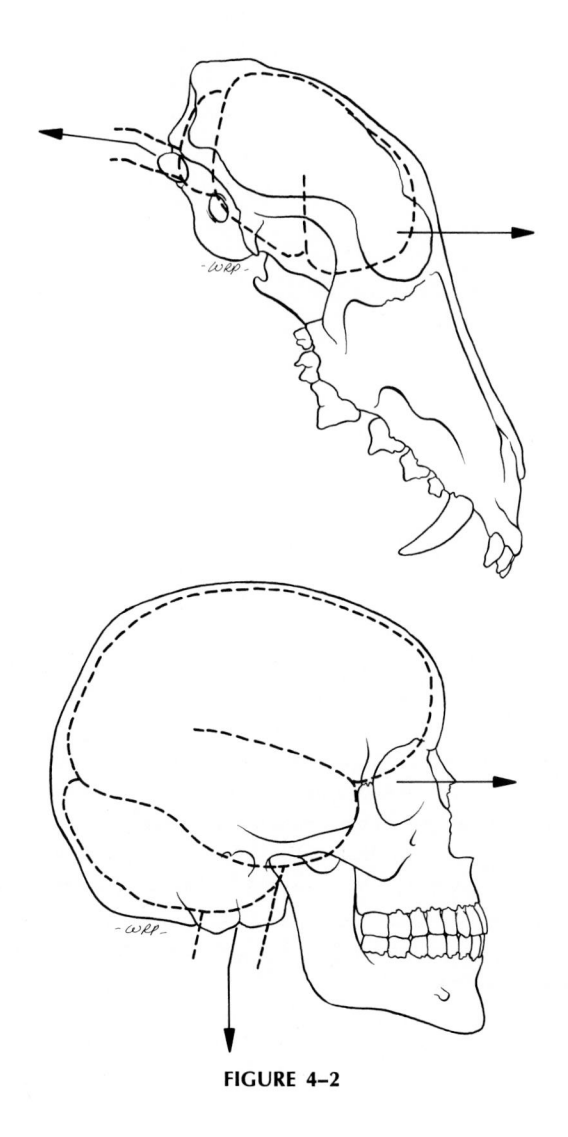

**FIGURE 4–2**

**FIGURE 4–3**

**Concept 3.**   The large size of the human brain has also caused the orbits to rotate toward the midline (or, hypothetically, the converse). This results in a binocular arrangement of the orbits, a feature that complements finger-controlled manipulation of food, tools, weapons, and so forth. The absence of a long, protrusive muzzle does not block the close-up vision of hand-held objects. The human **mind** directs the **free hands** that can work with **3–D** perspective in an **upright** stance on feet. The enormous size of the human brain and the human cranial base flexure are key factors, but **all** these changes are required, and they are all mutually interdependent. Which particular change was the actual start of the chain will likely always be speculative.

Complete orbital rotation into a forward-pointing direction, however, has also caused a reduction in the interorbital part of the face. This is significant, since the area involved is the root of the nasal region, and the result of man's close-set eyes is a **narrow nose.** Because the nose is so thin, it is also necessarily quite short. The much broader nasal base of most mammals supports a correspondingly much longer snout.

**FIGURE 4–4**

**Concept 4.**   The nasal region above and the oral region below are two sides of the same coin, that is, the palate. Reduction in nasal protrusion is accompanied by a more or less equivalent reduction of the jaws. The whole face has necessarily become reduced in horizontal length. However, the face has also been **rotated** into a nearly vertical alignment because of the massive enlargement of the brain and the flexure of the cranial base. The downward rotation of the olfactory bulbs and the whole anterior cranial floor has caused a corresponding downward rotation of the nasomaxillary complex.

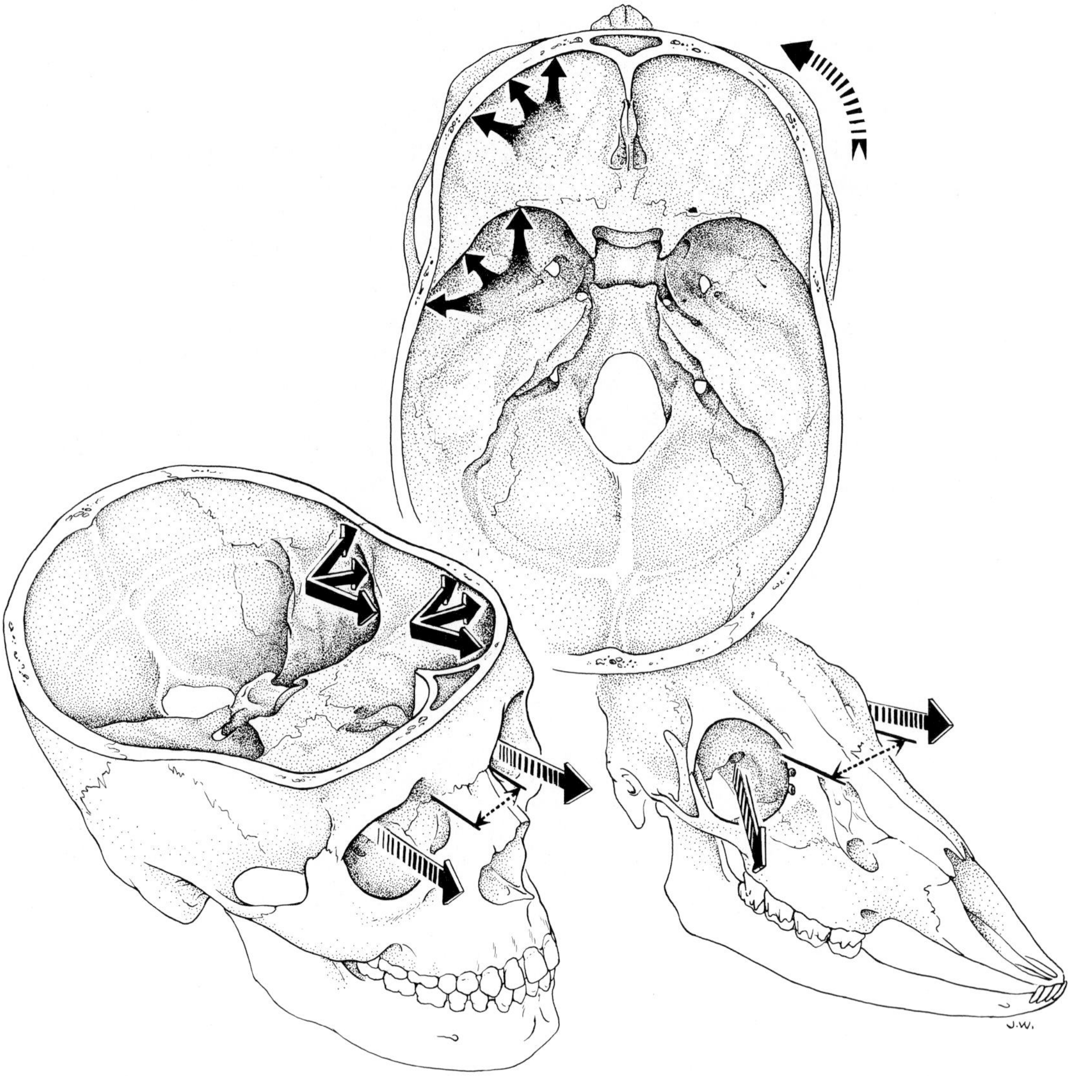

**FIGURE 4–3**

*(From Enlow, D. H.:* The Human Face. *New York, Harper & Row, 1968, p. 190.)*

**FIGURE 4–4**

*(From Enlow, D. H., and S. Bang: Growth and remodeling of the human maxilla. Am. J. Orthod., 51:446–464, 1965.)*

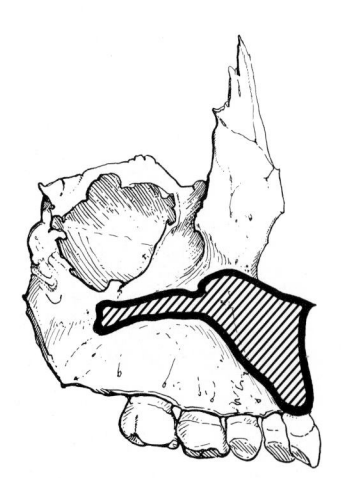

**FIGURE 4–5**

Facial rotation has led to the development of the **maxillary sinus** beneath the orbital floor and above the shortened maxillary arch. Because of its adaptation to facial rotation, the human maxilla is uniquely rectangular rather than triangular like that of most other mammals. It is a distinctively shaped upper jaw. An orbital **floor** has also been added to the human maxilla.

The nasal mucosa is ordinarily an active site involved in temperature regulation in most mammals. Vasoconstriction and vasodilation of the vessels in the massive spread of mucosa covering the turbinates controls the amount of heat loss. Because of marked nasal reduction in man, however, this function has been largely taken over by the relatively hairless and sweat gland–loaded integument. Control of blood flow in the dermis, combined with sweat gland activity, provides the equivalent for nasal thermoregulation. This is possible in man (and in a very few other species, such as the pig) because of a near-naked skin. In thick-furred animals, thermoregulation is carried out by regulating heat transfers in the nasal mucosa, panting to release excess heat, limited perspiration in hairless areas (such as the pads of the paws), and a fluffing of the fur to increase dead air insulation. The latter also makes the animal look larger to enemies, and it increases the nonvital part of his anatomy for enemies to bite. We have only the holdover: goose bumps.

**FIGURES 4–6 AND 4–7**

**Concept 5.** The human face is exceptionally wide because the brain and cranial floor are wide. However, the face has been almost engulfed by the massive brain behind and above it. Note the wondrously, incredibly colossal size of the human cranium, in comparison with that of the typical mammal. The expanded frontal lobes of the human brain lie **above** the eyes and almost the whole remainder of the face rather than behind, and a forehead has thus been added. This also relates to the rotation of the orbits into vertical, forward-facing positions.

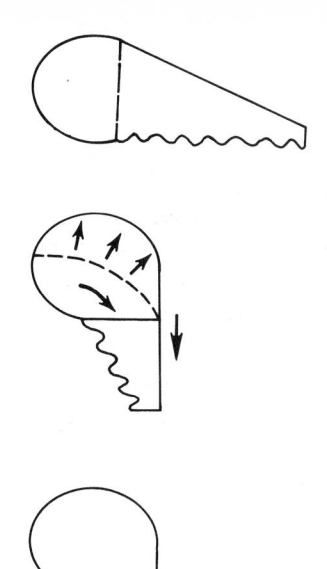
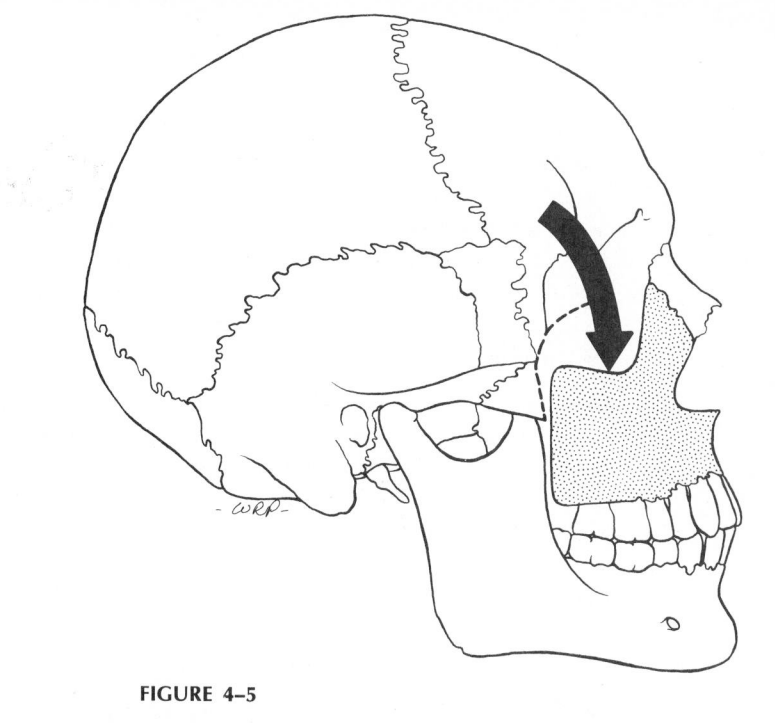

**FIGURE 4–5**

*(From Enlow, D. H.: Postnatal growth and development of the face and cranium. In Cohen, B., and I. R. H. Kramer (Eds.):* Scientific Foundations of Dentistry. *London, Heinemann, 1975.)*

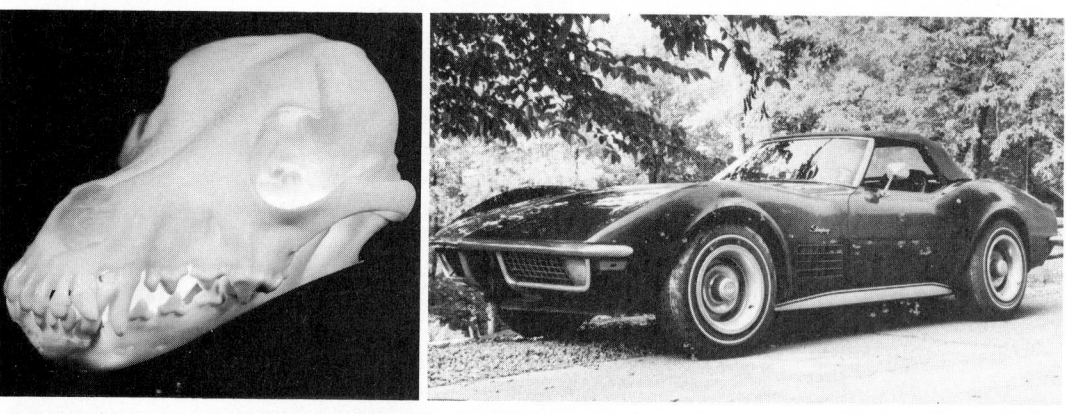

**FIGURE 4–6**

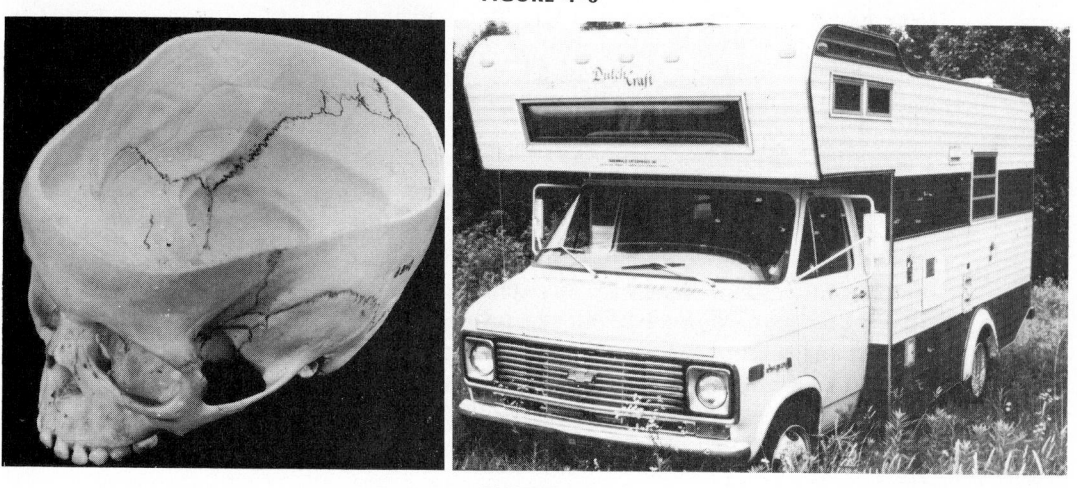

**FIGURE 4–7**

**FIGURE 4–8**

**Concept 6.** The expansion of the various parts of the cerebral hemispheres has created sizable pockets in the cranial floor. Each of these **endocranial fossae** relates to specific lobes of the brain on the inside of the cranial floor and to specific parts of the face, pharynx, and so forth on the outside. We can utilize our knowledge of these brain–cranial floor–facial relationships to advantage in analyzing the structure of the face and the basis for its many variations in form and pattern.

**FIGURE 4–9**

**Concept 7.** The nasomaxillary complex relates **specifically** to the anterior cranial fossa. The anterior boundary of this fossa establishes where the anterior boundary of the nasomaxillary complex will be. The posterior boundary of the anterior cranial fossa determines the corresponding posterior boundary of the nasomaxillary complex.

**FIGURE 4–10**

The pharynx relates specifically to the middle cranial fossa. Because of the human cranial floor flexure, the size of the middle cranial fossa in man determines the horizontal dimension of the pharyngeal space. The dimension of the middle cranial fossa **should** be equaled by the breadth of the mandibular ramus. The function of the ramus is to span the pharynx and middle cranial fossa in order to place the lower arch in occlusion with the upper. The length of the corpus of the mandible should also match the size of the bony maxillary arch. The mandible is a separate bone, however, joined to the skull by a movable articulation, and the size and the placement of its parts are independently variable. This is a major factor in the variations of facial form and profile. The maxilla, in contrast, is joined directly to the anterior cranial floor by sutures, and the growth of the cranium directly influences the corresponding growth of the midface, because common boundaries are shared by their respective growth fields.

**FIGURE 4–11**

**Concept 8.** One of the most basic and important planes in the whole head is the **"PM"** (**p**osterior **m**axillary) plane. This is a natural anatomic boundary that represents the contact interface among certain key facial and cranial sites of growth, remodeling, and displacement. It will be recalled that a vertical reference line was used in Chapter 3 to visualize various major growth changes. This is the **PM** line, and it is a fundamental axis of growth activity involved in many relationships that exist during the overall growth process.

The **PM** is a natural boundary line that separates *a, b,* and *c* in the diagram from *d, e,* and *f.* The **PM** line delineates and establishes the boundary for these "counterparts" of the face and cranium. Thus, parts *a, b,* and *c* are all structural and growth counterparts to one another. The same is true for parts *d, e,* and *f.* Note that the boundary between the anterior and middle cranial fossae is the **exact** posterior-superior corner of the midfacial compartment. The growth and size of the midface relate specifically to the frontal lobe of the cerebrum. They are counterparts. The floor of the anterior cranial fossa is the skeletal platform between them, and the anterior floor is thus a counterpart to the frontal lobe on one side and the upper part of the nasomaxillary complex on the other side.

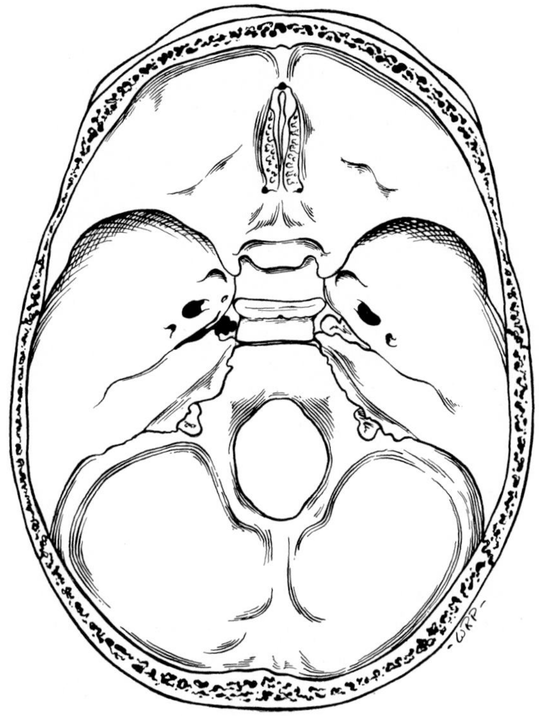

**FIGURE 4–8**

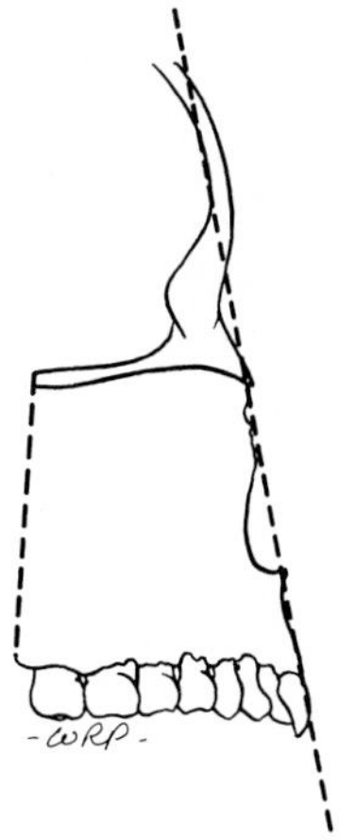

**FIGURE 4–9**

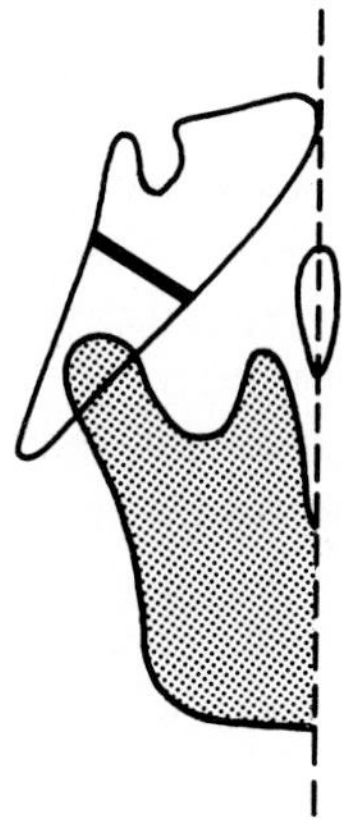

**FIGURE 4–10**

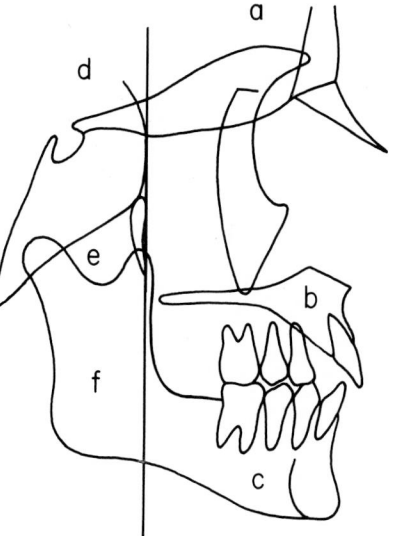

**FIGURE 4–11**

*(From Enlow, D. H.: Postnatal growth and development of the face and cranium. In Cohen, B., and I. R. H. Kramer (Eds.):* Scientific Foundations of Dentistry. *London, Heinemann, 1975.)*

**FIGURE 4–12**

**Concept 9.** Certain other boundaries of the brain are similarly shared by corresponding facial boundaries. These brain boundaries are established by growth fields, within which facial growth also takes place. The face has a prescribed perimeter of maximum growth, and this is the same as the growth perimeter for the brain. Some of the major **directions** of facial growth are established by the **special senses** located within the face itself (olfaction and vision). The two basic growth factors, amount and direction, constitute the growth "vector."

The anterior border of the nasomaxillary region corresponds to the anterior edge of the frontal lobe. The posterior margin of the nasomaxillary region corresponds to the posterior border of the frontal lobe (and in each case the anterior cranial fossa). The front and back vertical planes of the midface are **perpendicular** to the olfactory bulb and the orbit, respectively. The lower border of the midfacial compartment corresponds to the inferior level of the brain. All these various borders and planes represent the maximum normal growth boundaries for the nasomaxillary complex, and the growth directions relate to two major special senses that are directly involved in many functions of the face.

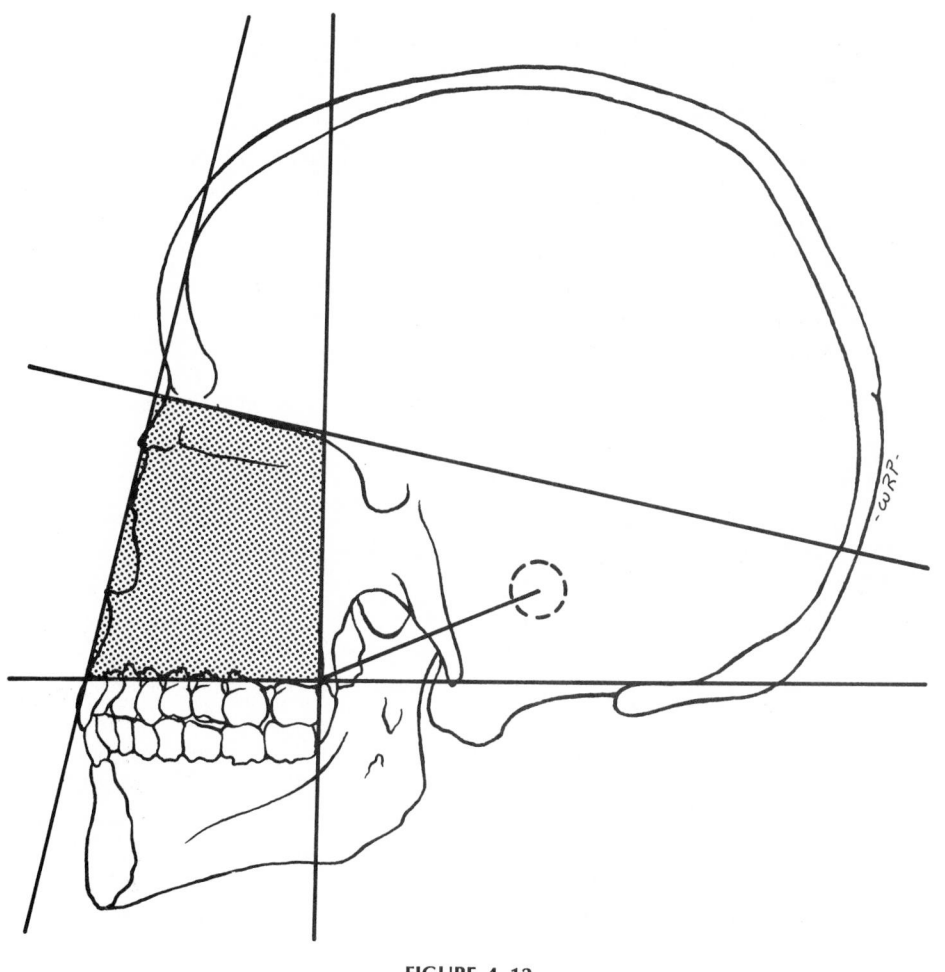

**FIGURE 4–12**

*(From Enlow, D. H., and M. Azuma: Functional growth boundaries in the human and mammalian face. In Bergsma, D. (Ed.):* Morphogenesis and Malformations of the Face and Brain. *Birth Defects Orig. Art. Ser., Vol. XI, No. 7. New York: Alan R. Liss, Inc. for The National Foundation—March of Dimes, White Plains, New York.)*

# PART TWO

**FIGURES 4–13 TO 4–16**

If a short piece of adhesive tape is affixed to a rubber balloon and the balloon then inflated, it will expand in a curved manner. The balloon bends because it enlarges around the nonexpanding segment. The enormous human cerebrum similarly expands around a much lesser enlarging midventral segment (the medulla, pons, hypothalamus, optic chiasma). This causes a bending of the whole underside of the brain. The **flexure** of the cranial base results. The foramen magnum in the typical mammalian skull is located at the posterior aspect of the cranium. In man, it is in the midventral part of the expanded cranial floor at an approximate balance point for upright head support on a vertical spine.

**FIGURES 4–17 AND 4–18**

The expansion of the frontal lobes displaces the frontal bone upward and outward. It becomes the distinctive, bulbous, upright "forehead" of the human face, although it is really part of the cranium and not the face proper. The frontal lobes also relate to a rotation of the human orbits into new positions. As the forehead is rotated into a vertical plane by the brain behind it, the superior orbital rim is carried with it. The eyes now point at a right angle to the spinal cord. The spine is vertical, and the orbital axis is horizontal. Vision is directed toward forward body movement.*

---

*In some anthropoids, such as the gorilla, the massive supraorbital ridges may also rotate vertically independent of the frontal lobe. In the human face, however, the orbits **must** rotate into a vertical alignment because of the expanded size of the frontal lobes.

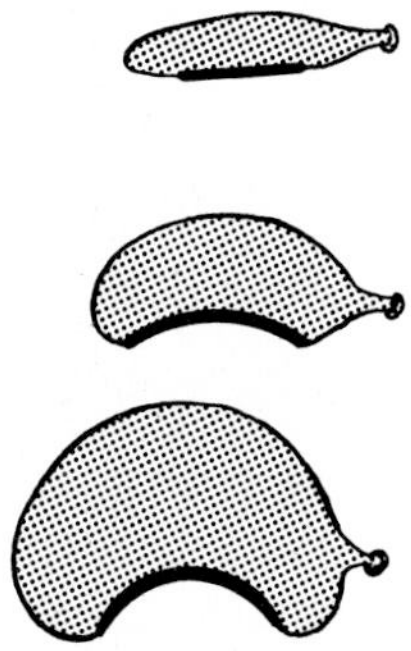

**FIGURE 4–13**

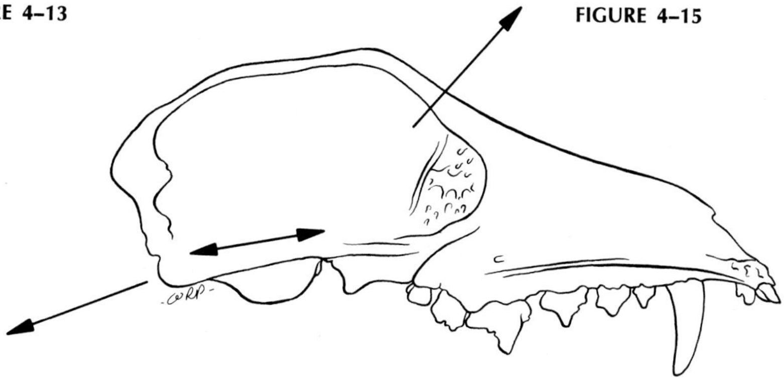

**FIGURE 4–15**

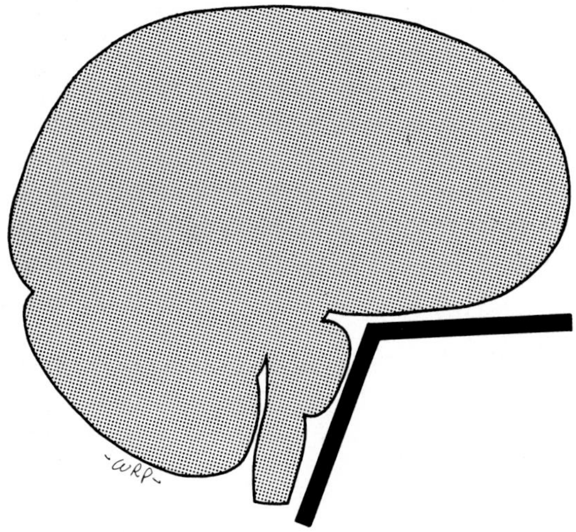

**FIGURE 4–14**

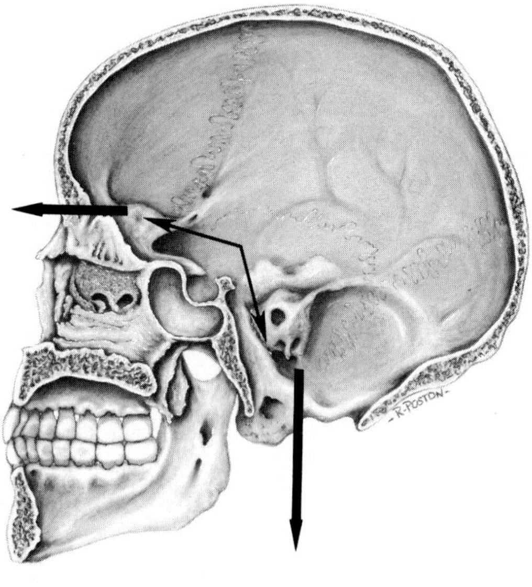

**FIGURE 4–16**

*(From Enlow, D. H.: Postnatal growth and development of the face and cranium. In Cohen, B., and I. R. H. Kramer (Eds.): Scientific Foundations of Dentistry. London, Heinemann, 1975.)*

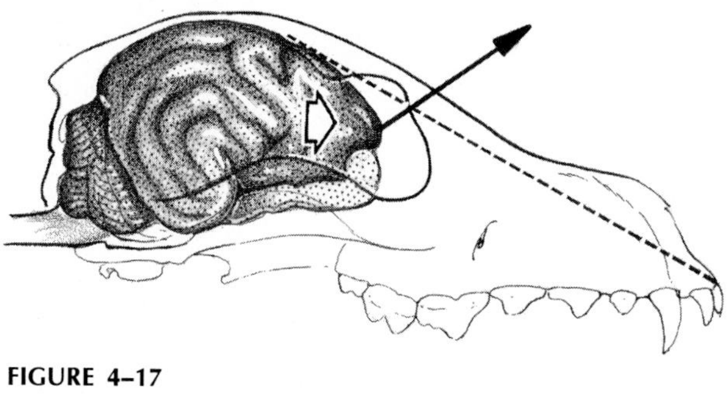

**FIGURE 4–17**

*(From Enlow, D. H., and J. McNamara: Angle Orthod., 43:256, 1973.)*

*(From Enlow, D. H., and J. McNamara: Angle Orthod., 43:256, 1973.)*

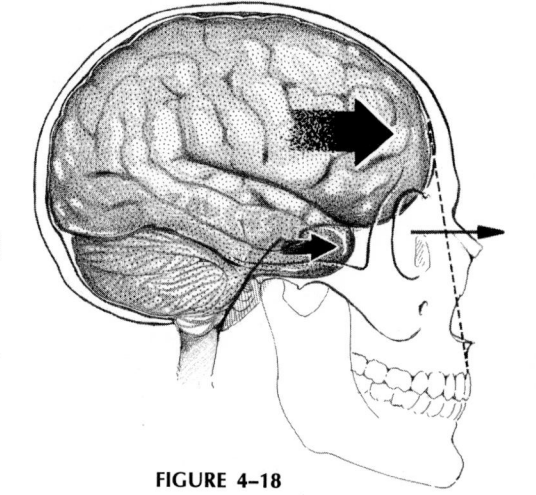

**FIGURE 4–18**

**FIGURES 4–19 AND 4–20**

The expansion of the frontal and, particularly, the temporal lobes of the cerebrum relates to a rotation of the orbits toward the midline. The eyes come closer together. Two separate axes of orbital rotation are thus associated with the massive expansion of the cerebrum. One displaces the orbits vertically, and the other carries them horizontally in medial directions into a binocular position.

**FIGURE 4–21**

Orbital rotation toward the midline, however, significantly reduces the dimension of the interorbital space. This is one of two basic factors that underlies reduction in the extent of snout protrusion in man. Because the interorbital segment is the root of the nasal region, a decrease in this dimension reduces the structural (and also the physiologic) base of the bony nose. A wide nasal base can support a proportionately longer snout. A narrow nasal base, however, reduces the architectural limit to which the bony part of the nose can protrude, and the snout is thereby shorter. The second basic factor involved in the reduction of nasal protrusion deals with the rotation of the olfactory bulbs.

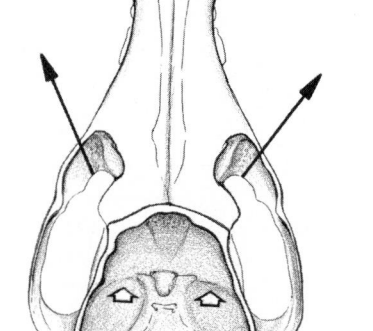

**FIGURE 4-19**

*(From Enlow, D. H., and J. McNamara: Angle Orthod., 43:256, 1973.)*

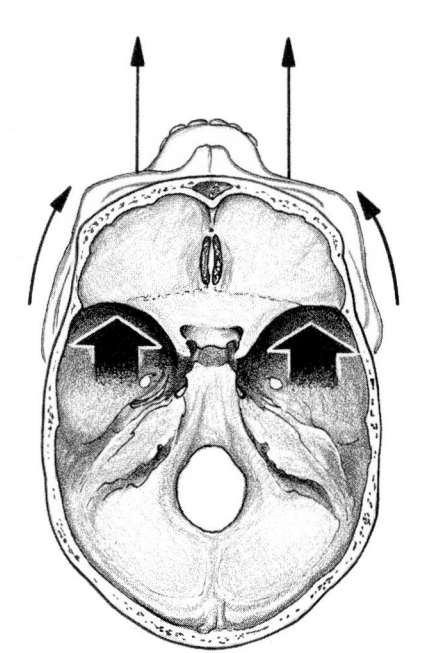

**FIGURE 4-20**

*(From Enlow, D. H., and J. McNamara: Angle Orthod., 43:256, 1973.)*

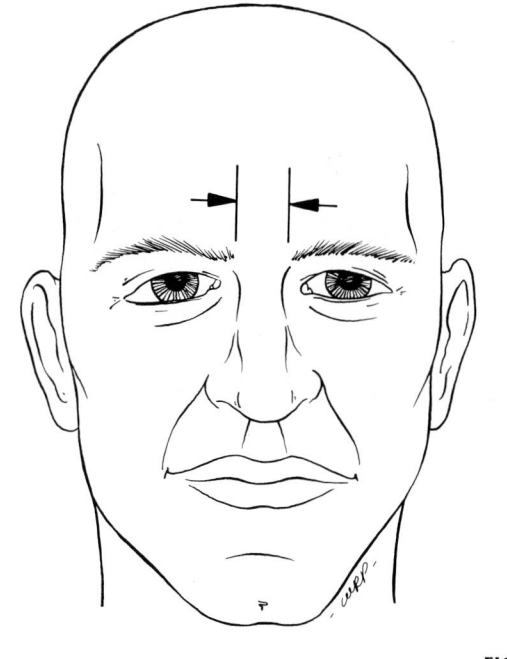

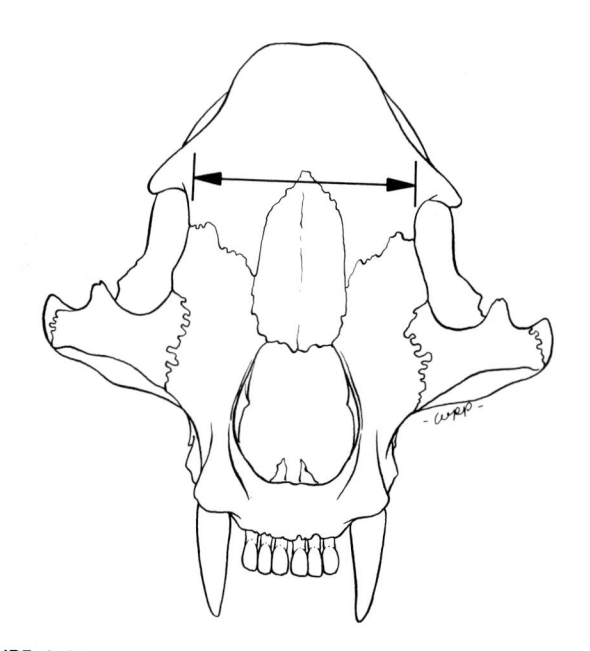

**FIGURE 4-21**

**FIGURE 4–22**

Note that the enlarged human cerebrum has caused a downward rotational displacement of the olfactory bulbs. In all other mammals, they are nearly upright or obliquely aligned, depending on the size and configuration of the frontal lobes. In man, the bulbs have been rotated into **horizontal** positions by the cerebrum. This is a significant factor in the basic design of the human face.

**FIGURE 4–23**

The olfactory bulbs relate directly to the alignment and the direction of growth of the adjacent nasal region. The long axis of the snout in most mammals is constructed so that it necessarily points toward the general direction of the sensory olfactory nerves within it. The plane of the nasomaxillary region is thereby approximately **perpendicular to the plane of the olfactory bulbs.** This is a major anatomic and functional relationship involved in the basic plan of the face in any mammal. As the bulbs become rotated progressively from a vertical position to a horizontal one because of increases in brain size or because of its shape (1, 2, 3), the whole face is similarly rotated from a horizontal to a vertical plane (1a, 2a, 3a). Or, stated another way, the face is rotated down by the expanded anterior cranial floor as it rotates downward as a result of the enlargement of the frontal lobes.

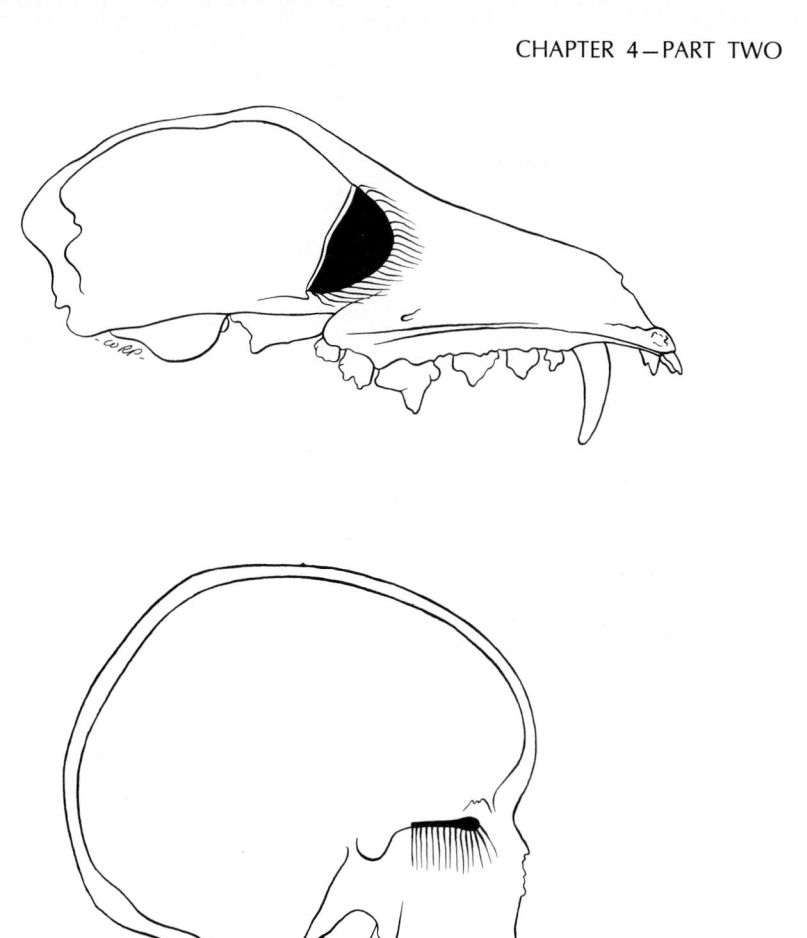

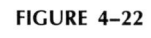

**FIGURE 4–22**

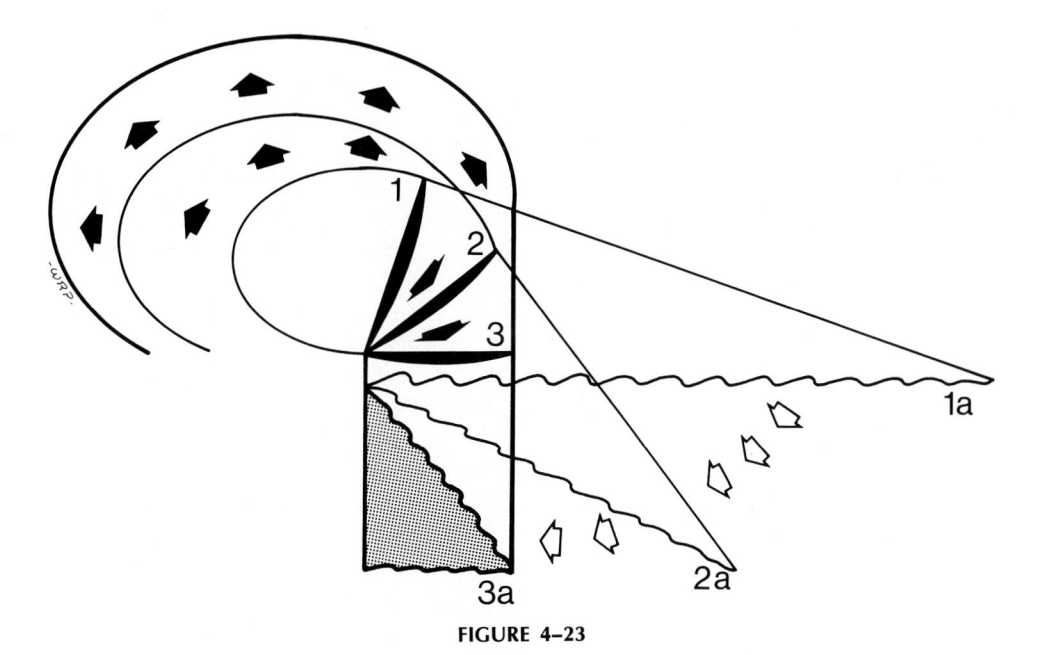

**FIGURE 4–23**

**FIGURES 4–24 AND 4–25**

The maxilla of most mammals has a triangular configuration. In man, it is uniquely rectangular. This is caused by a rotation of the occlusion into a horizontal plane to adapt to the vertical rotation of the whole midface. The occlusal plane in most mammals, including man, is approximately parallel to the Frankfort plane (a plane from the top of the auditory meatus to the inferior rim of the orbit). This aligns the jaws in a functional position relative to the visual, olfactory, and hearing senses. In the human maxilla, the design change that produced this resulted in the creation of a new arch-positioning facial region, the **suborbital** compartment. Most of this phylogenetically expanded area is occupied by the otherwise nonfunctional maxillary sinus (uses such as air warming, nasal drip, and voice resonance are secondary). An **orbital floor** was also newly created because of this added facial region. Compare also with Figure 4–23.

**FIGURE 4–26**

The nasal region is thus **vertically** disposed in the human face. The neutral axis of the spread of the sensory olfactory nerves is vertical, and the vertical vector of nasomaxillary growth has become a major feature of human facial development. The characteristic vertical human facial profile is a composite result of (1) a bulbous forehead, (2) rotation of the nasal region into a vertical plane, (3) reduction in snout protrusion in conjunction with medial orbital convergence, (4) rotation of the orbits into upright positions, (5) rotation of the maxillary arch downward and backward, and (6) bimaxillary reduction in the extent of prognathism matching nasal reduction. The face also becomes markedly widened because of the increased breadth of the brain and cranial floor and because the orbits and cheekbones are rotated into forward-facing positions. The face of man lies **beneath** the frontal lobe of the brain; in contrast, in other mammals the face is largely in front of the cerebrum. The nasal chambers are housed largely **within** the face, between and below the orbits, rather than projecting forward with a protrusive muzzle. The human snout itself houses very little of the mucosal part of the nasal chambers.

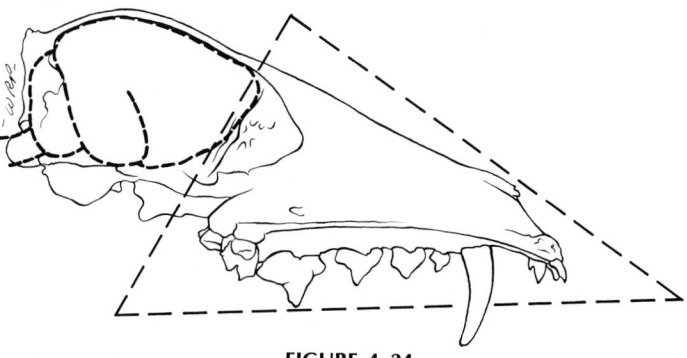

**FIGURE 4–24**

*(From Enlow, D. H.: Postnatal growth and development of the face and cranium. In Cohen, B., and I. R. H. Kramer (Eds.): Scientific Foundations of Dentistry. London, Heinemann, 1975.)*

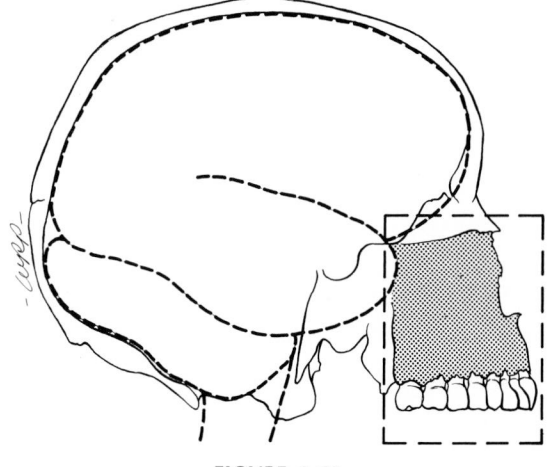

**FIGURE 4–25**

*(From Enlow, D. H.: Postnatal growth and development of the face and cranium. In Cohen, B., and I. R. H. Kramer (Eds.): Scientific Foundations of Dentistry. London, Heinemann, 1975.)*

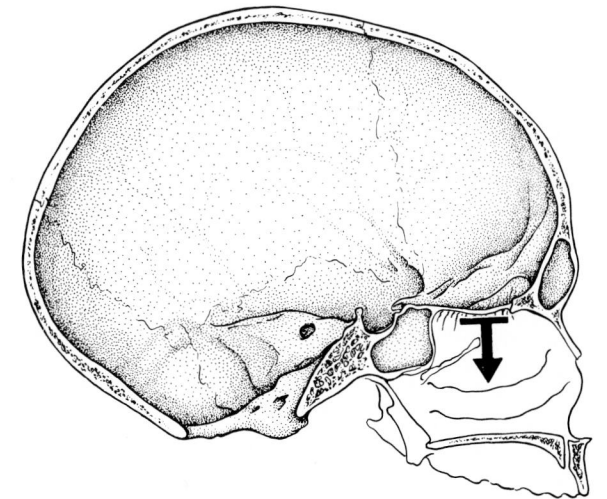

**FIGURE 4-26**

*(Modified from Enlow, D. H.: The Human Face. New York, Harper & Row, 1968, p. 187.)*

**FIGURE 4–27**

Reduction of the nasal region associated with orbital convergence must necessarily also be accompanied by a more or less equal reduction in maxillary arch length, because the floor of the nasal chamber is also the roof for the mouth. Only a relatively slight degree of horizontal divergence between the two can exist. The palate is shared in common by both regions. Whether the chain of events that led to nasal reduction "caused" a shortening of the maxillary arch, or whether, conversely, the chain went in the other direction, we shall likely never know. However, if either one becomes reduced in length, so must the other. This refers only to the bony part of the nasal region; some species have a fleshy proboscis protruding well beyond the jaws and palate (such as man and the elephant).

**FIGURE 4–28**

The protrusion of the cartilaginous and soft tissue portion of the nose provides for **downward**-directed external nares. This aims the inflow of air obliquely upward into the vertically disposed nasal chambers toward the vertically aligned sensory nerves of the olfactory bulbs located in the **ceiling** of the chambers. This is in contrast to the anteriorly directed external apertures of other mammals taking air into more horizontal nasal chambers having the cribriform plates located as part of the posterior wall.

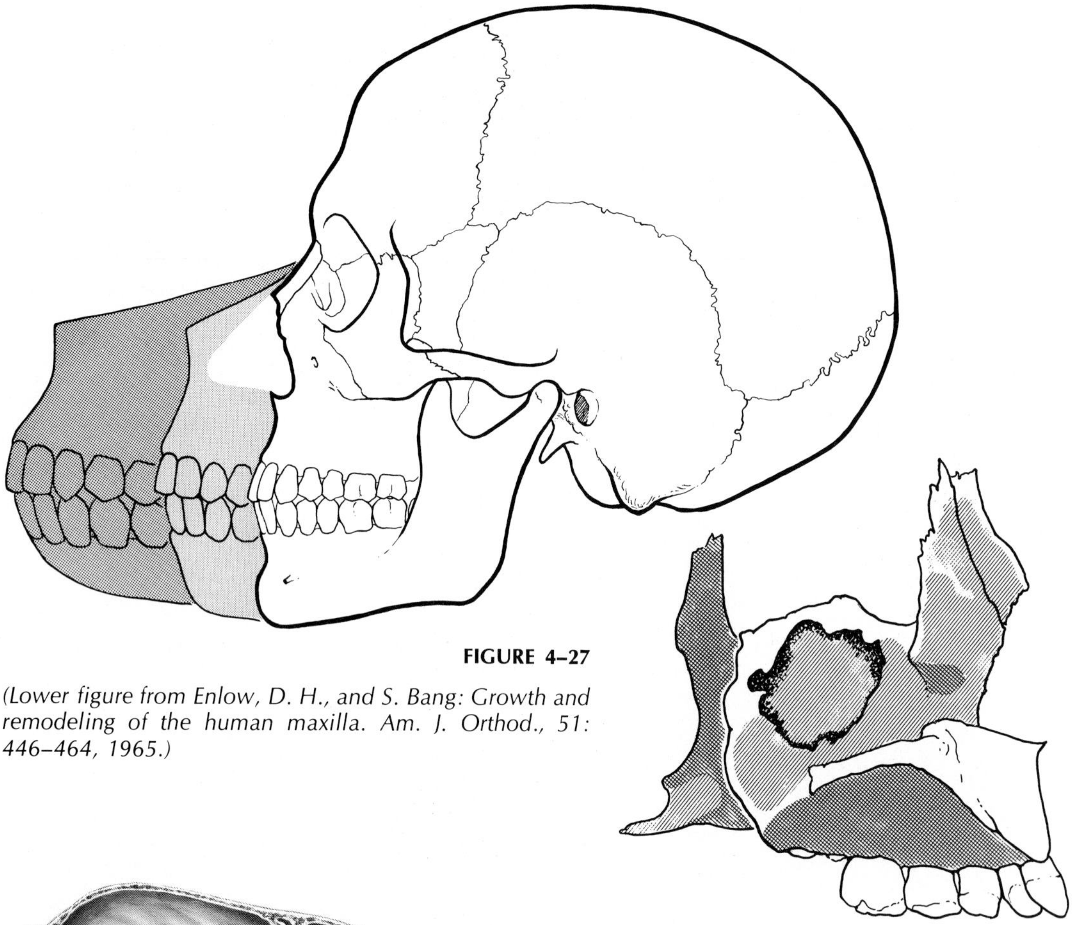

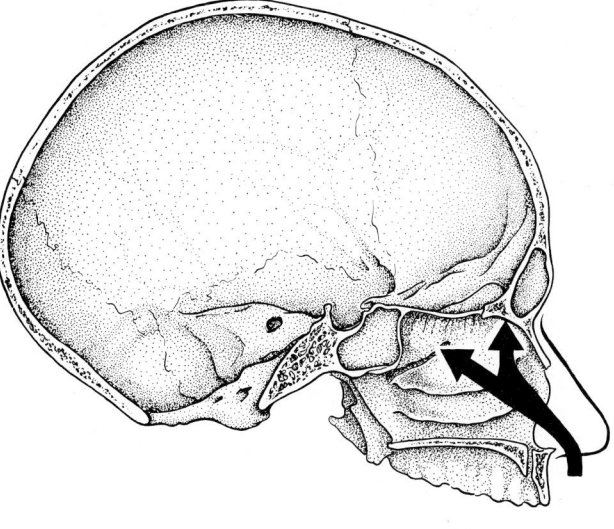

**FIGURE 4–27**

*(Lower figure from Enlow, D. H., and S. Bang: Growth and remodeling of the human maxilla. Am. J. Orthod., 51: 446–464, 1965.)*

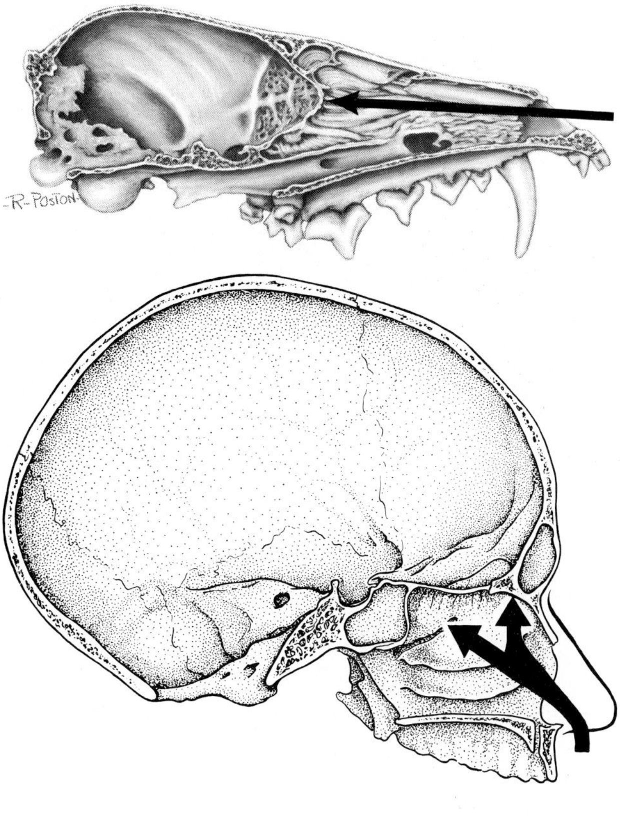

**FIGURE 4–28**

*(Lower figure From Enlow, D. H.:* The Human Face. *New York, Harper & Row, 1968, p. 188.)*

**FIGURE 4–29**

The rotation of the whole face downward and **backward** has resulted in a facial placement within the recess or pocket created by the cranial base flexure. What will happen if the brain **continues** to enlarge phylogenetically and thereby produce an even further extent of backward rotation? It has virtually no more room left to "rotate into," at least given the present design and arrangement of the various soft tissue and skeletal parts involved. Already the face has almost reached the airway because of rotation. The posterior cranial floor, vertebral column, and face, as in a closing vise, are coming together; there are important parts in between.

**FIGURE 4–30**

The growth of each part of the face involves two basic considerations. The first is the **amount** of growth, and the second is the **direction** of growth. These two factors constitute the growth "vector."

**FIGURE 4–31**

The amount of growth involves fields of growth and the boundaries of these fields. There is a prescribed perimeter of maximal growth capacity for each major part of the face, and growth does not ordinarily exceed this perimeter. Significantly, the forward, downward, backward, and lateral growth boundaries that exist for the face are **shared** by the brain. The perimeter for the field of brain growth and the perimeter for the field of facial growth have become established in common.

The reason for this is that the brain has evolved in conjunction with the cranial floor. The form, size, topographic features, and angular characteristics of one conform to the other. The floor of the cranium, in turn, is the **template** upon which the face is built. The junctional part of the face can be no **wider,** for example, than the maximal width of the cranium. There would be nothing to attach it to. Similarly, the **length** and **height** of specific parts of the cranial floor are expressed as equivalent dimensions for the face.

The face is **not** independent of the cranium structurally or developmentally. The whole face has often been described as a genetically and developmentally separate region, the only tie between them being that they happen to be placed in juxtaposition, as a picture hangs in juxtaposition to a wall. No cause and effect relationships were believed to exist between the cranium and the size or shape of the face. This is certainly not the case. **Many** structural features and dimensions of the face are based on brain–cranial base–facial relationships, as will be seen. This is an important concept because a number of normal and abnormal variations in facial form relate, at least in part, to underlying circumstances present in the cranial floor (see Chapter 5).

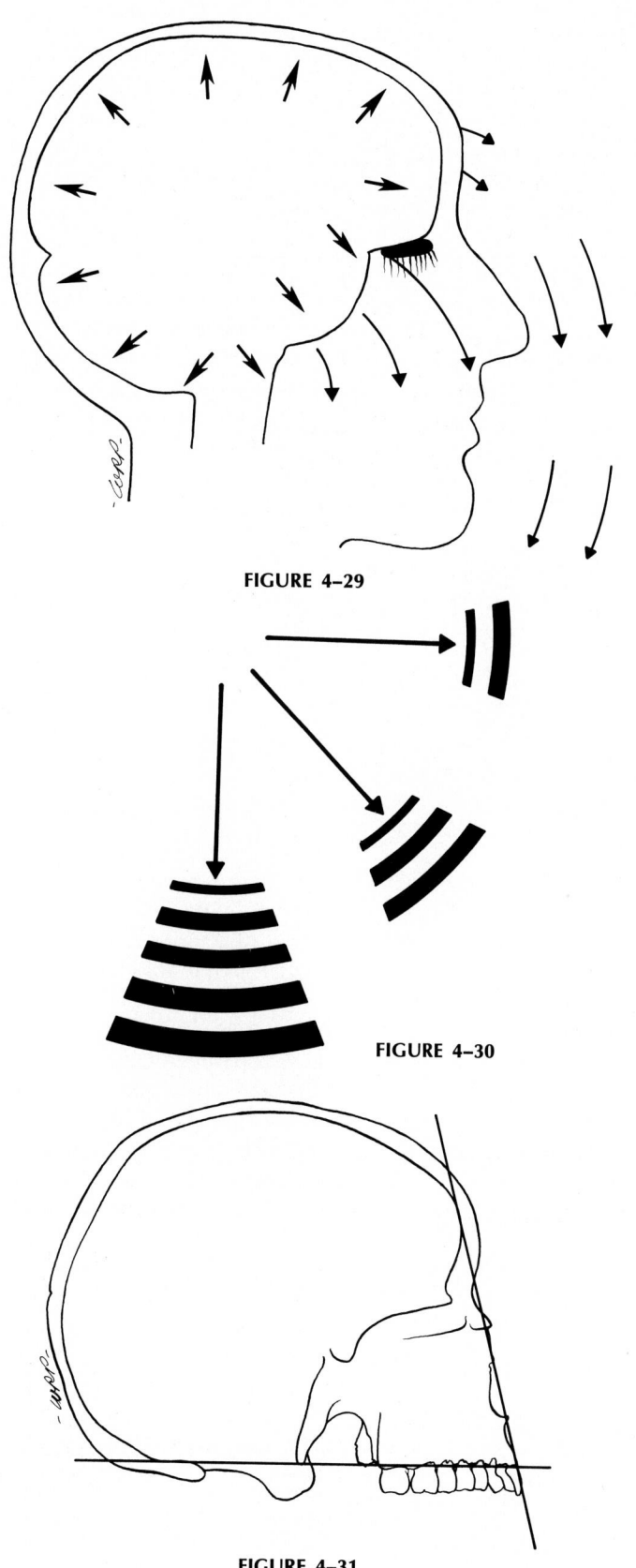

**FIGURE 4–29**

**FIGURE 4–30**

**FIGURE 4–31**

The floor of the cranium has developed in **phylogenetic** association with the brain. Whatever "independent" genetic control actually exists in the cranial base (this is controversial), the shape and size of the floor of the cranium have become established because its genes were especially adopted by the natural selection process to accommodate an **interdependent** association with the developing brain. Which came first will probably always be argued, although the brain gets the nod by most present day theoreticians. Thus established, however, the cranium presumably has a measure of genetic independence. The face similarly develops **in conjunction with** the cranial floor (and the brain), and the genetic control of facial development (by its soft tissues) has become established to accommodate its own functional association with the cranium. In all areas, however, a developmental **latitude** exists that provides **adjustments** during growth to accommodate variations in one part or another. This is involved in the operation of the "functional matrix," and it is the factor that allows for a unified coexistence of the many separate, developing parts all growing in relation to one another. There are regional differences, however, in the **capacity** for such developmental adjustment. Some areas, such as the bony alveolar sockets, are extremely labile and responsive to variable circumstances. Other areas, such as the cranial base, are much less sensitive and adaptable. The "intrinsic programming" for the latter is presumed to be greater than for the former because of their different levels of developmental independence and whatever the different, poorly understood factors are that determine and control this.

**FIGURES 4–32 AND 4–33**

The forward **boundary** of the brain is shared by the forward border of the nasomaxillary complex. The **direction** of growth by the nasal part of the face is established by the olfactory bulbs and the sensory olfactory nerves. These two factors underlie the "vector" of midfacial growth, that is, the amount and the direction. To show this, a line is drawn from the forward edge of the brain down to the anterior-most, inferior-most point of the nasomaxillary complex. This represents the **midfacial plane.*** Note that the midfacial plane is perpendicular to the **olfactory bulb** (or the cribriform plate, as seen in lateral headfilms). The long axis of the nasal region points in the same general direction as the neutral axis of its sensory nerve spread. The amount of growth is established by the prescribed perimeter of its growth field. **The nasomaxillary complex grows as far forward as the edge of the brain in a direction approximately perpendicular to the olfactory bulbs.**

---

*"Nasion" is often used in cephalometric studies as a point for drawing the facial plane, but this is a poor selection because nasion is so variable. Moreover, the purpose of the **midfacial** plane described above is to show the relationship between the **brain** and the **nasomaxillary complex.** Thus, the edge of the brain is used rather than nasion. Also, the above descriptions presume that the alignment of the nerves is the lead factor that determines the direction of midfacial growth. Of course, it may be the converse. Whichever, the important point is that they are established **together** in a constant relationship to the olfactory bulbs, which, in turn, are placed according to the size and shape of the brain.

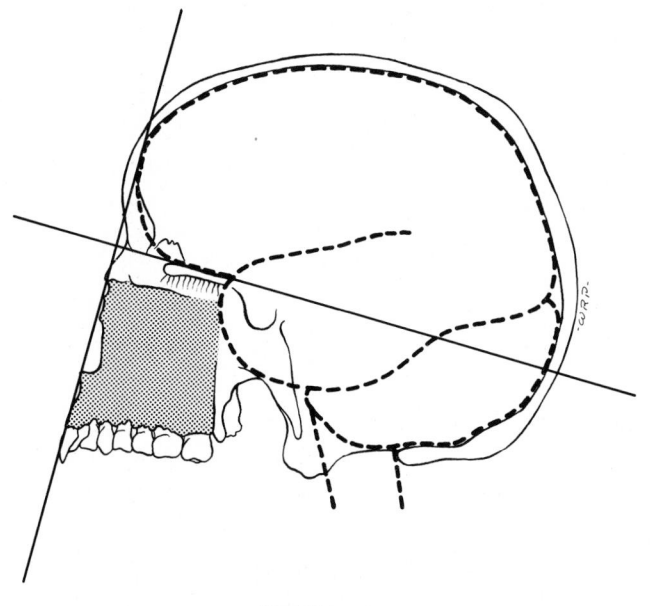

**FIGURE 4–32**

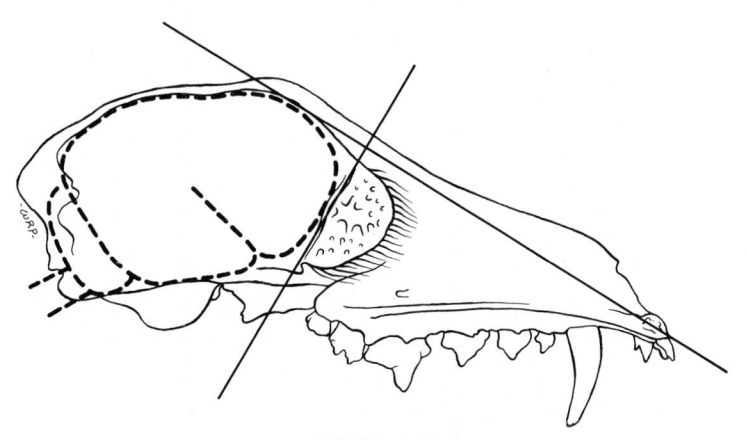

**FIGURE 4–33**

(From 'Enlow, D. H., and M. Azuma: Functional growth boundaries in the human and mammalian face. In Bergsma, D. (Ed.): Morphogenesis and Malformations of the Face and Brain. Birth Defects Orig.: Art. Ser., Vol. XI; No. 7. New York: Alan R. Liss, Inc. for The National Foundation—March of Dimes, White Plains, New York.)

**FIGURES 4–34 AND 4–35**

This relationship exists among mammals in general. In species or groups having a smaller brain and, as a result, a more upright olfactory bulb, the snout and muzzle tend to be correspondingly more horizontal and much more protrusive. As the olfactory bulbs become rotated downward because of increasing brain size (or shape, as in more round-headed species), the muzzle correspondingly rotates down with them. In man, the olfactory bulbs have become virtually horizontal because of the massive growth of the frontal lobes. The nasal part of the face is thus vertically aligned in conjunction with the neutral vertical axis of olfactory nerve distribution.

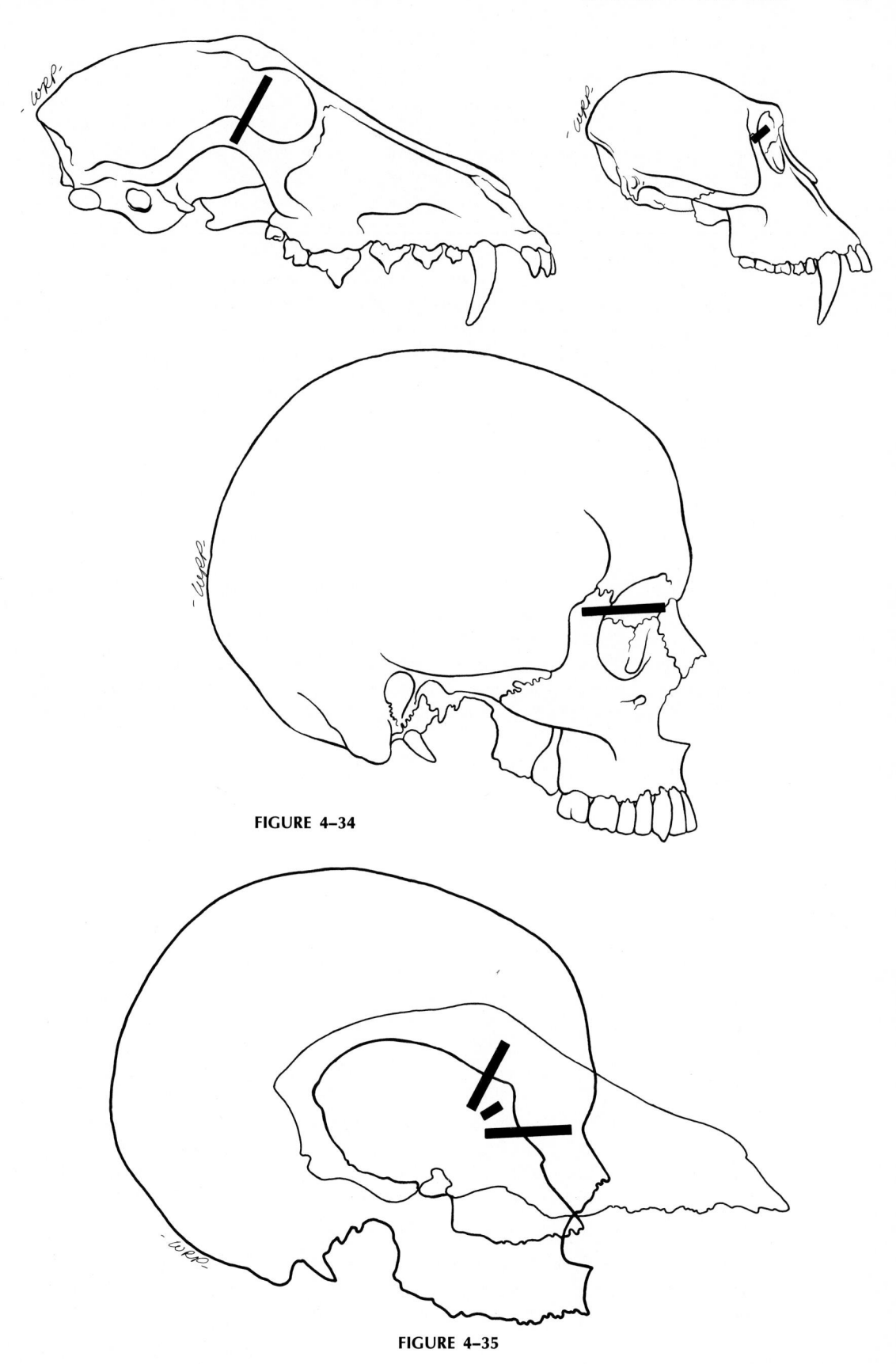

FIGURE 4–34

FIGURE 4–35

**FIGURE 4–36**

The nasomaxillary complex, as mentioned earlier, is specifically associated with the anterior cranial fossa. The posterior boundary of this fossa establishes the corresponding posterior boundary for the midface. This is essentially a nonvariable anatomic relationship. The **direction** of growth in this region is established by the particular special sense located in this part of the face, which is the visual sense. The posterior maxillary tuberosity is located beneath the floor of the orbit, and the orbital floor is the roof of the maxillary tuberosity. The tuberosity is aligned approximately perpendicularly to the neutral geometric axis of the orbit. **The posterior plane of the midface extends from the junction between the anterior and middle cranial fossae (and the inferior junction between the frontal and temporal lobes) downward in a direction perpendicular to the neutral line of the orbit.** This plane passes almost exactly along the posterior surface of the maxillary tuberosity.

**FIGURE 4–37**

The boundary just described represents one of the key anatomic planes in the face. This is the **PM** plane. There are many "cephalometric planes" in the face and cranium. Most of these, however, do not represent (and are not so intended) *(1)* key sites of growth and remodeling or *(2)* functional relationships among the various parts of the skull, including soft tissue associations. Most conventional cephalometric planes, such as sella-nasion, bypass the really important key sites of growth without recognizing them. The vertical **PM** boundary, in contrast, is a **natural anatomic and morphogenic** plane that relates directly to the factors that establish the basic design of the face. It is one of the most important developmental and structural planes in the face and cranium.

**FIGURE 4–38**

The **PM** (**p**osterior **m**axillary) plane delineates naturally the various anatomic **counterparts** of the craniofacial complex. The frontal lobe, the anterior cranial fossa, the upper part of the ethmomaxillary complex, the palate, and the maxillary arch are all mutual counterparts lying anterior to the **PM** line. All these parts have posterior boundaries that are placed along this vertical plane. Similarly, the temporal lobe, the middle cranial fossa, and the posterior oropharyngeal space are mutual counterparts located behind the **PM** plane. The anterior boundaries of these parts are precisely positioned along this vertical line. The **PM** plane is a **developmental interface** between the series of counterparts in front of and behind it. This key plane retains these basic relationships throughout the growth process.

The **corpus** of the mandible is a counterpart to those parts lying in front of **PM.** The **ramus** is a counterpart of the parts behind **PM.** The placement of the mandible and the size of its parts, however, are independently variable. The posterior boundary of

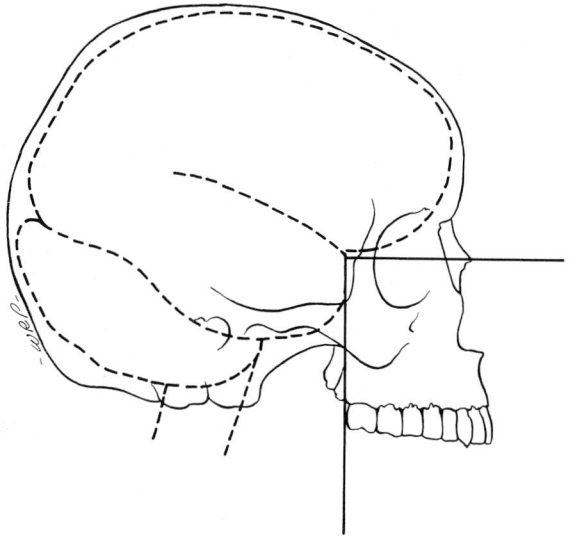

**FIGURE 4–36**

*(From Enlow, D. H., and M. Azuma: Functional growth boundaries in the human and mammalian face.* In *Bergsma, D. (Ed.):* Morphogenesis and Malformations of the Face and Brain. *Birth Defects Orig. Art. Ser., Vol. XI, No. 7. New York: Alan R. Liss, Inc. for The National Foundation—March of Dimes, White Plains, New York.)*

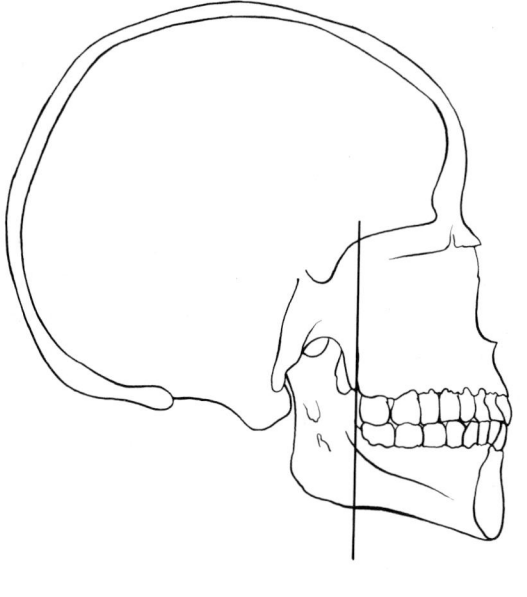
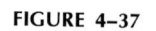

**FIGURE 4–37**

*(From Enlow, D. H., and M. Azuma: Functional growth boundaries in the human and mammalian face.* In *Bergsma, D. (Ed.):* Morphogenesis and Malformations of the Face and Brain. *Birth Defects Orig. Art. Ser., Vol. XI, No. 7. New York: Alan R. Liss, Inc. for The National Foundation—March of Dimes, White Plains, New York.)*

**FIGURE 4–38**

*(From Enlow, D. H.: Postnatal growth and development of the face and cranium.* In *Cohen, B., and I. R. H. Kramer (Eds.):* Scientific Foundations of Dentistry. *London, Heinemann, 1975.)*

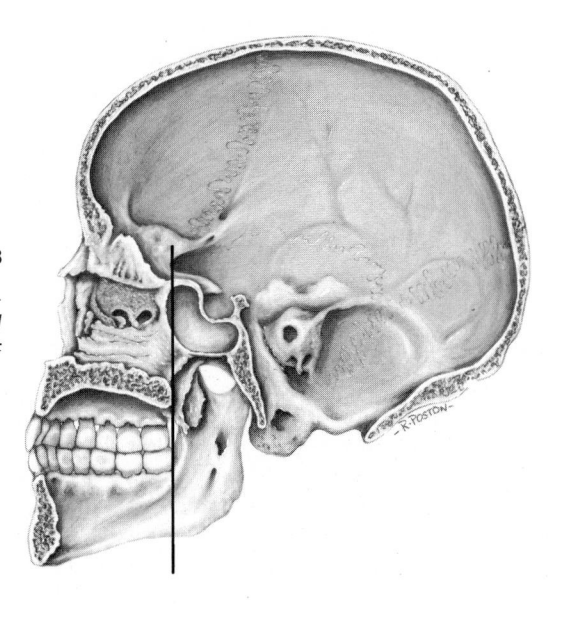

the corpus **should** lie on **PM**. This is the "lingual tuberosity," which is the direct mandibular equivalent of the maxillary tuberosity. The forward boundary of the ramus, where it joins the lingual tuberosity, should also lie on **PM.** (Note: The anterior edge of the obliquely aligned ramus overlaps the lingual tuberosity, but this edge does not represent the actual forward point of the effective ramus dimension; the lingual tuberosity itself is the functional junction between the corpus and ramus.) Because the mandible is a separate bone not attached directly to the cranium by sutures, its latitude for structural variation is not subject to the same degree of developmental and structural communality that occurs between the growth fields shared by the cranial floor and the maxilla. Independent variations can thus exist in the dimensions and the placement of both the ramus and the corpus. The ramus, for example, may fall short of **PM,** or it may protrude well forward of it. This variability feature is often **compensatory.** That is, a narrow or broad ramus can offset a tendency for an imbalance or a malocclusion caused by factors in **other** parts of the face and cranium. In any event, such mandibular variations can be recognized, and in analyzing the basis for any given individual's facial pattern, normal or abnormal, the part played by the mandible can be determined by noting where the lingual tuberosity is located with respect to the maxillary tuberosity (or with **PM**), as described in Chapter 5.

**FIGURE 4–39**

The above relationships hold for mammals in general. Thus, the posterior boundary of the midfacial compartment in most species coincides with the posterior boundary of the anterior cranial fossa (the frontal lobe–temporal lobe junction on the floor of the cranium). The posterior midfacial plane extends downward from this point in a direction approximately perpendicular to the neutral axis of the orbit. This line passes along the posterior surface of the maxilla.

**FIGURES 4–40 AND 4–41**

Just as the other boundaries of the midface coincide with respective brain boundaries, the inferior boundary of the nasomaxillary complex is established by the bottom-most surface of the brain and cranial floor. Note that a line from the floor of the posterior cranial fossa passes through, or very nearly so, both the inferior corner of the posterior maxillary tuberosity and the inferior corner of the front part of the bony maxillary arch (prosthion).

**FIGURE 4–39**

*(From Enlow, D. H., and M. Azuma: Functional growth boundaries in the human and mammalian face. In Bergsma, D. (Ed.): Morphogenesis and Malformations of the Face and Brain. Birth Defects Orig. Art. Ser., Vol. XI, No. 7. New York: Alan R. Liss, Inc. for The National Foundation—March of Dimes, White Plains, New York.)*

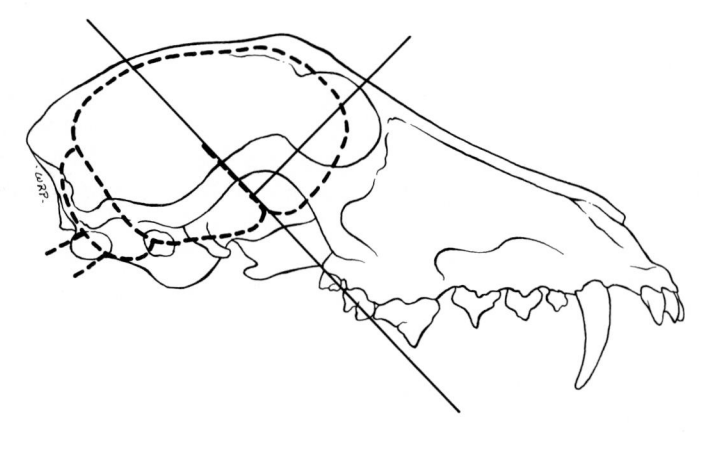

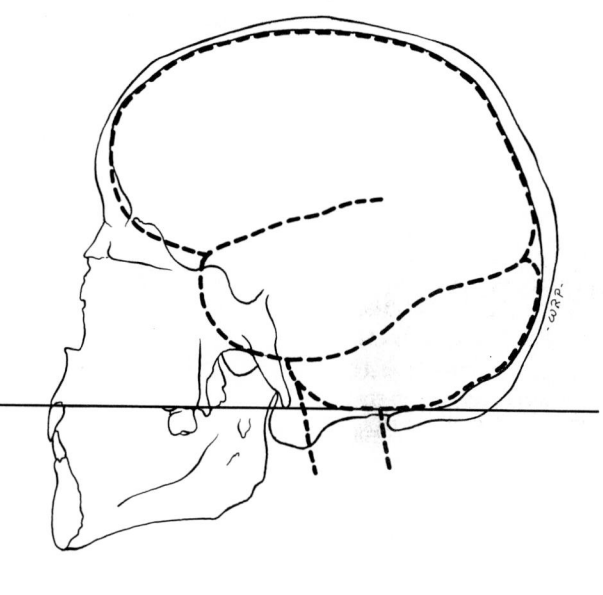

**FIGURE 4–40**

*(From Enlow, D. H., and M. Azuma: Functional growth boundaries in the human and mammalian face. In Bergsma, D. (Ed.): Morphogenesis and Malformations of the Face and Brain. Birth Defects Orig. Art. Ser., Vol. XI, No. 7. New York: Alan R. Liss, Inc. for The National Foundation—March of Dimes, White Plains, New York.)*

**FIGURE 4–41**

*(From Enlow, D. H., and M. Azuma: Functional growth boundaries in the human and mammalian face. In Bergsma, D. (Ed.): Morphogenesis and Malformations of the Face and Brain. Birth Defects Orig. Art. Ser., Vol. XI, No. 7. New York: Alan R. Liss, Inc. for The National Foundation—March of Dimes, White Plains, New York.)*

**FIGURES 4–42 AND 4–43**

The anatomic and functional plan of the face is also associated with another special sensory relationship, the sense of hearing. In man, the anthropoids, and the simians (and many other mammalian forms as well), the midpoint of the external auditory meatus forms an approximate 45° relationship between the midpoint of the orbital opening and the inferior corner of the maxillary tuberosity. The functional significance, if any, is not clear, but this arrangement does place the ear opening in an effective position with regard to both sight and the position of the jaws. All three are mutually involved in defense, offense, food-getting, and so forth. Unlike the direct 90° developmental and functional relationship between the nasal region of the face and the olfactory bulbs, this ear-eye-jaw arrangement may or may not represent an actual phylogenetically and developmentally **bound** interrelationship. That is, this alignment arrangement may be only incidental; it is, however, a phylogenetically successful morphologic pattern, and the relationship is useful in appraising the vertical growth of the face.

All the preceding angular and dimensional relationships have only a relatively small range of variation between individuals. The perpendicular, adult olfactory and orbital alignments to the anterior and posterior parts of the face, respectively, usually only vary within ±1° or 2° in the human head and in many other mammals as well (although some species show specialized variations). The adult 45° relationship of the auditory meatus also varies only about +2°. The adult inferior boundary of the midface is normally within 2 or 3 millimeters of the inferior plane of the brain.* While well within the latitude for normal anatomic and morphogenic relationships, these ranges of variation are too great for most kinds of cephalometric studies, in which accuracy to within the width of a pencil line is needed. Nonetheless, these relationships provide a useful and meaningful guide to understanding the fundamental plan of mammalian and human facial architecture and structural design and the basis for abnormal departures from it.

**FIGURE 4–44**

For example, if Class II and Class III headfilm tracings are superimposed on the cribriform plates (representing the olfactory bulbs), it is apparent that the anterior plane of the nasomaxillary region in both conforms to the normal, perpendicular olfactory relationship. Note the similarity of the midfacial plane alignments. In this particular Class II individual (and most others as well), it is not the maxilla itself that "protrudes"; rather it is the **mandible** that is actually retrusive. In the Class III individual, it is not the maxilla that is retrusive; the **mandible** is protrusive. In both individuals, the nasomaxillary complex is located where it is supposed to be, and its horizontal dimensions are not out of line as they relate to the brain.

---

*Except among simians and anthropoids, most of whom have an **established vertical hypoplasia** in the anterior part of the maxillary arch. In the rhesus monkey, for example, the premaxillary region is "high." A differentially greater extent of downward displacement takes place in the posterior part of the arch as compared with the anterior part. This in effect causes an "upward" rotation of the anterior region, and direct downward bone growth by this area does not move it fully to the inferior level attained by the posterior part of the arch. Anterior open bites are quite frequent, much more so than in the human being. A similar rotation does, in fact, occur in the human maxilla, but unlike in these other primates, the anterior part of the arch grows downward to an extent that fully offsets it. (See also Chapter 5.)

**FIGURE 4–42**

*(From Enlow, D. H., and M. Azuma: Functional growth boundaries in the human and mammalian face.* In *Bergsma, D. (Ed.):* Morphogenesis and Malformations of the Face and Brain. *Birth Defects Orig. Art. Ser., Vol. XI, No. 7. New York: Alan R. Liss, Inc. for The National Foundation—March of Dimes, White Plains, New York.)*

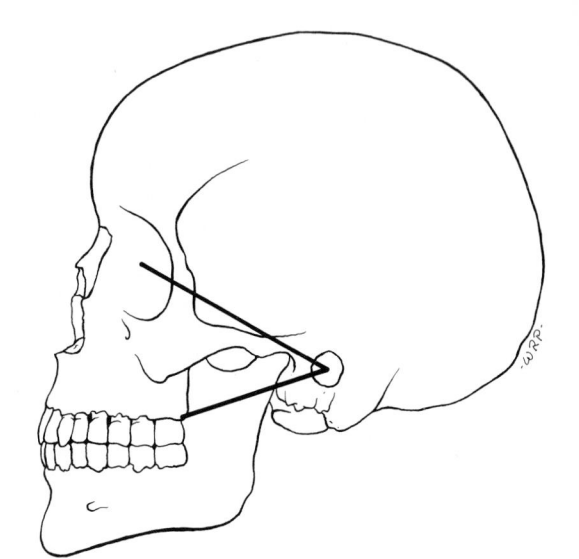

**FIGURE 4–43**

*(From Enlow, D. H., and M. Azuma: Functional growth boundaries in the human and mammalian face.* In *Bergsma, D. (Ed.):* Morphogenesis and Malformations of the Face and Brain. *Birth Defects Orig. Art. Ser., Vol. XI, No. 7. New York: Alan R. Liss, Inc. for The National Foundation—March of Dimes, White Plains, New York.)*

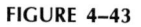

**FIGURE 4–44**

*(From Enlow, D. H., and J. McNamara: Angle Orthod., 43:256, 1973.)*

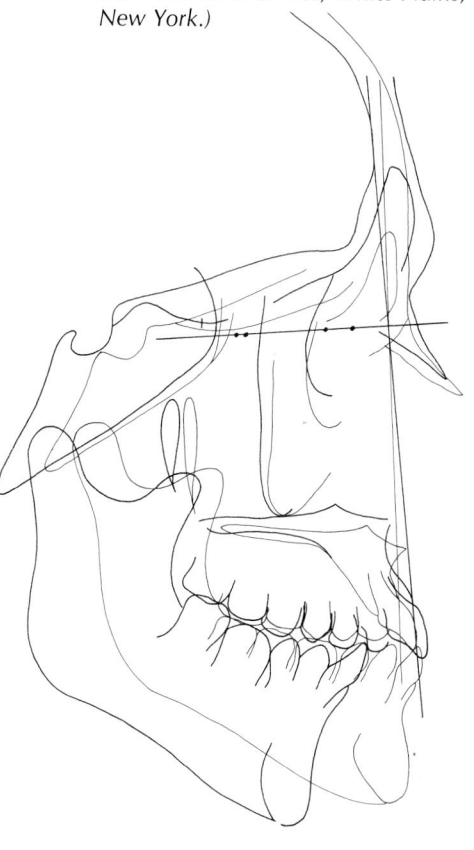

**FIGURES 4–45 AND 4–46**

In the first individual it is seen that the **anterior** part of the maxillary arch falls short of the inferior boundary down to which it is supposed to develop. The posterior corner of the bony arch falls on the 45° ear line intersection with the **PM** plane, which is where it should be. This vertical **hypo**plasia has contributed to the basis for the **anterior open bite** that exists. Or, in the second individual, the posterior part of the maxillary arch shows **hyper**plasia, and this has exceeded the capacity of the anterior portion of the arch to adjust. The posterior corner of the tuberosity is well below the normal limit set by the lower border of the brain and the 45° ear relationship (dashed lines), but prosthion itself is actually at its proper place on the horizontal dashed line. An anterior open bite is also produced, but the circumstances are different from those in the first individual. This "test" is usable only in older children when most or all of the vertical growth of the maxilla is complete.

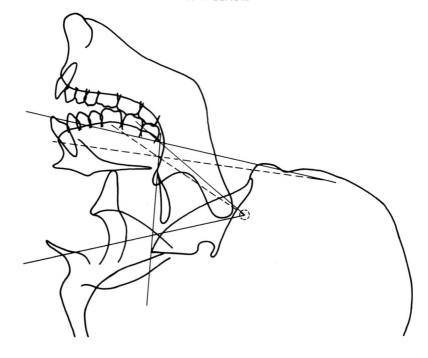

FIGURE 4-46

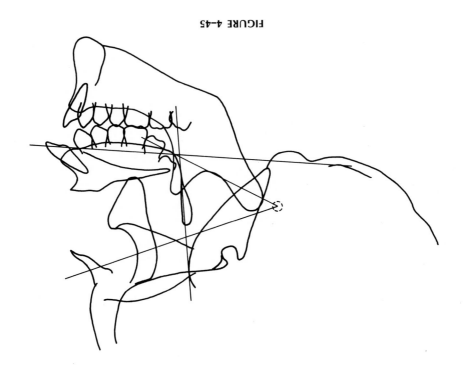

FIGURE 4-45

**FIGURES 4–47 AND 4–48**

In summary, the growth in each region of the face involves two basic factors: *(1)* the amount of growth by any given part, and *(2)* the direction of growth by that part. The brain establishes (or at least shares) the various **boundaries** that determine the amount of facial growth. This is because the floor of the cranium is the template upon which the face is constructed. The **directions** of regional growth among the different parts of the face are inseparably associated with the special sense organs housed within the face. These two factors establish a prescribed growth perimeter that defines the borders of the growth compartment occupied by the nasomaxillary complex. All the many components that constitute the midface, including the bones, muscles, mucosae, connective tissues, cartilage, nerves, vessels, tongue, teeth, and so on, contribute to a composite expression of growth the sum of which can produce enlargement up to the maximum, as determined by the midfacial growth boundaries. The growth of the **midface** is not limitless, and it is not independently and randomly determined entirely within itself.

Superior prosthion thus comes to lie in a predetermined position that has been programmed by the brain–cranial base–sense organ–soft tissue composite of developmental factors. Superior prosthion is comprised of alveolar bone, which is a highly labile and responsive type of bone tissue. Traditionally, this area of bone is regarded as quite unstable and subject to a wide range of variations according to the many forces that act on it. This is quite true, as will be seen below. However, prosthion has a specific target location that it will occupy if the growth process is not disturbed by intrinsic or extrinsic imbalances. The target point is not programmed within prosthion itself, or even just within the maxilla. It is determined, rather, by the composite of all the growth-establishing factors mentioned above. In most cases, prosthion will have settled in, when growth is complete, right on or very close to its target point.

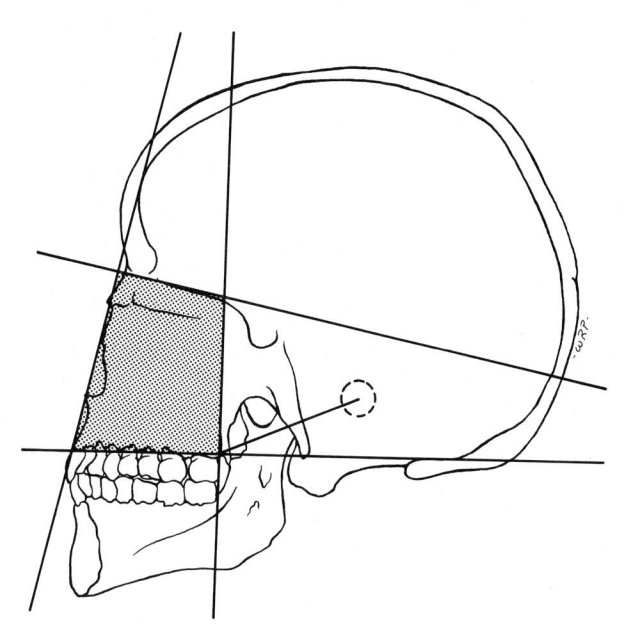

**FIGURE 4–47**

*(From Enlow, D. H., and M. Azuma: Functional growth boundaries in the human and mammalian face. In Bergsma, D. (Ed.):* Morphogenesis and Malformations of the Face and Brain. *Birth Defects Orig. Art. Ser., Vol. XI, No. 7. New York: Alan R. Liss, Inc. for The National Foundation—March of Dimes, White Plains, New York.)*

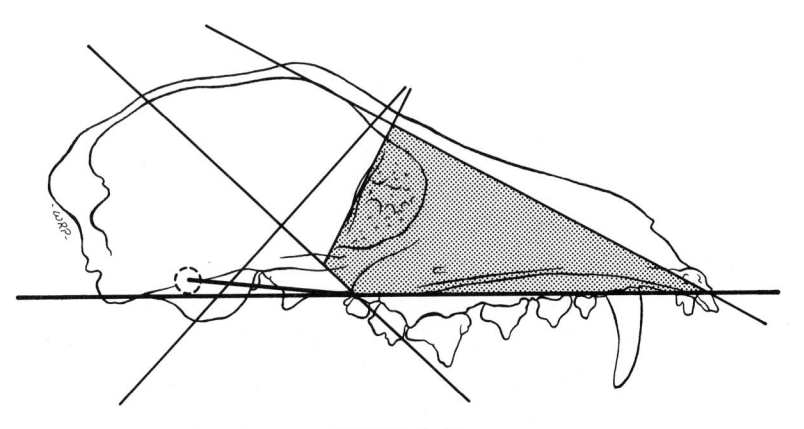

**FIGURE 4–48**

*(From Enlow, D. H., and M. Azuma: Functional growth boundaries in the human and mammalian face. In Bergsma, D. (Ed.):* Morphogenesis and Malformations of the Face and Brain. *Birth Defects Orig. Art. Ser., Vol. XI, No. 7. New York: Alan R. Liss, Inc. for The National Foundation—March of Dimes, White Plains, New York.)*

**FIGURES 4–49 AND 4–50**

In the headfilm shown in Figure 4–49 it is seen that prosthion falls short of the predetermined midfacial plane; growth is incomplete, however. In the same individual, when facial growth is largely completed, prosthion will have arrived at its place on the perpendicular adult (dashed) midfacial line. In Figure 4–50, the two headfilms are superimposed on the cribriform plane to show the before and after growth stages.

**FIGURE 4–51**

Can the brain–sense organ relationship with the face be violated? Of course; it frequently happens. For example, thumb-sucking and various developmental defects can move the teeth and alveolar bone to places that are out of bounds with respect to the normal growth process. The forces and factors of ordinary growth become overridden by extrinsic forces, and the prescribed boundary and the usual limit of growth are thereby overrun. However, this produces a structural and functional imbalance. If the overriding ectopic factors are removed, the normal balance of functional intrinsic forces work toward a greater or lesser return to the normal position, conforming with the natural anatomic boundary of the growth field.

Because many anatomic boundaries, large and small, exist throughout the face and cranium, the factor of boundary "security" is a major and important consideration to the clinician. If one given facial growth field is made to overrun the boundary of another field, either by clinical intervention or because of a developmental abnormality, one or the other will necessarily have to become compromised. A competition for the same space by the two overlapping growth fields occurs, and one field will necessarily become subordinate. This has great meaning with regard to the **stability** of a region and the functional "equilibrium" among different structural parts. If, for example, a given treatment procedure causes a violation of some growth boundary, will hard-earned treatment results be subsequently lost because functional stability and balance have been disturbed? Or, perhaps, will results be lost because the activity of a growth field that has been imposed upon subsequently causes a return ("rebound") toward the original structural pattern when treatment is stopped? Another similar question is whether a treatment procedure can actually **change** the long-term growth program. If it does not, subsequent growth, after treatment is ceased, can erase the treatment results because growth then proceeds along its original, unaffected course. These are fundamental clinical questions, and they have to do with priorities of growth control among the many fields of growth and the natural integrity of their boundaries. (Read the sections on "retention" in orthodontic texts.)

**FIGURE 4-49**

*(From Enlow, D. H., and J. McNamara: Angle Orthod., 43:256, 1973.)*

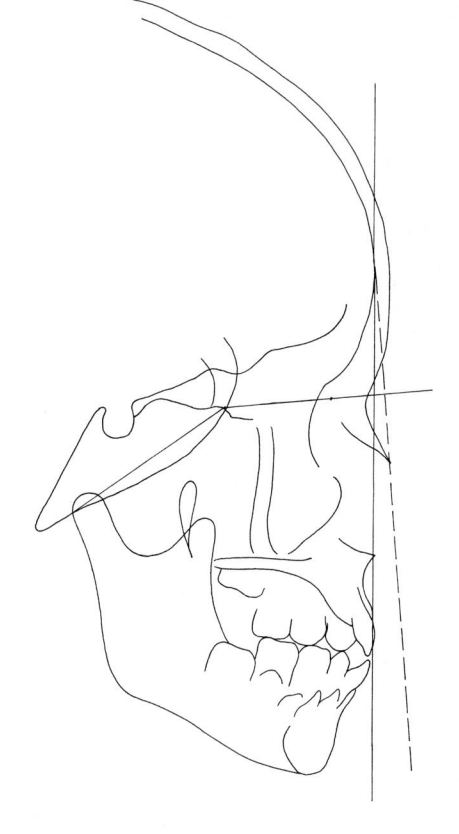

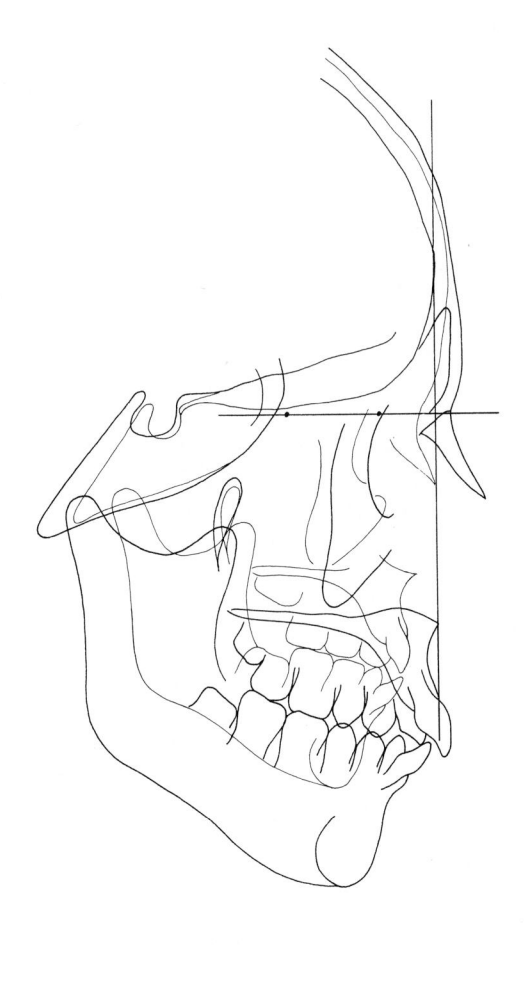

**FIGURE 4-50**

*(From Enlow, D. H., and J. McNamara: Angle Orthod., 43:256, 1973.)*

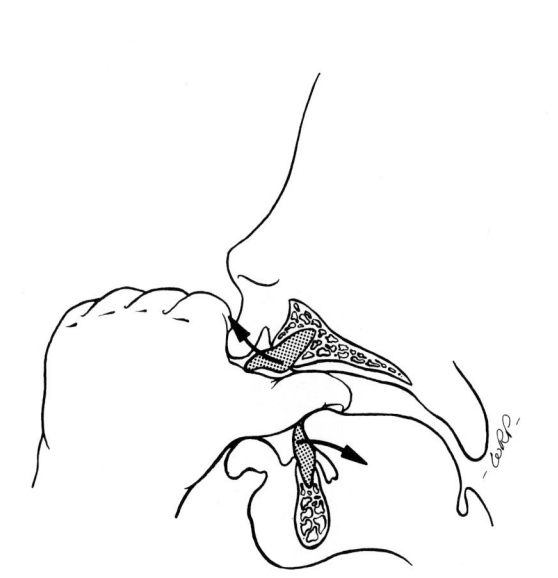

**FIGURE 4–51**

# Chapter 5

# Normal Variations in Facial Form and the Anatomic Basis for Malocclusions

# PART ONE

**Variation** is a basic law of biology. The pool of structural, functional, and genetic-based variations that are always present within a population of any species provides the capacity for adaptation to a changing environment. This increases the probability of survival for those individuals having features most suitable to the needs of the time. The human face, like most of our other "specialized" anatomic parts, certainly has its share of variations. Indeed, there are probably more basic, divergent kinds of facial patterns among humans than among the faces of most other species. This is because unusual facial and cranial rotations have occurred in relation to human brain expansion. A greater latitude for facial differences exists because the brain, proportionately, is so large and so variable in configuration. There is also a much greater likelihood for different kinds of malocclusions in the human face than in the faces of most other species for these same reasons. In fact, actual tendencies toward malocclusions are **built into** the basic design of our faces because of the unusual relationships that are inherent in its design.

**FIGURES 5-1 AND 5-2**

**Concept 1.** How does one "size up" a person's face to determine just what kinds and combinations of facial variations are present for that **individual** person? There are many sophisticated cephalometric "analyses" that can be used to appraise the structural details of any given face and cranium; these are briefly explained later. However, one quick and effective way to evaluate the general features of a face is simply to visualize that face as it would be represented by a skillful caricature artist. What are the topographic idiosyncrasies that the artist would seize upon to portray that person by **greatly exaggerating** these characteristics in cartoon style? The caricaturist always has a difficult time of it with a really well-balanced, attractive face, since there are no particularly special features to emphasize to make a caricature clearly recognizable. For most of us, on the other hand: Pug nose? Retrusive mandible? Sloping forehead? Heavy eyebrow ridges? Hollow cheekbone area? Moon-faced? Horse-faced (long)? Dish-faced (concave profile)? Mouse-faced (convex profile)? Flat-faced? Pointed mandible? Protruding teeth? Heavy chin? Aquiline nose? Narrow-set eyes? Criminal tendencies? High nasal bridge? Fat cheeks? And so on. Find a mirror.

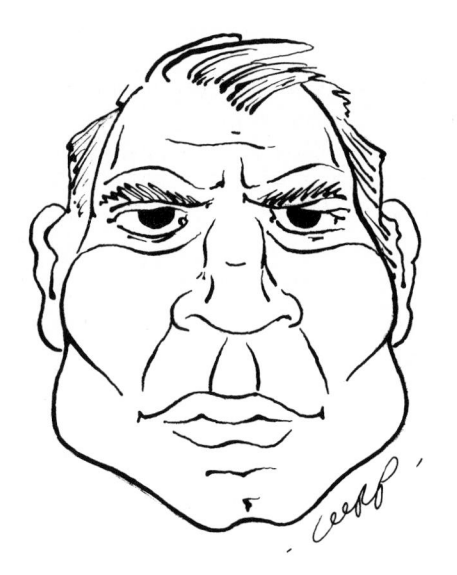

FIGURE 5–1

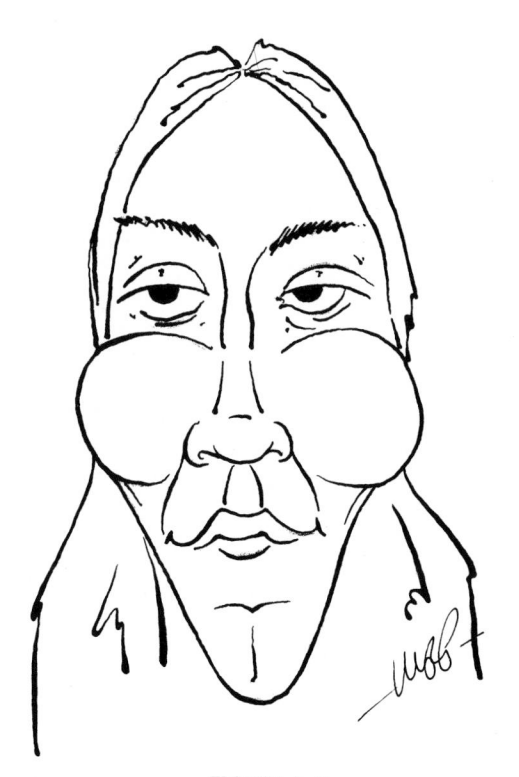

FIGURE 5–2

**FIGURES 5–3 AND 5–4**

**Concept 2.** There are two basic extremes in the shape of the head: *(1)* **dolichocephalic** and *(2)* **brachycephalic** (a third, mesocephalic, lies between). The oval-shaped dolichocephalic head form is horizontally long and relatively narrow, in contrast to the more rounded brachycephalic head form, which is horizontally shorter and broader. The **cephalic index** is the ratio between overall skull length and breadth: dolichocephalic, up to 75; mesocephalic, 75 to 80; and brachycephalic, over 80. Specific facial and occlusal types relate to these head form shapes, as explained below.

**FIGURE 5–5**

**Concept 3.** Three general types of facial profile exist: **orthognathic, retrognathic,** and **prognathic.** The orthognathic ("straight-jawed") form is the everyday standard for a good profile, and it is the type common to most Hollywood and TV big names. It is easy to "eyeball" a person's face, without actual need for headfilms or precision anthropometric measurements with instruments, to see what his or her profile type is. Simply visualize a line extending straight out from the center of the orbit looking straight forward with the head and this line angling neither upward nor downward (a). The head and body can be in any position, lying down, standing up, leaning over, and so forth, when visualizing this neutral orbital axis line. Now visualize a **vertical** line **perpendicular** to the orbital line extending down along the surface of the upper lip. This line will just touch the lower lip and the tip of the chin in a person with an orthognathic profile. (In many faces, also, the line will pass near the midpoint along the upper slope of the nose, but this feature is not required for an orthognathic type of face.) Time otherwise thrown away waiting around air terminals, sitting out classes, or standing in lines can be put to interesting use quietly studying people's profiles.

The retrognathic face has a characteristic convex-appearing profile. The tip of the chin lies somewhere behind the vertical line, and the lower lip is retrusive. The chin may be two or three centimeters behind the line in a severely retrognathic face (b). It is very common among Caucasians, however, to have about a half-centimeter or so of chin retrusion (c). The reason is explained later.

The prognathic face is characterized by a concave-appearing ("dished-in" midface) profile. The tip of the chin is protrusive and lies somewhere in front of the vertical line (d). The lower lip is forward of the upper. This kind of profile is far less frequent among Caucasians (in comparison with some other ethnic groups) than is the mandibular retrusive profile type.

**Concept 4.** There are four general categories of **occlusal** patterns: the "normal," the Class I, the Class II, and the Class III.

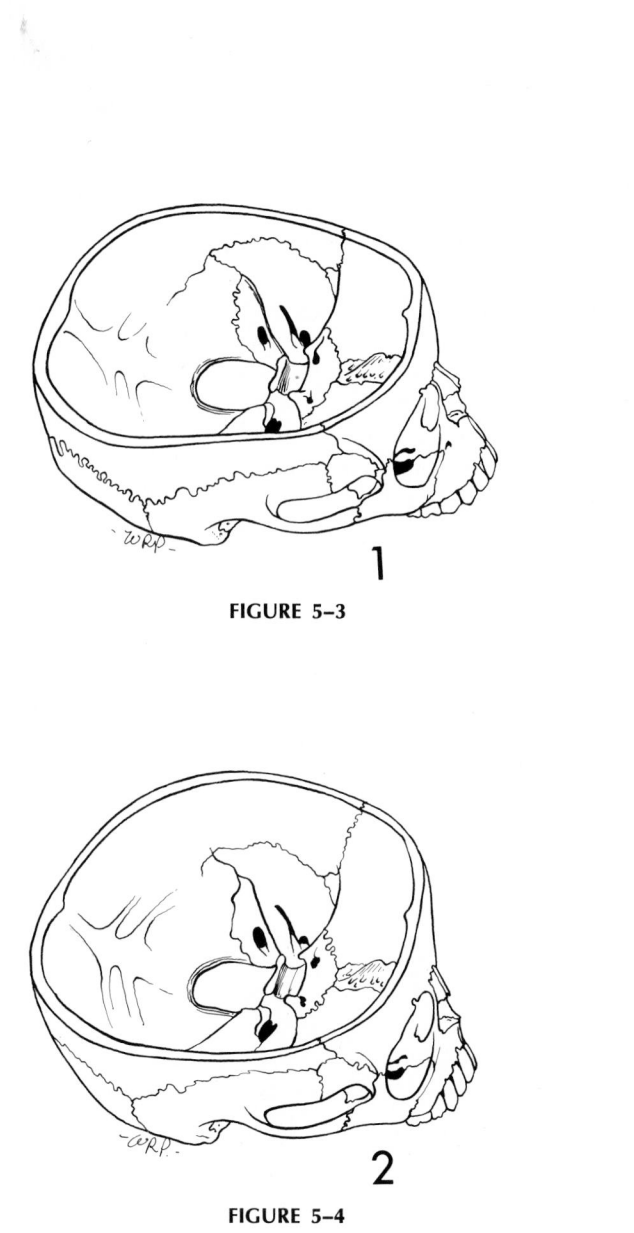

**FIGURE 5–3**

**FIGURE 5–4**

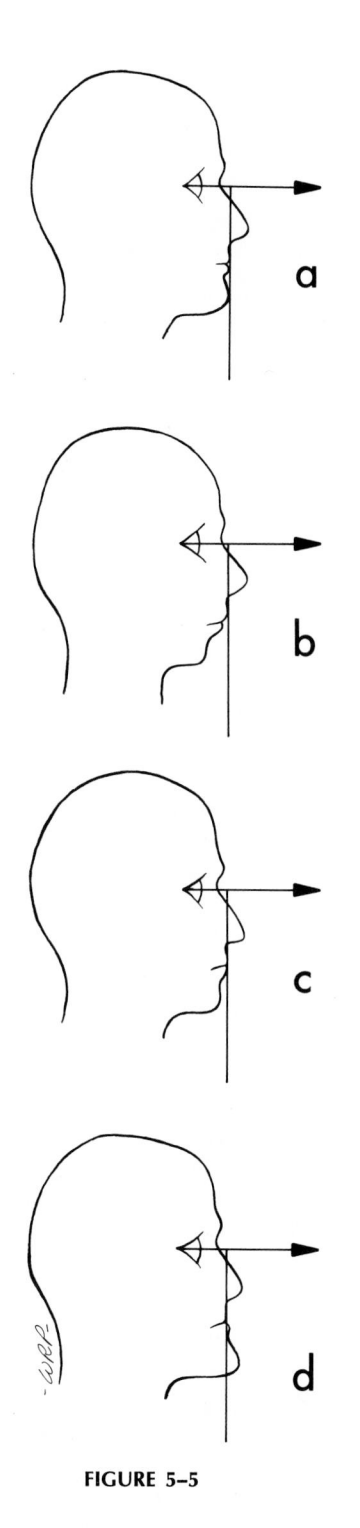

a

b

c

d

**FIGURE 5–5**

**FIGURE 5–6**

In the "normal" type of occlusion, all the many underlying skeletal and dental factors combine to place the upper and lower teeth in such a way that:

a. there is no undue amount of **overjet** (excessive maxillary incisor protrusion in front of the mandibular incisors);

b. all the teeth **interdigitate** perfectly (cusps fit with the grooves of antagonist teeth);

c. there is no undue amount of **overbite;** the maxillary front teeth should not overlap and cover the mandibular front teeth by more than about one-third the crown height of the lower incisors.

d. the maxillary canine is about one-half tooth width behind (distal to) the mandibular canine;

e. the maxillary first molar is about one-half cusp behind the mandibular first molar. That is, the mesiobuccal cusp of the upper should occlude with the mesiobuccal groove of the lower molar. This is the familiar "molar relationship" much used in appraising malocclusions. Note that overall mandibular arch length is shorter than maxillary arch length in the normal occlusion and that the more posterior positioning of the maxillary molars accommodates the larger size of the upper incisors. Normal differences also exist in root alignment; the incisor roots tip lingually, the canine roots tip distally, and the molar and premolar roots are essentially vertical.

All these dental and skeletal ("skeletal" because the various bones participate in placing the teeth) requirements are tough to meet. The majority of the population, in fact, does not have "normal" dentition. Most of us have one kind of malocclusion or another, however slight or severe in extent. There are various kinds of malocclusions which fall into one of three general categories. This system was first devised by Edward Angle and is thus called the **Angle classification.** Beginners can be pardoned for presuming that the word "Angle" refers to angles.

**FIGURE 5–7**

The **Class I** malocclusion is a less severe type, and it involves largely dental (rather than "skeletal") variations from the ideal. The molar relationship is normal (at least in the older child), and disharmonies usually involve the crowding of anterior teeth. The profile is usually good, although a few millimeters of retrognathia may normally be present. Overjet is not excessive. A variation of the Class I occlusion involves "bimaxillary protrusion." The mandible tends to be slightly long horizontally, and this causes a forward tipping (proclination) of the upper incisors. The result is a protrusion of **both** the upper and lower incisor regions. This gives a noticeably "full" appearance to the mouth.

**FIGURES 5–8, 5–9, AND 5–10**

The **Class II** malocclusion is skeletally as well as dentally based. The various bones (including those of the cranial base) cause a positioning of the teeth in such a way that a "Class II molar relationship" exists. The maxillary first molar lies either directly over or in front of the mandibular first molar, rather than slightly behind where it should be. In the most common variety ("Division 1") of the Class II type of malocclusion (Fig. 5–9), the maxillary incisors are quite protrusive with excessive overjet, and the profile is distinctly retrognathic. In another variety ("Division 2"), however, overjet is not pronounced, but a deep bite (overbite) and an outward flaring of the lateral maxillary incisors occur (Fig. 5–10). Both types have a Class II molar relationship to varying extents.

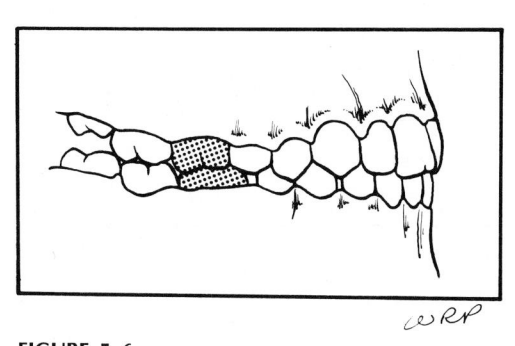

**FIGURE 5–6**

*(Adapted from Gianelly, A. A., and H. M. Goldman:* Biologic Basis of Orthodontics. *Philadelphia, Lea & Febiger, 1971.)*

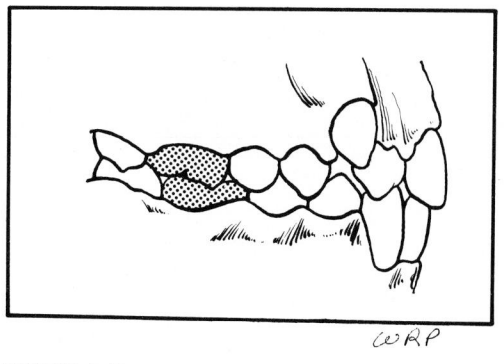

**FIGURE 5–7**

*(Adapted from Gianelly, A. A., and H. M. Goldman:* Biologic Basis of Orthodontics. *Philadelphia, Lea & Febiger, 1971.)*

**FIGURE 5–8**

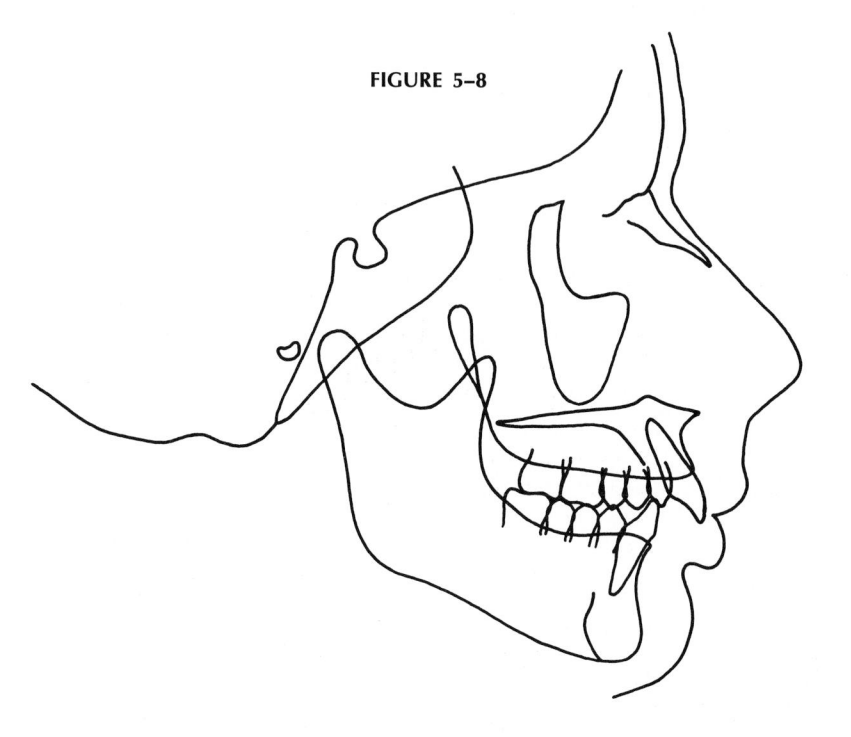

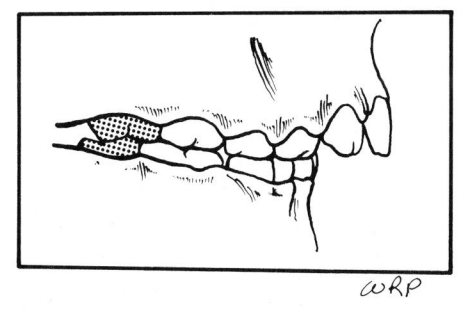

**FIGURE 5–9**

*(Adapted from Gianelly, A. A., and H. M. Goldman:* Biologic Basis of Orthodontics. *Philadelphia, Lea & Febiger, 1971.)*

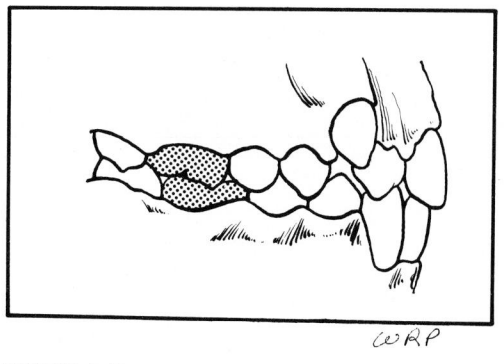

**FIGURE 5–10**

*(Adapted from Gianelly, A. A., and H. M. Goldman:* Biologic Basis of Orthodontics. *Philadelphia, Lea & Febiger, 1971.)*

**FIGURES 5–11 AND 5–12**

The Class III malocclusion is characterized by a marked protrusion of the mandible, a prognathic profile, and a molar relationship in which the lower first molar lies ahead ("mesial") of the normal position. This is largely a skeletally based type of occlusal variation.

**Concept 5.**   Three separate systems of classification have been described above. The first relates to the shape of the whole skull, the second to the facial profile, and the third to occlusion. Is there a direct developmental and structural relationship among them? Yes; the reasons are explained below. To introduce these correlations the following interesting questions are asked. First, Class II skeletal malocclusions are by far the most common among Caucasian individuals. Why? Most Caucasians have a tendency toward retrognathia. Why? Among Oriental individuals the Class III type is much more common. There is a stronger tendency toward an orthognathic profile. Why?

**FIGURES 5–13, 5–14, AND 5–15**

In individuals (or whole populations) having a **dolichocephalic** head form, the brain is horizontally long and relatively narrow. This sets up a cranial base that is somewhat more flat; that is, the flexure between the middle cranial floor and the anterior cranial floor is more open. It is also horizontally longer. These factors have several basic consequences for the pattern of the face. First, the whole nasomaxillary complex is placed in a more **protrusive** position relative to the mandible because of the horizontally longer anterior and middle segments of the cranial floor. Second, the whole nasomaxillary complex is lowered relative to the mandibular condyle. This causes a downward and **backward** rotation of the entire mandible. Third, the occlusal plane becomes rotated into a downward-inclined alignment. The **two-way** forward placement of the maxilla and backward placement of the mandibular corpus results in a tendency toward mandibular retrusion, and the placement of the molars results in a tendency toward a Class II position. The profile tends to be retrognathic. However, compensatory changes are usually operative, as explained later. Because of the more open cranial base angle and the resultant trajectory of the spinal cord into the cervical region, this type of face is associated with individuals having a greater tendency toward a somewhat stooped posture and anterior inclination of the head and neck.

**FIGURES 5–16 AND 5–17**

Individuals or ethnic groups with a **brachycephalic** head form have a rounder and wider brain. This sets up a cranial base that is more upright and has a more closed flexure, which decreases the effective horizontal dimension of the middle cranial fossa. The result is a relative retrusion of the nasomaxilla and a more forward relative placement of the entire mandible. This causes a greater tendency toward a prognathic profile and a Class III molar relationship. The occlusal plane as well as the ramus of the mandible may be aligned upward, but various compensatory processes usually result in either a perpendicular or a downward-inclined occlusal

**FIGURE 5–11**

*(Adapted from Gianelly, A. A., and H. M. Goldman:* Biologic Basis of Orthodontics. *Philadelphia, Lea & Febiger, 1971.)*

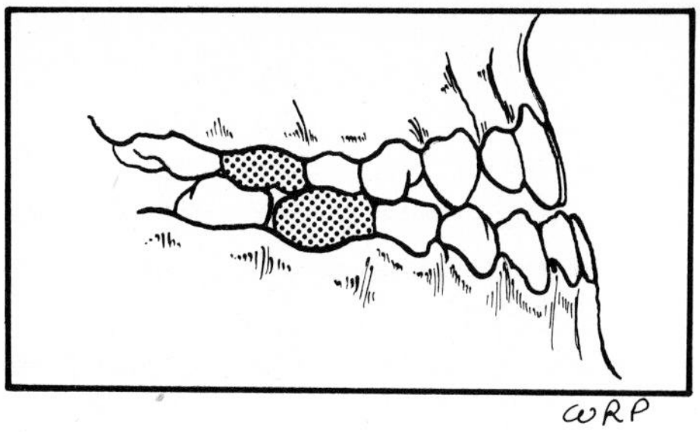

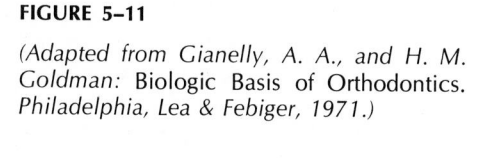

**FIGURE 5–12**

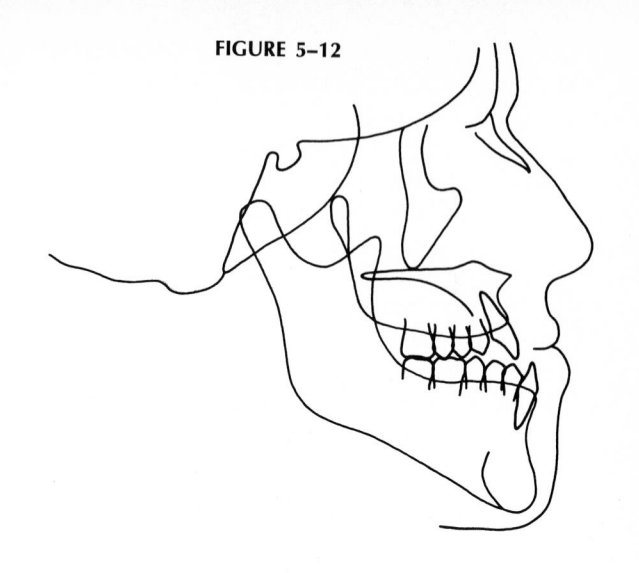

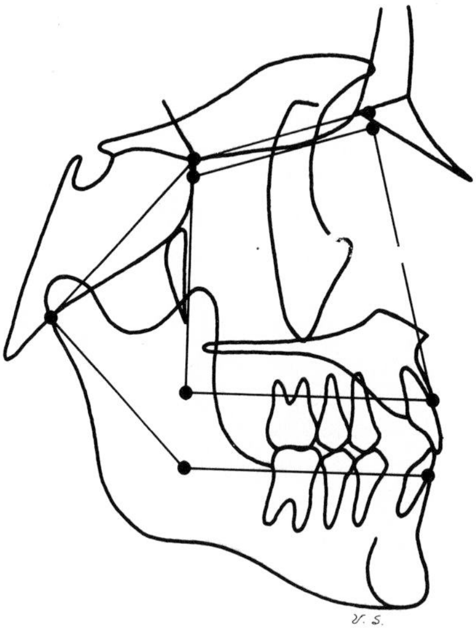

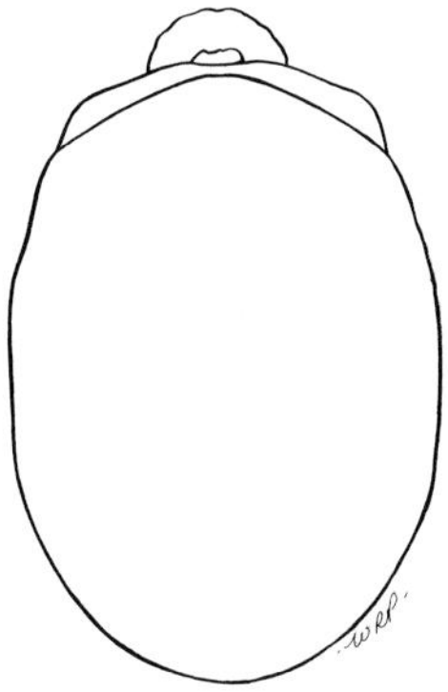

**FIGURE 5–13**

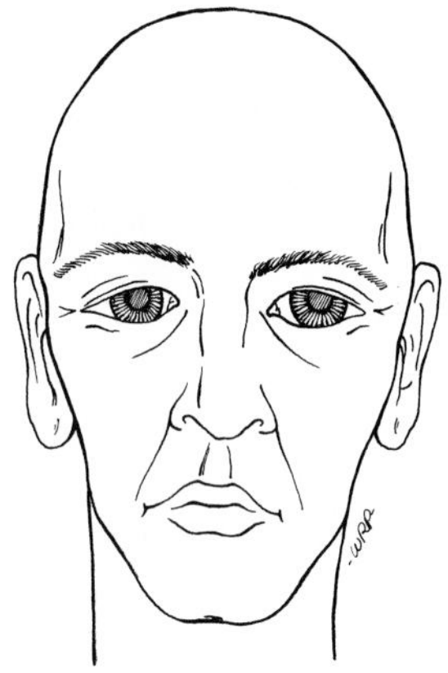

**FIGURE 5–14**

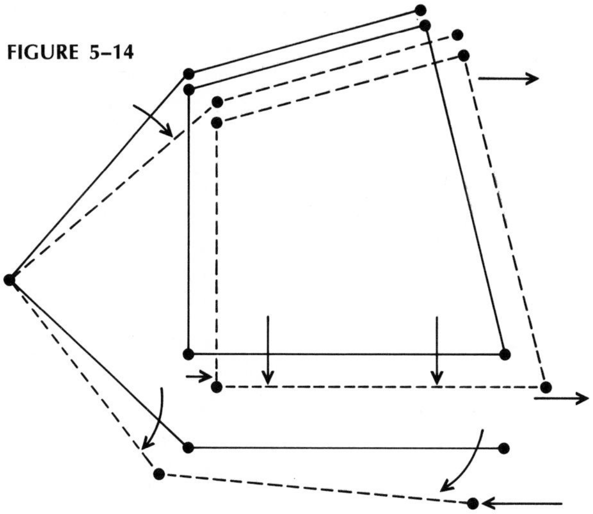

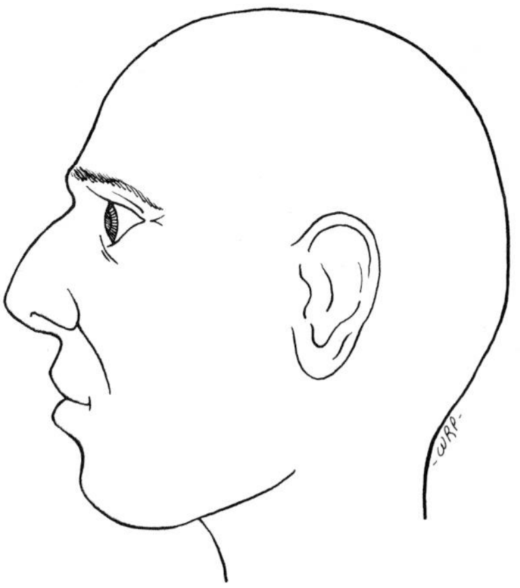

**FIGURE 5–15**

plane and slight backward rotation of the ramus. Other compensatory changes are also operative, as explained next, and these tend to counteract the built-in Class III tendencies. Because of the more upright middle cranial fossa and the more vertical trajectory of the spinal cord, individuals with all these various facial features also have a tendency for a more erect posture with the head in a more "military" (at braced attention) position.

**Concept 6.**    The basic nature of interrelationships among *(1)* brain form, *(2)* facial profile, and (3) occlusal type, as just seen, causes a predisposition toward character-istic facial types and malocclusions among different types of populations. Some Englishmen, for example, or the French and some other Europeans with a tendency for a dolichocephalic head form, have a corresponding **tendency** toward Class II malocclusions and a retrognathic profile. The Japanese, having mostly a brachyce-phalic head form, have a correspondingly greater tendency toward Class III malocclusions and a prognathic profile. These respective tendencies are built into the basic plan of facial construction. However, most of us also have intrinsic struc-tural features that have compensated for these tendencies. **If** we have such com-pensatory features, the built-in tendencies are offset, to a greater or lesser extent, and we thereby have at least reasonable facial proportions with a Class I occlusion, even though the underlying tendencies are still present. **If** these compensatory fea-tures do not develop, however, or if they are insufficient, the built-in tendencies then become expressed, and we have a more or less severe malocclusion and a greater extent of retrognathia or prognathia.

**FIGURE 5–18**

**Concept 7.**    How does a face undergo intrinsic compensations during its develop-ment? One example that is very common is shown here (see Part Two for others). In Concept 5 above, the mandible was placed in a retrusive (retrognathic) position owing to its downward and backward rotation resulting from the more open type of cranial base flexure. The mandibular **ramus,** however, can compensate by an increase in its horizontal dimension. This places the whole mandibular arch an-teriorly into a proper position beneath the maxilla, and it positions the teeth in a "normal" or a Class I type of molar relationship. The mandibular retrusion that would otherwise be present thus becomes partially or completely eliminated, and a profile in which the chin lies on or within a half-centimeter or so of the orthognathic profile line results. The downward mandibular rotation is compensated for by an upward drift of the anterior mandibular teeth and a downward drift of the anterior maxillary teeth. This causes a curved occlusal plane, the "curve of Spee" (see Part Two for details).

The face on each of us, virtually without exception, is the composite of a great many regional "imbalances." Some of these offset and partially or completely counteract the effects of the others. The wide ramus cited above, for example, is ac-tually an imbalance, but it serves to reduce, as a normal compensatory process, the effects of some other angular or dimensional imbalances caused by the built-in ten-dencies toward malocclusions. The particular feature of a wide ramus is very com-mon among dolichocephalic Caucasians. When this and other compensatory fac-tors are present, the underlying stacked deck toward retrognathia and a Class II malocclusion is removed or made less severe. Thus, many of us have a slightly retrognathic profile and a little anterior tooth crowding.

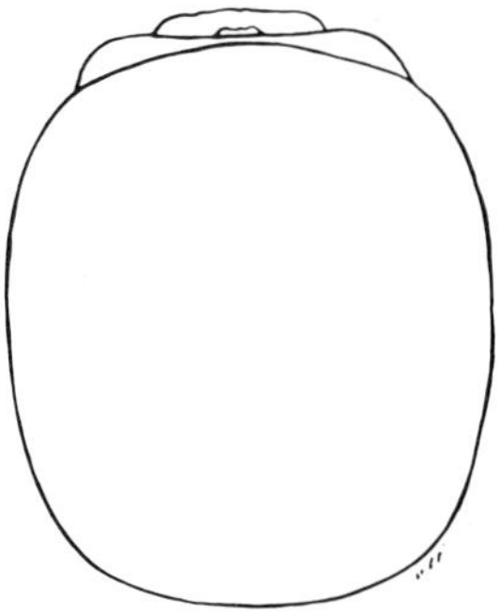

FIGURE 5–16

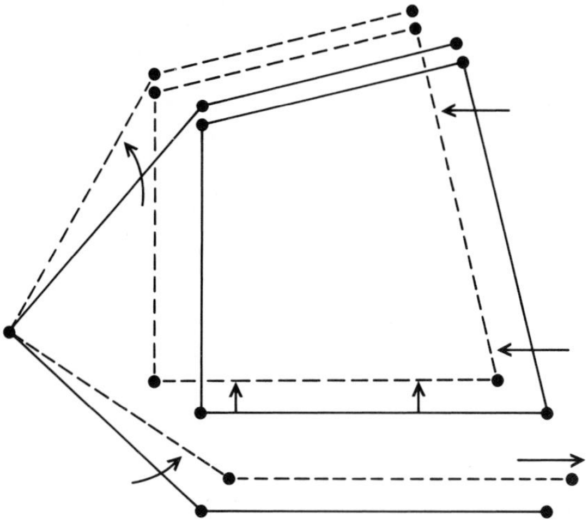

FIGURE 5–17

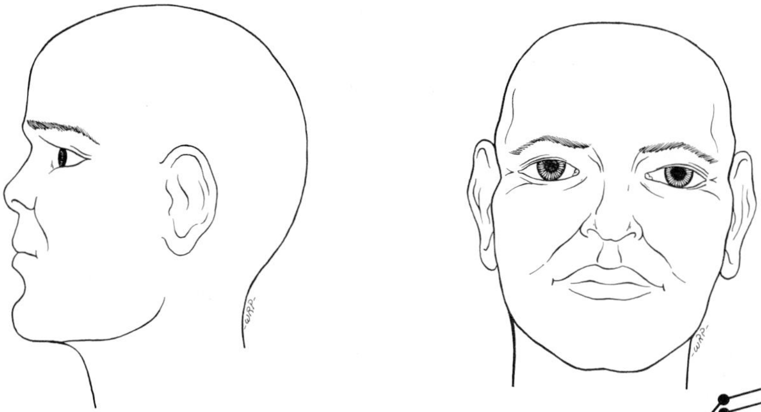

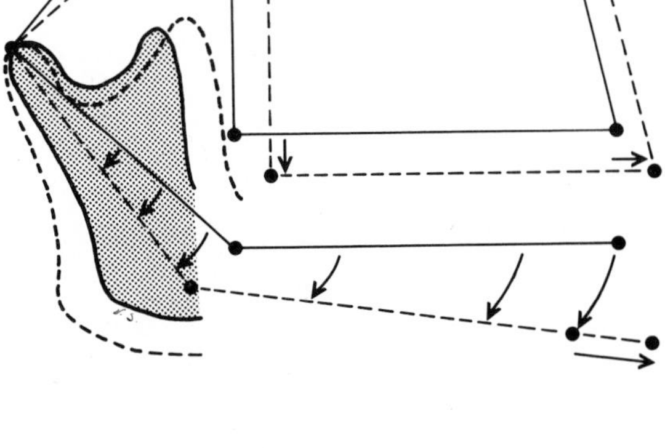

FIGURE 5–18

# PART TWO

In this section, specific cause and effect relationships underlying differences in facial pattern are explained. Each regional area throughout the face and cranium is considered separately. To evaluate the structural and developmental situation for each given region, a simple test is used: that region is compared with other regions with which it must "fit." If they do not have a good fit, the resultant effect is appraised by noting whether it causes *(1)* a mandibular retrusive or *(2)* a mandibular protrusive effect. As will be seen, imbalances in many parts of the head are passed on, region by region, and affect the placement of the jaws and the resultant nature of the occlusion.

### FIGURE 5–19

Importantly, two basic factors must be considered for each region. The first is the **dimension** of a particular part. Is it "long" or is it "short" with regard to its fitting with other parts? Parts *a* and *b* should "fit," as should parts *c* and *d*. If, however, part *b'* is short relative to *a'*, it causes part *d'* to be "retrusive," even though it actually matches the dimension of *c'*.

Great care must be used to evaluate only that particular span or dimension of a bone specifically involved in the actual, direct fitting of the bones one to another. This is the **effective dimension.** For example, bone *m* relates only to dimension *n*; dimension *o*, even though part of the same bone, is not involved in the counterpart relationship with *m*. Or, the "effective" straight-line dimension of oblique part *y* is actually the horizontal dimension *z*, which relates to the composite fit with separate bone *x*.

### FIGURE 5–20

The second fundamental consideration is the **alignment** of any given part. This must also be included (although many cephalometric studies do not), because any rotational change either increases or decreases the **expression** of a dimension. For example, separate parts *a* and *c* match, although their real oblique lengths are different; the straight-line parallel dimensions *b* and *d* are equivalent.

### FIGURE 5–21

Note what happens, however, if the **alignment** of bone *c* is changed to *d'*. Its actual dimension is exactly the same, but the **expression** of this dimension has been altered; the expressed horizontal dimension has been increased, and the expressed vertical has necessarily been decreased at the same time. Even though the actual dimensions have not been changed, parts *a* and *c* no longer fit because their effective dimensions do not now match. If part *c* is aligned (rotated) to *e*, the expressed horizontal dimension is decreased, but the expressed vertical is increased at the same time.

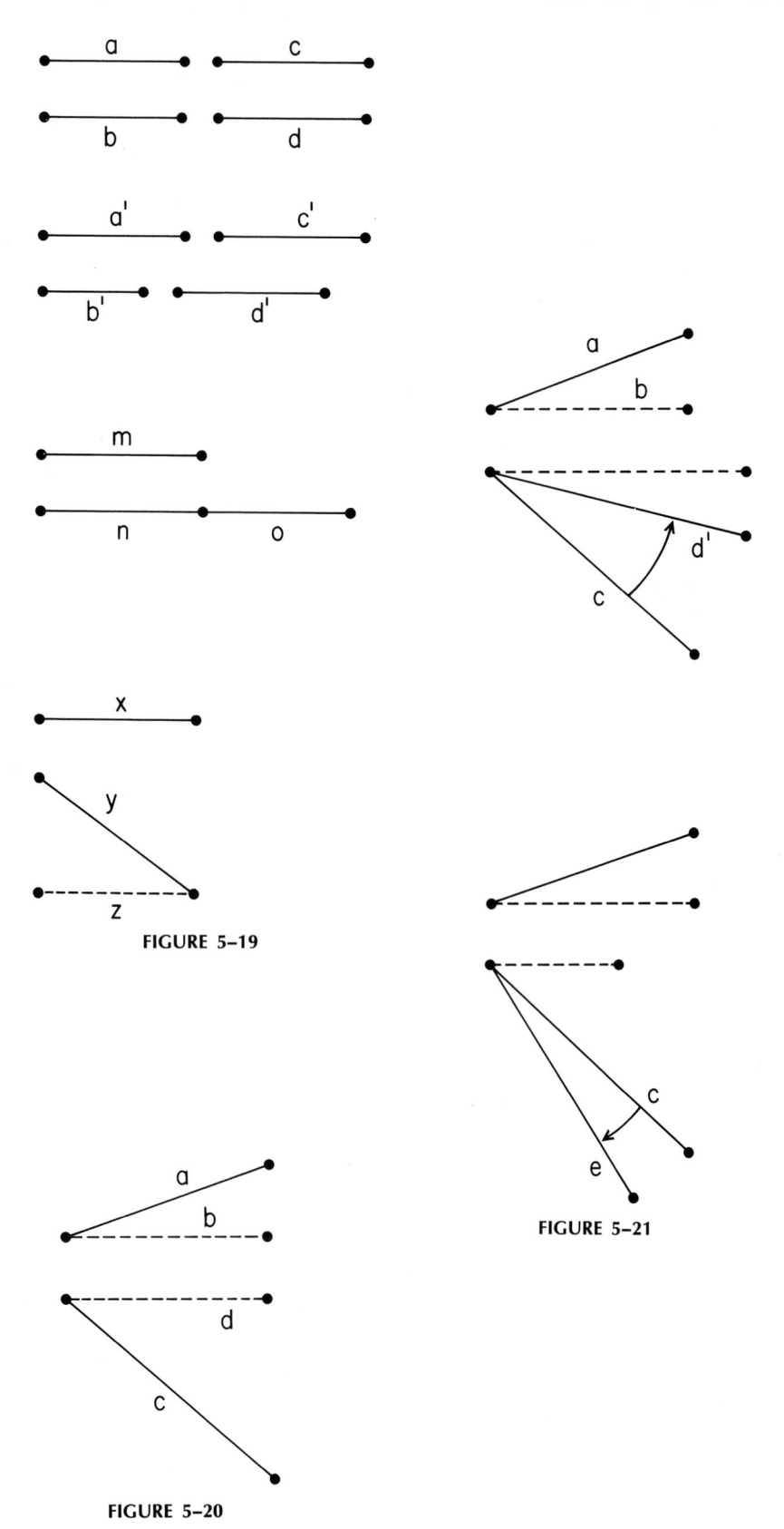

FIGURE 5–19

FIGURE 5–20

FIGURE 5–21

**FIGURE 5–22**

To illustrate the important effects of **alignment** as a basic factor involved in determining facial pattern, the alignment of the middle cranial fossa in this Class II child was changed (on paper) to a more upright position. All the other facial regions, including the mandible, maxilla, and the anterior cranial fossa, were then reassembled around the realigned middle cranial fossa. **No** changes in the actual dimensions of any of the parts were made. The horizontal and vertical **expression** of the middle cranial fossa dimension, however, resulted in a change from the Class II pattern into a Class I pattern, even though all the individual bones were exactly the same size.

**FIGURES 5–23 AND 5–24**

If horizontal dimension of the mandibular corpus *(b)* is short relative to its counterpart, the bony maxillary arch *(a)*, the effect is, of course, mandibular retrusion (probably with anterior crowding of the teeth). Note that this does not necessarily cause a Class II molar relationship, because the posterior parts of the upper and lower bony arches can still be properly aligned. It is emphasized that these are **relative** comparisons between two parts within the **same** individual. The mandible is not being compared with a norm or an average value derived from a population sample. Whatever the actual value of this mandibular dimension happens to be in millimeters, or regardless of how it compares to some statistical mean, it is short when compared with the dimensional value that really matters: its counterpart, the horizontal dimension of the maxillary body in that particular individual.

**FIGURES 5–25 AND 5–26**

If the mandibular corpus is dimensionally long, the effect, of course, is mandibular protrusion. A horizontally short maxillary arch has the same effect. (There are anatomic ways to tell which is long and which is short, as explained in Chapter 9.) Whether or not a long corpus produces a Class III molar relationship depends on whether it is long mesial or distal to the first molars.

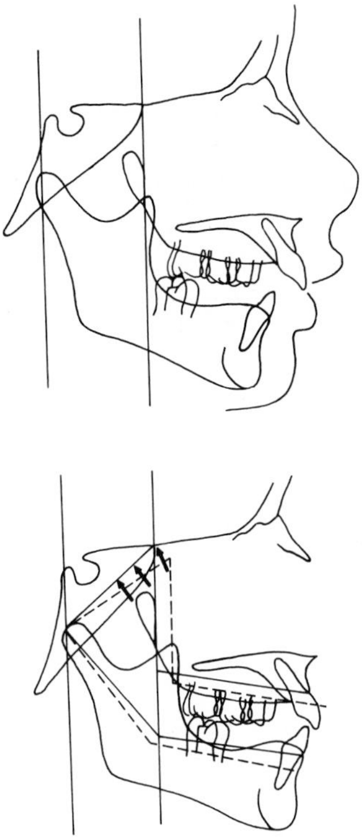

FIGURE 5–22

FIGURE 5–23

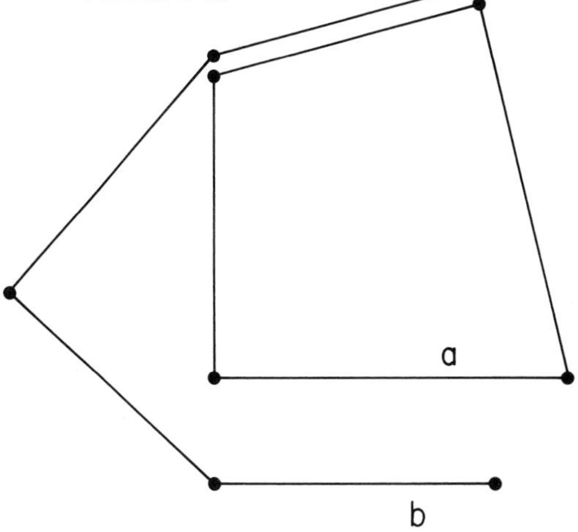

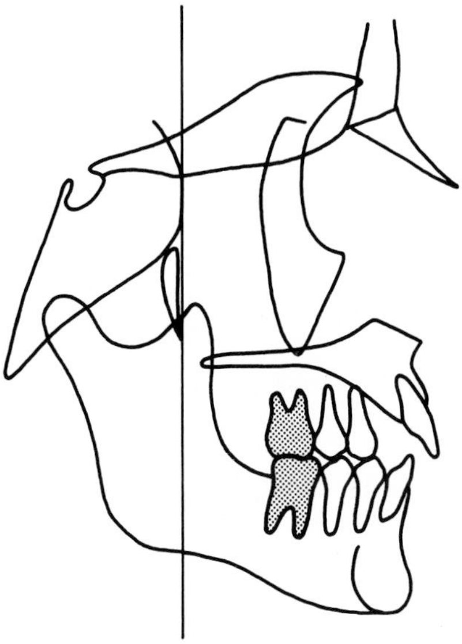

FIGURE 5–24

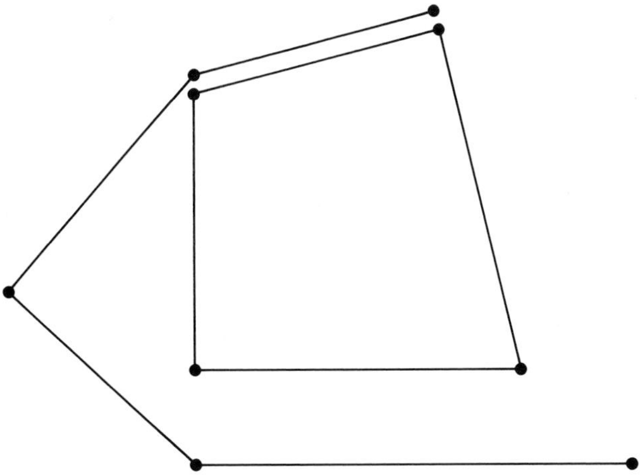

FIGURE 5–25

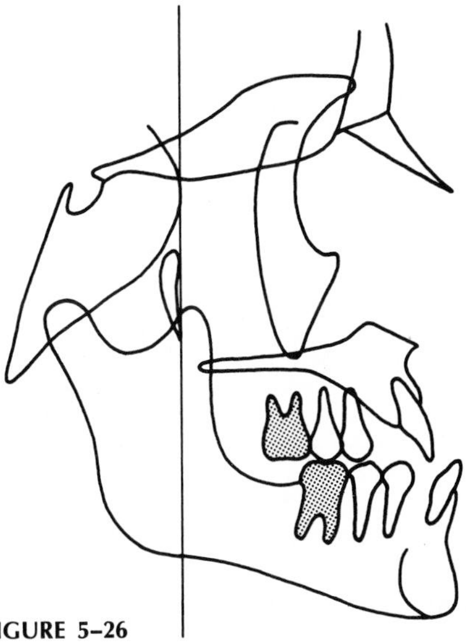

FIGURE 5–26

**FIGURE 5–27**

In this situation, the upper part of the nasomaxillary complex is horizontally long **relative** to its counterparts, the anterior cranial fossa, the palate, and the maxillary and mandibular arches. Note that this has no effect on the occlusion. The individual can **appear** retrognathic, but this is a result of the protrusive nature of the upper part of the face and not the jaws themselves. Because the superior part of the ethmomaxillary region is protrusive, the outer table of the frontal bone is carried with it. The result is a sizable frontal sinus, heavy eyebrow ridge and glabella, sloping forehead, high nasal bridge, and a long nose. The cheekbone area appears retrusive because of the prominent nasal region.

**FIGURE 5–28**

If this upper part of the nasomaxillary complex is quite protrusive, the upper edge of the nose will often be curved or bent into a classic aquiline (*L., aquila,* eagle, as for eagle beak), Roman nose, or Dick Tracy configuration **if** the nose is **also** vertically long. The longer the vertical dimension of the nose, the more its slope must bend. This nasal shape is quite common in some European population groups, such as the French, and typically has a rather narrow and sharp configuration. The ventral edge of this nose type may be horizontal but often has a tendency to tip downward, in contrast to the vertically shorter type of nose in which the lower margin can angle upward. In another type of nasal bending, the **lower** part of the nasal region, just above the maxillary arch, may be quite protrusive; this produces a characteristic and graceful recurved configuration of the nasal slope as the lower portion grades and curves back onto the less protrusive upper part. The cheekbone area in this type of face is often notably prominent because this entire level of the midface also tends to be protrusive. All the above facial features, in general, characterize the long, narrow-faced, dolichocephalic head form found among most (but not all) Caucasian groups, and the dinaric type of head form. The extent of some of these features is sex- and age-related, such as frontal sinus expansion and the slope of the forehead.

**FIGURE 5–29**

If the upper part of the nasomaxillary complex is **not** protrusive, so that its horizontal dimension more nearly matches counterpart dimensions in the anterior cranial fossa, palate, and maxillary and mandibular arches, quite a different facial profile results. The frontal sinuses are comparatively smaller, the forehead is more upright, the eyebrow ridges and glabella are not as prominent, the nose is not nearly as protrusive, and the nasal bridge is much lower. The jaws appear more prominent because the upper nasal region is less protrusive. The cheekbones also appear more prominent for the same reason. The whole face is much flatter. This composite of facial features is typically found in the broad-faced, brachycephalic type of head form that characterizes many Oriental individuals. Some Caucasian populations are also broad-faced with a shorter nose, more prominent mandible, lower nasal bridge, and so forth, including, for example, many individuals of southern Irish descent.

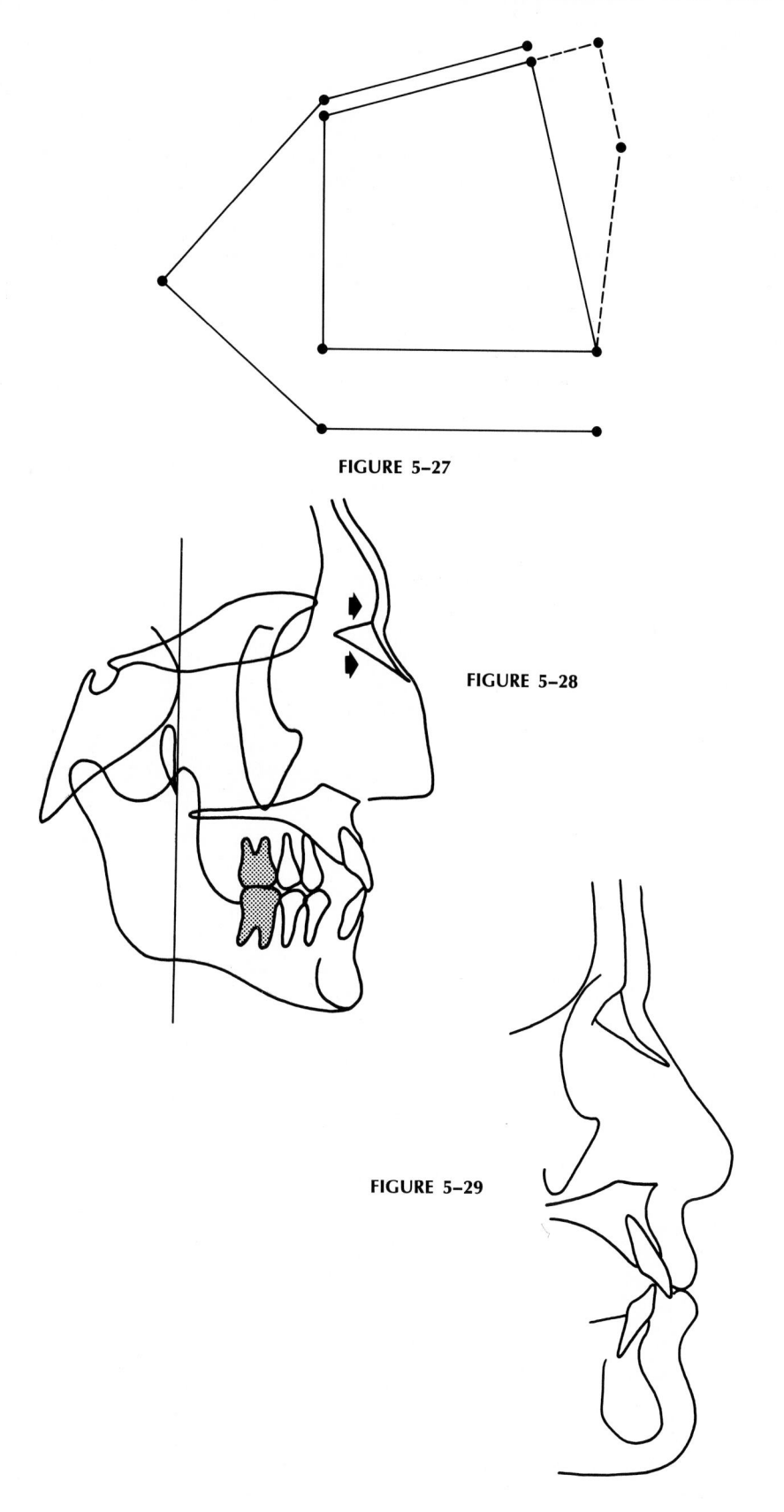

FIGURE 5-27

FIGURE 5-28

FIGURE 5-29

**FIGURE 5–30**

If the effective horizontal (not oblique) dimension of the ramus is narrow relative to its counterpart, which is the effective horizontal (not oblique) dimension of the middle cranial fossa, a mandibular retrusive effect is produced. Note that the mandibular arch lies in a resultant **offset position** relative to its counterpart, the maxillary arch. Even though the upper and lower arches themselves are actually matched in dimensions, the profile is retrognathic. The arches are in offset positions because the parts **behind** them are "imbalanced." Note that the posterior part of the maxillary arch lies well anterior (mesial) to the posterior part of the mandibular arch. This is one (of several) of the basic skeletal causes that underlie a **Class II molar relationships.** Remember, the "real" anatomic junction between the ramus and corpus is the lingual tuberosity rather than the oblique "anterior border" where it overlaps the corpus. Because this structure cannot be directly visualized in head films, it is not represented here. However, it is located distal to the vertical reference line because of the narrow ramus in this individual.

**FIGURE 5–31**

In this situation, the effective horizontal (not oblique) dimension of the ramus is broad relative to the middle cranial fossa. Or, the cranial fossa is horizontally narrow relative to the ramus (either way because this is a relative comparison). The effect is mandibular protrusion due to the resultant offset positions between the upper and lower arches, even though the horizontal dimensions of the arches themselves match. This is one (of several) of the basic skeletal causes for a **Class III molar relationship.** The lingual tuberosity (not shown) is mesial to the vertical reference line.

**FIGURES 5–32 AND 5–33**

If the ramus has a more upright alignment (as a result, for example, of a vertically long nasomaxillary region), the effect is mandibular retrusion. While this increases the expression of the vertical dimension, the horizontal is necessarily decreased at the same time. The whole mandible is rotated downward and **backward.** As a result, the mandibular arch becomes offset relative to the upper arch. The profile is retrognathic, and the offset placement of the arches causes a Class II molar relationship. Note that the mandibular corpus is rotated downward, causing a downward-inclined mandibular occlusal plane (see page 218 for an explanation of dental compensations).

**FIGURE 5–34**

If the ramus has a more forward-inclined alignment (as a result of a vertically short midface), the effect is mandibular protrusion because the expression of the horizontal dimension is increased. The vertical is decreased. Or, stated simply, the ramus rotates forward and upward, causing the mandible to protrude. The arches are offset, and the molars have a resultant Class III relationship. The occlusal plane has an upward inclination relative to the neutral orbital axis or to the maxillary tuberosity (that is, the vertical **PM** line). (The posterior teeth drift inferiorly and/or the gonial angle opens to provide proper occlusion.)

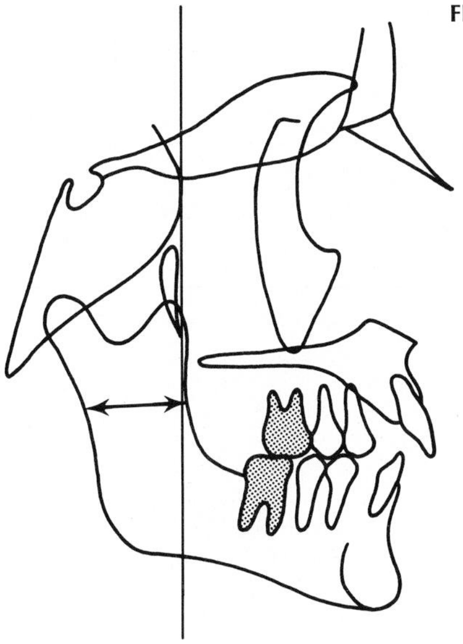

FIGURE 5–30

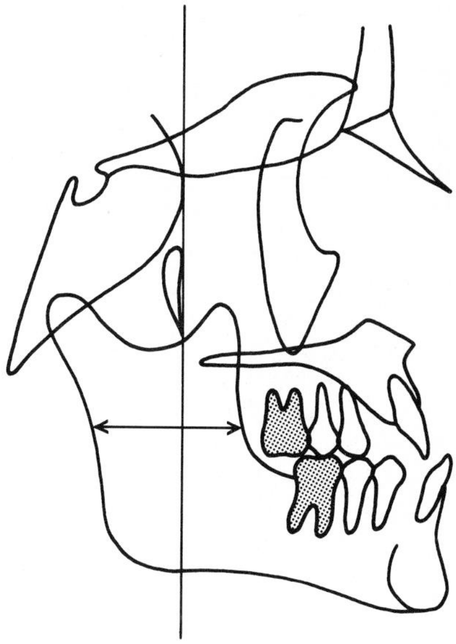

FIGURE 5–31

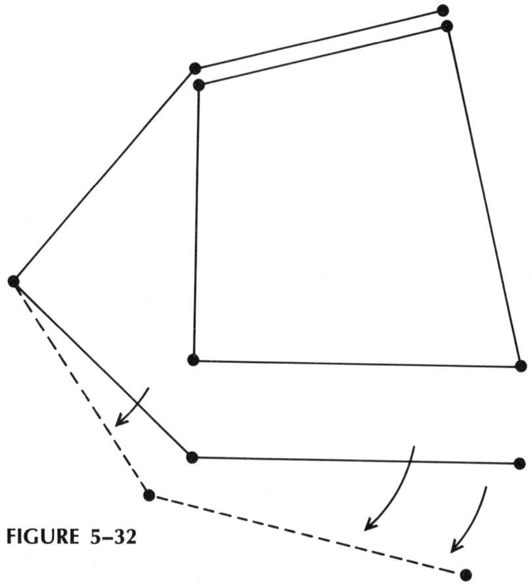

FIGURE 5–32

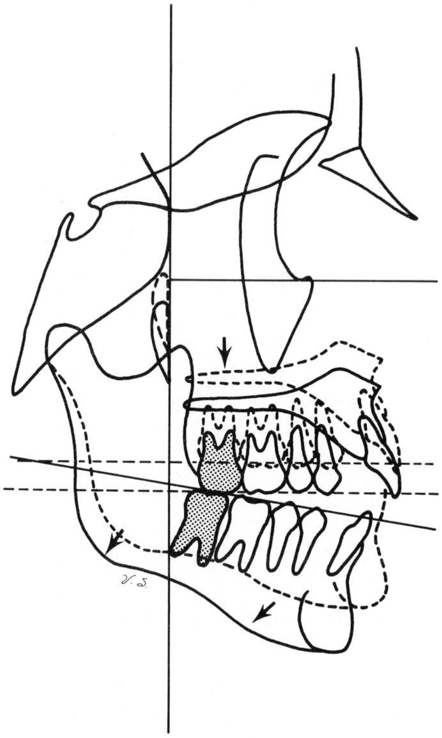

FIGURE 5–33

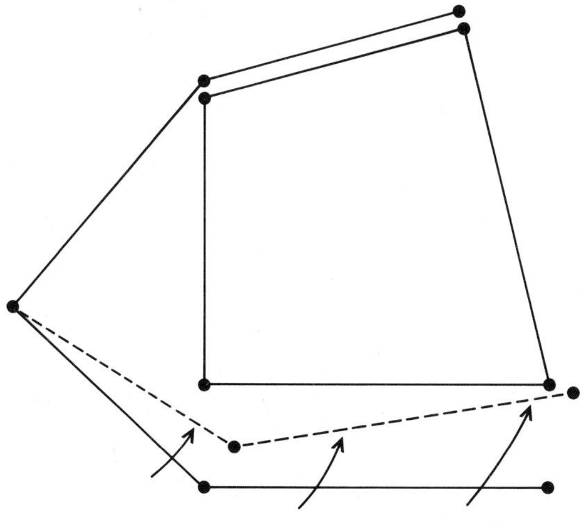

FIGURE 5–34

If the **corpus** (not ramus) has an upward alignment, a mandibular **retrusive** effect is produced. These relationships have often been misunderstood, and the whole subject of mandibular "rotations" has been perplexing to many workers. There are two basic and separate kinds of mandibular skeletal rotations (exclusive of dentition rotations, which will be described separately).

1. The **whole mandible** rotates up or down at the condylar pivot. This was described above as a "ramus" rotation because it involves changes in ramus alignment at the condylar pivot. The corpus is carried with it. The primary reason why this kind of rotation takes place is to adjust the ramus, and thereby the corpus, to whatever vertical position exists for the midface. The ramus rotates forward and upward to meet a short midface or an upright cranial base flexure, and it rotates down and back to accommodate a vertically long midface or a more open cranial base flexure.*

2. The angle between the ramus and corpus also can become increased or decreased as a separate kind of rotation. This does not refer merely to the conventional "gonial angle" but rather to the alignment between the whole of the ramus and the corpus. The axis of the ramus can thus be **more upright,** with the ramus-corpus angular relationship thereby "closed." Or, the converse can occur by an opening of the ramus-corpus angle. In either case, the corpus is aligned up or down **relative** to the ramus. Both parts can participate in the bony remodeling changes involved in the opening and closing of the angle between them, although it is necessarily the **ramus** rather than the corpus that carries out most of them because this is where most of the active remodeling processes occur. It would not be possible, for example, for the corpus to rotate upward by its own remodeling to close the gonial angle.

---

*Another type of whole mandible rotation has been reported by some investigators and is believed to involve a pivot located at some point along the occlusal plane. This axis of rotation may occur at the bicuspid level or the first erupted molar. The whole mandible thus "rocks" around this pivot as the maxillary arch grows downward. This will lower or raise the posterior and anterior ends of the mandible, depending on the direction of rotation. Compensatory remodeling changes by the mandible involve essentially the same combinations of resorption and deposition described in Chapter 3. That is, the ramus and condyle will grow either upward and forward or upward and backward in order to sustain proper articular position of the condyle. The mandibular condyle has a type of cartilage lacking the linearly oriented **columns** of proliferating chondrocytes. This is a special feature that gives the condylar cartilage a **multidirectional** growth capacity, in contrast to the restricted unilinear growth direction of a typical epiphyseal plate. This feature allows the condyle to adapt to the wide variety of rotational situations encountered among different individuals with different facial types, as well as the mandibular rotations that take place as a normal part of the growth process. Remember, the growth behavior of the condyle is **secondary;** it is not a pacemaker (see page 94). As the whole mandible becomes displaced by whatever vectors are involved at different ages and by whatever variations occur among different individuals, the condylar cartilage and the contiguous membranes forming the intramembranous bone of the condylar cortex grow in whatever directions and amounts are required to sustain constant functional position and articulation with the cranial floor. The variable capacity of condylar growth thus provides adaptation to different facial types, different occlusal patterns, and the normal structural changes occurring during progressive growth (such as the "rotations" that the ramus undergoes at different age levels).

There are two basic reasons why ramus-corpus rotations occur. The **first** was described on page 116 and deals with the need for a progressively more upright ramus to accommodate a vertically lengthening midface. The remodeling changes that carry this out were also outlined. The result is a ramus-corpus alignment that naturally and normally becomes more closed as the midface grows. The **second** reason is to accommodate the results of whole-mandibular rotation. When the entire mandible rotates forward and upward (see previous sections for the reasons), the mandibular corpus normally rotates downward to some extent in order to compensate. This helps to keep the mandibular arch in a constant or proper occlusal plane relationship. In addition, the posterior maxillary teeth may drift inferiorly. The occlusal plane can be brought to a perpendicular relative to **PM,** or it may still have a slight upward inclination. When the ramus (and whole mandible) rotates backward and downward, the ramus-to-corpus angle can also close to some extent, thereby compensating. The respective amounts of these counteracting rotations are not always exactly equal, however. If they are equal, or if no rotations at all occur, such individuals have an occlusal plane almost exactly perpendicular to the vertical **PM** plane. Often, however, the occlusal plane has a noticeable downward angulation because the amount of upward corpus realignment falls short of the downward rotation of the whole mandible. When you are observing the faces of your patients, you can easily "eyeball" how much downward occlusal plane rotation exists by visualizing it relative to the neutral horizontal axis of the orbit. If the two are parallel, the occlusal plane is perpendicular to the **PM** plane. In many of your patients the occlusal plane will angle downward, to a greater or lesser extent, and in a few it will angle upward. Persons with a vertically shorter nasal region tend to have a perpendicular or an upward occlusal plane alignment, or at least a much lesser amount of downward rotation. The occlusal plane in long-faced and long-nosed individuals tends to be downward-rotated to a greater extent.

**FIGURES 5–35 AND 5–36**

As just seen, an upward corpus alignment closes the ramus-corpus angle, and a downward alignment opens it. The former **shortens** overall mandibular length and thereby has a mandibular retrusive effect. The latter increases it and has a protrusive effect. There are two ways to illustrate why this occurs. First, the straight-line dimension (overall mandibular length) from $a$ to $b$ is decreased; the dimension from $a$ to $c$ is increased. Second, if the upper and lower arches $M$ and $N$ are aligned upward, $M$ protrudes beyond $N$ by the distance $x$ relative to the **occlusal plane** (not the vertical facial profile). When aligned downward, $N$ protrudes by the distance $y$ relative to the downward-inclined occlusal plane.

If the ramus-corpus angle is opened, the prominence of the **antegonial notch** is increased. This is caused by the downward angulation of the mandibular body at its junction with the ramus. If the ramus-corpus angle is **closed,** the size of the antegonial notch can be reduced or obliterated entirely because of the upward alignment of the corpus relative to the ramus. (See Stages 9 and 15 in Part Two of Chapter 3 for additional factors involved and for an account of the remodeling processes that produce these various rotations.)

**FIGURE 5–37**

Note especially that the effects of whole-mandible rotations and ramus-to-corpus rotations are **opposite.** When the entire mandible is aligned downward, a mandibular retrusive effect is produced, but when just the corpus is aligned downward, a mandibular protrusive effect results. An upward whole-mandible alignment is mandibular protrusive, and an upward corpus alignment is mandibular retrusive.

**FIGURE 5–38**

Can an individual have a retrognathic profile and **not** have a Class II malocclusion? Yes. Even though the underlying skeletal factors are the same for both, the effects for one may be more severe than for the other. This is because different planes of reference relate separately to the profile and to malocclusions. In the accompanying diagram, note that the profile (a) is retrognathic, but the upper incisor region, relative to the downward-inclined **occlusal plane,** is not protrusive (b).

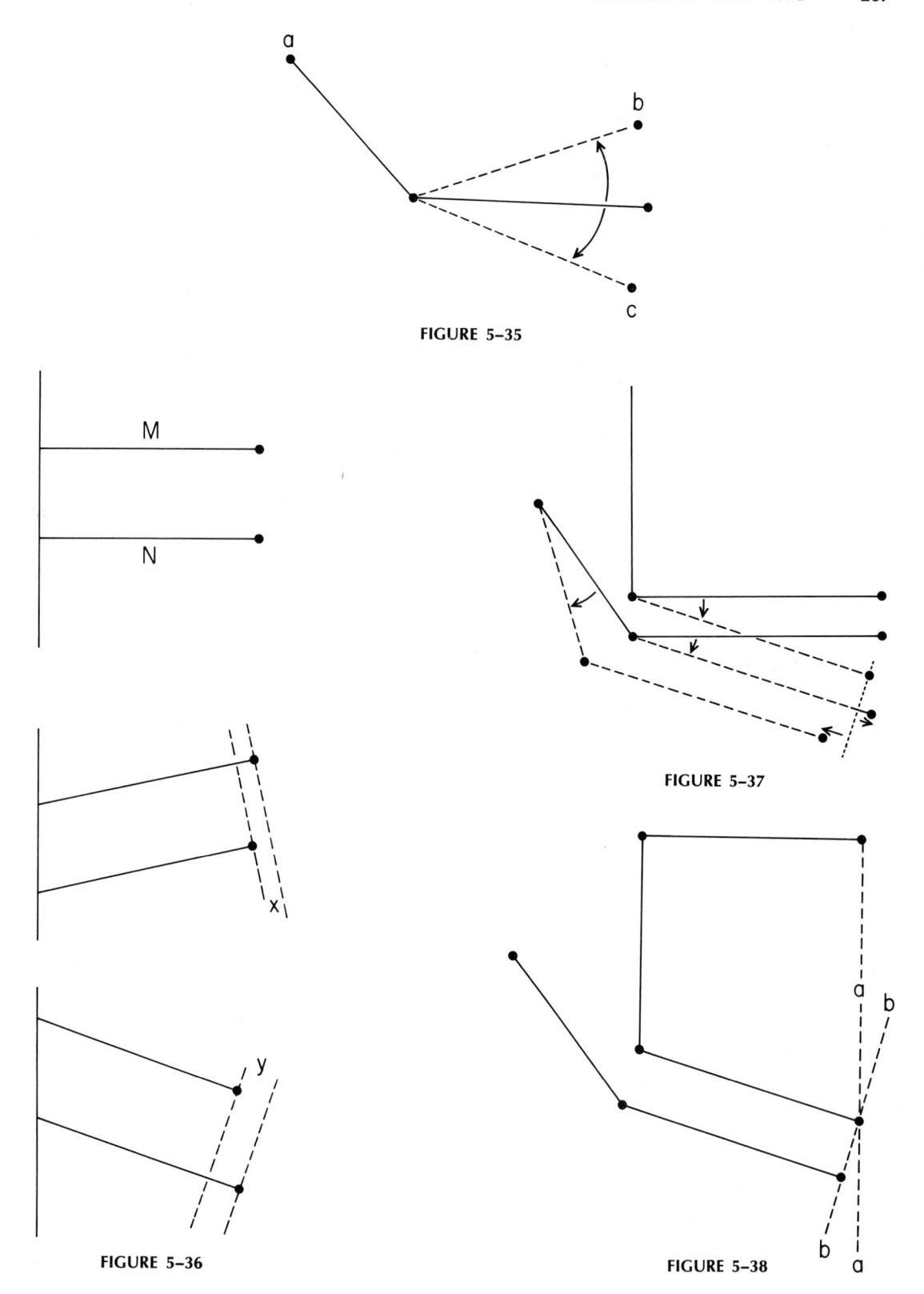

FIGURE 5–35

FIGURE 5–37

FIGURE 5–36

FIGURE 5–38

**FIGURES 5–39 AND 5–40**

A forward-inclined middle cranial fossa has a mandibular retrusive effect. Because the expression of the effective horizontal (not oblique) dimension of the middle fossa is thereby increased, the maxilla becomes offset anteriorly with respect to the mandibular corpus. The midface is also lowered, and this causes the whole mandible to rock down and back. The maxilla thus is carried forward, and the mandible is rotated backward in this composite, two-way movement. Mandibular retrusion results, even though the arch lengths of the upper and lower jaws can have equivalent dimensions, as shown here. These various changes in skeletal pattern cause a Class II molar relationship because the upper bony arch is anteriorly offset.

**FIGURES 5–41 AND 5–42**

A backward-inclined middle cranial fossa* has a mandibular protrusive effect. This contributes to a Class III type of molar relationship. The maxilla is placed backward, and the mandible rotates forward into a protrusive position. Note that the mandibular occlusal plane is rotated into an upward-inclined position. To compensate, the posterior maxillary teeth descend and/or the ramus-corpus angle is opened.

---

*Note this important point. The conventional way to represent the "cranial base angle" is by a line from basion to sella to nasion. This is not the anatomically meaningful way to do it. The **real** relationship (so far as the face is concerned) involves the contact between the condyle and the cranial floor (thus not basion), and the junction corner between the cranial floor and the nasomaxillary complex (thus not sella). **This** is the relationship that **directly** determines the anatomic effects of the three-point contact among the cranial floor, the mandible, and maxillary tuberosity. Basion-sella-nasion only indirectly reflects this. These three traditional landmarks have nothing to do with the actual anatomic fitting of the key junctions involved. They are removed **midline** structures that do not relate directly to the **lateral** positions of the upper and lower arches, the lateral contacts between the mandibular condyles and the cranial floor, and the lateral effects of the angle between the lateral parts of the floor of the middle and anterior cranial fossae relative to the maxillary tuberosities. Sella, basion, and nasion themselves can be almost **anywhere** along the midline axis, within normal variation limits, and not affect the "angle" that really counts: the angle from the condyle–glenoid fossa articulation to the point of junction between the middle and anterior cranial fossae, that is, the point where the nasomaxillary complex joins the cranial floor.

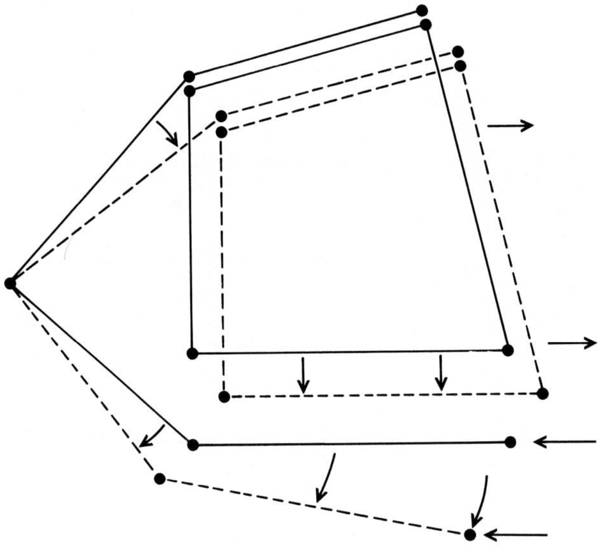

**FIGURE 5–39**

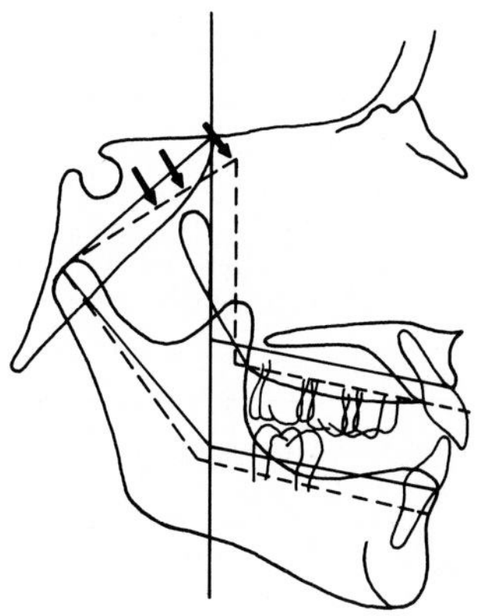

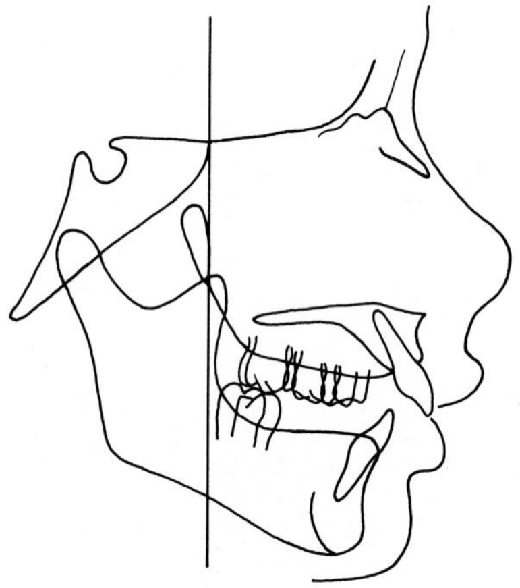

**FIGURE 5–41**

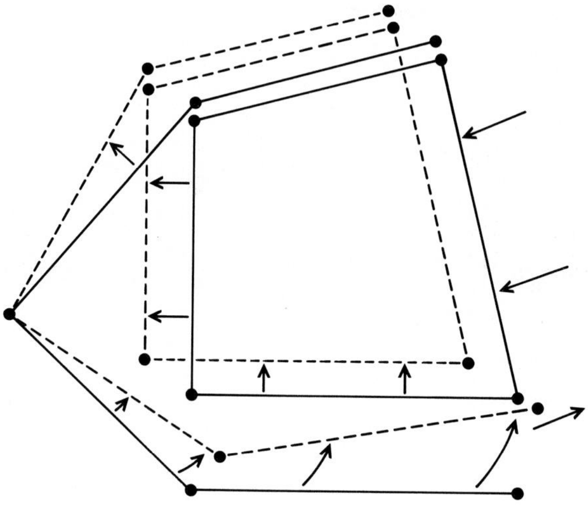

**FIGURE 5–40**

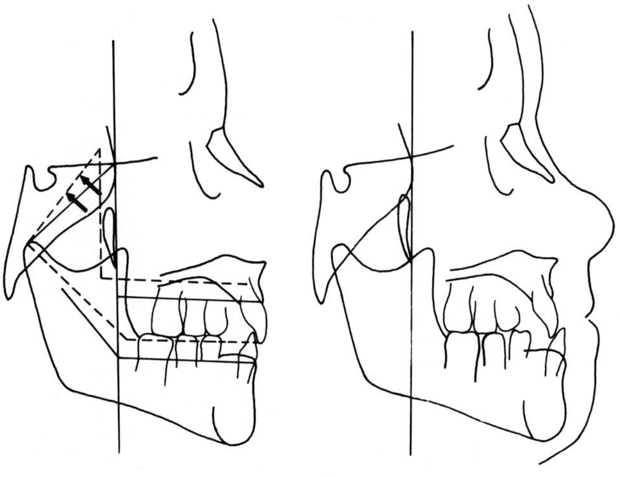

**FIGURE 5–42**

**FIGURES 5–43 AND 5–44**

The nasomaxillary region in most individuals tends to be vertically long relative to the ramus and middle cranial fossa. The result is a downward and backward placement of the whole mandible. Note the resultant mandibular retrusive effect, the retrognathic profile, and the skeletal basis for a Class II molar relationship. It will be recalled that a forward alignment of the middle cranial fossa also causes a similar kind of mandibular rotation. If **both** occur in the same individual, the total extent of mandibular rotation is the sum of the two. (Dental changes occur to preclude an anterior open bite; see under "curve of Spee" below.)

**FIGURE 5–45**

If the nasomaxillary region is vertically short, as noted earlier, a mandibular protrusive effect is produced. The mandible rotates forward and upward, and the resultant offset positions between the maxillary and mandibular arches can contribute to a Class III type of molar relationship. Note that a **vertical** imbalance has resulted in a **horizontal** structural effect. It is incorrect to assume, as many do, that malocclusions are based, essentially, only on horizontal dysplasias.

All the above relationships illustrate the various effects of changes in the dimensions or the alignment of any **one** given region, as for the ramus, middle cranial fossa, maxillary arch, and so on. The skull of any given individual, however, is a composite of many combinations of such relationships among **all** the regional parts. Outlined below are examples of several different combinations of various regional dimensional and alignment imbalances and balances.

**FIGURE 5–46**

In this particular combination, the horizontal dimension of the maxillary arch exceeds that of the mandibular arch (a). The middle cranial fossa has a forward-inclined alignment (b), and the midface (c) is also vertically long. The ramus is rotated backward (d). **All** these features have mandibular retrusive effects, and their combined sum (e) results in a severe Class II malocclusion and extreme retrognathia.

**FIGURE 5–47**

In this combination, the horizontal dimension of the middle cranial fossa exceeds its counterpart, the ramus. The mandibular corpus, however, is long relative to the horizontal dimension of the bony maxillary arch. The composite result actually produces a prognathic profile, but with a Class II molar relationship. Such individuals are encountered now and then.

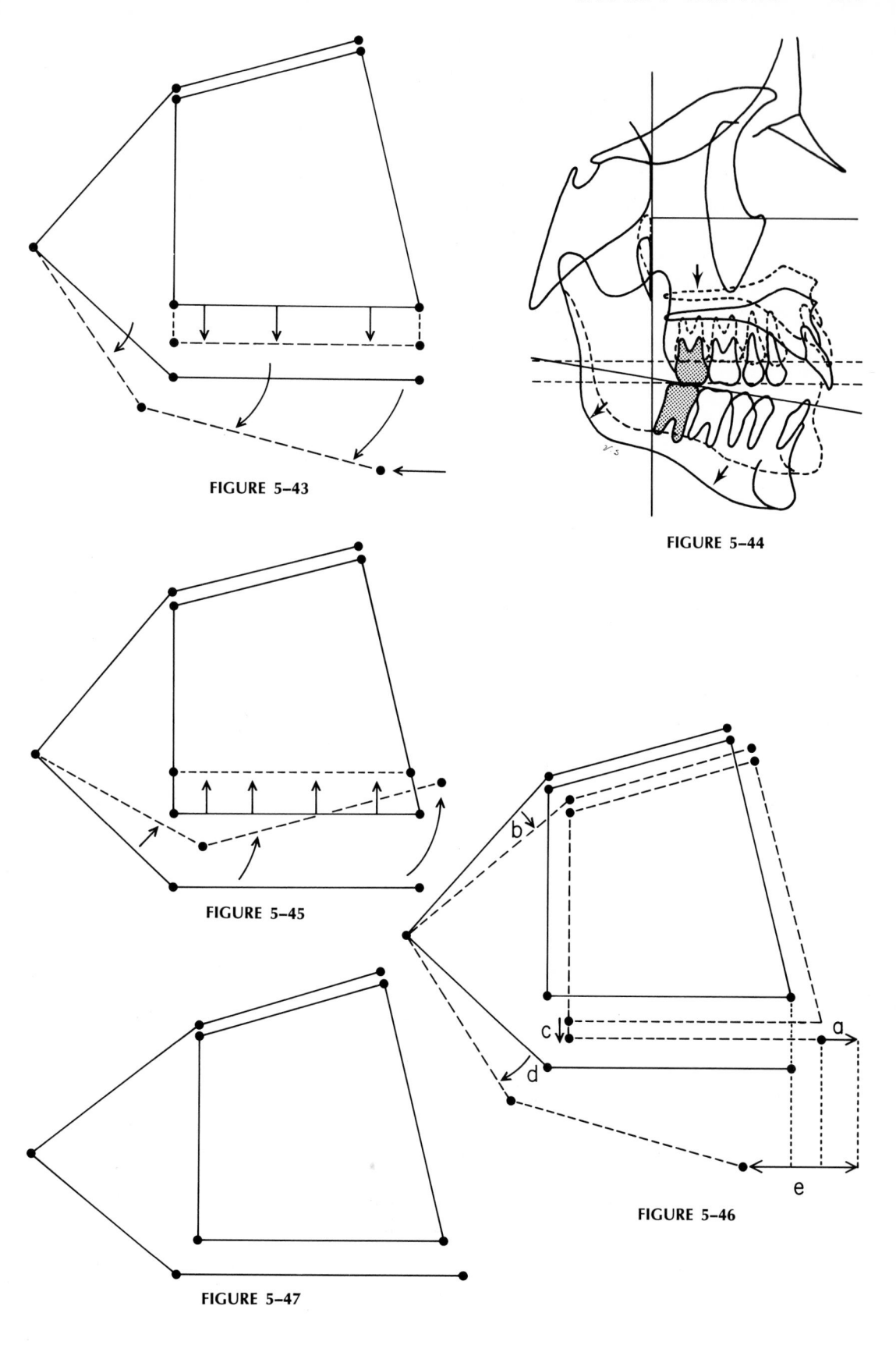

FIGURE 5–43

FIGURE 5–44

FIGURE 5–45

FIGURE 5–46

FIGURE 5–47

**FIGURE 5–48**

This combination illustrates a horizontally short mandibular corpus (relative to the individual's maxillary arch) in combination with a backward-rotated middle cranial fossa, a forward-rotated ramus, and a downward-rotated mandibular corpus. The composite result is an individual with a Class II type of upper arch, a Class III molar relationship, and a Class I (orthognathic) type of profile.

**FIGURE 5–49**

This combination involves a horizontally long mandibular corpus, a forward align-ment of the middle cranial fossa, and a backward rotation of the ramus. The cumulative result is a Class I orthognathic profile, a Class II molar relationship, and a Class III type of bony mandibular arch. The corpus-ramus angle has closed a bit, but there is still a downward-inclined alignment of the occlusal plane.

**FIGURE 5–50**

One of the most frequently encountered combinations involves a nasomaxillary complex that is vertically long and/or a middle cranial fossa that is obliquely forward. Both these features cause a downward-backward ramus rotation. However, the ramus is horizontally **broad,** and this counteracts the amount of its backward rota-tion. What would have been a Class II malocclusion and a retrognathic profile has thereby been converted into a Class I occlusion and an orthognathic (or nearly so) profile owing to the development of a wider ramus.

It is apparent that a Class II malocclusion is **not** merely caused by a "long maxillary arch." Malocclusions are multifactorial. Examples of Class II and Class III headfilm tracings are shown next for comparisons of their respective structural characteristics according to the different kinds and combinations of relationships outlined above.

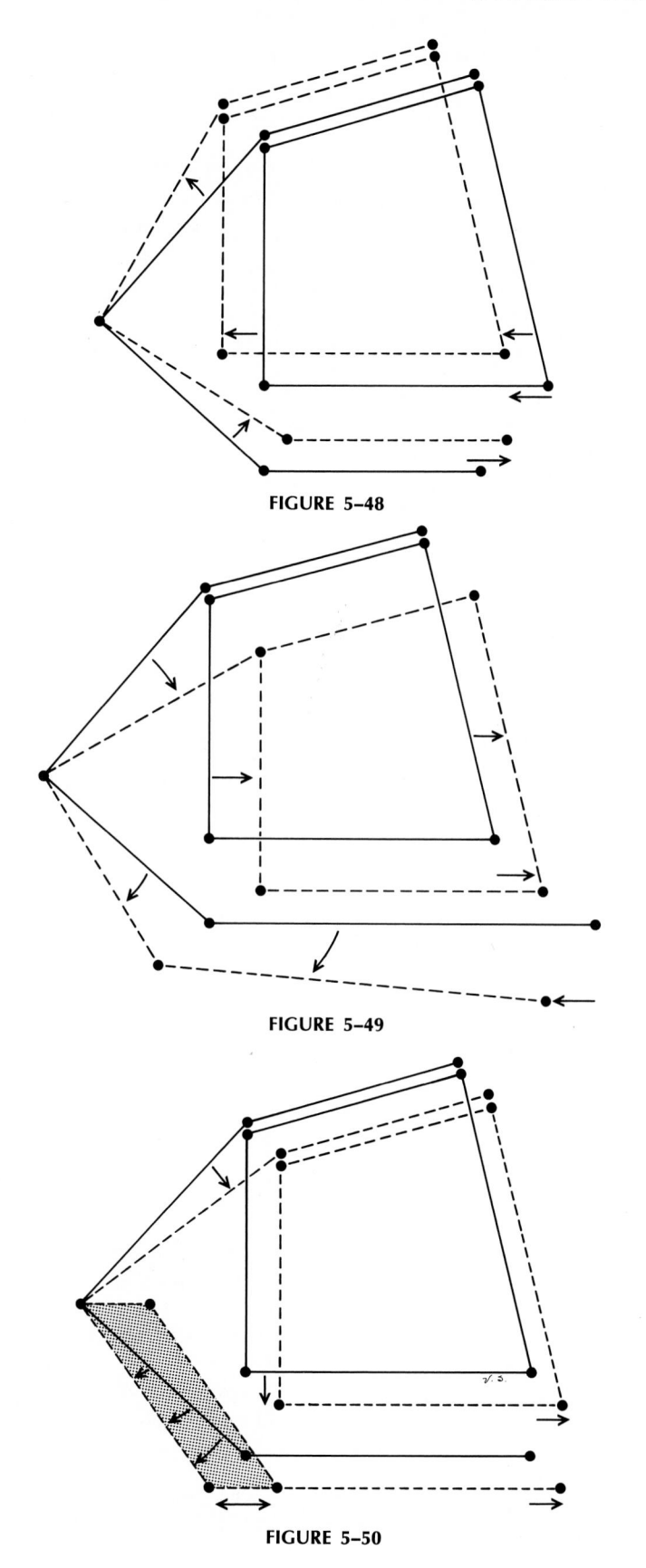

**FIGURE 5–48**

**FIGURE 5–49**

**FIGURE 5–50**

### FIGURES 5–51 AND 5–52

In this Class II individual, the mandibular arch is short relative to the maxillary arch. The mandibular body in the Class III individual, conversely, is long.

In the Class II individual, the corpus is rotated upward (closed ramus-corpus angle); it is rotated downward in the Class III individual. This causes the characteristic steeply angled mandibular corpus of Class III. Note, however, that the anterior mandibular teeth have risen considerably to compensate, so that the occlusal plane, unlike the corpus, is not angled as sharply downward. This causes the characteristically long, high alveolar region above the chin found in a typical Class III individual.

The nasomaxillary complex of the Class II individual is vertically "long" **relative** to the vertical dimension of the ramus and middle cranial fossa. Because of this, the ramus of the Class II individual is rotated downward and backward relative to the "neutral" alignment plane (dashed line).* The Class III ramus, conversely, is rotated forward in conjunction with a vertically short midface. While the face of a Class III individual **looks** quite long, it is usually the mandibular corpus (lower face), not the nasomaxillary region (midface), that causes this.

The middle cranial fossa in the Class II individual has a forward and downward-inclined alignment; it is aligned upward and backward in the Class III individual. These factors also contribute directly to the cause for the downward-backward and forward-upward rotations of the mandible, respectively.

To date, **all** these features contribute to the mandibular retrusion of the Class II individual and the mandibular protrusion of the Class III individual. However, note that the ramus in the Class II individual is horizontally broad. This is a **compensatory** feature, and it partially counteracts the other characteristics that combine to cause mandibular retrusion. Because of this, the resultant malocclusion is less severe than it would have been if the ramus were to have had a "normal" dimension, or especially if it were narrow, thereby **adding** to (rather than subtracting from) the composite basis for the Class II malocclusion.

Similarly, note the narrow horizontal breadth of the Class III ramus (compare it with the Class II ramus). This is also a compensatory feature that has reduced the severity of the malocclusion and the extent of prognathism.

Most Class II individuals have a horizontally short mandibular corpus, a vertically long nasomaxillary complex, a downward- and backward-aligned ramus, a forward middle cranial fossa alignment, an upward corpus rotation (ramus-corpus angle), and (in severe malocclusions) a narrow ramus and horizontally broad middle cranial fossa. The **converse** of all these regional relationships characterizes the Class III malocclusion. Each such feature occurs in about 70 percent or more of the Class II's and III's. What about the other 30 or so percent? This is where "offsetting penalties" come into play. Instead of a forward-inclined, dolichocephalic type of middle cranial fossa causing mandibular retrusion, for example, a given individual can have a **backward**-aligned, brachycephalic type of fossa. This feature may then combine with one or more other regional mandibular **protrusive** features, such as, perhaps, a broad ramus, or a long corpus, or a downward-rotated corpus, to par-

---

*See Enlow et al.: The morphological and morphogenetic basis for craniofacial form and pattern. Angle Orthod., 41:161, 1971, for the rationale underlying the various neutral alignment planes shown by the different dashed lines.

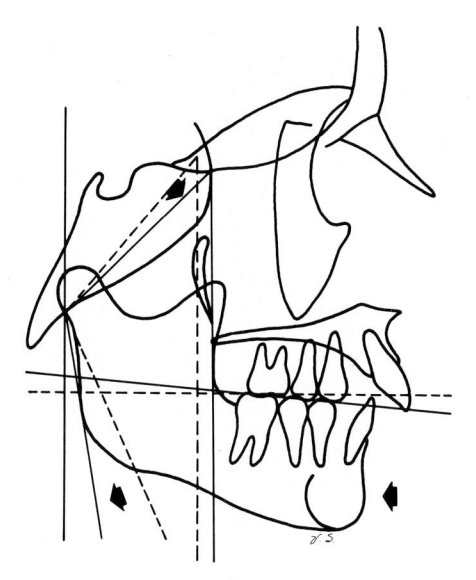

**FIGURE 5–51**

*(From Enlow, D. H., T. Kuroda, and A. B. Lewis: Angle Orthod., 41:161, 1971.)*

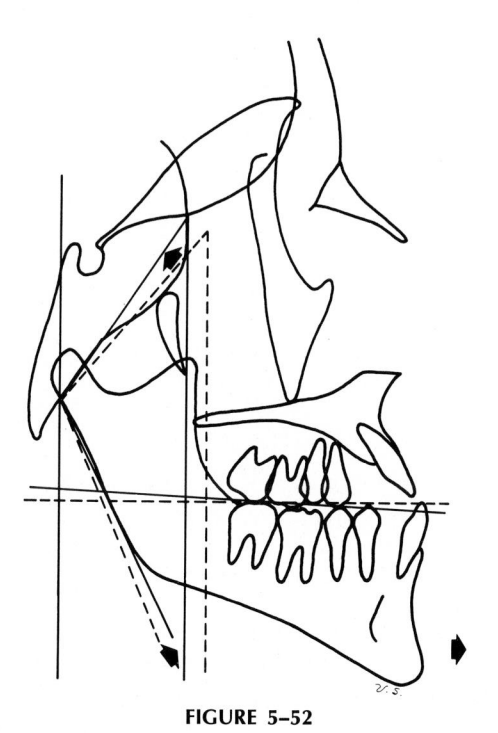

**FIGURE 5–52**

*(From Enlow, D. H., T. Kuroda, and A. B. Lewis: Angle Orthod., 41:161, 1971.)*

tially counteract the various mandibular retrusive factors that are also present. In any given individual, the sum of the dimensional values for all the mandibular protrusive features weighs against the sum of the values for all the mandibular retrusive features. Either they come into an effective balance, or one or the other wins. If the mandibular retrusive features dominate, the **severity** of the resultant Class II malocclusion and retrognathic type of face depends, first, on how much (in millimeters) the total of these retrusive features amounts to, and second, how much the counteracting features subtract from this total.

**Each** of us has a face and cranium that represents such a composite mixture of regional counteracting imbalances. Here and there in the different regions of the face and cranium, something can be balanced, but no one of us has a head that is regionally in balance throughout. Add to this the topographic facial features involving the frontal sinuses, upper face protrusion relative to the anterior cranial fossa, corresponding variations in the nasal bridge, the shape and size of the nose, and the narrowness or widening of the whole face in relation to brain shape and head form, and the almost limitless variations that occur in overall facial form can readily be appreciated.

**Each** of us has a natural, normal predisposition toward either mandibular retrusion (Class II) or protrusion (Class III). There is no such thing, in a sense, as a "separate" Class I facial category. All Class I individuals have a predominant tendency one way or the other toward a malocclusion. Most Class I Caucasians have the **same** underlying facial and cranial features that are present in the Class II Caucasians; the same 70 or so percent of the various mandibular retrusive relationships described above also occur in the Class I individual. This is why a Class II "tendency" usually exists to a greater or lesser extent. The difference between the Class I and II malocclusions, however, is the **extent** of the imbalances and the number and extent of counteracting features. If the compensating characteristics are adequate, a more or less normal face results. If they partially or totally fail, marginal to severe malocclusion and facial disproportion result. A person having an attractive, well-proportioned face, for example, with an orthognathic (or nearly so) profile and only relatively minor occlusal irregularities **also** has, unsuspected to him or her and deep within the face and cranium, the **same** underlying characteristics that caused a cousin to have a noticeably retrognathic profile and a Class II malocclusion. Our hero, however, has a particulary broad ramus and some other happy characteristics that are winners for him as an individual. Most of us have at least a reasonable-appearing face, although somewhat short of perfect, for the same reasons.

**FIGURE 5–53**

During the development and establishment of the occlusion, compensations occur involving dentoalveolar remodeling as well. The placement of the teeth interrelates with the many other skeletal and soft tissue growth processes taking place in the face and cranium. This series of illustrations explains some common changes involved.

In the first diagram, the vertical and horizontal dimensions among the various skeletal parts and counterparts are in balance. The alignments of all the parts also are in "neutral" positions. That is, the nature of the alignments is such that neither protrusion nor retrusion of the upper or lower jaws is produced; the angular relationships are balanced so as not to increase or decrease the "expression" of any of the various key dimensions. Note that the occlusal plane is perpendicular to the vertical reference line (the **PM** plane) and parallel to the neutral orbital axis (shown here below the orbit rather than within its geometric center).

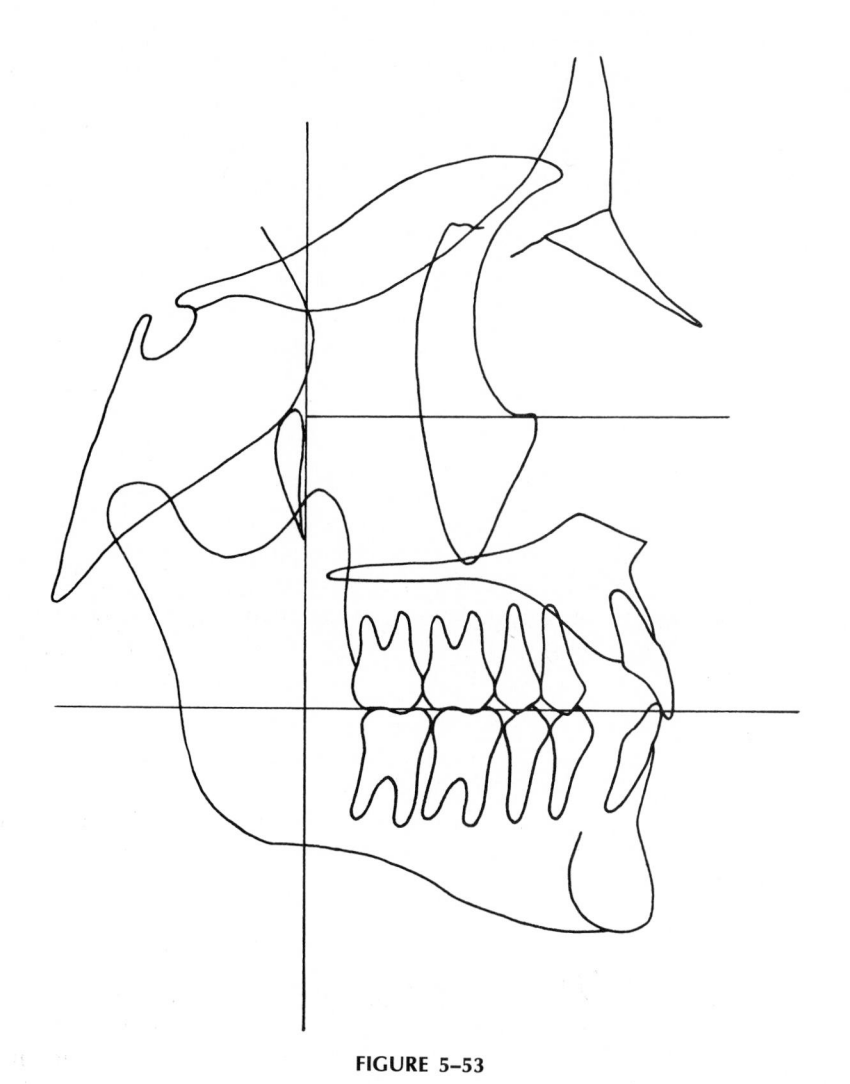

**FIGURE 5–53**

*(From Enlow, D. H., T. Kuroda, and A. B. Lewis: Angle Orthod., 41:161, 1971.)*

**FIGURE 5–54**

The nasomaxillary complex in this stage has become lengthened vertically to a disproportionate extent. This is common, as mentioned earlier. The amount of mid-facial growth has **exceeded** the vertical growth of the ramus–middle cranial fossa composite. The result is a downward and backward alignment of the whole mandible to accommodate the longer nasomaxillary complex. A vertical "imbalance" has thus been introduced, and the expression of the vertical ramus height has been increased to match it by a downward rotation. (This same effect on the mandible can also be caused by a proclination of the middle cranial fossa, as previously described). Note especially that the mandibular corpus, and with it the lower teeth, now has a consequent downward inclination relative to the vertical **PM** line. This "opens" the anterior bite; only the second molars (and perhaps the first molars because they are the first to erupt) are in occlusal contact. The amount of occlusal separation increases toward the incisors.

Note also the retrusion of the mandible, overjet, and the Class II molar relationship caused by the ramus rotation. These resultant effects, however, may be partially or completely offset by a widening of the ramus or other compensatory changes.

**FIGURE 5–55**

The upper teeth "drift" (**not** erupt) inferiorly until each comes into contact with its antagonist. The last molar was already in contact; the next molar must move downward only a short distance. The premolars drift inferiorly even more because of the greater gap potential involved. The central incisors move down the greatest distance. As a final result, full arch-length occlusal contact is attained. The occlusal plane is **straight** (not curved, as in other variations described below). The occlusal plane bisects the upper and lower incisor overlap, just as it did in the first, "balanced" stage. The occlusal plane, however, is now inclined obliquely downward.

**FIGURE 5–56**

Another remodeling combination may occur. The upper teeth drift inferiorly, but the premolars, canines, and incisors do not move down to the full extent needed to completely close the occlusion, only to about the extent that the first molar or premolars drift inferiorly.

**FIGURE 5–57**

The anterior **mandibular** teeth now drift superiorly until full arch occlusal contact is reached. The incisors must move upward much more, however, than the canines and premolars. Note that the cusps of the lower incisors and canines are **noticeably much higher** than the premolars and molars. Palpate your mandibular anterior teeth with the tongue to determine if this common growth and adjustment process has occurred in your own dentition.

There are two ways to represent the **occlusal plane.** The traditional method is to draw a line along the contact points of the posterior teeth and the midpoint of the overlap between the upper and lower anterior teeth. In the first two examples cited above, this line is straight. In the last, however, note how the line is curved as it exactly bisects the overlap of the upper and lower incisors. This is called the **curve of Spee,** and the reason for its development was outlined in the previous paragraphs. The second way to represent the occlusal plane is to run a line from the posterior-most molar contact point straight to the anterior-most premolar contact point. The

**FIGURE 5–54**

*(From Enlow, D. H., T. Kuroda, and A. B. Lewis: Angle Orthod., 41:161, 1971.)*

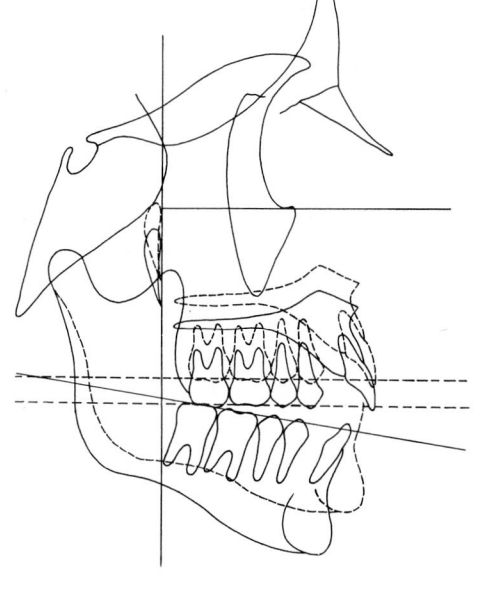

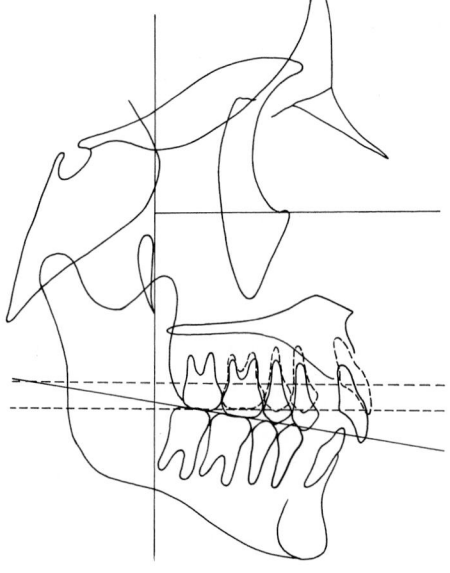

**FIGURE 5–55**

*(From Enlow, D. H., T. Kuroda, and A. B. Lewis: Angle Orthod., 41:161, 1971.)*

**FIGURE 5–56**

*(From Enlow, D. H., T. Kuroda, and A. B. Lewis: Angle Orthod., 41:161, 1971.)*

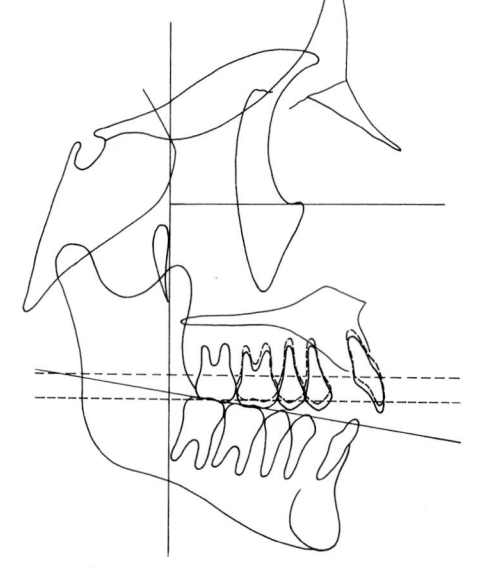

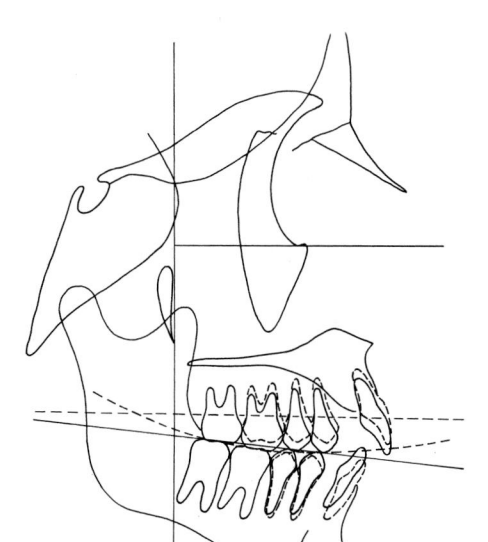

**FIGURE 5–57**

*(From Enlow, D. H., T. Kuroda, and A. B. Lewis: Angle Orthod., 41:161, 1971.)*

incisors are not considered. This is termed the "functional occlusal plane," and it is always a straight line whether or not a curve of Spee exists.

In the first and second examples of occlusal development outlined above, a curve of Spee has not developed, and the two methods for representing the occlusal plane result in the same line. In the last example, however, the curved occlusal plane bisecting the incisor overlap and the straight functional occlusal plane are divergent. Note how the mandibular incisors rise considerably above the level of the functional occlusal plane. The maxillary incisors, however, fall well short and do not even touch this straight-line functional occlusal plane. In individuals having a marked curve of Spee, the alveolar region of the mandible just above the chin is characteristically more elongate because the incisors have drifted superiorly for several millimeters or more.

If you have access to a darkroom try this test for the nature of the symmetry of your own face. Look at the two sides of your face in the mirror. They look about the same, right? Probably not. Prepare two frontal photographic prints of the face, but reverse the negative for one. Then, cut the prints into equal right and left halves and reassemble so that the two right halves and the two left halves each form a full face. Compare the two reassemblies. Two different faces? Which side, right or left, is the more "masculine," and which the more "feminine"? Which side of your own face would you want to favor in facing the camera to show your best profile? Are you right- or left-"faced" (similar to right- and left-handedness)?

Using the radiographic and anthropometric landmarks described in Chapter 9, study the **individualized variations** in facial structure on as many dry skulls as you can gather, on headfilms, on your classmates, and on yourself using the following as a guide.*

Observe the variations in different facial types between the profile plane (nasion to pogonion) and the denture base profile (A point to B point). Compare the relative height from A point to B point with total face height. Notice that small differences in the relative dimension of the denture base can cause marked changes in the proportions of the face, both vertically and horizontally. Compare between individuals: the depth from the denture base profile to the last molar; relative width across the molar region; protrusion angle of upper incisor to denture base profile; facial profile angle to Frankfort horizontal.

Compare the height of the face to its breadth, that is, nasion-gnathion to bi-zygion. Compare bi-zygion to bi-gonion and also bi-zygion to bi-frontotemporale.

Compare the nasion to pogonion profile plane in different individuals to determine the degree of convexity or concavity of the facial profile. See where nasospinale and B point lie with regard to the profile plane. Determine if nasal height (nasion to nasospinale) is greater or less than the 43 per cent "rule" of nasion-gnathion height, that is, a ratio of about 3:4 for the region above nasospinale (43 per cent) to the region below it (57 per cent). Is the individual vertically long-nosed or short-nosed?

Is the mandible heavily-built or relatively light? Note the effect of the canine

---

*Modified from Dempster (1960).

alveolar bone on the form of the face and lips. Is the mandible relatively broad or narrow? Pointed or more V-shaped? Is the gonial region broad, squared, and flared or the converse? Chin prominence?

Determine the extent of variations between face height (as from nasion to gnathion) compared with the height of the whole skull from bregma to gnathion. Is there a single frontal boss or two? Compare the prominence of the glabella and the supraorbital ridges. Note the transverse and horizontal nature of the curvature of the forehead. Compare minimum frontal breadth relative to other frontal bone dimensions. Where is the hairline with respect to metopion? Does the hairline come to a midline apex (with two lateral recesses) on the forehead or is it uniformly rounded? Male and female differences?

Compare the relative sizes of the orbits. Compare the degree of roundness. Obliquity of the superior margin? See if the eyebrow hairline follows the superior bony rim. Note the extent of protrusion of the lateral margin and the cheekbone and also the protrusion of nasion. Flat-faced or a peaked type of face? Compare the bi-dacryon breadth with the width across the nasal alae. Compare also the bi-canthus dimension with nasal width. Note whether the total bi-orbital width is greater or less than minimum frontal width (a diamond-shaped, heart-shaped, etc. face). Where does a vertical line from the pupil lie with respect to the angle of the mouth? The same with respect to the maxillary teeth? Where is nasion relative to the level of the pupil? Epicanthic fold (in Orientals)? Nordic lid fold? Senile lid fold?

Is the nasal bridge high or low? Narrow or broad? Does it look broad just because it is low? Is the nose shape curved, straight, bent, pug, flaring, turned up or down? Angulation of the upper edge of the nose? Compare nasal angle to forehead slope and to the profile plane. Compare height to width ratio of the nose. Extent of nasal protrusion? Looking from below, is the edge of the midline septum perpendicular to the lip or skewered to one side? How does it grade into the lip? Are the nasal openings equal in size and do they have the same symmetry?

Run a vertical line down along the surfaces of the upper lip at a right angle to the neutral orbital axis. Is the chin protrusive or retrusive relative to this line? Does the line intersect the upper nasal slope at about its midpoint? Compare this method of midfacial plane determination with the nasion-to-pogonion profile plane among different individuals with different facial types.

On dry skulls, construct a frontal triangle from superior prosthion to the two zygion points. Compare for flat, round-faced and for narrow-faced individuals. Between Oriental and Caucasian faces? Compare bi-zygomatic and minimum frontal widths. Compare bi-zygomatic and bi-gonial widths. How do these differences affect facial form? Note the angulation of the anterior surface of the malar region (that is, the cheek profile angle). Where does this plane fall with respect to the maxillary and mandibular arches?

Note the overall prominence of the maxillary "muzzle." Is it massive, reduced, or moderate? Are the upper incisors tipped out by the lowers to produce a bimaxillary protrusion? Compare the vertical span of the upper lip with the height of the nose. Compare the width of the maxillary arch with the height of the nose in different facial types. Compare maxillary arch width with bi-zygomatic width. Effect on facial appearance? Note the angle between nasion-prosthion and Frankfort and compare with the angle between nasion-pogonion and Frankfort. Prognathic? Retrognathic? Compare lip thickness in Caucasian, Oriental, and Black faces. Is the upper lip concave or flat? How prominent is the "cupid's bow" of the upper lip? Does one lip or the other protrude? Are they closed?

Compare the chin angle (infradentale or inferior prosthion to pogonion) in different profile types. Chin angle relative to Frankfort? Note the marked variations in chin shapes. Compare bi-condylar width with bi-gonial width; with the width at the trihedral eminences. Squared mandible, rounded, more pointed? Compare variations in the angle between the mandibular base plane and the occlusal plane and also with Frankfort. Variations in the angle between the mandibular plane and the posterior margin of the ramus and condyle? Is the whole ramus aligned more backward or forward relative to the occlusal plane? If the individual has a retrognathic profile, is he a Class I or a Class II? If a Class I, how do you explain this?

Is the head form elongate (dolichocephalic) or rounded (brachycephalic)? Note the degree of prominence of the parietal eminences. Narrow or broad forehead? Sloping or more upright? Estimate relative dome height (porion to vertex). Note the type of external contour of the occiput; flat or rounded? **Dinaric** type of skull form (narrow, slanting forehead, flat and slanting occiput, long face, long nose, a high nasal bridge, and an elongate, peaked skull dome)? Note whether the calvaria is bilaterally symmetrical.

Figures 5–58 through 5–61 catalogue a number of common facial variations. These are just a sample; there are, of course, many, many more. A good police department artist can reconstruct a "composite" facial representation of a given person by combining such feature variations (adding characteristics as hair color, type, and length, brow height, eyebrow form, age, wrinkles, and so forth). 1: Tarsal part of upper eyelid exposed; 2: laterally covered by an eyelid fold; 3: iris covered by upper eyelid; 4: most of iris exposed; 5: lateral corner of eye higher than medial corner; 6: lateral eye corner lower than medial corner; 7: top of nasal bridge (root) markedly indented; 8: high nasal root (so-called "Greek nose"); 9: narrow nasal root; 10: broad nasal root; 11: narrow nasal slope; 12: broad nasal slope; 13: concave nasal profile; 14: straight nasal profile; 15: convex nasal profile; 16: inconspicuous nasal wings; 17: prominent nasal wings; 18: V-shaped nasal wings; 19: rounded nasal wings; 20: arched nasal wings; 21: straight nasal wings; 22: narrow nasal tip; 23: broad, flattened nasal tip; 24: thick, fleshy nasal wing; 25: thin nasal wing; 26: asymmetrical nasal openings; 27: symmetrical openings; 28: **P-A** directed openings; 29: laterally directed openings; 30: narrow, elongate openings; 31: rounded nasal openings; 32: upward nasal inclination; 33: straight lower nasal border; 34: downward inclined nasal border; 35: vertically short upper lip; 36: long upper lip (check also to see if upper lip profile is straight or concave); 37: upper lip without midline "cupid's bow"; 38: deep midline notch in upper lip (look also for a more conspicuous philtrum above upper lip, and check for thinness or thickness of the red part of both the upper and lower lips); 39: acutely curved lower border (concavity) below lower lip; 40: lesser concavity below lower lip, and a greater distance between lip and mentolabial sulcus; 41: lower lip retrusive; 42: lips equally protruding; 43: lower lip protrusive; 44: pointed mandible; 45: squared mandible; 46: no chin cleft; 47: bifid chin; 48: retrusive mandible (and chin); 49: prominent chin; 50: slight rolling of upper border of ear helix; 51: pronounced helix rolling; 52: flat, shallow ear scapha; 53: pronounced, deep groove below scapha; 54: slight rolling of middle part of helix; 55: pronounced middle helix rolling; 56: short, low crus; 57: prominent, long crus; 58: dangling ear lobe; 59: ear lobe fused with facial skin; 60: slight ear protrusion; 61: marked ear protrusion; 62: diamond-shaped face; 63: long, narrow face; 64: round, short face; 65: oval face; 66: squared face; 67: egg-shaped face.

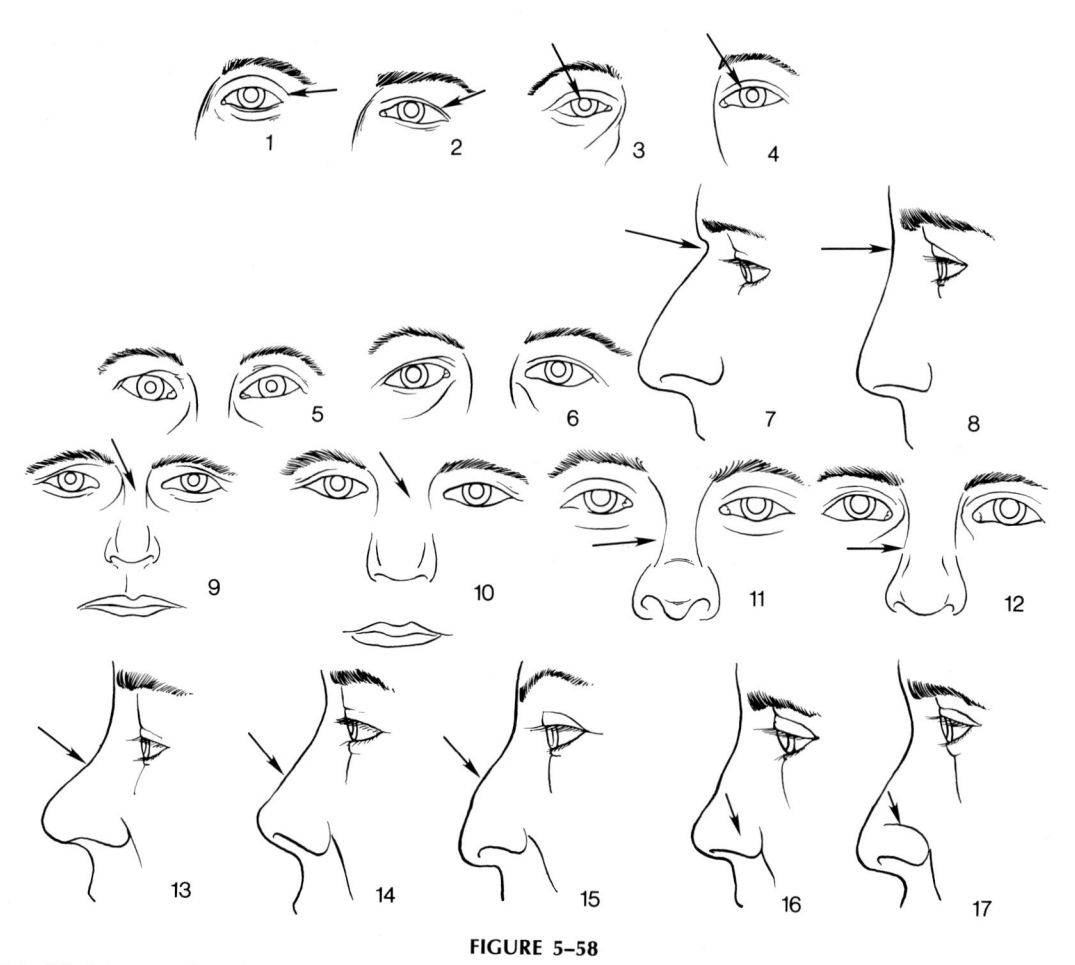

**FIGURE 5–58**

(Modified from Hulanicka, B.: Nadbitka Z Nru 86, Materialow 1 Prac antropologicznych Wroclaw, 115, 1973.)

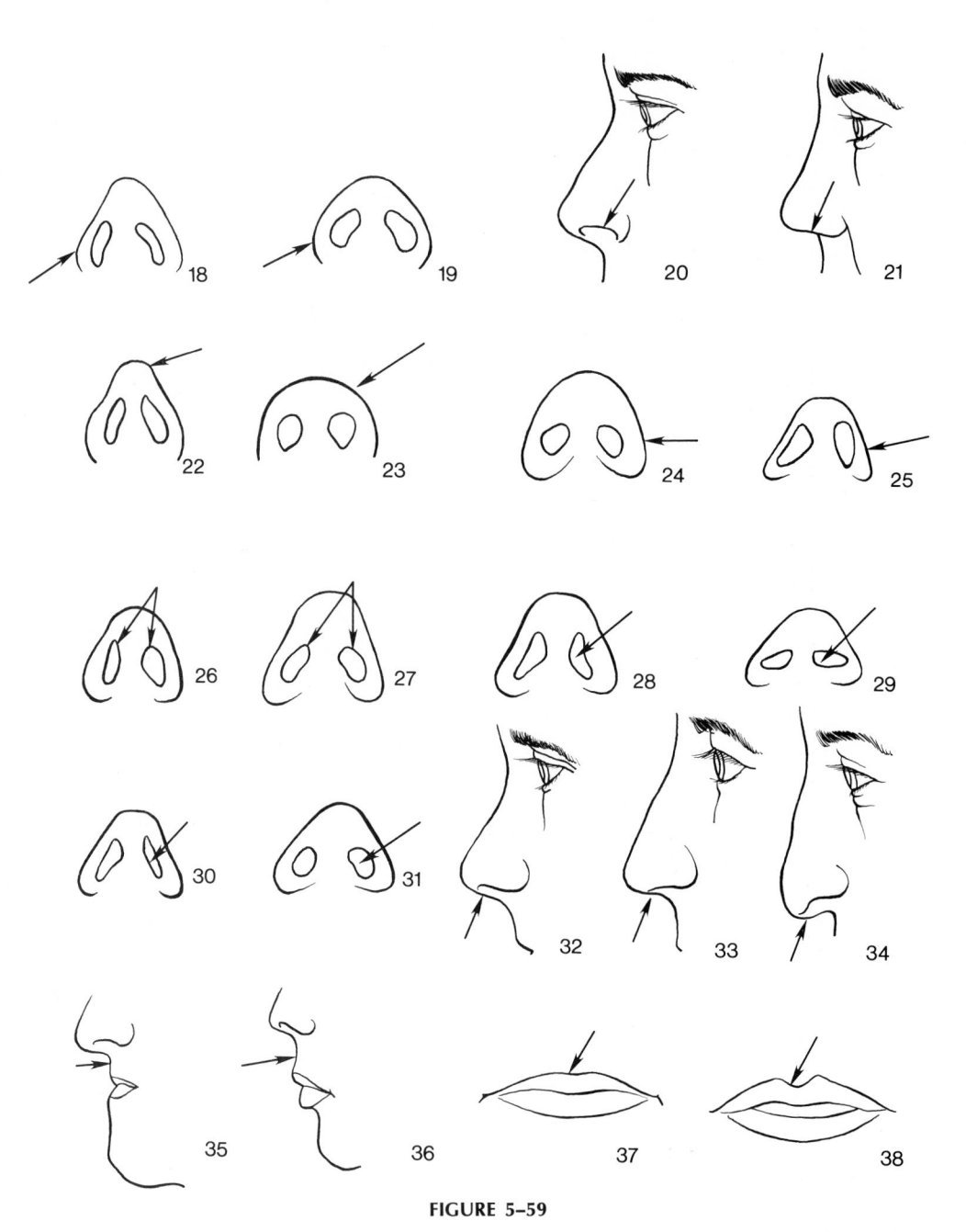

**FIGURE 5–59**

*(Modified from Hulanicka, B.: Nadbitka Z Nru 86, Materialow 1 Prac antropologicznych Wroclaw, 115, 1973.)*

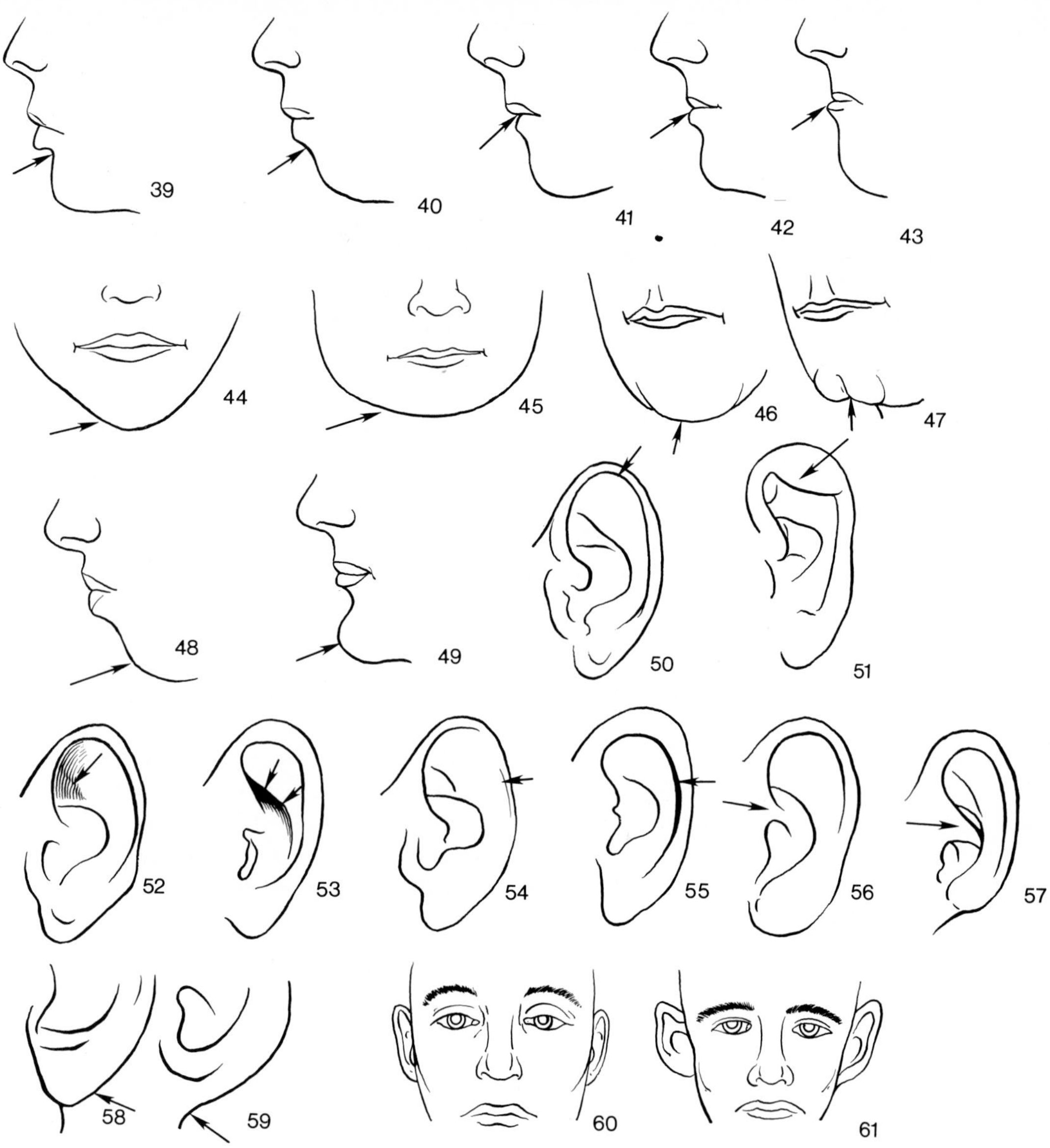

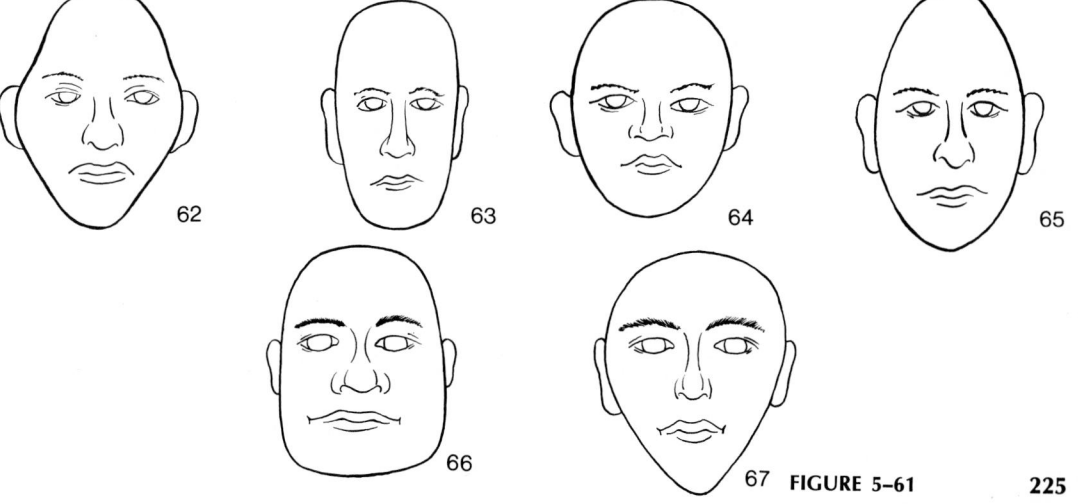

**FIGURE 5–60**

*(Modified from Hulanicka, B.: Nadbitka Z Nru 86, Materialow 1 Prac antropologicznych Wroclaw, 115, 1973.)*

**FIGURE 5–61**

## Chapter 6

# The Structural Basis for Ethnic Variations in Facial Form

Age, sex, and population differences in the pattern of facial structure have been pointed out in many different sections of the preceding chapters. The purpose of the present section is to summarize this information briefly and add to it as a separate topic. While this is an interesting subject in its own right, it is quite important for the clinician to realize that population norms derived from a given sample are not necessarily valid or accurate for other samples or groups, especially if ethnic variations are involved.

The phylogenetic basis for the form and pattern of the human face was outlined in Chapter 4. It will be recalled that both the shape and the size of the brain are key factors relating to the structure of the face. Because the cranial base is the bridge between them, and because the floor of the cranium is the template upon which the face is constructed, variations in the shape of the brain in **any** species are associated with corresponding variations in the form of the face. For example, the junctional part of the midface can only be as wide as the floor of the cranium. It cannot be wider because there is nothing to attach it to. Thus, narrow-brained species or subgroups are correspondingly narrow-faced. Compare the face of the long, narrow-brained collie dog with the short, round-brained boxer or bulldog.* Man has an exceptionally wide face, in comparison with the typical mammal, because of his colossal brain size and the shape of the brain. The various rotations of the olfactory bulbs, orbits, and so forth combine with the boundaries of the brain to establish, in all species, the amount and the principal directions of facial growth. Because of these factors, the shape and the size of the brain are involved in the variations of facial pattern **within** any given species as well as between species. There are, however, other factors that come into play, as will be seen.

Human population groups having a dolichocephalic head form naturally have a more narrow and a longer face than those with a brachycephalic type of head form. The wider brain (no special difference in overall volume) has the wider face. It has been claimed that there is an evolutionary (secular) trend toward the brachycephalic type among "long-headed" human groups. If this is happening, there will also be related long-term changes in facial structure, the nature of in-built tendencies toward malocclusions, and profile type.

---

*It has long been argued as to which is the pacemaker, the brain or the cranial floor (see Chapter 4). Regardless of which one is primary and which secondary, the effects on facial configuration are nonetheless real.

226

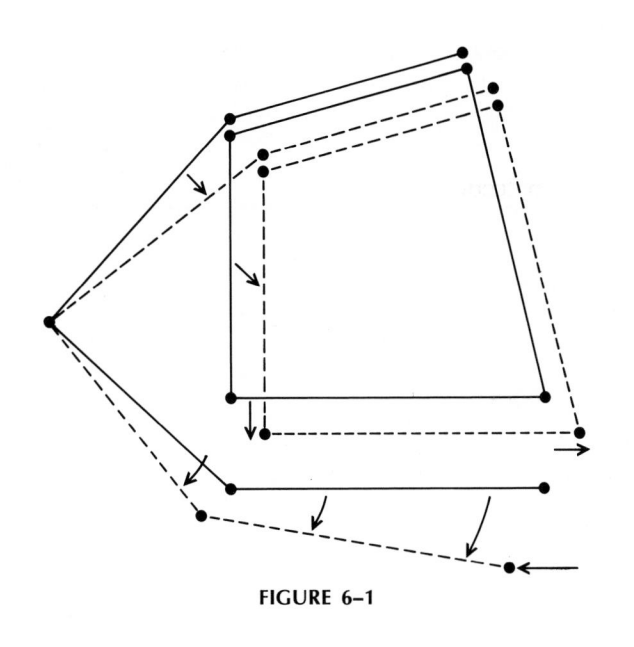

**FIGURE 6–1**

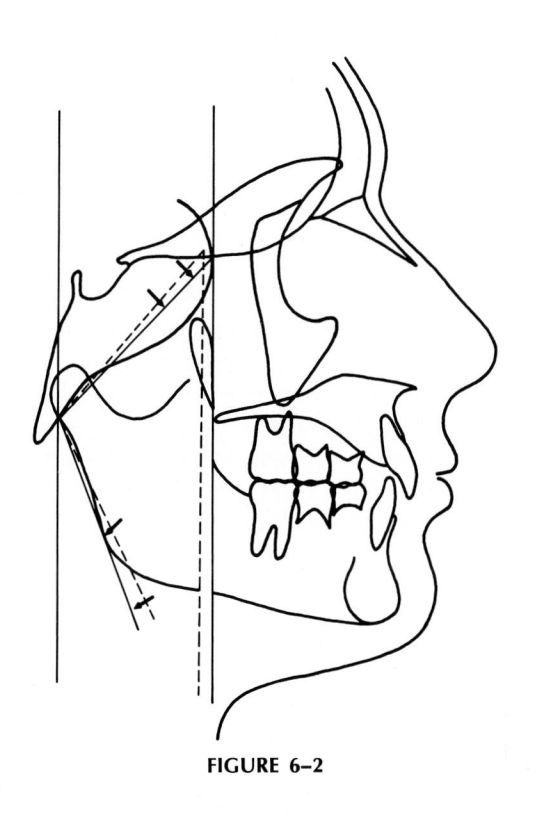

**FIGURE 6–2**

**FIGURES 6–1 AND 6–2**

The more open ("flat") cranial base flexure that usually characterizes the dolicho-cephalic head form in many Caucasian groups sets up a more protrusive upper face and a more retrusive lower face. The whole nasomaxillary complex is placed in a more forward position, and it is lowered relative to the mandibular condyle. Because the condyle is "higher," there is the tendency for a downward and backward rotation of the whole mandible. The posteroanterior dimension of the pharynx is relatively large because of the longer and more horizontally aligned middle cranial fossa. For these various reasons, there is a greater frequency of the retrognathic type of profile and a Class II tendency among groups with a dolichoce-phalic type of head. There is also a high incidence of a "broad" ramus to compen-sate for the built-in tendency toward mandibular retrusion.

See figures on preceding page.

**FIGURES 6–3 AND 6–4**

Among most dolichocephalic Caucasian groups, the upper part of the ethmomax-illary region characteristically is even **further** protrusive. This adds to the forward manner of upper face placement caused by the elongate and narrow brain. The an-terior cranial floor and the frontal lobe (a), the superior part of the ethmomaxillary complex (b), the palate and maxillary arch (c), and the mandibular corpus (d) are all structural "counterparts" to each other. Depending on (1) the actual horizontal dimension each part attains by its own regional growth, (2) the alignment of that part, and (3) the direction and the extent to which each part is **displaced** by other regions, the form and profile of the face is changed accordingly. Because the an-terior cranial fossa (a) is horizontally long and narrow, the upper face (b) is also correspondingly long and narrow. In many dolichocephalic Caucasians, however, part b adds to the extent of upper face protrusion by continued horizontal expan-sion beyond a and c. This causes a characteristically high, sharp nasal bridge and a sizable nose. If the extent of protrusion is quite marked, the nose **bends** in order to relate structurally with part c.* The outer cortical table of the forehead is carried an-teriorly with the nasal bridge, and a large frontal sinus is thereby formed between the inner and outer tables. The forehead is much more sloping as a result, and the glabella becomes noticeably protrusive. The cheekbones often appear less promi-nent and more "hollow" because the remainder of the upper and the middle face are so protrusive. Because the mandible is rotated posteriorly, it tends to be re-trusive, and the whole profile takes on a characteristic convexity for all these reasons. A Class II tendency is **built-in.**

---

*In another type of facial configuration, the **lower** part of the nasal region anterior to the cheek-bones, rather than the upper part, is the more protrusive. This also causes a bent nasal shape, and the slope is characterized by a graceful recurved configuration. The whole cheekbone area is also pro-trusive. In either case, the curved type of nose is **vertically** long. The longer it is, the more sharply it must curve. The lower border of the nasal profile in a vertically long nose sometimes angles downward, while the ventral edge of the shorter nose in other facial types sometimes angles upward.

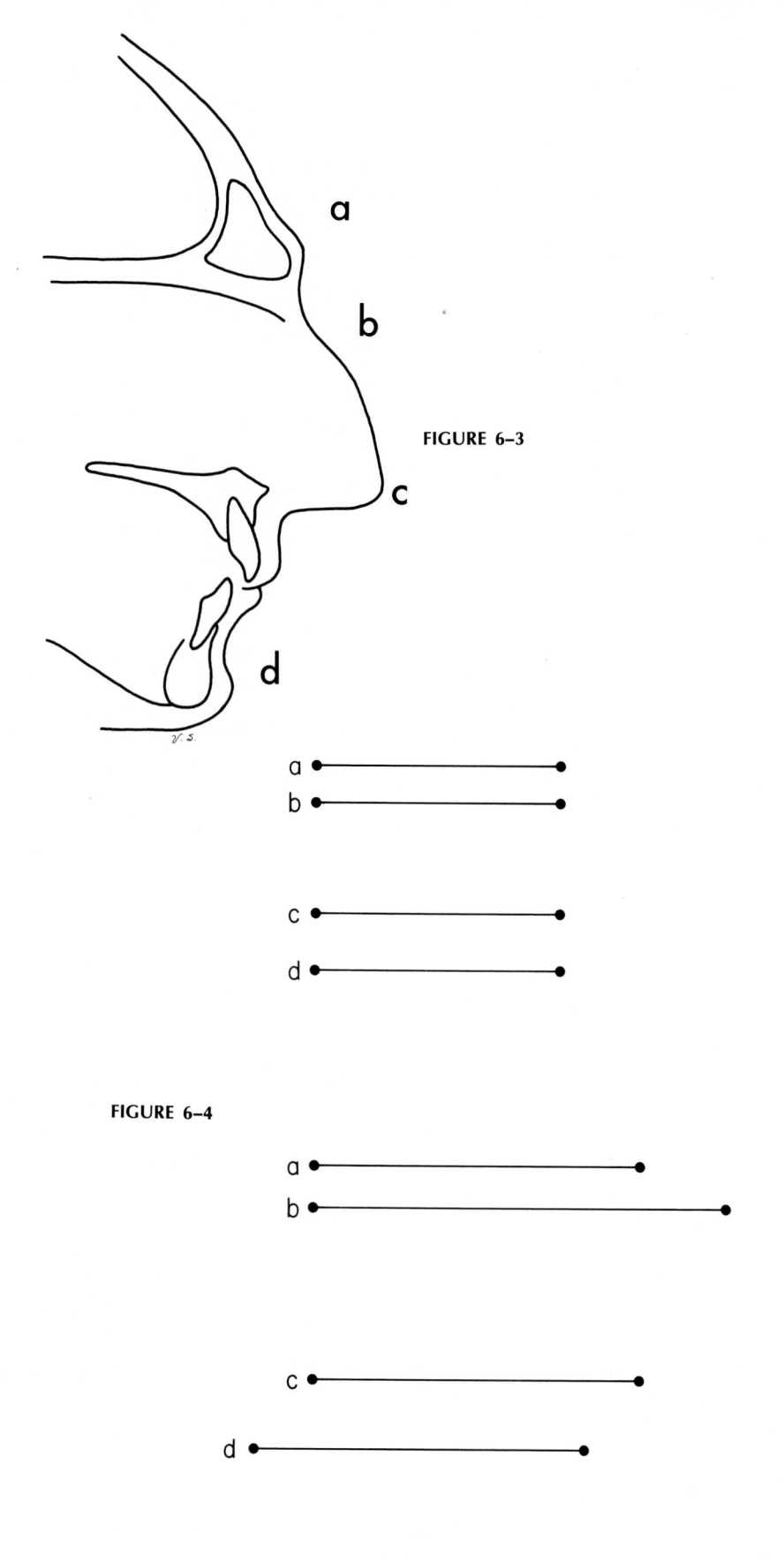

FIGURE 6–3

FIGURE 6–4

**FIGURES 6–5, 6–6, AND 6–7**

The more closed, upright cranial base flexure that usually characterizes the brachycephalic head sets up a correspondingly wider, flatter, more upright type of face. The rounder, horizontally shorter brain and correspondingly shorter anterior cranial floor limits facial size and places the whole upper face in a less protrusive position. The middle cranial fossa is horizontally shorter for the same reason and, also, because it is much more vertically inclined. In addition, the upper part of the ethmomaxillary complex does not expand anteriorly to nearly the same extent described for the previous facial type. The composite result is a more upright and bulbous forehead, a lesser protrusion of the glabella and eyebrow ridges, a smaller frontal sinus, a much lower nasal bridge, a shorter pug-type nose, and a tendency for a forward rotation of the entire mandible (unless offset by a vertical lengthening of the midface, which is a common feature). These features give a quite "vertical" character to the whole face, and the face appears much flatter, broader, and squared. The cheekbones are more prominent-appearing because the remainder of the upper and middle face is not as protrusive. There is a greater likelihood for an orthognathic (straight) profile, and the chin appears prominent and the mandible quite full. A greater tendency for a Class III type of malocclusion and a prognathic mandible exists. However, the nasomaxillary complex can also be vertically long, and the ramus can thus rotate well downward and backward as a result. The face does not "look" as long, however, because it is wider. The eyes "look" more wide-set because the nasal bridge is low. The mandibular corpus tends to be horizontally shorter relative to the maxillary arch, as compared with the situation in the dolichocephalic type, and this factor, together with the backward ramus rotation, contributes to a compensation for the built-in tendency toward prognathism and bimaxillary protrusion.*

---

*The information in this section is based on previously unpublished work carried out by the author in collaboration with Dr. Takayuki Kuroda of the Tokyo Medical and Dental University.

The above features characterize the Oriental face. Some Caucasian groups also have a rounder mesocephalic or brachycephalic type of head form, and many of these same facial features are thereby present. The face is wider, the nasal bridge lower, the nose flatter and shorter, the midface vertically shorter, the forehead more upright, and the mandible more prominent. There are fewer underlying Class II tendencies in this basically different type of Caucasian face. Class I individuals having this composite facial structure tend toward a more orthognathic type of profile. When a Class II malocclusion does develop, however, it is a different kind (see Enlow et al.: Angle Orthod., 41:3, 161, 1971). Care must be taken by the orthodontist because there are often stronger mandibular protrusive factors within this face, and Class II treatment procedures can sometimes produce unexpected and undesired results. However, this type of Class II malocclusion is characteristically less severe than the others, and proper treatment results are often gratifying.

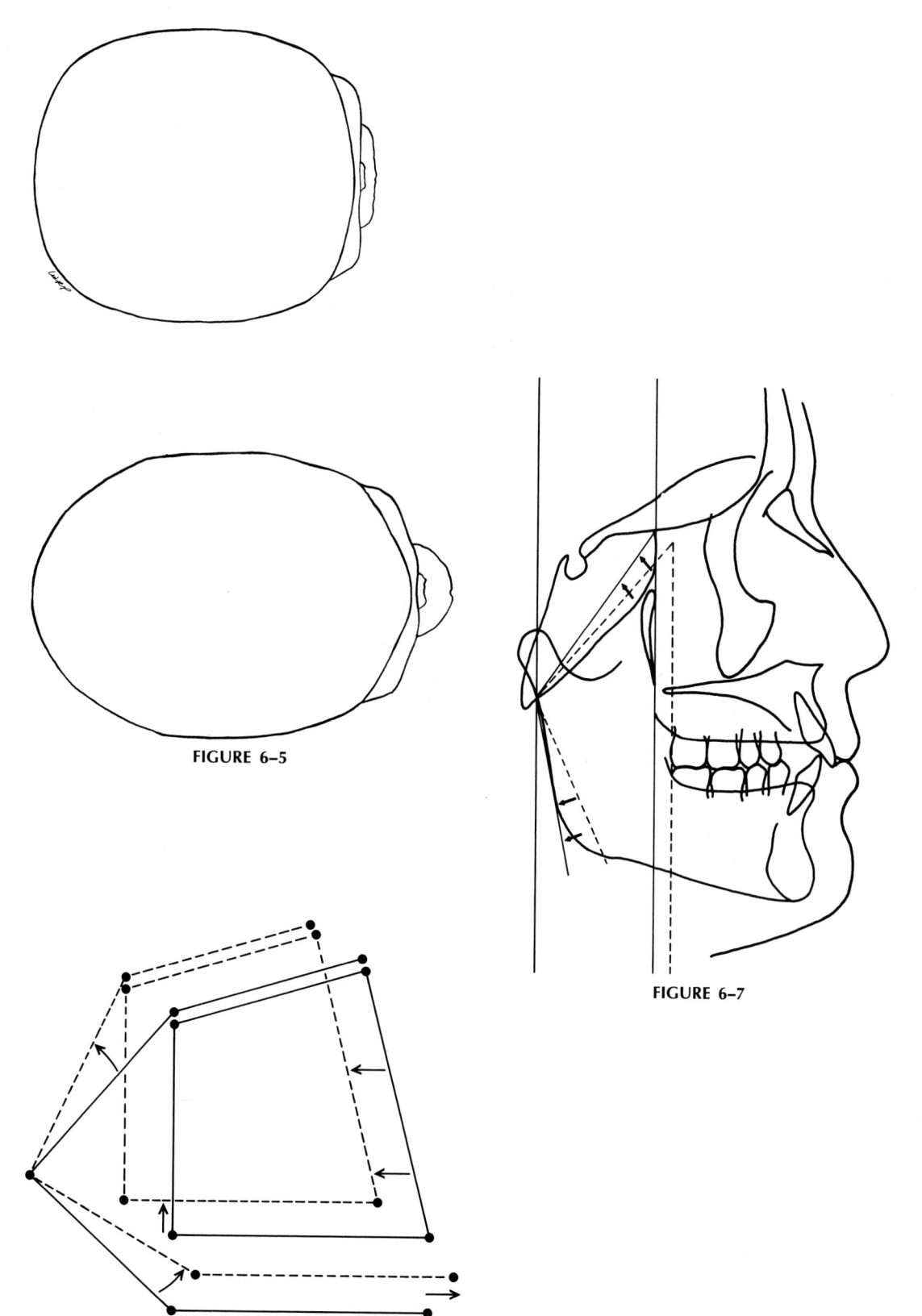

FIGURE 6-5

FIGURE 6-7

FIGURE 6-6

**FIGURE 6–8**

Black individuals,* like many Caucasians, tend to have an elongate, dolichocephalic head form. The middle cranial fossa has an anteriorly inclined (open) alignment. This factor, together with a vertically long nasomaxillary complex in some Black groups, causes the ramus (and whole mandible) to rotate down and back. The mandibular corpus tends to be horizontally long relative to the bony (not dental) maxillary arch. This is similar to the Caucasian pattern but unlike the Oriental. Unlike the typical "long-headed" Caucasian facial type, however, the upper part of the face in the Black expands much less and is thereby not nearly so protrusive. In this respect, the face of the Black corresponds to that of the Oriental. The forehead is more upright and bulbous than in most Caucasians, the frontal sinus proportionately less expanded, the nasal bridge lower, the nose flatter, wider, and less protrusive, and the cheekbones more prominent. One special feature characterizes the Black face; the mandibular **ramus is quite broad**. In a previous chapter, it was pointed out that the horizontal dimension of the ramus is a site that commonly participates in compensations for structural imbalances in other parts of the face and cranium. The forward inclination of the middle cranial fossa that characterizes many Caucasian groups, for example, is partially or completely counteracted by the development of a wider ramus, thereby offsetting an intrinsic tendency for mandibular retrusion and a Class II malocclusion. The mandible of the Black also has this feature, but the amount is characteristically **much** greater. The very broad ramus places the mandibular corpus (which can also be long relative to the bony maxillary arch) in a resultant protrusive position. This, in turn, causes the maxillary incisors to tip labially, and a **bimaxillary** protrusion is thereby produced. This is an advanced feature that, for the dolichocephalic Black, often forestalls severe Class II malocclusions. They are usually of the Class II "B" type. That is, mandibular *B* point lies well ahead of maxillary *A* point, in contrast to the Class II "A" type, in which *A* point is the more protrusive relative to the occlusal plane (see Enlow et al.: Angle Orthod., 41:3, 161, 1971).

*The information in this section is based on previously unpublished work by the author in collaboration with Dr. Elisha Richardson, Meharry Medical College, School of Dentistry, and Dr. Takayuki Kuroda.

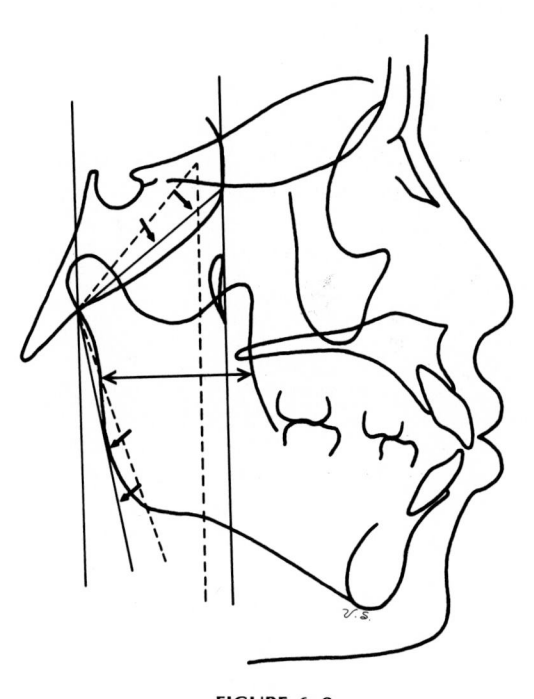

**FIGURE 6–8**

# Chapter 7

# Control Processes in Facial Growth

Many students of facial growth feel that we are in the midst of a great conceptual revolution dealing with the basic processes that govern morphogenesis. It is believed that we are now at the threshold of major, exciting breakthroughs into one of the most important biological problems of our time: how the local growth control process actually operates at the tissue and cellular level.

Until fairly recently, some explanations for the growth control process were regarded as more or less complete, with the theories underlying them sound and secure. This has all changed. We are now beginning to be able to recognize just what we don't really understand at all, and we think we can define the problems involved. This in itself is a breakthrough. So many advances in so many diverse, interdisciplinary fields relating to the general subject of growth control have been made in the past decade that we are dazzled by what is happening. Is there clinical significance? Can you **imagine** what could be done with treatment procedures if we could effectively control, at the most basic level, the local growth control process itself. This is the great goal. However, we must understand the overall process much better than we presently do. Major progress has already been made.

The explanations of the growth control process that prevailed until just a few years ago were straightforward, easy to understand, and so plausible that they became adopted and used for many years as the basis for a number of clinical concepts. If one did not question certain basic premises, the whole process of growth control was no particular puzzle and was readily explainable. First, the growth of all bones having cartilage growth plates was presumed to be regulated entirely and directly by the intrinsic genetic programming within the cartilage cells (see pages 79, 92, 102, and 110). This, of course, has now become a controversial point, and the matter is presently far from settled. Intramembranous bone growth, however, was believed to have a different source of control. This type of osteogenic process is particularly sensitive to biomechanical stresses and strains, and it responds to tensions and pressures by either bone deposition or resorption. Tension, as traditionally believed, specifically induces bone formation. Pressure, if it exceeds a relatively sensitive threshold limit, specifically triggers resorption. When tension is exerted on a bone, as at places of muscle attachment, the bone grows locally in response. Thus, sites of muscle insertion are usually marked by tuberosities, tubercles, and crests which form because of direct, localized fields of muscle traction. Since many muscles attach near the ends of a bone rather than on its shaft, the epiphyses are much larger than the diaphysis, because this is where the muscles

234

apply the most tension and where the bone thereby expands. As long as a muscle continues to grow, the bone is also stimulated to grow. This is because of the continuing biomechanical imbalance between them due to the expansion in muscle mass and resultant increasing force. The growing muscle exceeds the capacity of the bone to support it, and the osteoblasts are thereby triggered to form new bone in response. When muscle and overall body growth is complete, the bones attain biomechanical equilibrium with the muscles (and body weight, posture, and so forth). The forces of the muscles are then in balance with the physical properties of the bone. This turns off osteoblastic activity, and skeletal growth ceases. If any future circumstances cause departure from this sensitive state of bone–soft tissue equilibrium, such as major changes in body weight, loss of teeth, or the fracture of a bone, the process is revived until once again mechanical equilibrium subsequently becomes attained.

It is easy to understand why the above explanations were attractive and almost universally adopted by earlier workers. These concepts served to explain most everything then known about bone and its growth.

More recently, the realization that a number of shortcomings exist with regard to these reasonable concepts led to a reevaluation of the whole process of growth

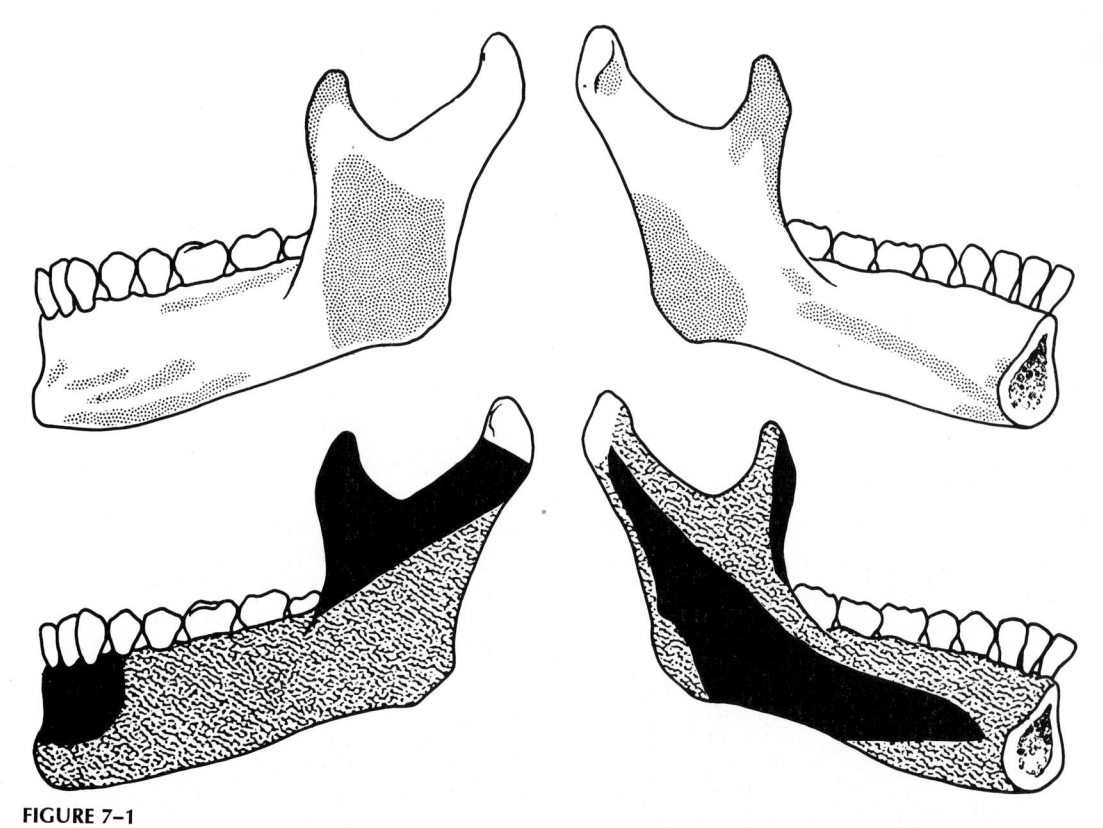

**FIGURE 7–1**

*The top figures show the distribution of muscle attachments on the buccal and lingual sides of the mandible. The bottom figures illustrate the pattern of surface resorptive (dark) and depository (light) growth and remodeling fields. Note that there is no one-to-one correlation between these respective patterns. As described in the text, this does not mean that muscle forces are not involved in growth control; it does show, however, that the old "muscle tension–direct bone deposition" concept is invalid. (From Enlow, D. H.: Wolff's law and the factor of architectonic circumstance. Am. J. Orthod., 54:803–822, 1968.)*

control. The subject has become a "new" frontier in facial biology. It is perhaps the most important problem that now faces us.

There is not a one-to-one correlation between places of muscle attachment and the pattern of distribution of resorptive and depository fields (Fig. 7–1). Growth control is more complex than this. Moreover, it is now known that there is **not a** direct, one-to-one correlation between tension-deposition and pressure-resorption (this pressure-tension concept is greatly oversimplified; see pages 332 and 342). This is important. About half of all bone surfaces to which muscles attach are actually **resorptive**, not depository. Many muscles have widespread attachments, and within these surface areas, some growth fields are resorptive and others are depository. Yet, these different surfaces are subject to the same pull by the same muscle, supplied by the same blood vessels, and innervated by the same nerves. The temporalis muscle, for example, inserts onto the coronoid process of the mandible (Fig. 7–2). As shown in Chapter 3, parts of this mandibular region have external surfaces that are resorptive. The muscle exerts tension, but the bone to which it directly attaches undergoes resorption. Other surfaces of temporalis muscle attachment are characteristically depository. Further, some muscles pull in one direction, but the bone surfaces into which they insert grow in other directions. The pterygoid muscle, for example, attaches onto the posterior part of the ramus. The muscle pulls anteriorly, but this part of the bone grows posteriorly.

Growth control involves graded feedback chains from the systemic down to the local tissue, cellular, and molecular levels. The problems at hand deal with the **local** control process.

With the development of the **functional matrix** principle, a number of important considerations began to receive attention. One of these is that the "bone" does not regulate its own growth. The genetic and epigenetic determinants of skeletal development are in the functional soft tissue matrix, not within the bone (page 80). The bone tissue itself is essentially passive and secondary in regard to growth control. It is important to understand that the functional matrix principle describes **what** happens during growth. It does not account for **how** it happens. The term is a title for a biological process; how the process actually operates is not explained. Wolff's law,* similarly, is the title for a biological process, but it does not explain the mechanism that underlies it. What, then, carries out the actual operation of this growth control process? Historically, two general explanations for the regulation of skeletal growth have existed: the genetic and the biomechanical. It is, in a sense, the old question of heredity versus environment, and there are advocates for both camps. Many investigators hold that skeletal development is regulated largely by genetic determinants and that the actual seat of this control lies directly in the genes of the soft tissue matrix. Many others believe that the play of mechanical stresses on a developing bone regulates the details of its progressive growth and differentiation. The variable stresses presumably trigger **selective** genetic expressions or induce direct physiologic changes to provide control. Whatever level of genetic control exists, however, pressures and tensions are held to be **the** overriding factor. Many experiments have been carried out in which muscles were severed, or the soft tissues otherwise altered, and in which artificial mechanical forces were ex-

*Wolff's law, as it has customarily come to be interpreted, simply maintains that the form and structure of a bone represent direct developmental adaptations to the bone's composite of functions. The bone's morphology becomes progressively structured to accommodate the sum of all the changing mechanical forces exerted on it during growth and development.

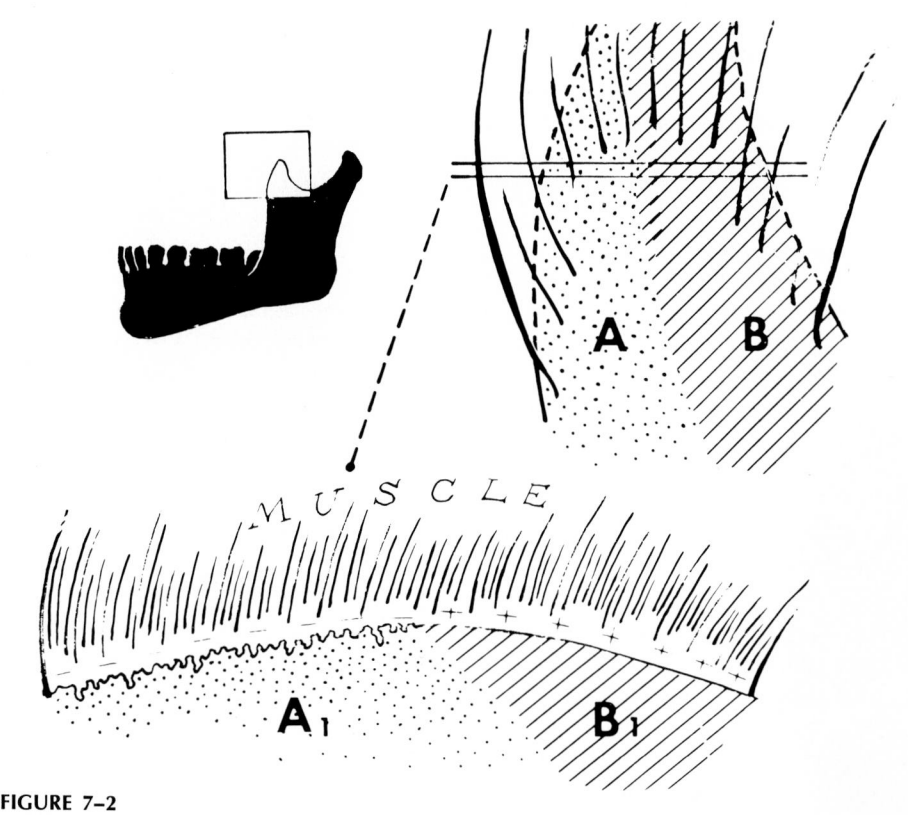

**FIGURE 7–2**

*The temporalis muscle attaches to surfaces* A *and* B *on the lingual side of the coronoid process. In microscopic sections, it is seen that the attachment on* $A_1$ *involves a resorptive bone surface; the same muscle is also inserted on surface* $B_1$, *which is depository. (From Enlow, D. H.: Wolff's law and the factor of architectonic circumstance. Am. J. Orthod., 54:803–822, 1968.)*

erted on a living bone. Because such procedures always result in some kind of response with a resultant change in the form of the bone, it has often been concluded that **stress** is therefore the principal factor controlling bone growth. Such experiments, however, do not "prove" such a role for mechanical forces, since certain critical variables necessarily exist that cannot be controlled in the experimental design. These include vascular interruption, nerve severance, temperature changes, alterations in pH and oxygen tension, and so on, all of which are known to affect bone growth. The fundamental question must be asked concerning whether or not extrinsic or unusual factors that can affect the course of bone development **also** necessarily represent the same factors that actually carry out the direct, primary control of the basic processes of growth and differentiation.

A great many factors have been shown or are presumed to be involved in growth control. Two of these, as just mentioned, are genetic influences and biomechanical forces, but the nature of the balance between them is still, at best, uncertain. Other factors include, for example, the neurotrophic effect, the piezo effect, and the activity of inductor substances.

The **neurotrophic factor** involves the network of nerves (all kinds, motor as well as sensory) as links for feedback interrelationships among all the soft tissues and bone. The nerves provide pathways for stimuli that trigger bone and soft tissue responses. It is not believed, however, that this process is carried out by actual ner-

vous impulses. Rather, it appears to function by transport of neurosecretory material along nerve tracts (analogous perhaps to the neurohumoral flow from the hypothalamus to the neurohypophysis along tracts in the infundibulum), or by an axoplasmic streaming within the neuron. In this way, feedback information is passed, say, from the connective tissue stroma of a muscle to the periosteum of the bone associated with that muscle, and the "functional matrix" thereby operates to govern bone development.

The **piezo factor** has been one of the great hopes of the 1960's and 1970's and promises to explain just how muscle action can be translated into bone remodeling responses. The idea, in brief, is that distortions of the collagen crystals in bone, caused by minute deformations of the bone due to mechanical strains, produce bioelectric changes (that is, the piezo effect). These altered electrical potentials appear to relate, either directly or indirectly, to the triggering of osteoblastic and osteoclastic responses. The link between the bioelectric changes and the cellular responses, as of this writing, has not yet been fully explained, and the role of the piezo effect in the overall picture of growth control has not yet been fully nailed down. However, at present this is a most active and exciting area of research.

It is to be realized that growth control is essentially a **localized** process because we are dealing with specific, restricted **fields** of growth activity (page 10). Bone resorbing and depositing cells respond to chemical, electric, and mechanical changes ("first messengers") in their **immediate** environment. The plasma membrane of each cell contains sensitive **surface receptors** that are responsive to such extracellular stimuli and which, in turn, activate a membrane-bound enzyme (adenyl cyclase). This substance accelerates the transformation of ATP to cyclic AMP (a "second messenger") within the cytoplasm, which then activates the synthesis of other specific enzymes relating selectively to bone resorption or bone deposition. Ionic calcium is also believed to be involved as a second messenger. The chain from receptor site to adenyl cyclase to cyclic AMP mobilizes $Ca^{++}$ from storage in the mitochondria. Changes in cell and organelle membrane permeability result, and this selectively controls the flux of other ions involved in the synthesis of those materials secreted by the cell (ground substance, fibers, enzymes, and so forth).

During bone formation, the osteoblast takes in amino acids, glucose, and sulfate for the synthesis of the glycoproteins and collagen in the organic part of the bone matrix. The cytoplasmic organelles within the osteoblast participate in the formation, storage, and secretion of tropocollagen, the ground substance, and also ions which form the inorganic (hydroxyapatite) phase of the bone matrix. Alkaline glycerophosphatase is related to bone formation (in contrast to acid phosphatase, which relates to resorption) and is associated with the collagen fibril as it is released from the osteoblast. High levels of alkaline phosphatase are also involved in the formation of the hydroxyapatite. The citric acid cycle and glycolytic enzymes provide generalized energy sources for all these activities.

The osteoclast contains an abundance of mitochondria in addition to lysosomes and an extensive endoplasmic smooth membrane system. The osteoclast produces, stores, and secretes enzymes (such as collagenase) and acids that relate to the breakdown of both the organic and inorganic components of bone. The lysosomes are involved in acid phosphatase storage and transport. First messengers, such as parathyroid hormone or bioelectric changes, stimulate receptor sites on the cell membrane. This activates adenyl cyclase, which in turn causes increases in cytoplasmic cyclic AMP. The latter then increases the permeability of the lysosomal membrane. By an exocytosis of the lysosomal contents, the resorption of both the

organic and inorganic parts of the bone is carried out through the activity of the acid hydrolases, lactates, and citrates. The endoplasmic smooth membrane system is also involved in this process of enzyme transport and release.

The preceding accounts provide a thumbnail, introductory outline of where we stand today in our general understanding of the control of bone growth and remodeling. Overall, it is a fairly complex picture. It is one of the most important of all our problem areas. A great many laboratory studies are now being conducted in attempts to clarify the relationships between, for example, cyclic AMP and biomechanical forces. Other studies are now dealing with the role of important substances such as the prostaglandins and neurotrophic agents. Still other work is under way seeking out possible chalone-like agents in bone (that is, localized tissue-level, hormone-like substances that are believed to accelerate or retard cell divisions). Day by day, new information and new links are being added. To help place all these old and new factors in some kind of meaningful perspective, one can use the following, relatively simple "test" to classify their respective roles:

1. Is a given factor the sole, primary agent that is directly responsible for the master control of growth? Of course, such a single, ubiquitous agent does not exist. Historically, bone investigators have searched for the master control factor, such as a special "hormone" or inductor agent, that does it all. Based on our present knowledge, however, it is now realized that the control process is multifactorial. Control involves a **chain** of regulatory links. Not all the individual links are involved in all types of growth changes. Rather, a **selected** combination exists for different specific control pathways that can follow many routes and involve different agents.

2. Does a given factor function as a "trigger" that induces or turns on other specific, selected agents that then launch the process of control response? Is it the first link in the chain? Is it the initial agent in the process of "induction"? Is it a "first messenger"? Biomechanical forces presumably represent such a trigger. It is still justifiably believed that pressures and tensions are indeed among the basic agents involved in growth control (although not following the traditional but oversimplified concept). Different kinds of bone, however, have variable thresholds of response to physical forces (e.g., basal bone versus alveolar bone in the mandible and maxilla, which are relatively nonsensitive and labile, respectively, to physical strains). It is also evident that biomechanical forces are not the **only** agents of control. Even when they are involved, many other links are also required as second and third level messengers.

3. Is a given factor, in effect, the **title** for some biological process without accounting for the actual functional mechanism involved? This is an important category. Such a title describes **what** happens but not **how** it happens. It does not explain the operation of the control process it represents. It is, in effect, a synonym for "control process" without explaining how it works. The reason why an awareness of this category is so important is because we often tend to use such titles as though they do, indeed, explain the mechanism involved. With continued use, we delude ourselves into believing that we actually understand the real basis for the control process. "Wolff's law," the "functional matrix," and the process of "induction" are all such descriptive labels for biological control systems. They do not, of course, explain how the system works (they were actually never intended for this ultimate purpose, even though many have used them as such). Be ever alert for this conceptual pitfall.

4. Does a given factor function in a supportive role or as a catalyst? Many nutrients would be an example of this category.

5. Does a given factor accompany the control process but not actually take part? It has been necessary for laboratory researchers studying the piezo effect, for example, to establish whether bioelectric potentials are actually first messengers or whether they simply "occur" in a nonparticipating role.

6. Is a given factor an actual **cause** rather than an effect in the growth control process? The piezo factor also serves as an example for this category. Do bioelectric changes in bone directly trigger remodeling responses or are they merely the incidental result of them? Present evidence is believed to support the former.

One major, fundamental feature of the control process is now clear. Any given tissue, such as bone, does not grow and differentiate in an isolated, independent manner by a wholly intrinsic regulatory process. Control is essentially a system of feedback pathways, informational interchanges, and reciprocal responses. Tissues develop in conjunction with one another. A given bone and all its muscles, nerves, blood vessels, connective tissues, and epithelia represent an interdependent, developmental **composite**. Bones have specific mechanisms for increasing in **length** (e.g., an epiphyseal plate, synchondrosis, condyle, suture), and they have another specific mechanism for increasing in **width** (subperiosteal, intramembranous growth). Correspondingly, muscles **also** have a specific growth mechanism for increases in length and another for increases in width. Both of these muscle growth processes proceed in concert with respective growth mechanisms for the bone. There are reciprocal feedback interrelationships between the muscle and bone, as well as the various other tissues, and they all enlarge together, not as "separate" and independent units. For example, input to the sensory nerves in the periodontal membrane can trigger growth responses to occlusal signals from the teeth. These signals can be passed on through an arc to motor nerves supplying the muscles of mastication. In conjunction with muscle adaptations to the individualized nature of the occlusion, thus, the bones of the face can remodel in association with the muscle and soft tissue matrix that encloses them and controls their course of growth. The old concept that a "growth cartilage" serves as the primary regulator for the overall development of a musculoskeletal composite is now regarded as an incomplete and unacceptable explanation because many more input factors are now known to be involved. However, we still have a long way to go in understanding the whole of the growth control system. History will likely judge this one of the great problems of our time.

# Facial Growth and Development in the Rhesus Monkey

Primates have become increasingly popular, in recent years, for laboratory studies of the facial growth process. Other animals, of course, have been used quite profitably for basic research, but it is felt by many investigators that the macaque (and some other simian species as well) is particularly appropriate for certain clinically related types of experimental studies. The purpose of this short chapter is to introduce the researcher to the elements of craniofacial growth in the monkey and briefly to outline major similarities and differences in comparison with the human facial growth process.

The basic plan of growth in the monkey's mandible, nasomaxillary complex, and cranium generally parallels that found in the human face and cranium. While there are certain specific and important differences (Fig. 8–1), as pointed out below, the overall distribution and pattern of resorptive and depository growth fields and the nature of displacement movements are similar. The simian mandible grows predominantly posteriorly, and the ramus lengthens in a superoposterior direction (Fig. 8–2, 8–3, and 8–4). As in man, a continuous sequence of remodeling changes is involved in backward and upward ramus growth and the progressive remodeling conversion from ramus to corpus. The same basic rationale exists for the various regional growth fields in both the ramus and corpus. As in man, the monkey's mandible becomes displaced anteriorly and inferiorly as it enlarges. The nasomaxillary complex in the monkey also grows posteriorly as it is simultaneously displaced anteriorly. The maxillary arch grows downward by the same pattern of deposition, resorption, sutural growth, and displacement as in the human face. The floor of the monkey's cranium, similarly, shows remodeling changes resembling those of the human cranial base, although differences in extent are involved in relation to the respective sizes of the brain (Fig. 8–10).

Because there are differences in the gross topographic shape in several regions of the simian skull, there are corresponding differences in the growth and remodeling of these particular parts. One notable difference exists in the mental region of the mandible (Fig. 8–5). In the human mandible, a prominent chin marks this region, a distinctive feature that characterizes the face of modern man (and also, for reasons yet to be studied, the elephant). This structure is lacking in the monkey as well as in the mandibles of extinct species of *Homo*. The mandible of the monkey, however, has a unique "simian shelf" on the lingual side in the mental region. This is not present in man. In the human mandible, the chin is formed by *(1)* variable amounts of bone deposition on the mental protuberance, together with *(2)* surface resorption in the alveolar area just above it. The lower incisors tip lingually, and

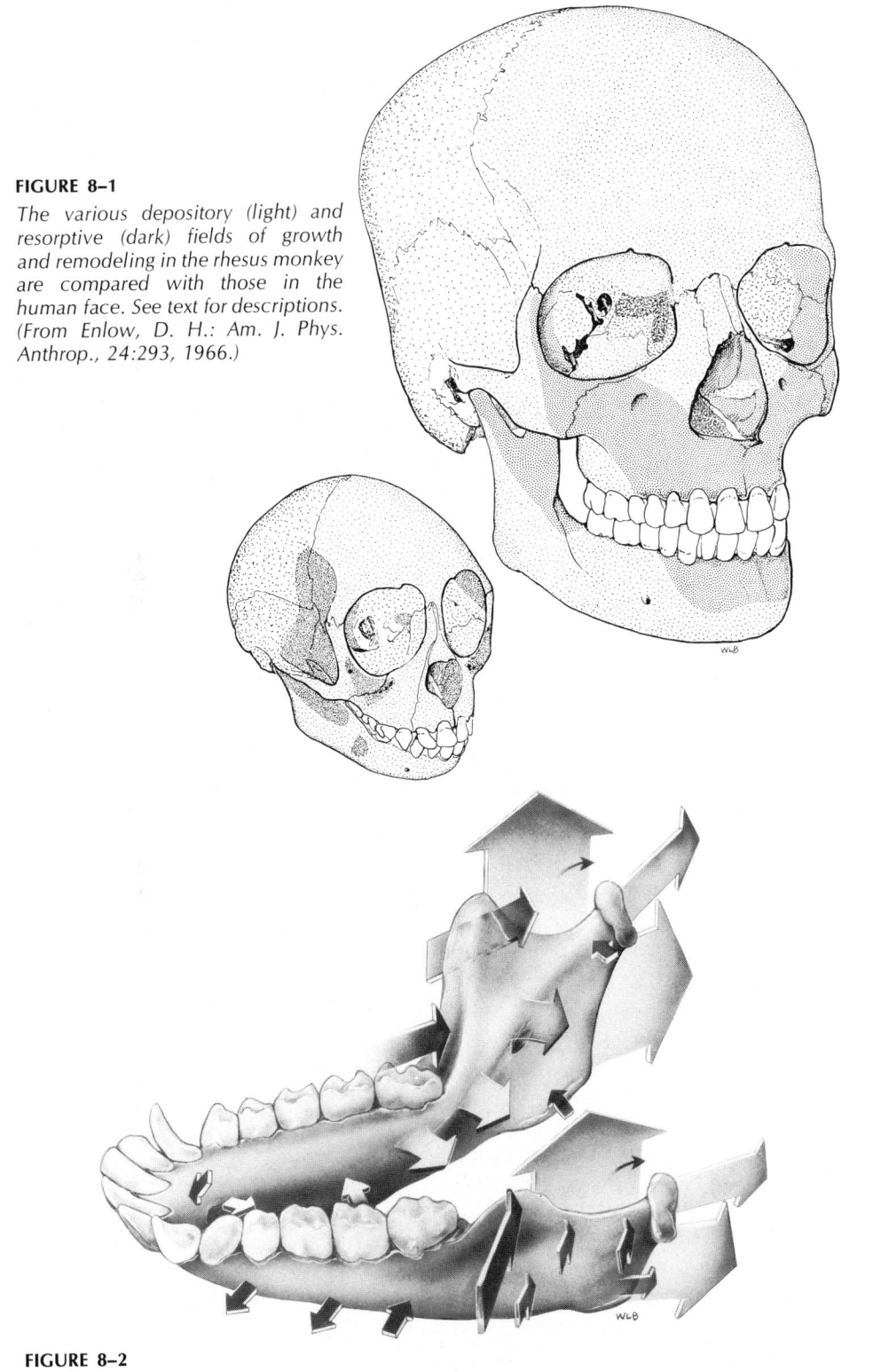

**FIGURE 8–1**

*The various depository (light) and resorptive (dark) fields of growth and remodeling in the rhesus monkey are compared with those in the human face. See text for descriptions. (From Enlow, D. H.: Am. J. Phys. Anthrop., 24:293, 1966.)*

**FIGURE 8–2**

*The growth and remodeling of the rhesus monkey's mandible is shown by resorptive arrows (going into the bone surface) and depository arrows (coming out of the bone surface). Compare with the human mandible in Figure 3–150. (From Enlow, Donald H.:* Principles of Bone Remodeling, *Courtesy of Charles C Thomas, Publisher, Springfield, Illinois.)*

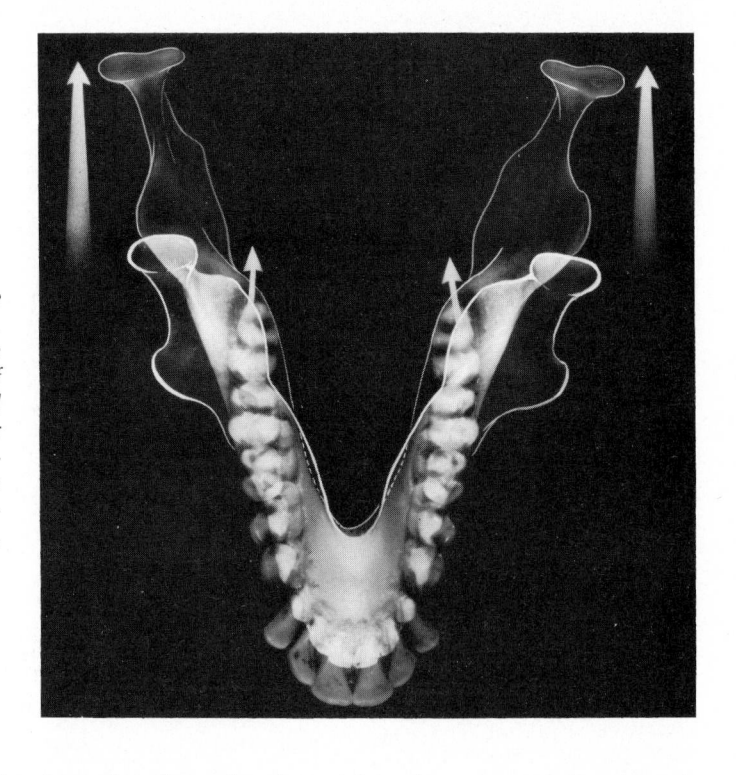

**FIGURE 8–3**

*Superimposed growth stages of the mandible in the rhesus monkey. Posterior growth of the ramus from the lingual surface (principle of the "V") is indicated by the small arrows, and the direction of condylar growth is shown by the larger arrows. (From Enlow, Donald H.: Principles of Bone Remodeling, Courtesy of Charles C Thomas, Publisher, Springfield, Illinois.)*

"overjet" and "overbite" result in lieu of end-to-end tooth contact. In the pointed, chinless mandible of the monkey, the entire labial surface of the mental region undergoes progressive anterior bone growth (Fig. 8–6). The incisors do not drift lingually (posteriorly). Rather, the mandibular corpus of the monkey grows forward in the anterior part as well as backward in the posterior part. The result, in conjunction with an equivalent mode of maxillary growth, is a more protrusive muzzle as compared with that of man. In mammalian species having markedly protrusive lower and upper jaws, the extent of such forward growth is even more than in the monkey. This growth and remodeling pattern produces an elongate, angular, chinless mandible in contrast to the short, broad, rounded, regressive, flattened, and chin-tipped arch of the human mandible.

On the lingual side, a contrasting pattern of growth and remodeling is present between the human and the monkey mandibles. In man, the entire lingual surface shows a marked accumulation of periosteal bone. This lingual surface in the

**FIGURE 8–4**

*This diagram illustrates the growth of the monkey's mandible by the V principle. The condyle and ramus do not simply grow backward along the principal axis, as represented by B. Rather, the enlarging V involves deposition on the lingual side, with some areas of resorption on the buccal side, because it increases in length more than in width. (From Enlow, Donald H.: Principles of Bone Remodeling, 1963. Courtesy of Charles C Thomas, Publisher, Springfield, Illinois.)*

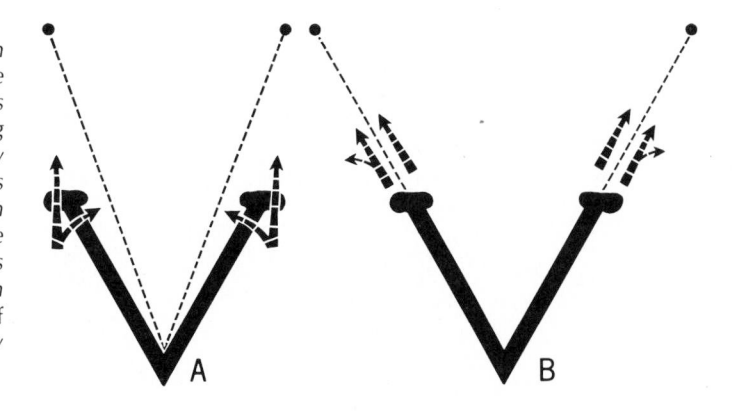

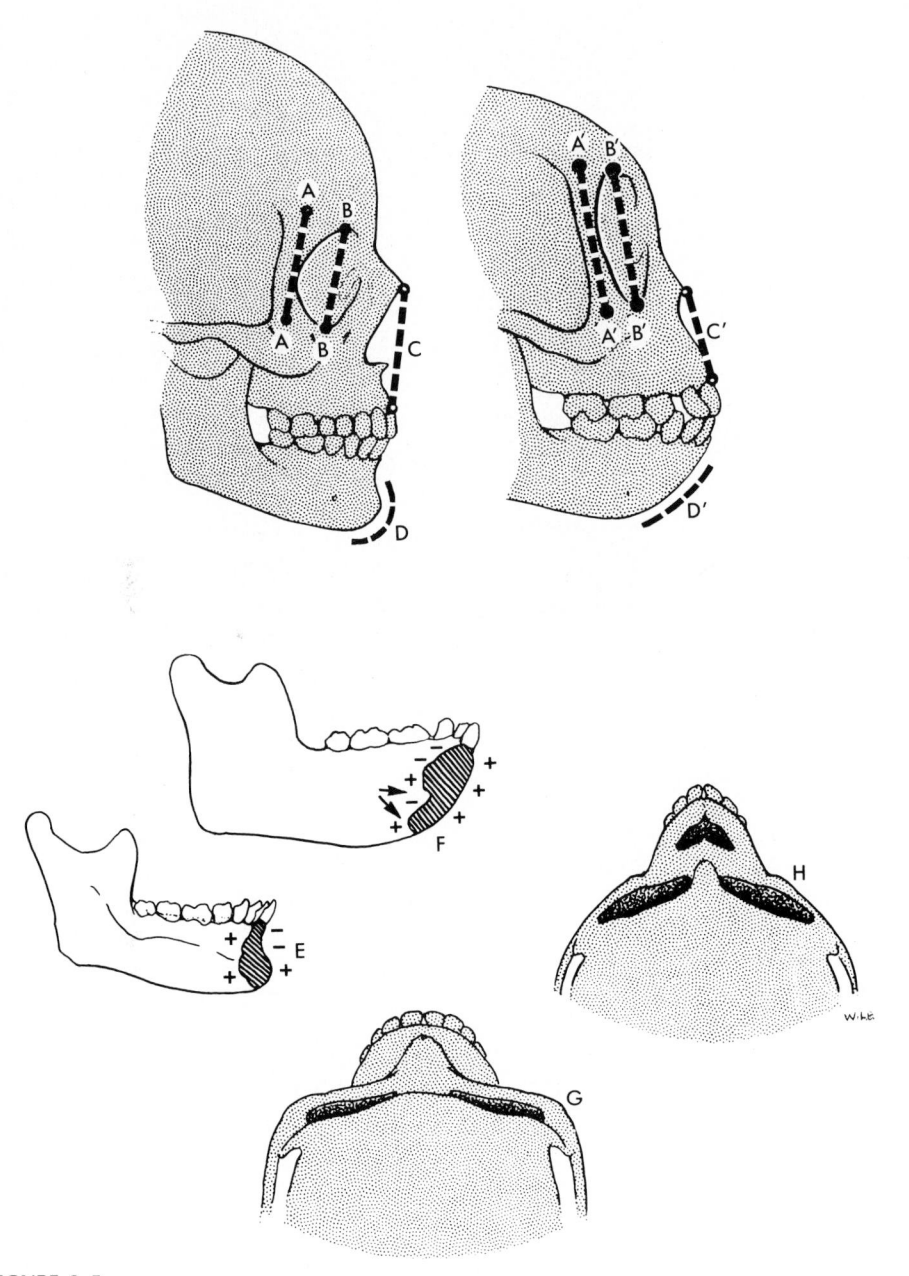

**FIGURE 8–5**

*Because of anatomic differences between the human and monkey facial parts, corresponding differences are involved in their respective growth and remodeling processes. Compare the size and slope of the frontal prominence, the relative positioning of the upper and lower orbital rims (B), the angle of the lateral orbital rims (A), the relationships between the tip of the nasal bone and the premaxilla (C), the structure of the mandibular mental region (D), the presence of a simian shelf in the monkey's mandible and a chin on the human mandible (E and F), and differences in the contour of the malar region (G and H). (From Enlow, D. H.: Am. J. Phys. Anthrop., 24:293, 1966.)*

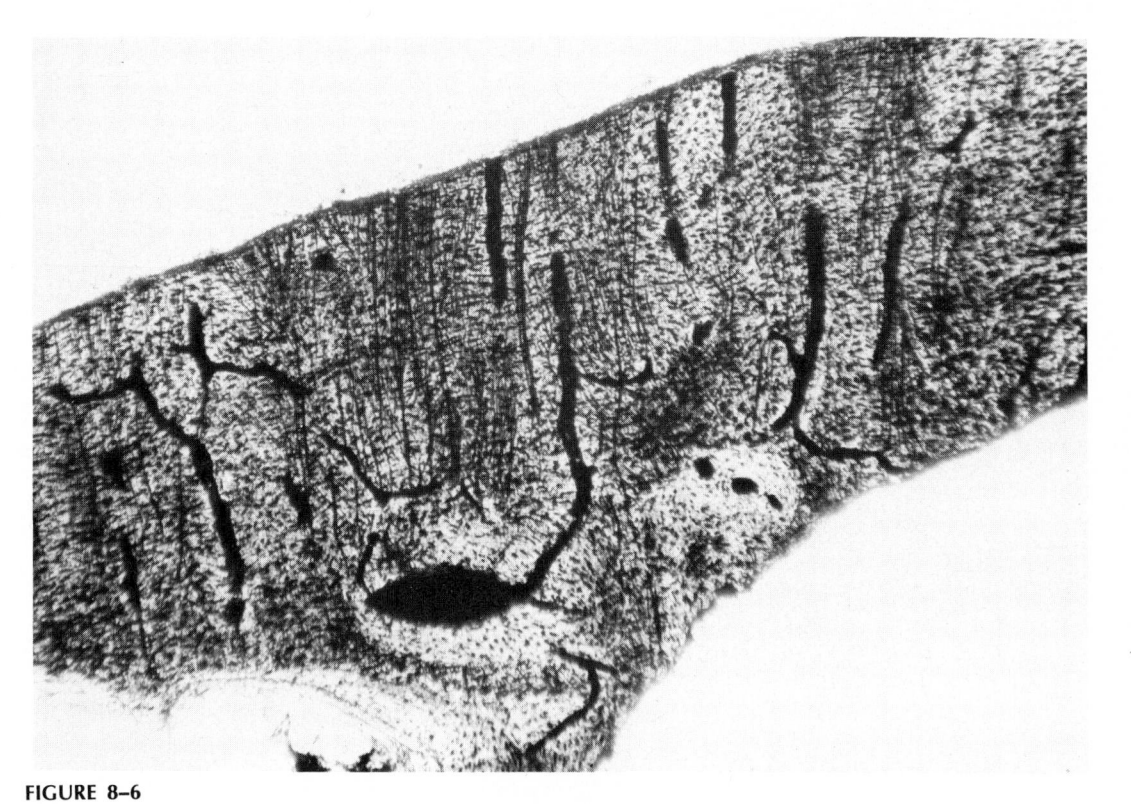

**FIGURE 8–6**
*This cortical section from the monkey's mandible was taken from the labial side of the cuspid region. It shows a depository periosteal surface and a resorptive endosteal surface. Note the attachment fibers embedded in the cortex. (From Enlow, Donald H.:* Principles of Bone Remodeling, *1963. Courtesy of Charles C Thomas, Publisher, Springfield, Illinois.)*

monkey, however, is largely resorptive, a feature that complements the forward, protrusive growth in the anterior part of the bony arch. But at the genial crest, a reversal line is present, and below this line part of the lingual surface is depository, not resorptive. This remodeling combination produces the prominent, characteristic **simian shelf** of the monkey's mandible. These contrasting remodeling differences between man and monkey in the forward part of the mandible result in a chin on one side and a simian shelf on the other, respectively (Fig. 8–5).

Because the relative dimensions of the human brain and its cranial floor are a great deal wider in man, the bi-condylar dimension of the human lower jaw is proportionately more broad. Together with the proportionately shorter arch, the **U**-shaped human mandibular configuration is characteristically much more rounded. In contrast, the **V**-shaped mandible of the monkey is proportionately longer, more narrow, and quite angular (Fig. 8–2.) The massive trihedral eminence on each side of the human mandibular corpus contributes to its rounded form. In the monkey a distinctive resorptive field, not usually present in man, occurs just anterior to the trihedral region (Fig. 8–1). This produces a lingual shift of the corpus and contributes to the more angular shape of the monkey's bony arch (Fig. 8–2).

A major growth difference exists in the anterior part of the maxillary arch. The outside (labial) surface of the human "muzzle" is characteristically resorptive. This is a unique human facial growth feature associated with the markedly decreased extent of muzzle protrusion and the essentially straight-down growth direction of the maxilla. In the monkey (Figs. 8–1, 8–7, and 8–8) the labial surface of the entire

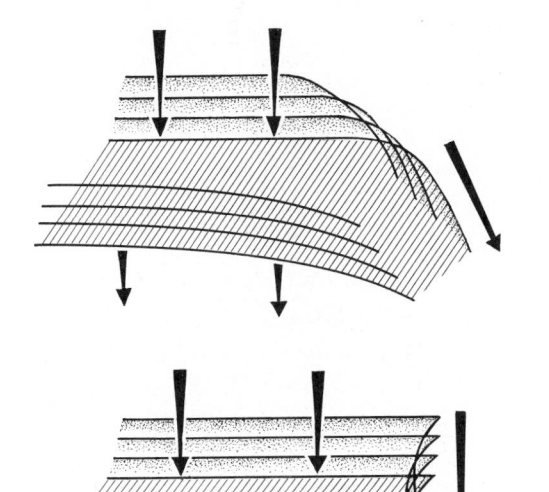

**FIGURE 8-7**

*The monkey's palate (top) grows downward by resorption on the nasal side and deposition on the oral side. The human palate (bottom) has an equivalent mode of growth. Note that the protrusive premaxillary region in the monkey, however, grows* **forward** *by external deposition, in contrast to the human maxilla, in which this region grows straight downward (after the primary teeth are established) by external resorption. (From Enlow, D. H.: Am. J. Phys. Anthrop., 24:293, 1966.)*

anterior part of the arch, including the separate premaxillary segments, is depository (as it is in other mammalian species as well). It grows **forward** as well as downward and is thereby more protrusive than the human maxillary arch. Note that the external contour of the monkey's premaxillary region is convex in contrast to the concave nature of the forward part of the human arch. Because of this, the downward mode of growth by the arch in man requires resorption on the labial side of the alveolar cortex (the portion below A point). In the **postnatal** human face, the loss of the premaxillary sutures, a lack of forward premaxillary cortical growth, the downward and backward rotation of the whole facial complex, and the reduced extent of anterior displacement all combine to result in a marked regression in the extent of maxillary prognathism. The anterior mode of cartilaginous and soft tissue growth by the overlying nasal region, although reduced in extent, results in the formation of the distinctive human "nose," which protrudes well forward of the short maxillary arch. This provides for **downward**-directed external nares which aim the inflow of air vertically, in the vertically-aligned nasal chamber, toward the sensory olfactory nerves in the **roof** of the chamber (in comparison to the posterosuperior wall in other species).

In the face of both man and the monkey, the outer surface of the cheekbone region is resorptive. The extent, however, is much less in the monkey (Fig. 8-9). There are two reasons. First, because the maxillary arch in the monkey grows anteriorly as well as posteriorly, the amount of backward relocation needed by the malar protuberance is not as great. In man, the direction of bony maxillary arch growth is posterior (after the primary dentition is established), and the malar region must necessarily grow backward to a greater extent than in the monkey in order to keep proportionate position relative to it. Second, the topographic configuration of the malar region is different in the two species. The more forward-rotated orbits in man produce a distinct squaring of the cheekbone and a flattening of the whole

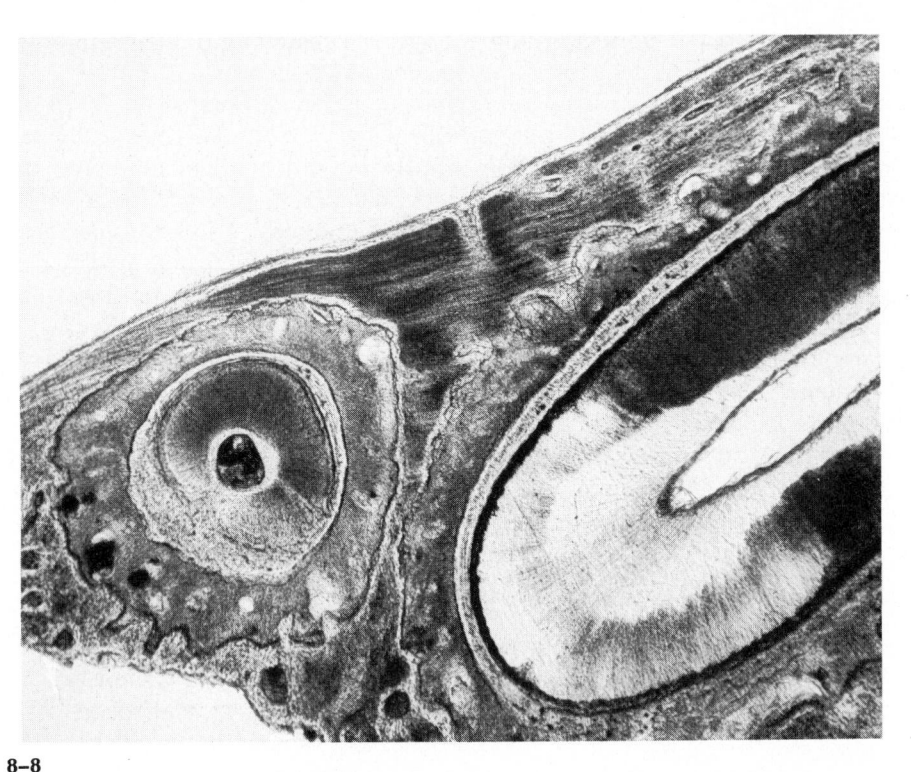

**FIGURE 8-8**

*This photomicrograph shows the external depository nature of the monkey's premaxillary region. The forward alveolar lining surface is resorptive as the teeth drift forward in the protrusively growing muzzle. Compare with the converse pattern in the human maxilla in Figure 3–157. (From Enlow, D. H.: Am. J. Phys. Anthrop., 24:293, 1966.)*

**FIGURE 8-9**

*Remodeling differences in the malar region between the monkey (a) and the human (b) faces are shown. Note that the anterior surfaces of both are resorptive (−) but that the squared configuration of the human cheekbone involves more extensive surface resorption. The rounded contour of the monkey zygomatic region, in contrast, has a lateral depository surface (+) that extends farther onto the anterior face of the malar. These differences also relate to the lesser extent of backward malar relocation in the more protrusive forward-growing muzzle of the monkey. (After Enlow, D. H.: Am. J. Phys. Anthrop., 24:293, 1966.)*

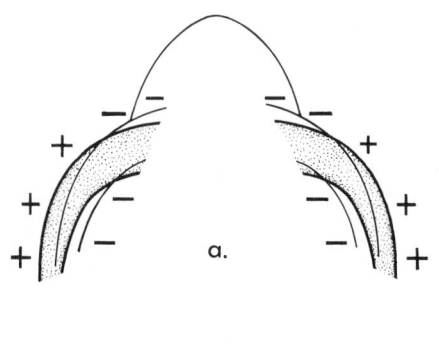

face. The malar region of the monkey is more rounded. Because of this, the lateral drift of the monkey's zygoma requires periosteal deposition that extends farther around onto the anterior-facing surface of the cheek, as shown in Figure 8–9.

Another basic difference occurs in the growth patterns of the lateral orbital areas in the human and the simian skulls. The orbital rims in man are much more vertically disposed because of the larger frontal lobes and upright forehead. The anterior edge of the lateral rim is entirely resorptive, and the postorbital surface is depository. This combination moves the lateral rim backward as the contiguous malar region beneath it grows posteriorly. The backward growth of the malar together with the forward growth of the forehead and superior orbital rim result in a human lateral orbital rim that is inclined obliquely forward (Fig. 8–5). In the monkey, conversely, the anterior edge of the lateral orbital rim is depository, and the postorbital side is resorptive. In conjunction with a much lesser extent of frontal lobe and forehead growth, the lateral rim remains inclined obliquely backward. Its inferior part grows forward and laterally rather than backward and laterally, as in the human face. The upper orbital rim in the human face is rotated so that it protrudes forward of the lower rim. In the monkey, the superior rim remains at a level posterior to the inferior rim.

As in the human calvaria, the individual bones of the skull roof in the monkey enlarge in perimeter by sutural bone growth. Both the inner and outer surfaces are depository except for limited remodeling near the sutures. The bony cortex is usually composed of a single, thin layer of lamellar bone. Where diploic spaces exist, the cortex has two tables, and the endosteal surfaces of both are resorptive. Two conspicuous differences between the human and simian calvariae exist, however. First, the proportionately deeper temporal fossa, medial to the zygomatic arch, has a medial wall that is resorptive in the monkey, and this sizable remodeling field covers parts of the temporal, parietal, and frontal bones. Second, the nuchal region has a large resorptive field on the exterior of the monkey skull. This provides an inward rotation of the inferior part of the occipital bone to sustain contact with the slower-growing cerebellum as the whole occipital bone is being displaced outward by the cerebrum.

The occipital part of the clivus is resorptive on the endocranial surface in both species (Fig. 8–10). This moves it forward and downward as the clivus lengthens at the spheno-occipital synchondrosis and at the rim of the foramen magnum. Unlike in man, however, the sphenoidal part of the clivus is always depository in the monkey (this area is variable in man). It flares posteriorly much more and moves markedly upward and backward during growth, together with the posterior lining wall of the sella turcica, which is resorptive. In man, the superoposterior wall of the sella turcica also moves back, but apparently not to the same relative extent. The floor of the monkey's pituitary fossa is depository (see page 106). This important region, however, needs further basic study in both species; the growth process here is complex and relates to differential extents of growth among the midline and lateral parts of the brain, the hypophysis, the extents of direct bony growth, and the extents of displacement of the various bony parts involved. Differences at various prenatal and postnatal age levels are also likely to be involved because of differentials in growth rates among the different parts of the cranial floor and among the lobes of the brain.

In the monkey, a sizable ring of resorption occurs on the floor of the cranium surrounding the endocranial side of the foramen magnum. This serves to lower as well as enlarge its diameter. Unlike the endocranial surface of the human cranial floor in general, only scattered patches of resorption exist in the various en-

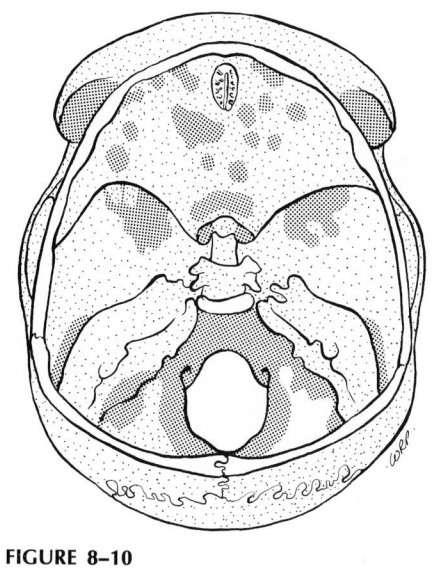 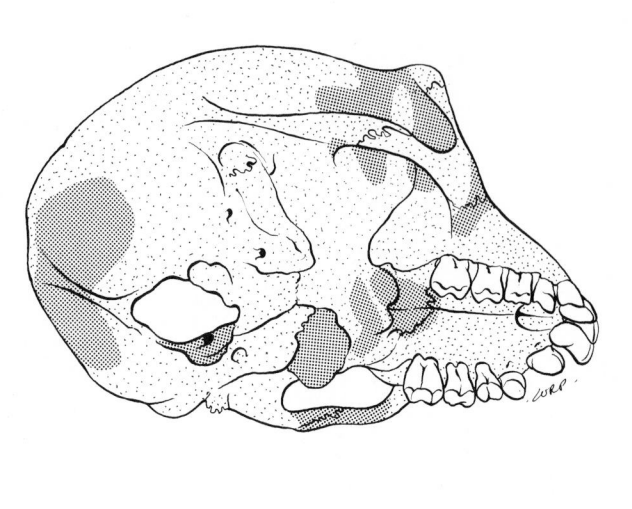

**FIGURE 8–10**

*Growth and remodeling fields in the cranium of the rhesus monkey. Resorptive (dark) and depository (light) surfaces are indicated. See text for descriptions. (Adapted from Duterloo, H. S., and D. H. Enlow: Am. J. Anat., 127:357, 1970.)*

docranial fossae of the monkey. This relates to the much more shallow nature of the cranial fossae, in contrast to the deep endocranial pockets present in man, and to the much lesser extent of cranial base flexure.

The entire superolateral half of the orbital roof in the monkey is resorptive. This same resorptive growth field is present in the human orbit but is more restricted in extent (postnatally). It is confined to that part of the roof beneath the overhanging supraorbital ridges. The roof of the orbit is also the floor of the anterior cranial fossa. In man, the endocranial side is resorptive in conjunction with the massive enlargement of the cerebrum and the flexure of the cranial base (see section on early fetal orbital growth, page 310). The opposite (orbital) side of the bony plate receives deposits as the entire anterior cranial fossa grows downward. The situation in the monkey is different. The anterior cranial floor slopes upward much more, and there is a lesser extent of cranial base flexure. The simian face lies essentially **in front of** rather than beneath the floor of the anterior cranial fossa. The surface of the bony floor of the anterior cranial fossa is largely depository rather than resorptive, in conjunction with a much lesser extent of growth expansion by the brain. Sutural growth provides a proportionately greater amount of the bony enlargement involved. This depository surface is complemented by a resorptive surface on its opposite (orbital) side, a pattern that also provides for orbital expansion.

The pterygoid plate in the human skull is typically depository on the surface lining the troughlike pterygoid fossa, and a large part of the external side is resorptive, at least during the latter period of childhood growth when middle endocranial fossa growth is complete. This combination produces growth enlargement essentially straight downward. In the monkey, the pterygoid plates are characteristically resorptive on the fossa side and depository on the external sides. This produces a downward and also **forward** growth expansion in conjunction with the more protrusive nasomaxillary complex and the lesser extent of cranial base flexure.

In man, the bony lining of the auditory canal is usually resorptive on the anterior surface and depository on the posterior surface. This results in an anterior drift of the entire canal. The lining of the monkey's auditory canal, in contrast, is

**FIGURE 8–11**

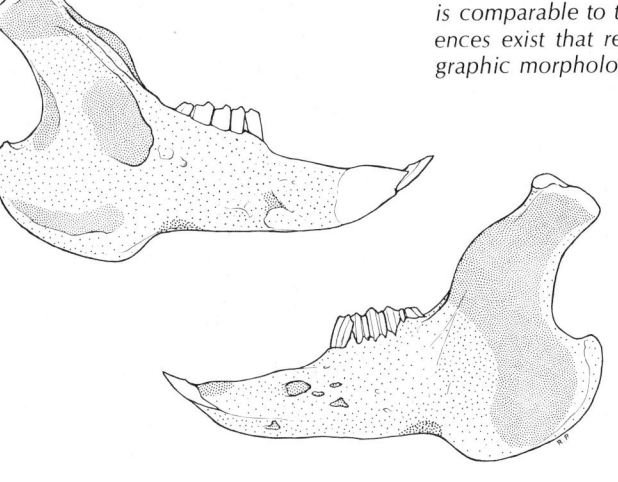

*The depository (light) and resorptive (dark) fields of growth and remodeling in the mandible of the rabbit, a popular laboratory animal in bone research, are shown. Although the overall plan is comparable to that of the human mandible, specific differences exist that relate to corresponding differences in topographic morphology. While growth takes place posteriorly, a major direction of growth also occurs anteriorly in the protrusive face of the rabbit. As the condyle and ramus grow posteriorly, sequential relocations are involved, just as they are in man. The lingual side of the ramus is predominantly resorptive, and the buccal side is largely depository. Note that a coronoid process is poorly represented, as is the lingual tuberosity. The patches of resorption on the lingual surface of the ramus relate to corresponding surface hollows. The resorptive zone on the dorsum of the alveolar bone just behind the incisors is involved in a progressive downward relocation of this area as the incisors move mesially with the lengthening mandible. (Adapted from Bang, S., and D. H. Enlow: Arch. Oral Biol., 12:993, 1967.)*

entirely resorptive. This enlarges the canal to its definitive size, but a forward drift movement is not involved.

The oral side of the horizontal shelf of the palatine bone in man is depository, and the nasal side is resorptive. This produces the downward growth of this part of the palate in conjunction with corresponding growth movements by the other bony palatal parts. In the monkey, however, the oral side of the palatine bone is often **resorptive** and the nasal side depository, although variations occur. The remainder of the bony palate is basically similar to that in man with regard to the growth pattern. The meaning of this difference is not known but probably relates to the counterclockwise displacement rotation of the whole maxilla. This should be taken into account when using the monkey in cleft palate research. The anterior part of the nasomaxillary complex is not displaced inferiorly to the same extent as the posterior portion (in relation to sutural growth). This has the effect of a counterclockwise "rotation," which is masked because the anterior part of the midface grows downward by direct deposition and resorption (remodeling growth) to a greater extent than the posterior part, as shown by the studies of McNamara. A similar combination exists in the human face as well, but the extent of direct inferior cortical growth in the anterior part of the maxilla in man is notably greater (as shown by Bjork). In many simians and anthropoids, there is a vertical hypoplasia of the anterior midface as an established anatomic feature, a characteristic not shared by the face of man. The posterior part of the anthropoid and simian maxilla appears quite large and "low," but it is actually of proper dimension (relative to the brain) and in its correct anatomic position (relative to the senses, as described in Chapter 4). It is the **anterior** part of the maxilla that is vertically shallow. Anterior open bites in the monkey are common. Moreover, an end-to-end type of incisor contact exists rather than an overbite, as in the human dentition. This also relates to the absence of a chin on the monkey's mandible. The human chin, which is formed in association with the development of overbite and overjet, may thus relate to the more extreme extent of maxillary rotation caused by frontal lobe expansion and the greater extent of anterior maxillary remodeling compensation.

# Cephalometrics*

## PART ONE

This chapter is intended to be an introductory overview of the science of radiographic cephalometry, or "cephalometrics." It is to provide a basic understanding of the techniques and principles involved in utilizing oriented head radiographs in the study of craniofacial morphology, growth, and treatment results. Since its introduction by Broadbent in 1931, cephalometrics has grown to be an integral part of orthodontic research, education, and clinical practice. Understanding human facial growth is basic to the development of clinical skills by all medical and dental practitioners who treat in this area, and application of the principles of cephalometrics is a necessary aid in this fascinating study.

Simply stated, cephalometrics involves making measurements from lateral and frontal head radiographs taken with the head held in a fixed position in a **cephalostat.** The head is held in this position by means of ear rods which are aligned on the central axis of radiation from the X-ray tube. Thus, for a profile view the sagittal plane of the head is at right angles to the direction of the X-rays and for a frontal or anteroposterior view is parallel to the flow of radiation (Fig. 9–3).† A standard distance of 60 inches from tube head to cephalostat is maintained to eliminate variation in degree of magnification resulting from divergence of the X-rays. Standardization of technique is necessary to minimize error when taking serial radiographs of the same individual at different times and to permit universal use of cephalometric data obtained from many different sources.

Measurements are made from the oriented radiographs, utilizing identifiable anatomic landmarks. To accomplish these procedures, tracings of basic skeletal, dental, and soft tissue structures are made on thin matte acetate with a fine pointed pencil. The tracing should include the soft tissue profile, outlines of the anterior and posterior cranial base, the orbit, the maxilla and mandible, the pterygomaxillary fissure, the first permanent molars, and the most anterior incisors.

Certain landmarks (see Glossary, page 288) are connected by lines or planes, and systems of angular and linear relationships may be employed to relate spatially

---

*By William W. Merow, D.D.S., Professor and Chairman, Department of Orthodontics, West Virginia University School of Dentistry.

†All illustrations are in Part Two of this chapter.

various structures within the complex. Figure 9–5 illustrates a standard tracing with basic cephalometric landmarks identified.

Over the years certain measurements have become standardized and have been applied to selected population samples to develop statistical means or averages. This approach has provided useful data in studying morphologic growth changes in the head, evaluating dentofacial abnormalities, and assessing response to orthodontic treatment procedures. The data have been particularly useful to the clinician in determining the timing and type of procedure he selects to treat individual problems. The majority of morphologic abnormalities occur in the sagittal plane, and the clinician's potential for correction is limited by the degree of severity of the problem and the pattern of growth associated with it. As a result, the measurements and analyses utilized are primarily profile-oriented and provide both anteroposterior and vertical relationships of the various parts of the dentofacial complex. In other words, cephalometrics has developed primarily in response to the clinician's demand for useful diagnostic and treatment planning criteria. For example, the clinician wants to know where the chin is in relation to the cranial base. How are the maxilla and mandible related to each other horizontally and vertically? How are the teeth positioned in relation to their supporting structures? What direction will these parts move in future growth change?

In answer to these questions, several standard cephalometric measurements have become widely adopted. The facial plane, drawn from nasion to pogonion and related angularly to the Frankfort horizontal plane, provides an assessment of the anteroposterior position of the chin (Fig. 9–7). The mean or average facial angle is 87.8°. High facial angles are associated with mandibular prognathism and low angles with a retrognathic profile.

Anteroposterior relation of the maxilla and mandible to each other and to the cranial base is established by the SNA and SNB angles and by the difference between them, the ANB angle (Fig. 9–8). The angles are formed between a line from sella to nasion and lines from nasion to points A and B, respectively. Normal readings for the angles are an SNA of 82°, an SNB of 80°, and an ANB of 2°. An ANB angle significantly higher than 2° means that the maxilla is protrusive, the mandible is retrusive, or a combination of these. A negative ANB angle means that point A is behind point B, and is associated with a concave facial profile.

The mandibular plane angle, formed between the mandibular plane and Frankfort horizontal plane (Fig. 9–11), provides an assessment of lower face morphology in the vertical dimension. The average for the angle is 21.9°. High mandibular plane angles mean that the posterior face height is deficient, the anterior face height is excessive, or a combination of these.

The Y-axis (Fig. 9–11) is a line drawn from the point sella to gnathion. Its average angular relation to Frankfort horizontal is 59°. High Y-axis angles are associated with vertically growing faces, while low angles indicate more forward growth in the lower face. High and low angle patterns are illustrated in Figures 9–12 and 9–13, respectively.

The most significant measurements of tooth position in profile analysis involve relating the incisor teeth anteroposteriorly to the face. The extent of dental protrusion may be determined by measuring horizontally from the tips of the incisors to the facial plane (Fig. 9–14) or to the A-Po line (Fig. 9–15). In both instances the tip of the lower incisor crown should be within approximately 2 mm. of the reference line. Angular measurements relating the long axis of the lower incisor to the mandibular plane, the long axis of the upper incisor to the Frankfort plane, and upper to lower as an interincisal angle provide additional useful information on

dental posture (Fig. 9–16). Low interincial angles are usually associated with dental protrusion.

Combinations of these and other measurements have been made to form "analyses" of dentofacial morphology. Several examples of these analyses applied to a sample patient are outlined in Part Two of this chapter. The individual measurements described in previous paragraphs were taken from Downs', Steiner's, and Ricketts' analyses, which are illustrated in Figures 9–18, 9–20, and 9–24, respectively. Most of these analyses, with the most notable exception being the archial (Sassouni), are based on established norms statistically derived from population samples. Their primary use is to provide a means of comparing an individual's pattern with a population average to locate areas of significant deviation. Some analyses, such as the Tweed analysis, stress particular relationships (lower incisor) as treatment planning criteria. Others provide a more complete craniofacial assessment. The Bjork analysis as adapted by Jarabak utilizes the configuration of a craniofacial polygon to evaluate and forecast direction of facial growth change.

In the assessment of general growth and treatment change, head radiographs of the same individual taken at separate times are traced and the tracings superimposed to ascertain the changes which have occurred. A common method is to register the two tracings at the point sella with the sella-nasion lines superimposed (Fig. 9–31). This method provides a gross overview of dentofacial and soft tissue change, but it is useful only in evaluating what has already occurred.

Clinical interest in changes that will occur during future growth has led to computerized programs of growth forecasting, and diagnosis-treatment planning assistance. These programs are, like the cephalometric analyses, based on comparions with accumulated information from large population samples having similar patterns and are largely in response to an increased awareness by the clinician of the significance of growth change in the relative success of treatment procedure.

Unfortunately, cephalometrics, while providing an excellent assessment of morphologic change, does not identify sites of growth and remodeling change, nor does it provide basic understanding of how or why certain growth patterns develop. The principal author of this text provides this basic information and understanding and, complementary to it, presents a form of cephalometric analysis outlined in Part Two of this chapter. Dr. Enlow's analysis, while perhaps not intended for the routine clinical use described above, will enhance the student's understanding of how and why craniofacial patterns develop as they do.

# PART TWO

Cephalometric radiography, or "cephalometrics" as it is more frequently called, is a technique employing oriented radiographs for the purpose of making head measurements. It has found wide use in growth research and in orthodontic diagnosis and treatment evaluation. The principles of cephalometrics are patterned closely after the science of craniometry, which had long been used in anthropology in the quantitative study of the skull.

Historically, the study of the human body began in the dissecting room and was primarily a descriptive discipline. Biologic and genetic variability in size and shape stimulated a desire to measure the bony skeleton and resulted in the science of osteometry, of which craniometry is a subdivision.

Craniometry was initially applied in measuring dried skulls. Standard landmarks and measurements were developed, and much useful information was obtained. The technique, however, had the obvious shortcoming of being a one-time-only evaluation of a static subject with frequently an unknown clinical history. Serial study of growth changes was not possible. When the technique was transferred to living subjects for growth measurement, accuracy was lost because landmarks were obscured, the soft tissue covering was of varying thickness, and there was no access to deeper structures.

The advent of radiography presented a method for recording shadow images of hard and soft tissues on living, growing, changing subjects. In 1931, Dr. B. Holly Broadbent introduced the basic technique of cephalometric radiography. In the years since, the technique has been developed and refined primarily as an orthodontic procedure, and the cephalometric radiograph is now universally adopted as an integral part of orthodontic record systems. More recently, its potential in the study and treatment of the whole range of craniofacial structural abnormalities has become increasingly apparent. The dental profession has indeed inherited a reasonably precise system for performing craniofacial measurements.

It is the purpose of this chapter to provide a basic overview of the technique and principles of cephalometric radiography and the manner in which it is used. It is intended to augment this text on dentofacial growth, on the assumption that students of medicine and dentistry will find that the principles of cephalometrics will further their understanding of the development and function of the human face.

## TECHNIQUE AND METHODS

The most important aspect of cephalometric radiography is standardization. It is necessary that the position of the patient and the orientation of the X-ray beam are established in such a way that repeated exposures may be made on successive occasions under the same conditions. The profession has gone to considerable lengths to assure standardization of the equipment and technique employed in acquiring satisfactory radiographs.

254

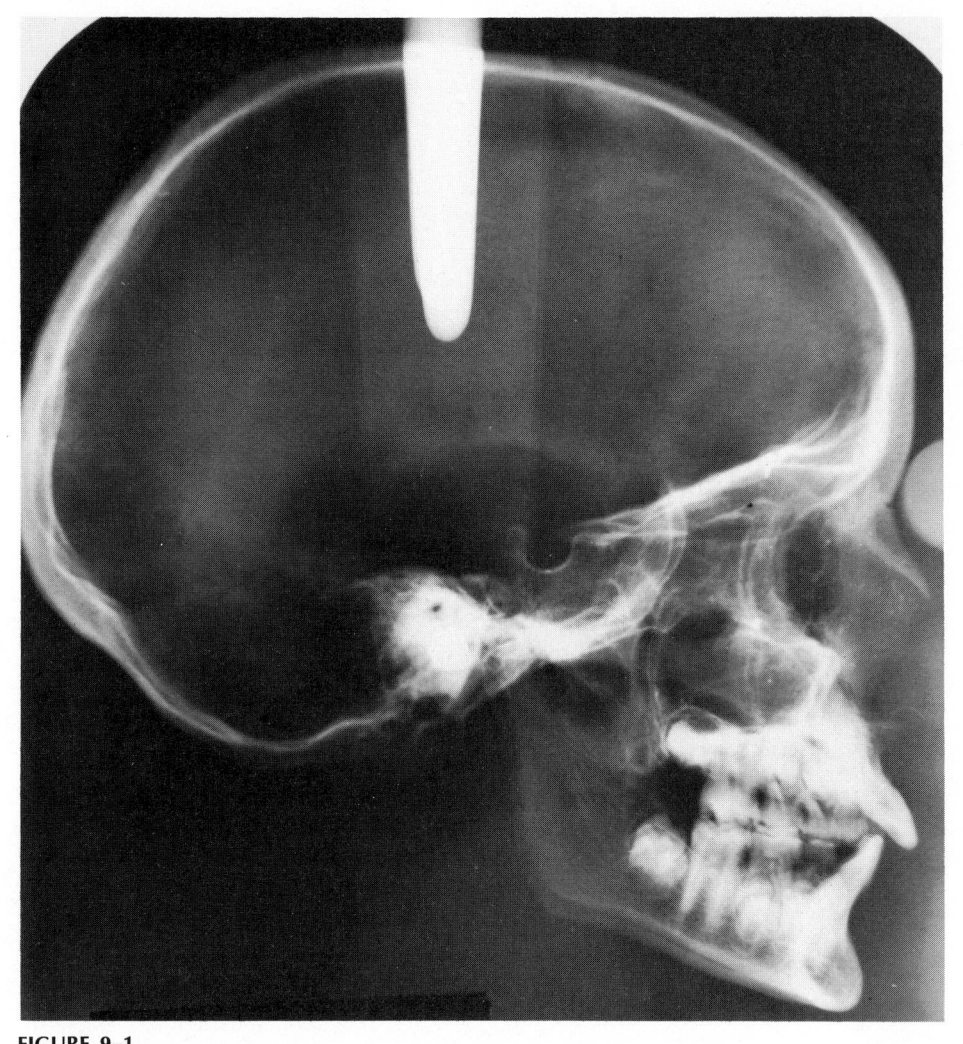

**FIGURE 9–1**
*Lateral cephalometric radiograph. (From Enlow, D. H.:* The Human Face, *New York, Harper and Row, 1968, p. 270.)*

Two radiographic views are commonly used: a lateral or profile view (Fig. 9–1) and a frontal or posteroanterior (P-A) view (Fig. 9–2). The accepted practice for the lateral view is to place the left side closest to the film cassette, and for the P-A view to place the face closest to the film (Fig. 9–3).

Equipment design has progressed through a series of modifications, but the essentials are a holder to position the X-ray tube and a **cephalostat** or head-positioning device which is precisely located in relation to the tube head. The most common contemporary design is a wall-mounted unit featuring a horizontal bar which supports the X-ray tube head on one end and the cephalostat on the other end. This assembly travels vertically in a counterbalanced system to adjust for variations in patient height (Fig. 9–4).

Standardization and accuracy in the relationship of tube head to cephalostat are most important. The cephalostat locates patient head position by means of laterally adjustable ear posts. Incorporated in the ear posts are small metal rings which are

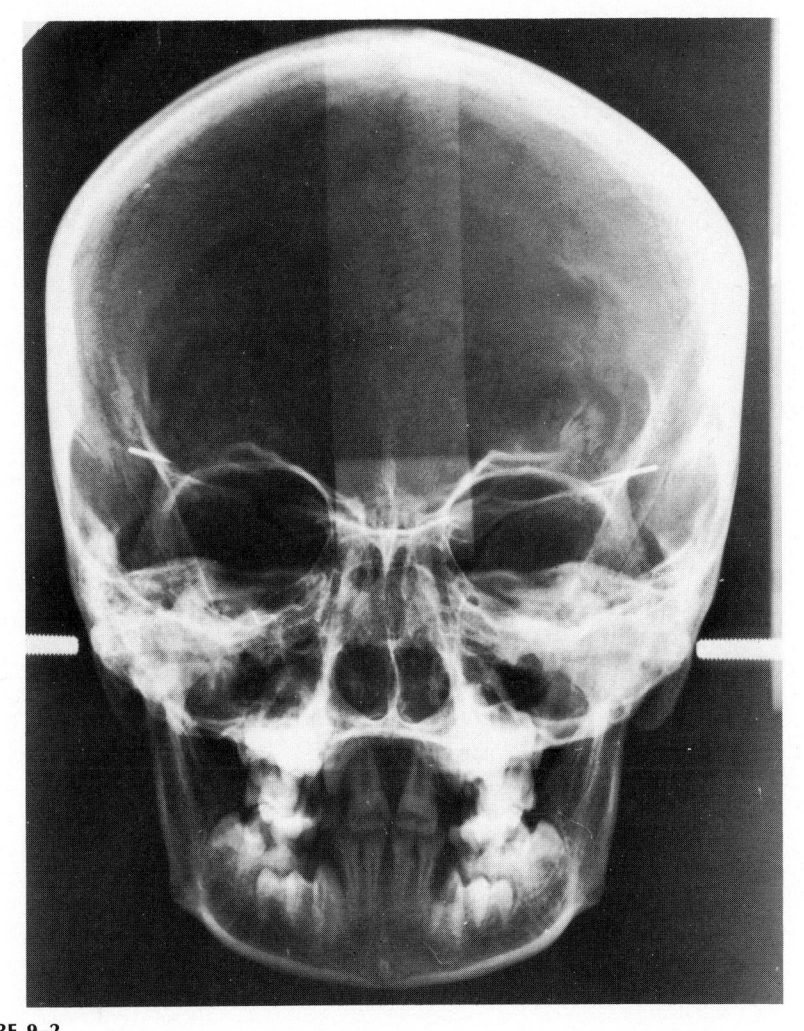

**FIGURE 9–2**
*Frontal or posteroanterior radiograph. (From Enlow, D. H.: The Human Face, New York, Harper and Row, 1968, p. 271.)*

useful in aligning the device. The cephalostat must be positioned in such a way that the ear posts are aligned to the X-ray tube head so that the upper edges of their images are superimposed on the film. In other words, the ear post axis must be centrally aligned to the source of radiation.

The linear distance from tube target to subject is standardized at 5 feet from target to the sagittal plane of the patient's head. Figure 9–3 illustrates diagrammatically these relationships. The film is positioned perpendicular to the ear post axis and as close to the patient's head as possible.

In the interest of good radiation hygiene, it is necessary that certain limitations be incorporated into the system. The cone of radiation is collimated to just cover the area of the film (usually 8 × 10 inches). In order to keep exposure at the lowest possible level consistent with high quality radiographs, high speed intensifying screens are built into the film cassettes. These screens function by fluorescing when exposed to radiation, and the effect is to markedly reduce the amount of radiation

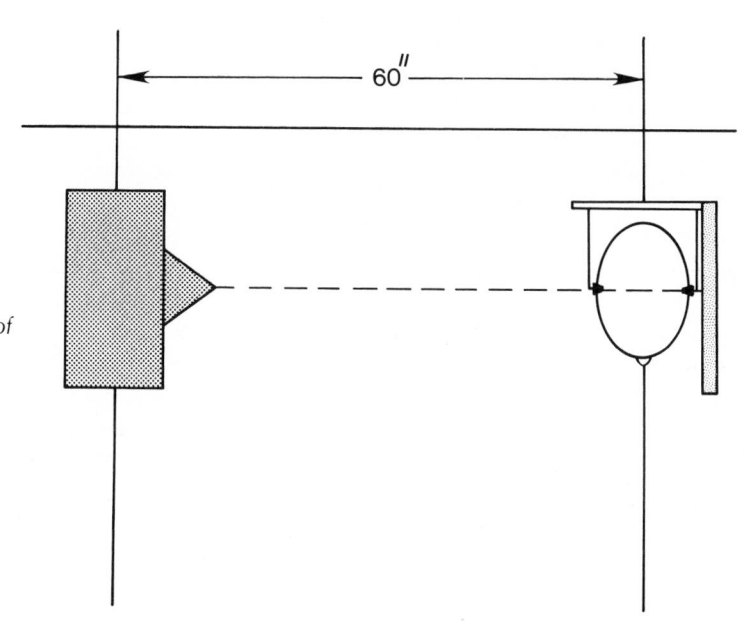

**FIGURE 9–3**

*Diagram of standard design of radiograph cephalometer.*

necessary to produce an image on the film. It also permits shorter exposures, thus reducing the possibility that patient movement will produce a blurred image.

Most orthodontic record films require an outline of the soft tissue profile. The film range may not be adequate to provide sharp skeletal contrast and soft tissue outlines at the same time, in which event it is necessary to employ additional means for the latter. This may be accomplished by placing an aluminum screen on the film cassette over the profile area, outlining the profile with a radiopaque material, or employing a second film especially formulated for this purpose.

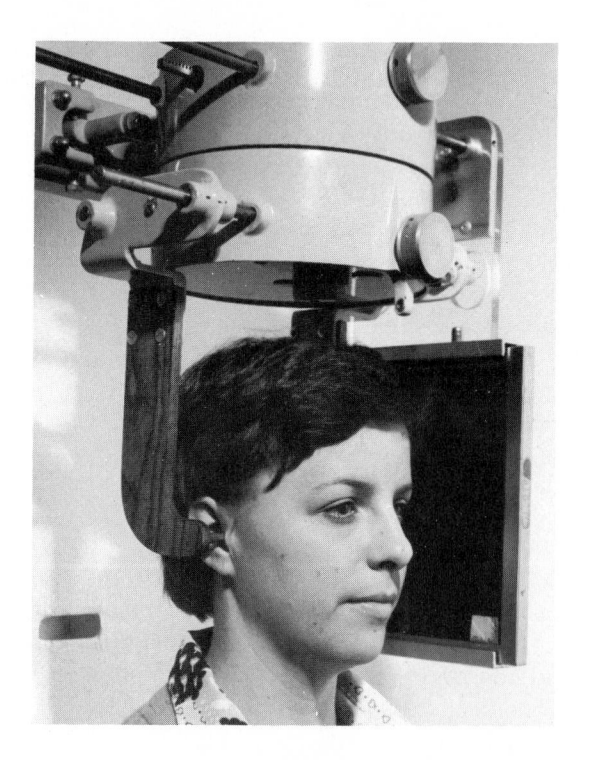

**FIGURE 9–4**

*Patient positioned in cephalostat for lateral film exposure.*

It must be understood that, because the X-rays emanate from a point source and are therefore divergent, a certain amount of magnification of image is always present. This is where standardization of equipment is important. If the unit is properly built and adjusted to these standards, the magnification effect is the same for all and therefore can be canceled out. This divergence of rays also results in a double-image effect in showing bilateral structures, such as the orbits, lower border of the mandible, and posterior teeth. This effect is most evident in peripheral areas of the film where the degree of ray divergence is more pronounced.

The lateral or profile headfilm is obtained by positioning the patient's head in the cephalostat with the right side of the face toward the X-ray tube. The ear posts are located in the ear openings and moved together until the head is firmly positioned. The cephalostat should then be moved upwards slightly to create firm contact between the ear posts and the bony outline of the patient's ear opening. This is to reduce error caused by variation in soft tissue thickness. The head should be upright, with the patient looking straight ahead. Exposure is usually made with the teeth in complete centric occlusion, although for certain special applications rest position or maximum open position may be used. The film cassette is moved as close to the face as possible, and the exposure is made. For the frontal (P-A) film the cephalostat is rotated 90° and the patient placed in the ear posts with the face toward the film cassette. More exposure is necessary for the P-A view than for the lateral view. After processing and drying the exposed film, it is ready for examination and tracing.

When the student starts studying the oriented radiograph, he is apt to be confused by what appears to be a crowded assortment of overlapping shadows with varying intensities and contours which are difficult to follow. A clear knowledge of anatomy, especially osteology of the skull and the pharyngeal soft tissue, is a prerequisite for reading and tracing the film. For the beginner it is helpful to have a skull available while tracing, and the dental casts should always be at hand while tracing the dentition.

Equipment and materials necessary for tracing include a view box, preferably with variable light intensity; 0.003-inch thick tracing acetate with one matte surface; a millimeter rule; protractor; compass; two draftsman's triangles; and a well-sharpened medium-hard (No. 3) pencil. The matte acetate is customarily attached to the film with two small pieces of masking tape along one edge.

Accuracy and consistency in tracing technique are essential and are developed only through practice. It is usually helpful in accurately locating landmarks to trace with reduced room illumination, which enhances the contrast in the film. If the film is not overexposed, the soft tissue profile can be seen by blocking out all peripheral light except in the area of the profile.

Since the lateral film is more widely used and all the analyses to be discussed in this chapter are based on it, similar emphasis will be followed in this discussion. Most clinical concern in the diagnostic sense as well as in the appraisal of treatment change is related to the height and depth proportions seen in the lateral film. The frontal (P-A) film is useful for determining breadth and symmetry, but it is more difficult to interpret because of the structures that are superimposed.

An example of the lateral tracing is shown in Figure 9–5. Definitions of the various landmarks and points used in most contemporary analyses are provided in the Glossary on page 288. The lateral tracing should include the soft tissue profile, the bony profile, the outline of the mandible, the posterior outline of the brain case, the odontoid process of the axis, the anterior lip of the foramen magnum, clivus, planum, and sella turcica outline, the roof of the orbit, the cribriform plate, the lat-

**FIGURE 9–5**

*Lateral tracing with standard cephalometric landmarks.
See Glossary of Terms at end of chapter for definitions.*

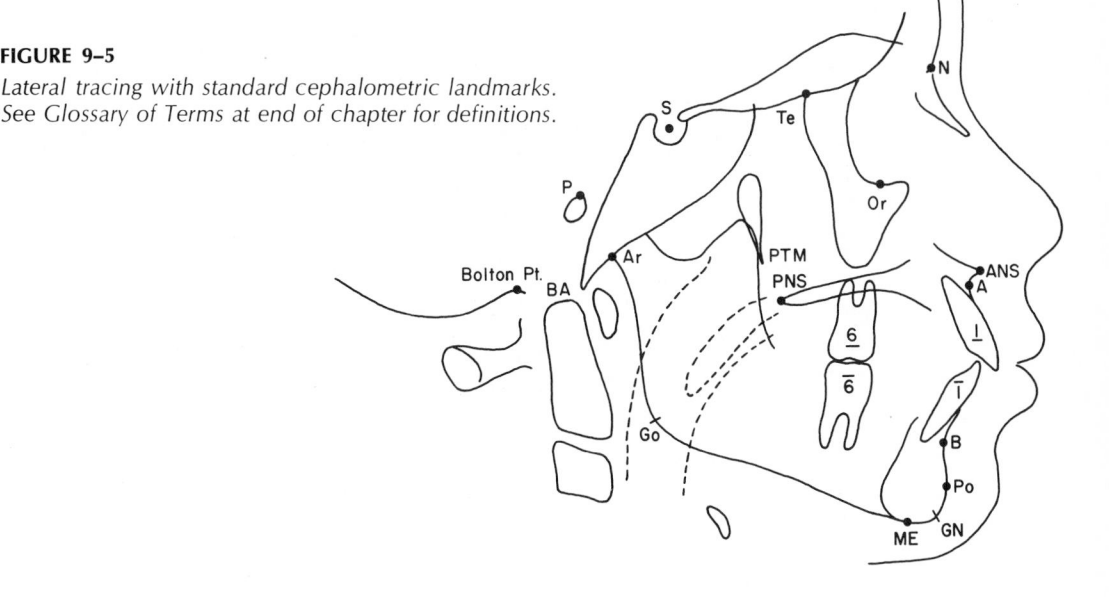

eral and lower borders of the orbit, the outline of the pterygomaxillary fissure, floor of nose, and roof of palate, the soft palate, the root of the tongue, the posterior pharyngeal wall, and the body of the hyoid bone. Minimum tracing of teeth should include the first permanent molars and the most anterior incisors.

In the case of bilateral structures, an attempt to distinguish right from left sides is difficult and may result in error. For that reason it is recommended that a midline between right and left structures be used.

All cephalometric analyses attempt to establish a spatial relationship between various parts of the craniofacial and dental structures. As previously stated, primary emphasis is on the lateral view, thus providing an assessment of these relationships horizontally and vertically in the sagittal plane. Specifically, the evaluation of the lateral film involves the flexure of the cranial base, the relation of the maxilla and mandible to each other and each to the cranial support, and the position and posture of the dentition in relation to facial structures. In order to measure these relationships, a system of anatomic points and landmarks has been developed, and by connecting certain of these with lines to form reference planes, angular and linear measurements can be made. It is not within the scope of this chapter to discuss the development of this system or to debate the relative validity of different measurements. Let it suffice that the reference planes and measurements to be discussed have been widely adopted in the field of cephalometrics for some time.

## NORMS, VARIABILITY, COMPARISON

Before entering into a description and discussion of the various cephalometric measurements and some of the more frequently used analyses, it seems appropriate to attempt to define a concept of how and for what purpose we are to use this technique.

It has already been stated that measurements made on oriented headfilms provide useful information in both research and clinical practice. In the research

application three primary efforts have been made: first, the accumulation of data related to craniofacial growth changes; second, the establishment of statistical norms for numerous cranial and dentofacial dimensions; and third, the evaluation of response to various treatment procedures. All these approaches have taken into account the obvious variables of age, sex, and racial and ethnic background. In growth research much has been learned of the morphologic and dimensional patterns of skull growth. However, the information is limited to just those aspects: change in shape and size. While it is true that change in rate, direction, and pattern of growth have been recorded, the cephalometric technique does not locate the sites of growth or measure the contribution of growth sites. Realistically, the single film provides only a static evaluation of that individual's size and shape at that point in time. A subsequent film on the same individual permits an evaluation of size and shape changes which occurred in the interval between films but does not tell exactly where the growth occurred.

In orthodontic clinical application, the common practice is to make any number of the prescribed measurements on the film and to compare these to established norms. When the word "normal" is used, the researcher and clinician alike may ask, "What is a normal face—or a normal head?" or they may ask, "Normal with reference to what?" Many thousands of cephalometric headfilms have been traced and as many comparisons made in a massive effort to find this elusive "norm." But when we combine the obvious differences in age, sex, race, and environment and add to these the vicissitudes of biologic and genetic variation, it becomes evident that variation is indeed the name of the game.

Realistically, the clinician works with one patient at a time, and when he compares that individual's measurement to statistically derived norms, he is comparing his patient to a generalization drawn from a large group of individuals. Or he may be attempting to forecast the pattern of facial growth of a young patient by comparing his static measurements to those of a large group of similar individuals whose patterns of facial growth have been statistically evaluated. In the world of biology where variability is still the rule, these efforts, while meeting with increased success, continue to entail a certain amount of hazard.

In practice, the orthodontist compares his patient's measurements with the norms and notes areas of deviation. These norms are calculated means or averages of many equivalent measurements. Along with the mean, a standard deviation (S.D.) is usually calculated. In clinical use, this S.D. might be called an acceptable range of variability. In other words, when there is a variation within one S.D., treatment by conventional orthodontic means should yield a good result. As the degree of deviation increases, the relative success of treatment decreases. This may appear to be an oversimplification, but essentially it means that, to some extent, the treatment is the victim of the individual pattern presented.

On the positive side, cephalometric-aided research has identified growth patterns and predictable responses to certain treatment procedures, either of which may assist the treatment planning process. The extent to which deviations from the norm influence the evaluation of the individual patient is a judgment made by the clinician, which he incorporates with all the other diagnostic information he has accumulated.

These comments are certainly not intended to downgrade the value of cephalometric research or invalidate the use of cephalometric norms in orthodontic diagnosis and treatment planning. Rather, they are to emphasize that cephalometric data, while valuable, are only a part of the information that must be employed in research and clinical applications.

## LINES, PLANES, ANGLES

In an effort to provide the beginning student with a basic understanding of cephalometric evaluation, several of the lines, planes, and angles which are common to several of the analyses will be diagrammed and discussed separately.

Figure 9–6 illustrates the most frequently used horizontal planes and lines. The sella-nasion (SN) is drawn from the selected point sella to nasion. It is described as representing the anteroposterior extent of the anterior cranial base and serves as a reference line when relating facial structures to the cranial base.

The Frankfort horizontal plane is drawn tangent to the superior outline of porion and extends through orbitale. It is widely accepted as **the** horizontal plane of the head. Some researchers feel there is a postural significance in the position of this plane.

The palatal plane is drawn by extending a line through and connecting the anterior nasal spine (ANS) and the posterior nasal spine (PNS). By relating palatal plane to Frankfort horizontal, postural tilt of the maxilla can be measured.

The occlusal plane bisects the incisor overbite (or open bite) and passes over the distal cusps of the most posterior teeth in occlusion.

The mandibular plane (MP) is drawn tangent to the inferior border of the symphysis outline and extending posteriorly is tangent to the inferior border of the mandible posterior to the antegonial notch. Relating the mandibular plane to the sella-nasion or Frankfort horizontal planes provides an assessment of vertical proportion in the lower face.

The planes and lines just described serve as reference planes for other measurements or may be related to each other, as will be discussed later. All tracings used in the illustrations are of the same patient and will also be used later in discussing several analyses.

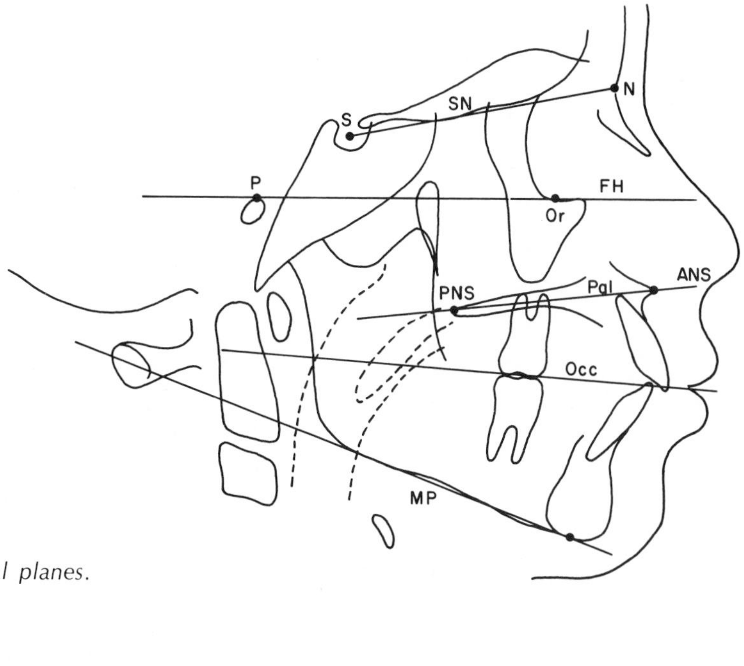

**FIGURE 9–6**
*Most frequently used horizontal planes.*

## SKELETAL ASSESSMENT

Profile assessment includes determining anteroposterior position of the chin, the maxilla, the anterior teeth, and the soft tissue. The facial angle is used to determine anteroposterior chin position (Fig. 9–7). It is the angle between the Frankfort horizontal plane and the facial plane. The facial plane is a line drawn from nasion through pogonion. The mean value for this angle is 87.8°, with a range from 82° to 95°. Values larger than these would indicate a lower face prognathism and Class III malocclusion, while smaller values would be associated with a retrognathic mandible and Class II malocclusions. The tracing used in the illustration (Fig. 9–7) shows a facial angle of 90°, which indicates an acceptable anteroposterior position of the chin.

The maxilla and mandible may be related to each other anteroposteriorly by the SNA-SNB angles (Fig. 9–8). The angles are read between the SN and the NA and NB lines, respectively. Although points A and B, by definition, appear to represent maxillary and mandibular basal structures, respectively, some authors question their validity on the basis that they may be influenced by incisor tooth movement during treatment. In any event, their mean values, 82° and 80°, respectively (at ages 12 to 14), are used to assess the anteroposterior position of the maxilla and mandible with respect to the anterior cranial base. Of perhaps more interest to the clinician is the difference between the angles—the ANB angle. The mean value of the ANB angle is 2°, and significant deviations from this mean indicate an anteroposterior discrepancy of those basal structures which support the dentition. A high ANB angle indicates a maxilla which is forward, a retrognathic mandible, or a combination of these deviations (Fig. 9–8). The tracing used in this illustration shows a high ANB angle (8°) but the SNB angle is normal (80°). It must be assumed, then, that the maxilla, or point A at least, is positioned anteriorly. Large ANB devia-

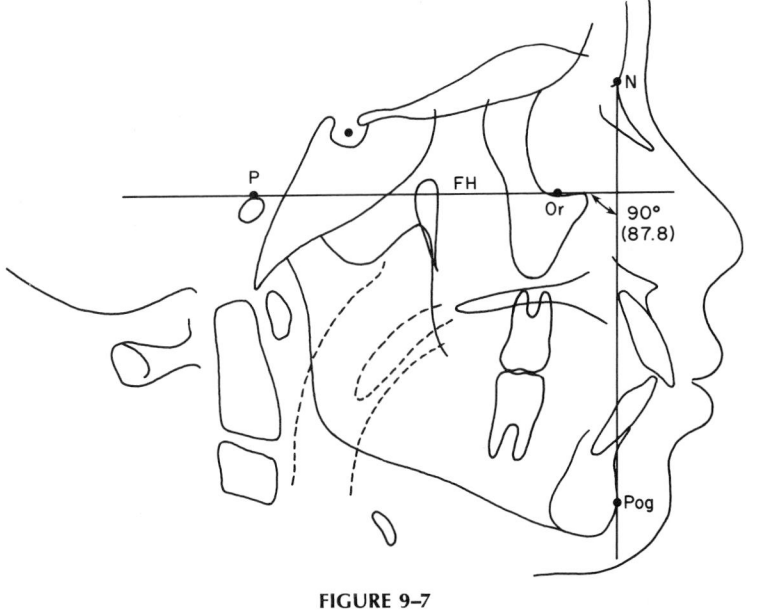

**FIGURE 9–7**
*Facial angle.*

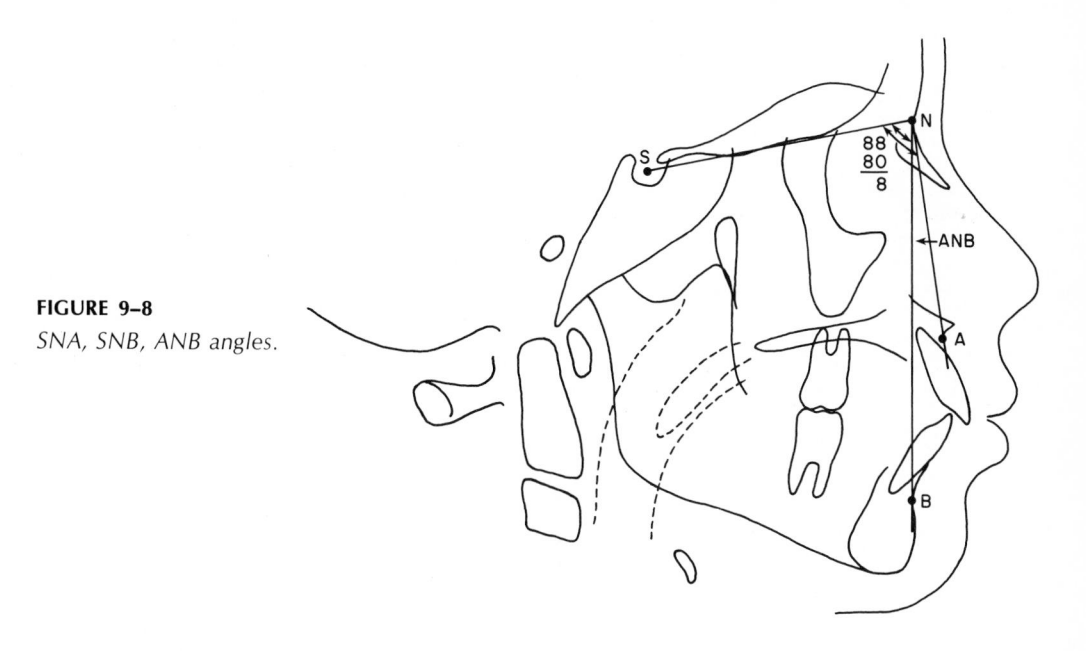

**FIGURE 9–8**

*SNA, SNB, ANB angles.*

tions in either direction indicate to the clinician that the problem to be treated has an orthopedic significance, and it may not be responsive to treatment by tooth movement alone.

Anteroposterior variations in facial profile may be assessed by the angle of convexity (Fig. 9–9). The angle of convexity provides information similar to that from the ANB angle, but in this case it takes into account the influence of the "chin button" or prominence of pogonion. The mean value of the angle of convexity is 0°, with a range of −8.5° to +10°. In the tracing used in the illustration, the high angle of 17° indicates a very convex facial profile and agrees with the information previously obtained from the SNA-SNB angles.

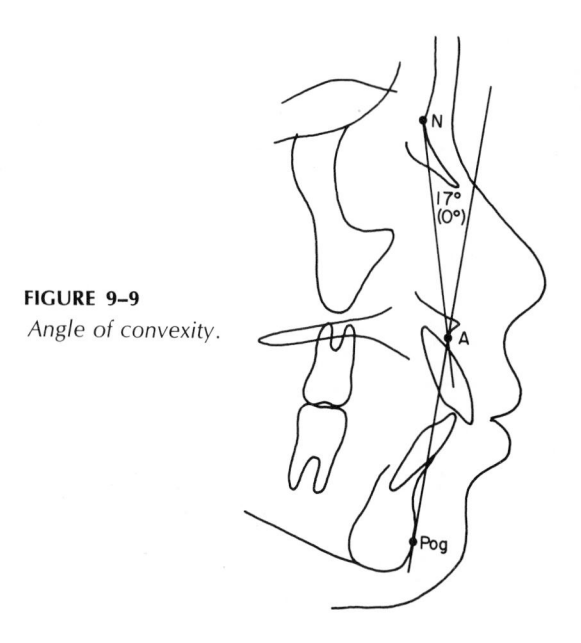

**FIGURE 9–9**

*Angle of convexity.*

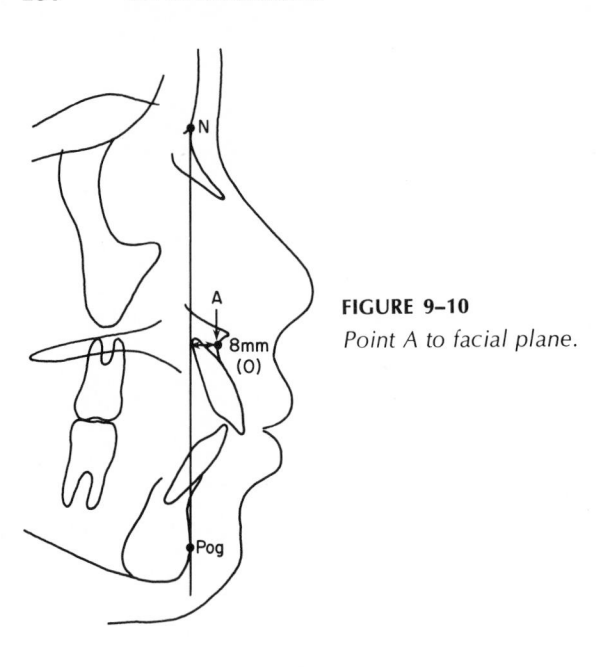

**FIGURE 9–10**
*Point A to facial plane.*

Another means of measuring convexity is by the relationship of point A to the facial plane (Fig. 9–10). The measurement is made horizontally from point A to the facial plane and recorded in millimeters. The mean value is 0, with a range of − 3 mm. to +4 mm. Deviations greater than 5 mm. in front of the facial plane or 3 mm. behind it are suggestive of an orthopedic problem in anteroposterior skeletal relation. In the illustration, the deviation is 8 mm. in front of the facial plane, agreeing with previous assessments of profile convexity.

The measurements outlined above are primarily related to anteroposterior assessment of skeletal profile. The mandibular plane angle provides a means of as-

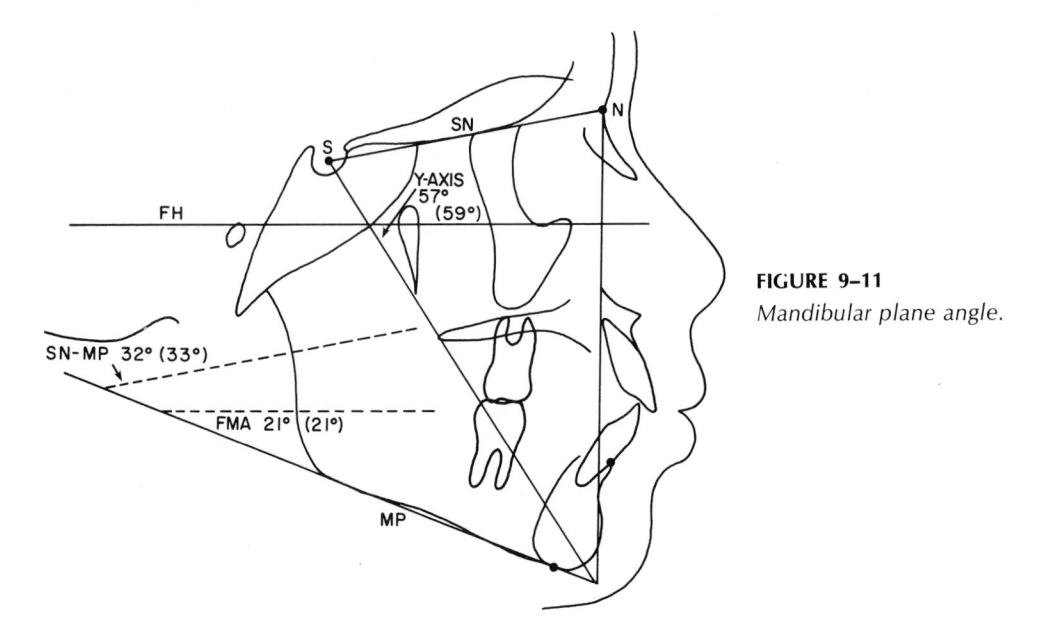

**FIGURE 9–11**
*Mandibular plane angle.*

**FIGURE 9–12**

*Steep mandibular plane angle. Note disproportion of anterior to posterior face height and shallow face depth.*

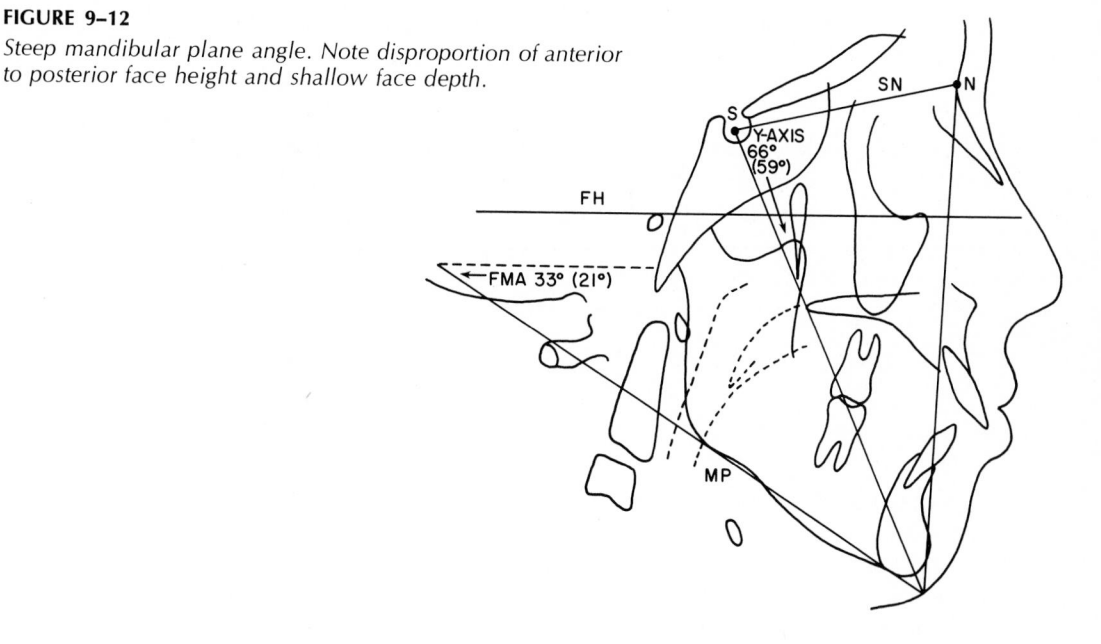

sessing vertical relation and the morphology of the lower third of the face. The mandibular plane angle may be measured in relation to the Frankfort horizontal plane (FMA) or in relation to the SN line (SN-MP). The mean FMA is 21.9° (Fig. 9–11), and the mean SN-MP angle is 33°. High mandibular plane angles may result from a short ramus, obtuse gonial angle, high position of the glenoid fossa, long anterior face height, or any combination of these. High mandibular plane angles are frequently associated with anterior open bites and vertically growing facial patterns. Conversely, low or flat mandibular plane angles, frequently associated with deep anterior overbite and horizontal mandibular growth patterns, are the result of a long ramus, acute gonial angle, short anterior face height, or any combination of these.

Figures 9–12 and 9–13 illustrate tracings showing high and low mandibular plane angles, respectively, and the differences in facial pattern associated with

**FIGURE 9–13**

*Low mandibular plane angle. Note relative decrease of anterior face height and increased depth of face.*

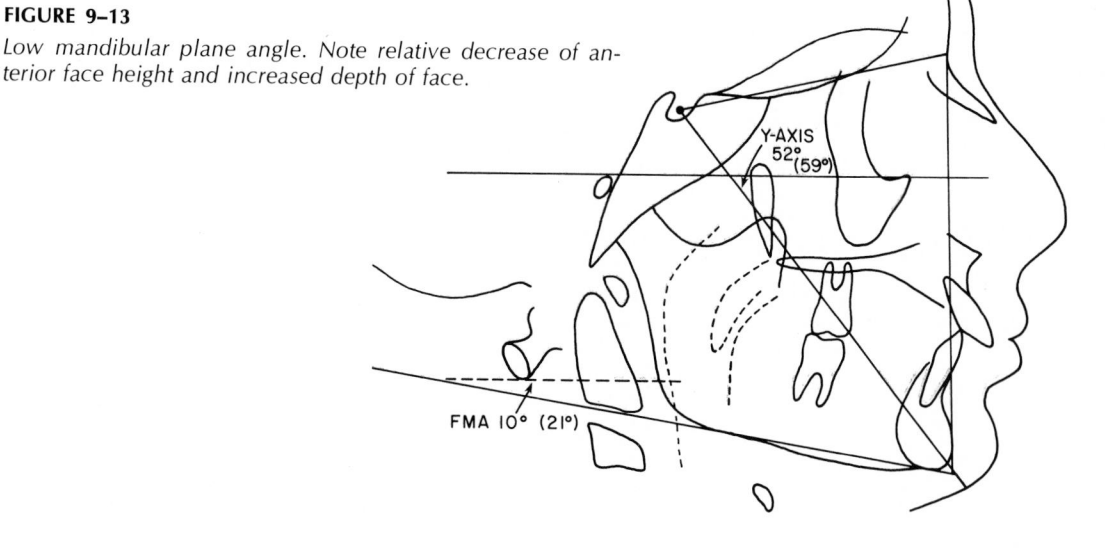

them. Note the differences in the ratio of facial height to depth and in the ratio of posterior face height to anterior face height.

The Y-axis, drawn from the sella point through gnathion and related angularly to the Frankfort horizontal plane, is judged to be indicative of the direction the mandibular symphysis will follow in future growth. The mean value of the Y-axis is 59°. The tracing of the sample patient (Fig. 9–11) shows a measurement of 57°. Angles significantly larger than the mean indicate a greater proportion of vertical growth at the symphysis, while smaller angles indicate a relatively more forward pattern. Figure 9–12 shows a 66° Y-axis and Figure 9–13 a 52° measurement. These measurements are consistent with the respective vertical and horizontal growth patterns of these two patients.

## DENTAL ASSESSMENT

The dental assessment is made by a combination of several measurements, both angular and linear, primarily involving the incisors. Anteroposteriorly, the incisor crowns may be related to the facial plane (Fig. 9–14), with the ideal position of the lower incisor crown being right on the plane or within an acceptable range of −2 mm. to +3 mm. The illustration (Fig. 9–14) would be described as protrusive, with the maxillary incisor being 15 mm. in front of the facial plane and the mandibular incisor 10 mm. forward.

The incisors may be related anteroposteriorly in the same general manner with reference to the A-Po line (Fig. 9–15). Again, the ideal for the lower incisor is right on the line, with an acceptable range of −2 mm. to +3 mm. It must be pointed out that in this instance the reference line (A-Po) is related to the maxillary denture base and the chin rather than to the facial plane and will thus vary with anteroposterior deviation between maxilla and mandible. In the patient illustrated in Figure 9–15, the maxilla is markedly in front of the mandible, and the lower incisor is 6 mm. in front of the A-Po line, compared with the 10-mm. relationship to the facial plane.

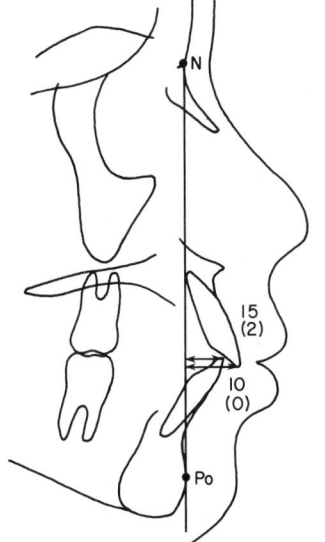

**FIGURE 9–14**
*Upper and lower central incisors to the facial plane.*

**FIGURE 9–15**

*Upper and lower central incisors to the A-Po line.*

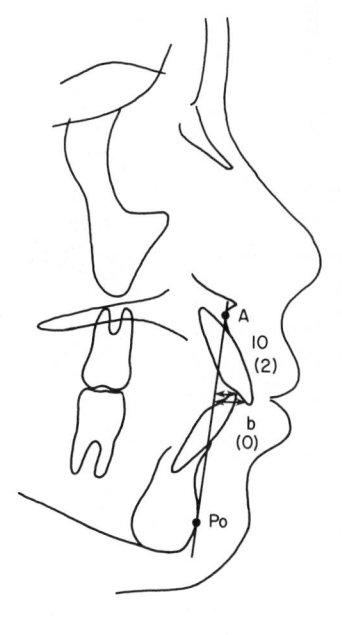

The most commonly used angular measurements are the interincisal angle, the long axis of the lower incisor to the mandibular plane, and the long axis of the upper incisor to the Frankfort horizontal plane. Figure 9–16 illustrates these measurements. Various authors place the mean interincisal angle between 125° and 135°. Larger angles result from very upright incisors and are frequently associated with deep overbite. Small angles occur in dental protrusion, as illustrated in Figure 9–16.

The ideal lower incisor to mandibular plane angle is 90° and would be expected to occur with an average mandibular plane angle (FMA). Some variations in

**FIGURE 9–16**

*Interincisal angle, lower incisor to mandibular plane, and upper incisor to Frankfort horizontal plane.*

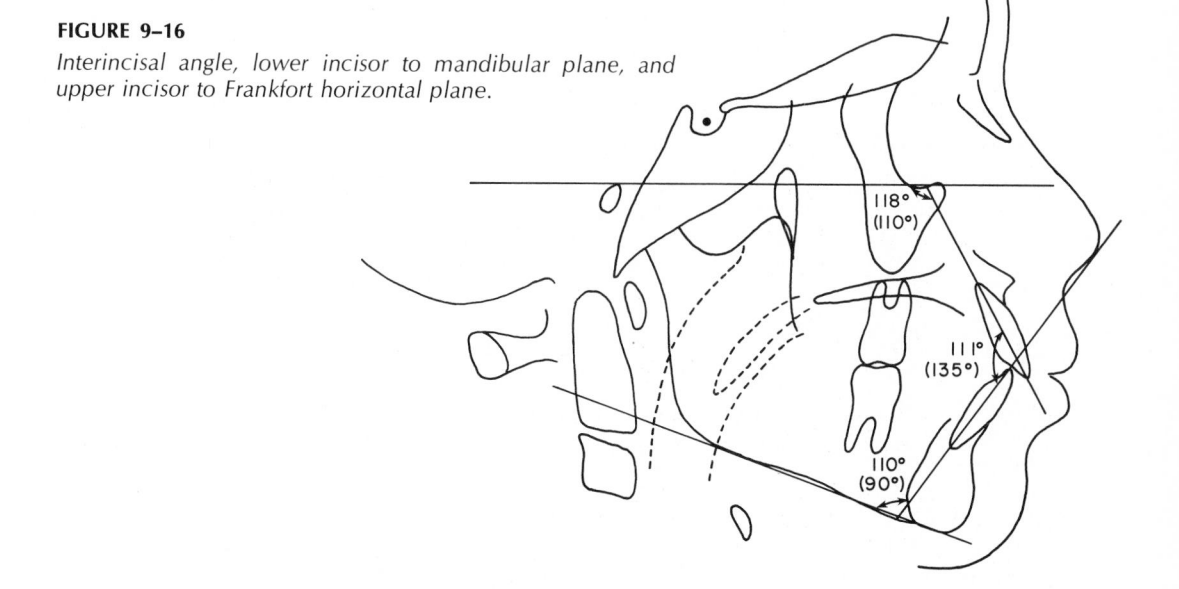

lower incisor posture according to mandibular posture are mentioned later in the discussion of the Tweed analysis.

The upper incisor angulation to the Frankfort horizontal plane is also illustrated in Figure 9–16 and has a mean value of 110°. Larger angles usually indicate maxillary incisor protrusion.

## SOFT TISSUE ASSESSMENT

Perhaps the most commonly used evaluation of soft tissue profile is the relationship of the lips to the esthetic plane (Ricketts), as shown in Figure 9–17. While this is primarily an esthetic evaluation, it is based on the fact that lip posture is influenced directly by anteroposterior position of the teeth behind the lips. As a criterion, the lower lip position should be within a range of 2 mm. behind the E-plane to just touching it. In the patient used in the illustration, both lips are well in front of the plane and reflect the extent of dental protrusion already documented.

The measurements just described are but a few of those used in cephalometrics and are illustrated here not because they are more significant than others, but because they are representative of the type of measurement made and provide a good initial understanding of this type of evaluation. The following pages will illustrate several of the more commonly used analyses incorporating these and similar measurements. Again, there is no suggestion that these are the most significant or the most accurate, but they are selected as being representative of a cross section of contemporary cephalometric thought. Each analysis is based on a background of research and hypothesis, and its author presents it as an attempt to provide a practical clinical application of the research effort.

## ANALYSES

In an effort to illustrate the techniques of cephalometric analysis, seven different analyses will be illustrated. Some of these are "total" in the sense that they make an effort at "total facial analysis." Others are limited to emphasis on a particular area or dimension. A single patient's lateral tracing will be used to illustrate each analysis, and, where appropriate, comments are made pertaining to interpretation.

The subject used is a Caucasian male, aged 13 years, 7 months, in good health and with no previous orthodontic treatment. An earlier headfilm, taken 3 years previously, will be used as a final illustration to demonstrate the technique of superimposing tracings to evaluate growth changes.

## DOWNS' ANALYSIS

Downs' analysis is based on a sample of 20 children, aged 12 to 17 years, with excellent occlusions. The diagram (Fig. 9–18) consists of the lines Na-Pog, Na-A, A-B, A-Pog, S-Gn, the occlusal plane, mandibular plane, the long axis of upper and lower incisors, and the Frankfort horizontal plane.

An addition to Downs' analysis by Voorhies and Adams is a chart providing a graphic portrayal of the ten measurements in the analysis (Fig. 9–19). The line of

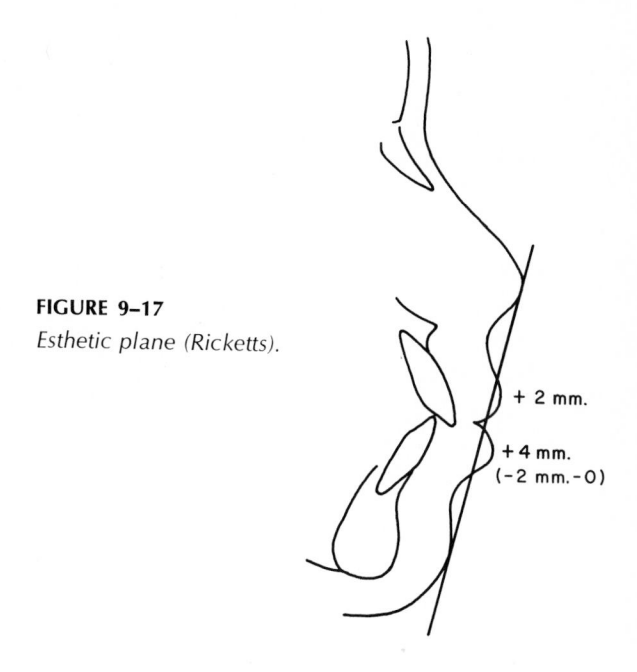

**FIGURE 9–17**
*Esthetic plane (Ricketts).*

+ 2 mm.

+ 4 mm.
(−2 mm.−0)

small arrows down the center of the diagram identifies the mean figure for each measurement, while the extent of the polygon outlines the range of each measurement. The dotted line in the illustration is the plot of the measurements of the subject used in this discussion.

The top half of the diagram charts those measurements related to skeletal configuration, while the lower half shows denture relationships. The interpretation for the patient illustrated shows that the chin is well situated anteroposteriorly, and the mandibular plane and Y-axis measurements forecast continued normal downward and forward growth in this area. The high angle of convexity and A-B plane angle

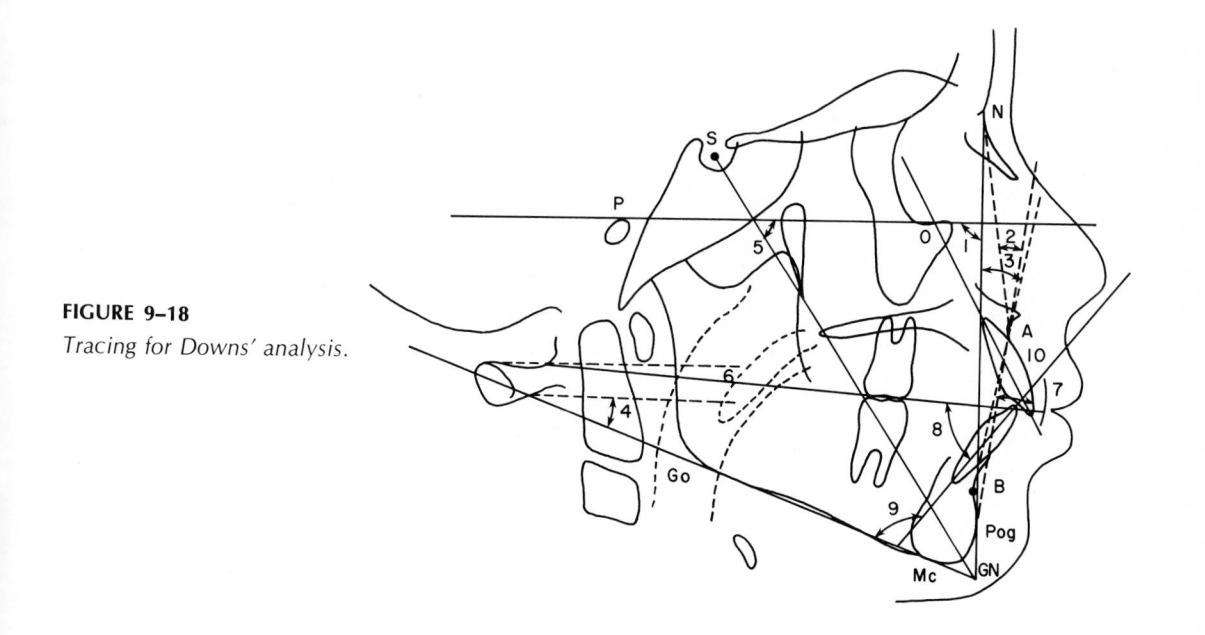

**FIGURE 9–18**

*Tracing for Downs' analysis.*

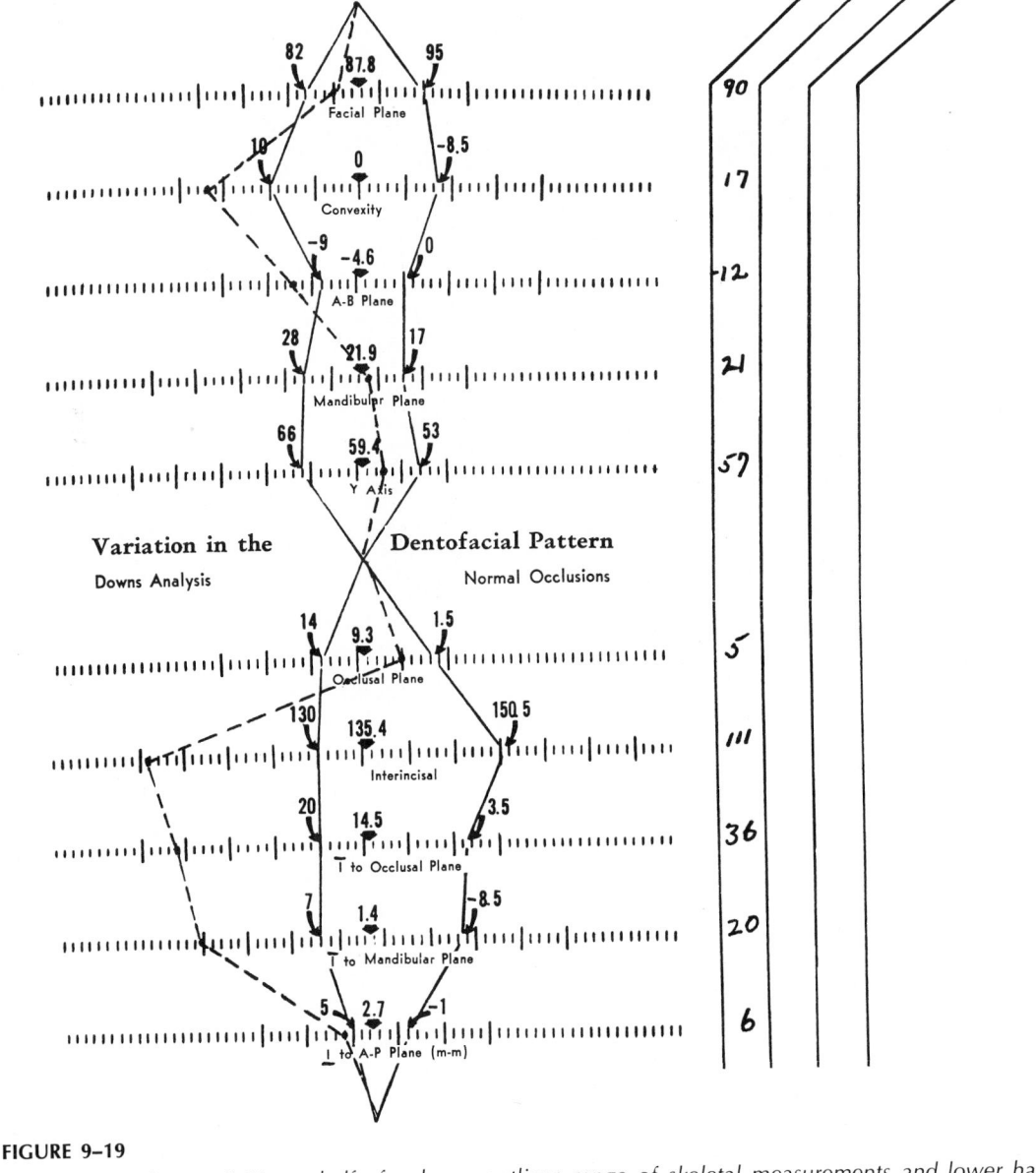

**FIGURE 9–19**

*Downs' "Wigglegram." Upper half of polygon outlines range of skeletal measurements and lower half the range of dental measurements. Dotted line shows the relation of the sample patient's measurements to the acceptable ranges. (Adapted from Voorhies, J. W., and J. W. Adams: Polygonic interpretations of cephalometric findings. Angle Orthod.; 21:194, 1951.)*

confirm the midface convexity. If the chin position is normal, the convexity must be due to midface or maxillary prominence.

The lower half of the diagram charts the marked bi-dental protrusion of this patient, as indicated by the deviations in interincisal, lower incisor to occlusal plane, and lower incisor to mandibular plane angles.

The anterior prominence of point A, which moves the A-Po line forward at its upper end, masks the protrusion of the upper incisor.

In summary, this patient would be described as having normal skeletal arrangement, except for the midface convexity, and a superimposed bi-dental protrusion.

Downs' analysis might be described as being profile-oriented. The primary reference plane is the Frankfort horizontal. Vertical assessment is only with the mandibular plane and Y-axis.

## STEINER ANALYSIS

The Steiner analysis is actually a composite of measurements from several other sources (Margolis, Thompson, Riedel, Wylie, and Downs). It is based primarily on a single plane of reference—the S-Na line—and does not take into account variations in the length or cant of this reference plane. A particular feature of the analysis is the linear as well as angular relation of the incisors to reference lines (Na-A and Na-B).

Lines to be drawn are S-Na, Na-A, Na-B, Go-Gn, occlusal plane, and long axis of the upper and lower incisors (Fig. 9–20). The Steiner analysis form (Fig. 9–21) lists the reference norms and the measurements made from the subject.

The appraisal of this patient (Fig. 9–21) reveals the chin well related anteroposteriorly, the maxilla forward, and a high ANB angle confirming the basal discrepancy. The upper incisor is well related linearly and angularly to its base, but the lower

**FIGURE 9–20**

*Tracing for Steiner analysis.*

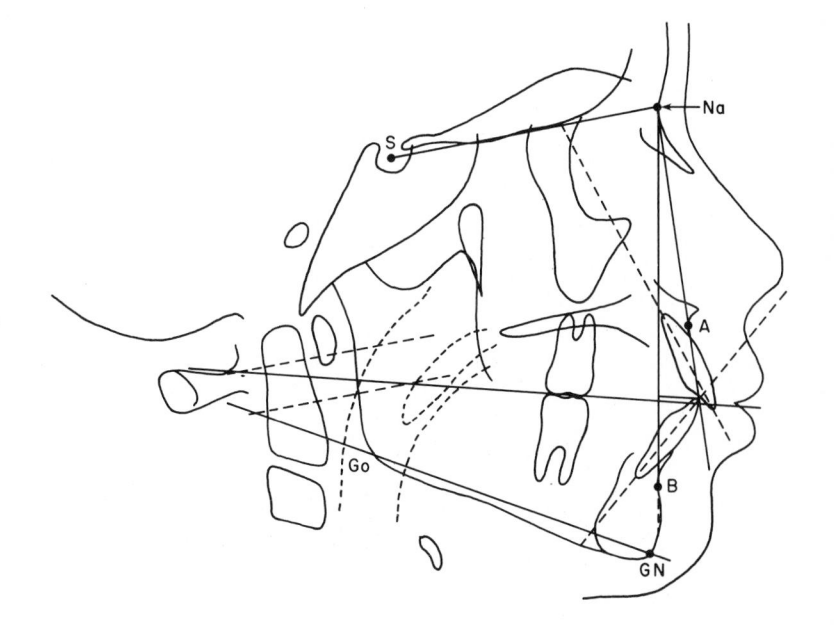

## STEINER ANALYSIS

Ref. Norm.

| | | Ref. Norm. | | | | | | | |
|---|---|---|---|---|---|---|---|---|---|
| SNA | (angle) | 82° | 87 | | | | | | |
| SNB | (angle) | 80° | 79 | | | | | | |
| ANB | (angle) | 2° | 8 | | | | | | |
| SND | (angle) | 76° or 77° | | | | | | | |
| 1 to NA | (mm) | 4 | 4 | | | | | | |
| 1 to NA | (angle) | 22° | 20 | | | | | | |
| 1̄ to NB | (mm) | 4 | 11 | | | | | | |
| 1̄ to NB | (angle) | 25° | 40 | | | | | | |
| Po to NB | (mm) | not established | 1 | | | | | | |
| Po & 1̄ to NB | (Difference) | | 10 | | | | | | |
| 1 to 1̄ | (angle) | 131° | 111 | | | | | | |
| Occl to SN | (angle) | 14° | 14 | | | | | | |
| GoGn to SN | (angle) | 32° | 30 | | | | | | |
| Arch length discrepancy | | | -3 | | | | | | |

| (mm) | + | − |
|---|---|---|
| Correcting Arch Form Moves 1̄ | | |

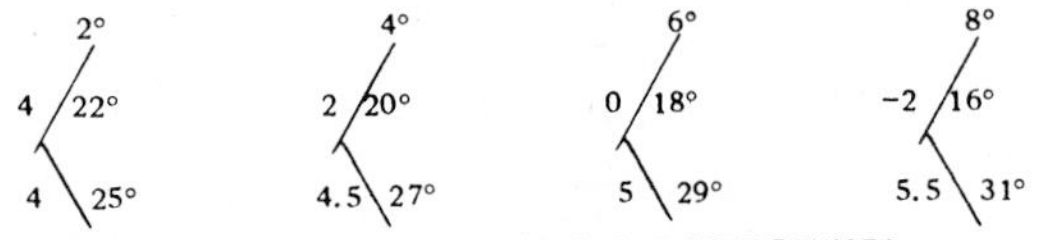

| LOWER ARCH | + | − |
|---|---|---|
| Discrepancy | | |
| Expansion | | |
| Relocation 1̄ | | |
| Relocation 6̄ | | |
| E Space | | |
| Intermaxillary | | |
| Extraction | | |
| | | |
| Total Net | | |

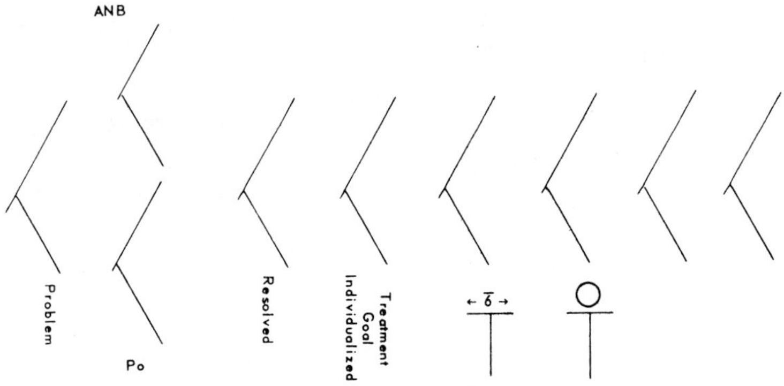

```
              2°          4°          6°          8°
           4 /22°      2 /20°      0 /18°     -2 /16°
           4 \25°    4.5 \27°      5 \29°    5.5 \31°
             IDEAL        ACCEPTABLE COMPROMISES
```

ANB

Problem    Resolved    Individualized    Treatment Goal Individualized    ← 6̄ →

Po

* *These estimates are useful as guides but they must be modified for individuals.*

Dental Corporation of America    •    P.O. Box 1011    •    Washington, D. C. 20013

**FIGURE 9–21**

*Recording and treatment planning form for Steiner analysis, with measurements of sample patient, R. B. (Courtesy of Dental Corporation of America, Washington, D.C.)*

incisor is markedly procumbent by both forms of measurement. The low interincisal angle is consistent with the bi-dental protrusion.

Summary appraisal of the patient shows normal chin position, forward maxilla, and bi-dental protrusion. The analysis does show that, while the lower incisor posture can be corrected by simple uprighting movement, it would be necessary to move the maxillary incisor bodily to preserve its normal angular posture.

An additional feature of the Steiner analysis is the provision made on the form (Fig. 9–21) for a treatment planning procedure. It is not within the scope of this

chapter to go through this procedure in detail, but briefly it is a means of synthesizing treatment objectives, taking into account the original cephalometric data, the arch length discrepancy, and selected treatment objectives.

The Steiner analysis is also profile-oriented and provides excellent visualization of incisor position and anterior facial profile detail. With the addition of its treatment planning rationale, it has enjoyed wide clinical use in the field of orthodontics.

## WYLIE ANALYSIS

In the following paragraphs, only the "assessment of anteroposterior dysplasia" as presented by Wylie in 1947 will be presented. The averages were derived from a sample equally divided between males and females with an average age of 11 years, 6 months.

All measurements except mandibular length are made parallel to the Frankfort horizontal plane from projections to the following points: posterior border of the condyle, sella, PTM, upper first molar, and the anterior nasal spine. The mandibular plane is drawn and perpendicular projections made to the posterior border of the condyle and pogonion for mandibular length.

Figure 9–22 shows the diagram and measurements on the sample patient. Comparison with the means is shown in Table 9–1.

Basically, the analysis provides a means of evaluating anteroposterior size and position of the maxilla and mandible. For maxillary measurements below the norm, the difference is put in the prognathic column, and for values above the norm in the orthognathic column. For mandibular measurements above the norm, the difference is put in the prognathic column, and in the orthognathic column when the measurements are below normal.

In the appraisal of this patient, there is a net difference, indicating the mandible is 5 mm. smaller than the maxilla. Actually, the individual measurements show

**FIGURE 9–22**

*Tracing for Wylie's assessment of anteroposterior dysplasia. Mean values for each measurement are shown in parentheses.*

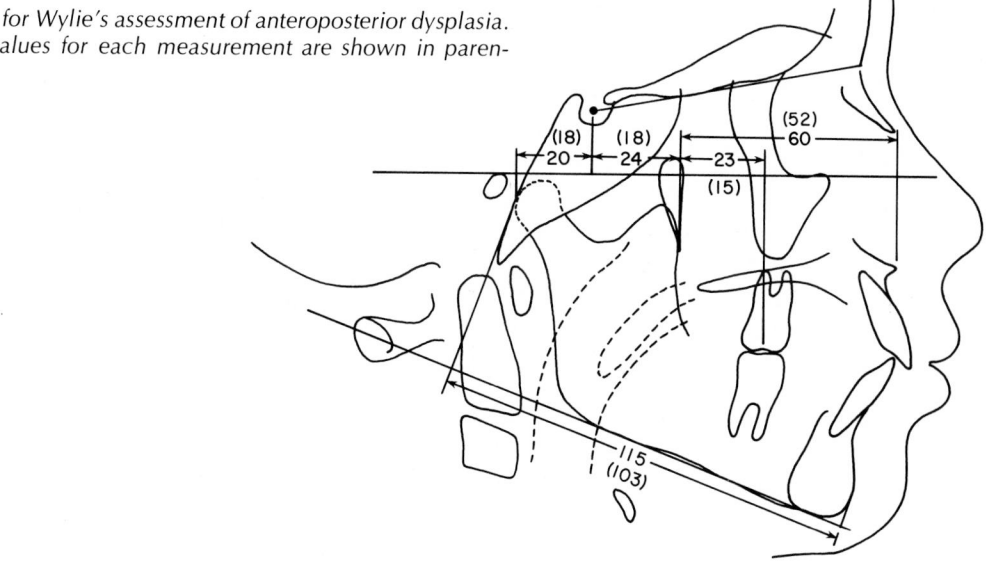

**TABLE 9–1**

| Dimension | Males | Females | R.B. | Ortho | Prog |
|---|---|---|---|---|---|
| Glen. fossa to S. | 18 | 17 | 20 | 2 | |
| S. to PTM | 18 | 17 | 24 | 6 | |
| PTM to ANS | 52 | 52 | 60 | 8 | |
| PTM to 6 | 15 | 16 | 23 | 1 | |
| Mand. length | 103 | 101 | 115 | | 12 |
| Totals: | | | | 17 | 12 |

that this is a very deep face with greater depth in the maxilla than in the mandible. This analysis is particularly useful in evaluating Class III skeletal patterns.

## TWEED ANALYSIS

It should be stated at the outset that the Tweed analysis is not a total facial analysis. Dr. Tweed did not intend it to be a total analysis, and it is unfortunate that some practitioners have used it in this manner. The analysis is based primarily on the deflection of the mandible, as measured by the Frankfort-mandibular plane angle (FMA), and the posture of the lower incisor. The objectives of the analysis appear to be twofold: first, to determine the position the lower incisor should occupy at the end of treatment. Predetermination of this relationship provides useful treatment planning information, especially in regard to the extraction decision. Second, Dr. Tweed established a prognosis on the treatment result, based on the configuration of the triangle.

Basically, the analysis consists of the so-called Tweed triangle, formed by the Frankfort horizontal plane, the mandibular plane, and the long axis of the lower incisor (Fig. 9–23). The three angles thus formed are the Frankfort-mandibular plane (FMA), lower incisor to mandibular plane (IMPA), and the lower incisor to Frankfort horizontal (FMIA). The basis is the FMA angle, as the following norms and prognoses indicate:

1. FMA 16° to 28°: prognosis good
   at 16°, IMPA should be $90° + 5° = 95°$
   at 22°, IMPA should be 90°
   at 28°, IMPA should be $90° - 5° = 85°$
   Approximately 60% of malocclusions have FMA between 16° and 28°
2. FMA from 28–35°: prognosis fair
   at 28°, IMPA should be $90° - 5° = 85°$, extractions necessary in majority of cases
   at 35°, IMPA should be 80–85°
3. FMA above 35°: prognosis bad
   extraction frequently complicates problem

Sometime after introducing the original figures shown above, Tweed began stressing the importance of the FMIA angle, recommending that it be maintained at 65 to 70°.

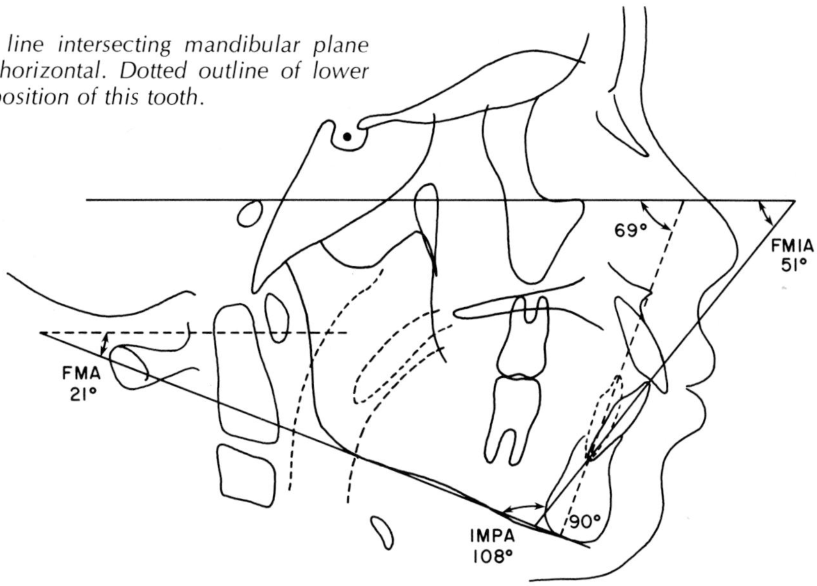

**FIGURE 9–23**

*Tweed triangle. Dotted line intersecting mandibular plane is parallel to Frankfort horizontal. Dotted outline of lower incisor represents ideal position of this tooth.*

In the sample tracing (Fig. 9–23) the FMA is 21°, FMIA 51°, and IMPA 108°. According to the analysis, with the FMA of 21°, the IMPA should be 90°, as shown by the dotted line on the tracing. With this change, the FMIA would be 69°, which is within the recommended range. To attempt to achieve this relationship would obviously necessitate removal of dental units.

The Tweed analysis is primarily for clinical treatment planning by establishing a position the lower incisor should occupy, with provision made for variations in mandibular position.

## RICKETTS' ANALYSIS

Ricketts' analysis has progressed through a series of modifications and has now burgeoned into a detailed evaluation of craniofacial and dental morphology. It has been adapted to a computer-based diagnostic and treatment forecasting service. Continued refinements of growth and treatment forecasting aspects may be anticipated as additional data are accumulated in the computer. Since this chapter is only providing an overview of several of the lateral or profile analyses, only the lateral summary analysis will be presented.

The lines traced are Frankfort horizontal, facial plane, occlusal plane, mandibular plane, esthetic plane (tip of nose to tip of chin), N-Ba, Pt-vertical (tangent to posterior outline of PTM and perpendicular to Frankfort), facial axis (upper margin of foramen rotundum to gnathion), and the long axis of the incisors.

From this tracing (Fig. 9–24) eight relations are measured to provide an overall appraisal of the case. The means and age changes from age 9 are shown in Table 9–2. The eight measurements are as follows:

1. Facial axis: the angle between the Ba-Na plane and the line from the foramen rotundum to gnathion. This gives the direction of the growth of the chin and is a modification of the Y-axis from Downs' analysis.

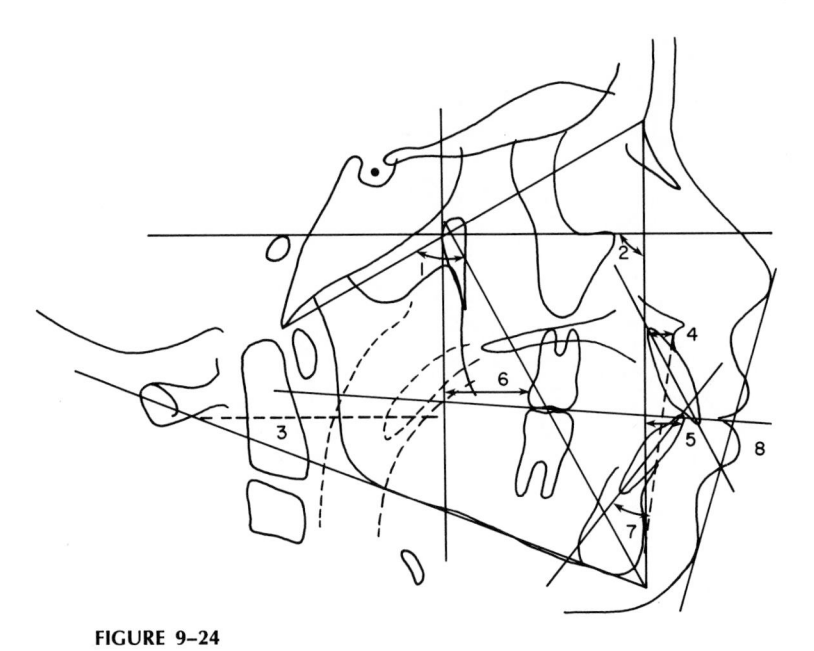

**FIGURE 9–24**
*Tracing for Ricketts' analysis (summary profile analysis only).*

2. Facial depth: the angle between the Frankfort plane and the facial plane.
3. Mandibular plane to Frankfort plane angle.
4. Convexity: the horizontal distance between point A and the facial plane. This measures anteroposterior deviation between maxilla and mandible.
5. Lower incisor to A-Po: locates the lower teeth anteroposteriorly in the mandible.
6. Upper molar position: the horizontal distance from PTV to the distal surface of the upper first molar.
7. Lower incisor inclination: the angle between the lower incisor axis and the A-Po line. This is a refinement of lower incisor to NB line (Steiner), which takes into account basal relationships.
8. Esthetic plane: lower lip relation anteroposteriorly to the esthetic plane.

This portion of the analysis provides an excellent initial survey of the case under study. For the deeper, more sophisticated analysis and the mechanism of growth forecasting, reference to the syllabus, *An Orthodontic Philosophy,* by Carl Gugino, is recommended.

**TABLE 9–2**

|  | Means | For 9 Yr. Old + Change | R.B. |
|---|---|---|---|
| 1. Facial axis | 90° ± 3° | No change | 90° |
| 2. Facial depth | 86° ± 3° | +1°/3 yrs. | 90° |
| 3. Mand. plane | 26° ± 6° | −1°/3 yrs. | 21° |
| 4. Com. of pt. A | 2 ± 2 mm. | −1 mm./3 yrs. | 8 mm. |
| 5. T to A-Po | +1 ± 2 mm. | No change | 10 mm. |
| 6. Upper molar to PTV | Age +3 mm. ± 2 mm. | 1 mm./yr. | 23 mm. |
| 7. T to A-Po | 22° ± 4° | No change | 32° |
| 8. Lower lip to E plane | −2 mm. ± 2 mm. | Decrease | 4 mm. |

For an evaluation of the patient R.B., the measurements (Table 9–2) show that the skeletal configuration (facial axis, facial depth, and mandibular plane) is essentially normal, taking into account the age difference between R.B. and the 9 year old norms. However, the last five measurements are all indicative of the midface convexity and dental protrusion evident in this patient.

This basic summary analysis is somewhat similar to Downs' analysis, except that convexity is measured by direct linear relation of point A to the facial plane. While Ricketts uses Frankfort as the horizontal reference plane, it should be pointed out that anatomic porion is used instead of "machine" porion, as described earlier. Also the facial axis, which replaced the Y-axis, is quite different, as described.

## BJORK ANALYSIS

Bjork has been an outstanding researcher in the field of cephalometrics, and his work, "The Face in Profile," is certainly recommended reading for those interested in cephalometric studies. His research was based on a study of 322 Swedish boys 12 years of age and 281 conscripts 21 to 23 years of age, and included almost 90 different measurements. Since it is not within the scope of this brief overview to attempt to abstract his massive work, only the principal portions of Bjork's analysis, as adapted and modified by Jarabak, will be presented.

The profile analysis is similar to Steiner's in that S-Na is the reference line, and SNA-SNB, along with Go-Gn, provides basic skeletal evaluation. The incisor axis and incisor to A-Po relate the denture to the skeletal base.

Lines to be drawn are: S-Na, S-Ar, Ar-Go, Go-Gr, Na-Pog, S-Gn, Na-Go, Na-A, Na-B, A-Po, occlusal plane, and long axis of the incisors (Fig. 9–25).

A feature of the analysis is the use of the polygon N-S-Ar-Go-Gn to assess anterior and posterior face height relationships and to predict direction of growth change in the lower face. The basis of this approach is the relationship of three angles —saddle angle (Na-S-Ar), articular angle (S-Ar-Go), and gonial angle (Ar-Go-Me)— and the lengths of the sides of the polygon.

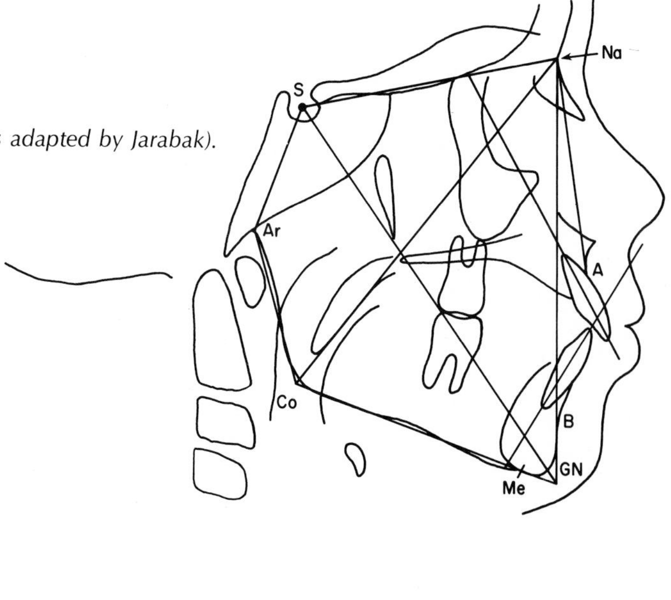

**FIGURE 9–25**

*Tracing for Bjork analysis (as adapted by Jarabak).*

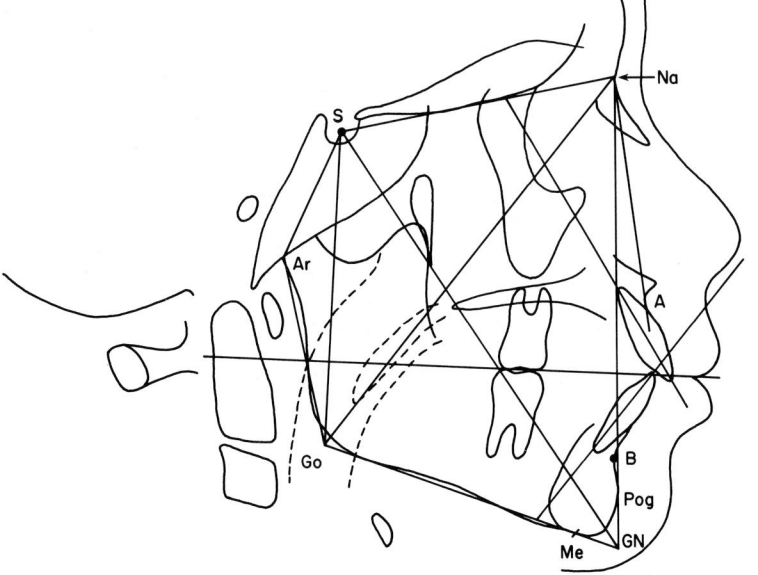

**FIGURE 9–26**

*Bjork analysis tracing on the patient shown in Figure 9–25 after 3 years' growth change.*

In the example illustrated (Fig. 9–25), a lateral film of R.B. taken in 1971 was traced and measured. Another film, taken in 1974 (Fig. 9–26), was also measured, and the values together with the means for each are shown in Table 9–3.

Briefly, at age 11 the anterior cranial base (S-Na) should equal mandibular body length (Go-Me). An ideal ratio of posterior cranial base length (S-Ar) to ramus height (Ar-Go) is 3:4. If the sum of the three angles described earlier is greater than 396°, there would be a tendency toward "clockwise" growth change in the mandible. The reverse (counterclockwise) would be the case in angles with a sum less than 396°. A ratio of posterior face height (S-Go) to the anterior face height (Na-Me) of 56 to 62 per cent would indicate a clockwise pattern of growth change in the mandible, while a ratio of 65 to 80 per cent would indicate counterclockwise change. Clockwise change means that anterior face height is increasing more rapidly than posterior face height and would be associated with downward and backward growth change at the symphysis and anterior open bite tendency. Counterclockwise change indicates more rapidly increasing posterior face height, forward growth of the chin, and anterior deep bite tendency.

In the example illustrated the sum of the posterior angles (390°) and the 65 per cent ratio of posterior to anterior face height suggest a closing or counterclockwise pattern. However, an examination of individual components shows the ramus height to be short in its ratio to posterior cranial base length, which would counteract this tendency. The prediction would be a normal downward and forward growth pattern with the mandibular plane dropping in a parallel manner, which is what occurred (Fig. 9–27). The mandibular corpus length, which was 3.5 mm. shorter than the anterior cranial base length, grew to the normal 1:1 ratio.

Figure 9–28 shows a female patient, aged 10, who was observed for 1 year of growth change. In this example the high posterior angle sum (401°), short ramus and posterior cranial base, and low posterior to anterior face height ratio (59 per cent) were combined to predict the clockwise pattern of change which occurred (Fig. 9–29).

The overall appraisal of the patient R.B. by this analysis agrees with the other analyses in identifying the midface convexity and the bi-dental protrusion.

**TABLE 9–3**

| Measurement | Average | R.B. 8-3-71 | R.B. 6-6-74 |
|---|---|---|---|
| Saddle angle | 123 ± 5 (Bjork) | 123 | 126 |
| Articular angle | 143 ± 6 (Bjork) | 141 | 142 |
| Gonial angle | 130 ± 7 (Bjork) | 127 | 122 |
| Sum | 396    (Bjork) | 391 | 390 |
| Ant. cranial base length | 71 mm. ± 3 (Bjork) | 71.5 | 74 |
| Post. cranial base length | 32 mm. ± 3 (Bjork) | 35 | 36 |
| Gonial angle | | | |
| Upper | 52°–55° | 56 | 53 |
| Lower | 70°–75° | 71 | 69 |
| Ramus height | 44 mm. ± 5 (Bjork) | 43 | 49 |
| Body length | 71 mm. ± 5 (Bjork) | 68 | 74 |
| Man. body to | | | |
| ant. cranial base:ratio | 1:1 | 1:1 | 1:1 |
| SNA | 80° | 86 | 87 |
| SNB | 78° | 80 | 80 |
| ANB | 2° | 6 | 7 |
| SN–MP | | 31 | 31 |
| Y–Axis | | 57 | 57 |
| Ant. face height | | 114 | 122 |
| Post. face height | | 74 | 81 |
| Post. face-ant. face:ratio | | 65% | 66% |
| 56–62% clockwise | | | |
| 65–80% counterclockwise | | | |
| Facial angle (SN–Po) | | 86 | 87 |
| | | | |
| *Denture* | | | |
| Occ. P1–M-P1 | | 13 | 15 |
| 1 to M-P1 | 90° ± 3 | 100.5 | 110 |
| *1* to SN | 102° ± 2 | 106 | 108 |
| *1* to facial plane | 5 mm ± 2 | 14 | 15 |
| *1* to facial plane | −2 mm ± 2 mm. | 9 | 10 |
| 1 to *1* | | 121 | 112 |

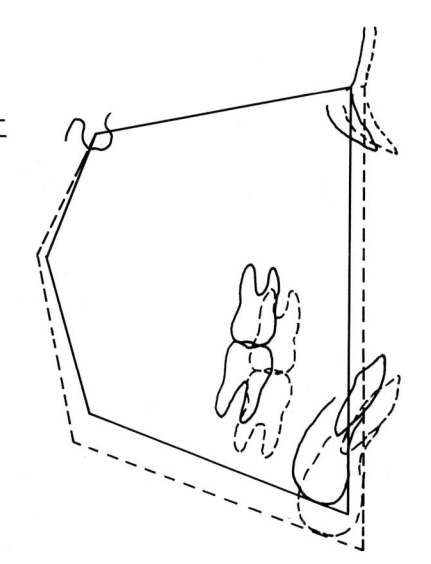

8-3-71 ———
6-6-74 ------

**FIGURE 9–27**

*Bjork polygons from tracings in Figures 9–25 and 9–26 superimposed to show growth change from age 10⁹ to age 13⁷. The SN lines are superimposed, registered at sella.*

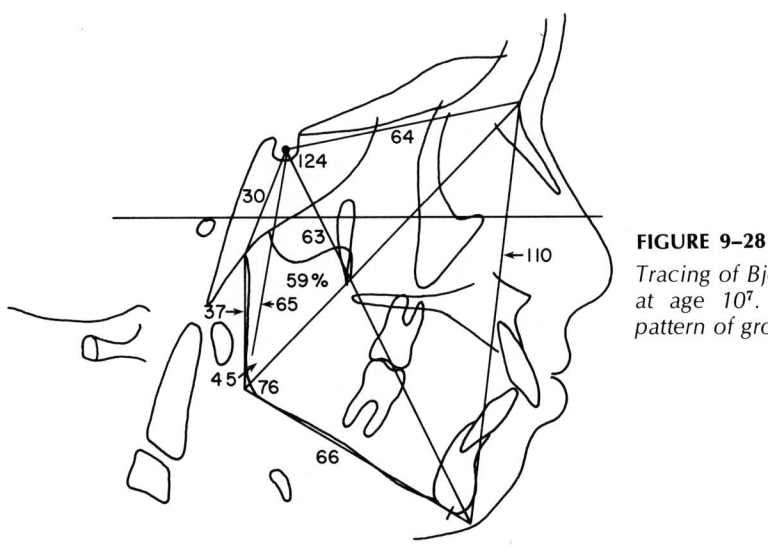

**FIGURE 9–28**

*Tracing of Bjork polygon on female patient at age 10[7]. Data indicates a clockwise pattern of growth change in the lower face.*

## SASSOUNI ANALYSIS

The Sassouni, or "archial," analysis is unique in that it does not employ a set of established norms, but rather defines relationships within the individual pattern which are judged "normal" or "abnormal."

Lines and points not previously described include the supraorbital plane: tangent to the anterior clinoid process and the most superior point on the roof of the orbit; Si: the lowest point on the contour of the sella turcica; Sp: the most posterior point on the sella turcica outline; Te: the intersection of the cribriform plate and the anterior wall of the infratemporal fossa; and point "O": the center of the convergence where the four horizontal planes tend to intersect.

A plane is drawn parallel to the supraorbital plane, tangent to Si, and the occlusal, palatal, and mandibular planes are drawn. These four planes should con-

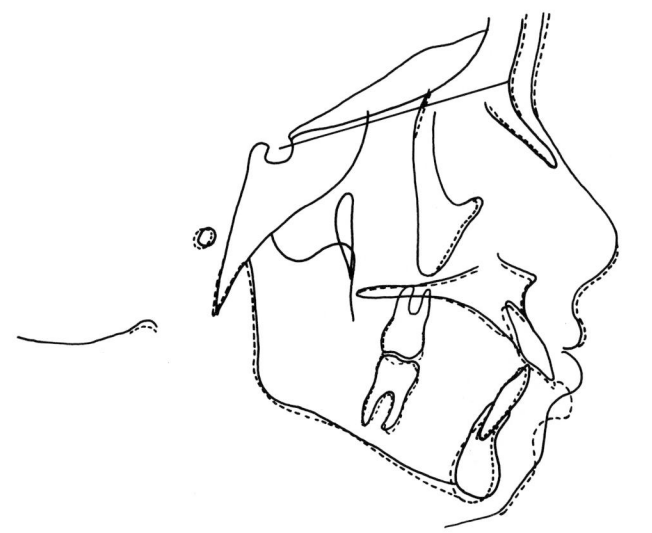

**FIGURE 9–29**

*Tracings of patient shown in Figure 9–28 superimposed, showing one year of clockwise growth change in mandible, with the symphysis dropping down and back.*

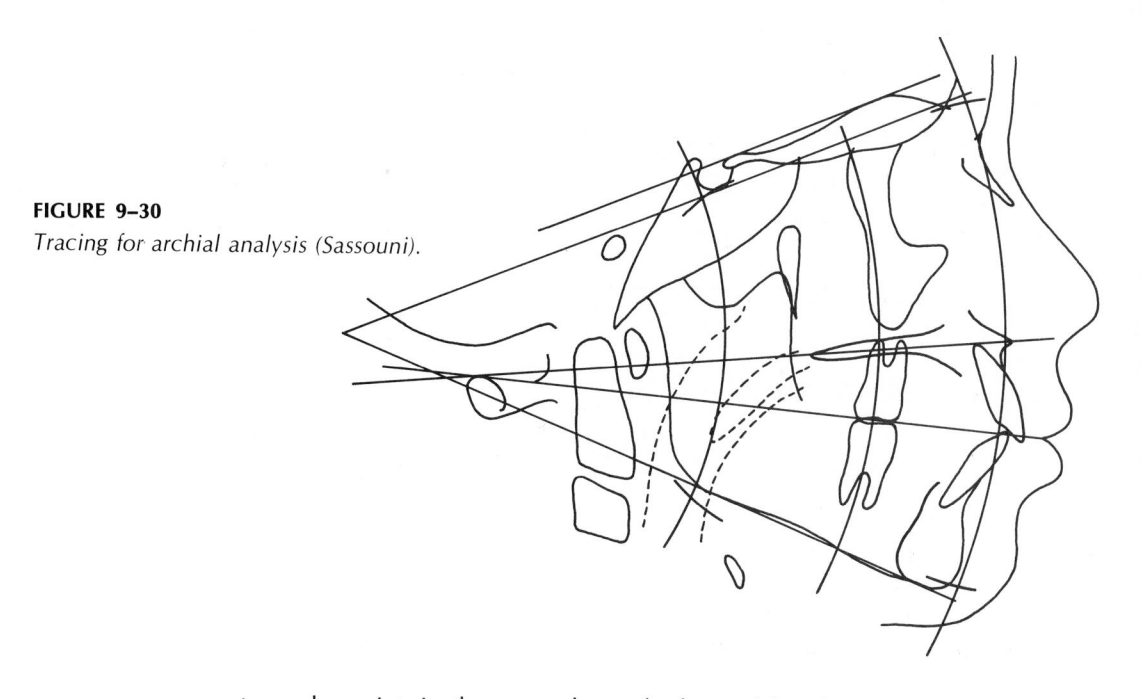

**FIGURE 9–30**

*Tracing for archial analysis (Sassouni).*

verge toward a point. In the event that only three of the planes converge toward a point, the fourth plane is divergent from the facial pattern. If only two of the planes converge, as in Figure 9–30, the junction of cranial base plane and mandibular plane is used as point O. From point O, four arcs are drawn: from nasion, from point B, from Te, and from Sp. These arcs are called the anterior, basal, midface, and posterior arcs, respectively.

General evaluation of the planes shows that the more they are parallel, the greater the tendency to skeletal deep bite, and the more they are steep to each other, the greater the tendency to skeletal open bite.

The anterior arc from nasion should pass through ANS, the tip of the maxillary incisor, and pogonion. The basal arch from point A should pass through point B. The midfacial arc from Te should pass tangent to the mesial surface of the maxillary first molar when ANS is on the anterior arc. If ANS is not on the anterior arc, the maxillary first molar relationship should be adjusted in the same amount and direction as the ANS deviation. The posterior arch from Sp should pass through gonion. If pogonion is on the anterior arc and gonion is on the posterior arc, it means that the mandibular corpus length is equal to the anterior cranial base length and is the normal relationship at age 12 years.

Vertically, the upper and lower face heights should be equal, both anteriorly and posteriorly. The anterior measurement is made by placing the point of the compass on ANS and striking an arc at supraorbitale. That dimension is then transferred by rotating the compass and intersecting the anterior arc at the symphysis area. For the posterior measurement, the tip of the compass is placed at PNS and an arc is struck at the intersection of the parallel plane (cranial base plane) and the posterior arc. This measurement is transferred inferiorly by striking an arc intersecting the posterior arc in the area of gonion.

In Figure 9–30 the anterior arc passes through ANS and is slightly forward of pogonion. Both incisors are forward of the arc, indicating incisor protrusion. Point B is behind the basal arc, indicating a basal discrepancy.

The maxillary first molar mesial contour is forward from the midfacial arc, indicating forward position of the maxillary teeth.

Gonion is behind the posterior arc slightly more than pogonion is behind the anterior arc, but corpus length in relation to anterior cranial base length is essentially normal for the age of this patient.

Vertically, the palatal plane is deviant, appearing elevated in front and depressed posteriorly. This results in anterior lower face height being slightly excessive and posterior lower face height slightly deficient.

In summary, this analysis indicates that the subject's general facial pattern is well coordinated, except for the tilted palatal plane, the protrusion of the dentition, and the mildly retrognathic mandible.


## ANALYSES SUMMARY

In comparing the seven analyses used on the sample patient, there is little serious disagreement in the appraisals. There was general agreement that the chin was well located anteroposteriorly, the face was convex, and a bi-dental protrusion was present. According to the archial analysis, the facial skeletal profile was well related, with some convexity resulting from a slightly retrognathic mandible, the remaining convexity being dental in nature. The other analyses found point A forward. The extent to which point A is influenced by incisor position is likely a factor in this difference. In a general sense, Wylie's assessment of anteroposterior dysplasia agreed with the archial analysis in that it demonstrated great depth to the face and showed the mandible to be 5 mm. shorter than the upper base.

All analyses agreed on the existence and magnitude of the bi-dental protrusion, with the exception of Wylie's, and a portion of that analysis did not measure dental protrusion.

In general, all analyses were quite close in their evaluation of growth direction with regard to mandibular change. This would not always be the case if many different patterns were subjected to this same overview approach.


## SUPERIMPOSING FOR GROWTH CHANGE EVALUATION

One of the most useful procedures to the clinician is acquiring headfilms of individual patients at periodic intervals and comparing them to obtain a general view of growth changes. As mentioned at the beginning of this chapter, this technique will not locate the sites of growth but will provide a quantitative directional appraisal of changes that have occurred.

There are numerous planes and landmarks that may be superimposed to view growth change. Probably the most common is to superimpose two serial tracings with point registration at sella and the S-Na lines superimposed one over the other. This provides a composite view of growth change during the period between the two films and is reasonably accurate as long as the growth change at nasion follows an extension of the original S-Na line. Most of the change in this area is due to frontal sinus growth, and an upward or downward migration of the frontonasal suture would result in error. However, it remains a most common method of superimposing serial headplate tracings for this purpose. An example, using patient R.B., is shown in Figure 9–31 and illustrates an essentially normal growth pattern. Downward and forward progress of both maxilla and mandible have proceeded in parallel fashion. The chin has followed an extension of the Y-axis very closely. In the almost 3 years of growth, the anterior cranial base length increased 2.5 mm. and the mandibular body length 6.0 mm. The cranial base flexure angle opened slightly,

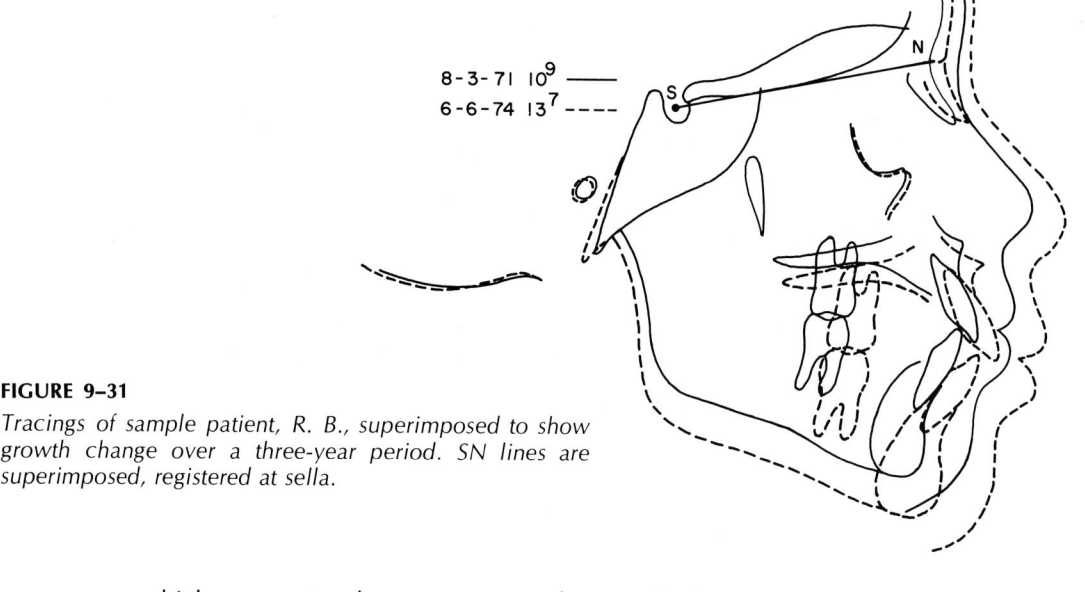

8-3-71 10⁹ ———
6-6-74 13⁷ - - - -

**FIGURE 9–31**

*Tracings of sample patient, R. B., superimposed to show growth change over a three-year period. SN lines are superimposed, registered at sella.*

which counteracted to some extent the mandibular growth in terms of anteroposterior chin position.

Figure 9–29 illustrates another patient in whom chin position has moved downward and backward, increasing facial convexity.

### SUMMARY

This discussion of cephalometrics has been admittedly a very brief overview of the technique and principles used in this form of craniofacial evaluation. The technique of acquiring headfilms and tracing and measuring them is straightforward and has demonstrated good accuracy. Over the past 40-odd years, various measurements and systems of analysis have been developed and currently are widely used, especially in the orthodontic field. Cephalometric research and its clinical application have centered on facial pattern evaluation, growth prediction, and treatment evaluation, all of which are external morphologic appraisals. The widespread clinical interest in cephalometrics has undoubtedly resulted in the development of measurements and analyses that evaluate areas of primary interest to the clinician, with emphasis on locating and quantifying external positional deviations rather than pin pointing sites of growth change or treatment response. The immediate concern of the clinician was, "What is the problem? Where is it? How severe is it? What can I do to correct it?" Cephalometrics has helped answer these questions, but it has not added significantly to an understanding of the true nature of the deviation, how it got that way, and by what mechanism it will respond to treatment.

In defense of cephalometrics it must be stated that many of the landmarks and lines were carry-overs from preradiographic craniometric techniques, and while they bore little relationship to sites of growth and/or treatment change, they were easy to locate and lent themselves well to the kind of geometric evaluation in current use.

Deviation was recognized by comparison of patient measurements with statistically derived means obtained by study of selected population samples. Perhaps it is an overstatement to say that this procedure appears to have the potential error of

not recognizing biological variability, and there is an implication of trying to make everybody look alike.

Gross evaluation of growth and treatment change, which was illustrated earlier, is of limited value in this context because it records only what has happened without divulging the true nature and location of the change. This is not to say that cephalometrics is not a useful tool for the clinician and researcher alike, as it is quite likely that its present mode of use will continue to be an important part of craniofacial research and clinical evaluation.

More recently, growth and development research has shifted emphasis to a more complete understanding of individual patterns of growth change and of what is happening in this complex anatomic area to cause these patterns to develop. The principal author of this text, as a counterpoint to the system and philosophy of analysis just described, visualizes use of the cephalometric principle and technique for a quite different purpose. While the cephalometric study to be described by Dr. Enlow in subsequent pages may not be designed for routine clinical use, an understanding of it may help to close the "gap" between the questions, "What is the problem?" and, "How did it get that way?"

## THE COUNTERPART ANALYSIS (OF ENLOW)

This is a method in which the various facial and cranial parts are compared with each other to see, simply, how they fit. The individual is measured against himself rather than compared with population standards and norms. Most conventional methods of analysis and cephalometric growth studies are intended essentially to determine **what** a particular growth or form pattern is. This procedure was developed to explain **how** such a pattern was produced in any given person. The "ANB" angle, for example, tells one the nature of the positional relationship between the anterior part of the upper and lower arches and provides an index to gauge the extent of malocclusions. The counterpart procedure is intended to account for the composite of the anatomic and morphogenic factors that **produced** the particular ANB angle (and other measurements) found in a given person.

Most conventional cephalometric planes and angles are not intended to coincide with or indicate actual sites and fields of growth and remodeling, and they are thus not appropriate for the essentially anatomic purposes just described. Because most standard planes and angles do not represent the patterns and distribution of growth fields, comparisons of the individual with population standards are required; there is usually no other basis for interpretation due to the nature of the planes themselves. However, if planes are constructed so that the activities of the growth and remodeling fields are in fact directly represented, a built-in and morphologically natural set of "standards" is identifiable, which allows meaningful evaluation of overall craniofacial form and pattern without population comparisons.

The analysis is based on the **counterpart principle.** This is the actual design basis upon which the face is constructed and which underlies the plan of its intrinsic growth process. The counterpart concept was described in Chapters 3 and 4, and it was used as the working basis for explaining how the face grows. The counterpart analysis is, in effect, the same thing. It shows **where** imbalances exist, **how much** is involved, and what the **effects** are. Refer to pages 49 and 154 to review just what a "counterpart" is.

In Figure 9–32, construction lines have been drawn on a headfilm tracing to represent several key fields and sites of growth. These include the maxillary

tuberosity, the mandibular condyle (using articulare for convenience rather than condylion), the ramus-corpus junction, the posterior border of the ramus, the anterior surfaces of both the maxillary and mandibular bony arches, the occlusal plane, and the junction between the middle and anterior cranial fossae (the anterior-most extent of the great wings of the sphenoid where they cross the cranial floor). Other planes may be added to represent other major growth areas, if desired, such as the zygomatic arch, the palate, the olfactory plane, and the anterior-vertical plane of the midface.

Note that the **PM** vertical plane is represented. This is the important boundary that separates the anterior cranial fossa and nasomaxillary complex from the middle cranial fossa and pharynx. The ramus relates to the latter and the corpus to the former (see page 154).

Two basic factors are important in evaluating the role of any bone or part of a bone in a composite assembly of several different bones. The first is the bone's **size** (horizontal and vertical), and the second is its **alignment** (rotational position). In this analysis, all must be considered. The reason is that the nature of the alignment of any bone affects the **expression** of its various dimensions, as explained in Chapter 5. The determination of a bone's dimension alone is not enough (and can be misleading); its alignment must also be known in order to see just how this factor affects its actual dimensions. In the counterpart analysis, both are determined for all the various bony parts and counterparts.

The rationale, in brief, is that the vertical and/or horizontal size of one given part is compared with its specific counterpart(s). If they exactly match, or nearly so, a dimensional "balance" exists between them. If one or the other is long or short, however, the resulting imbalance can cause either **protrusion** or **retrusion** of the part of the face involved and thereby affect the profile, either directly or indirectly (Fig. 9–32). The various parts and counterparts are then checked for their alignment to see if each, independently, has a protrusive or a retrusive effect, regardless of the nature of the dimensions. Then, all the regional part-counterpart relationships are added up to see how the combined, aggregate sum of all underlies the face of any given individual. This may be done on a single headfilm tracing at any age, or serial headfilms can be used to determine the progressive effects of age changes or of treatment results.

Figure 9–33 shows a Class II individual in whom major variations and imbalances are present for the different horizontal and vertical dimensions and for the alignment relationships (compare with the Class III individual described in the next paragraph). Note that (1) the mandibular corpus is short **relative** to its counterpart, the maxillary bony arch (both skeletally and dentally in this individual); (2) the corpus is aligned (rotated) upward (that is, the "gonial angle" is more closed); (3) the middle cranial fossa is aligned obliquely more forward (the dashed lines represent "neutral" alignment positions); (4) the ramus is aligned more backward; and (5) the nasomaxillary complex is vertically long (resulting in a downward and backward ramus rotation). **All** these features are either mandibular retrusive or maxillary protrusive, and they have combined to produce the multifactorial basis for a Class II malocclusion and retrognathic profile. Note, however, that the horizontal breadth of the ramus **exceeds** its counterpart, the horizontal (not oblique) dimension of the middle cranial fossa. This is a compensatory feature that has partially offset the aggregate effects produced by the other features and thereby reduced the severity of the malocclusion. If desired, the actual amounts of each and all of these effects can be measured.

Figure 9–34 shows an individual in whom the dimensions and alignments combine to produce the composite, multifactorial basis for a Class III malocclusion.

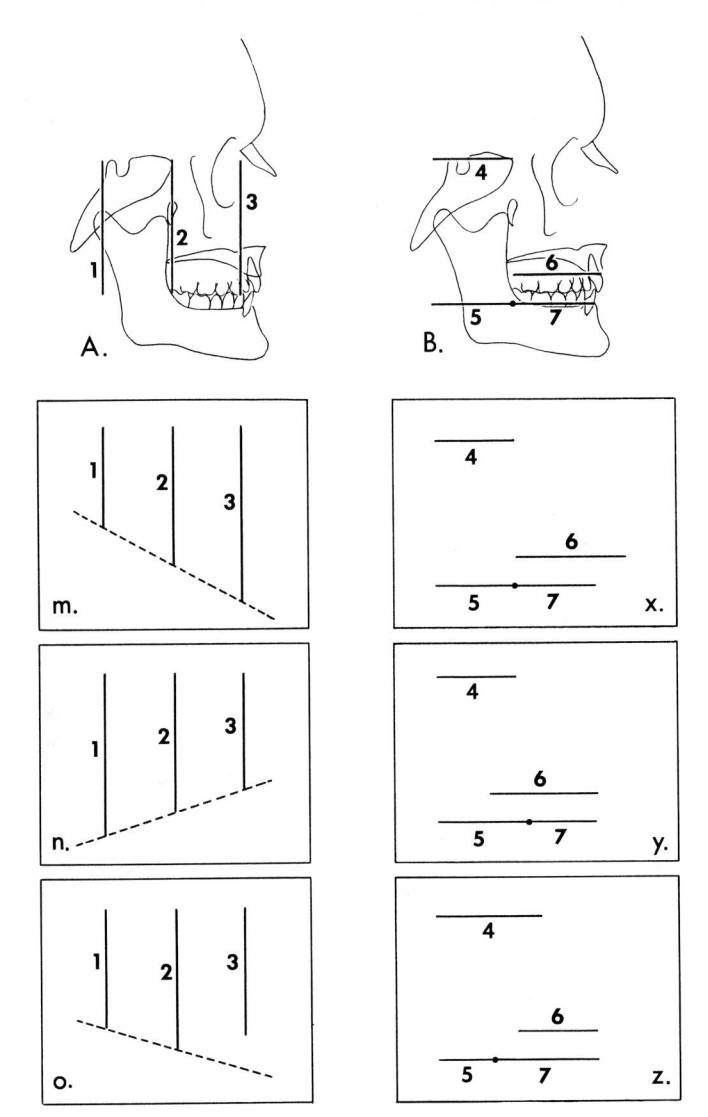

**FIGURE 9–32**

*Diagram A illustrates three vertical architectural counterparts: the cranial floor–ramus vertical (1); the posterior nasomaxilla (2); and the anterior nasomaxilla (3). As shown in A, they are all in exact dimensional balance, and the functional occlusal plane therefore coincides with the neutral occlusal axis, which is perpendicular to these three vertical planes. If vertical dimensional imbalance occurs, however, downward occlusal rotation (m), upward occlusal rotation (n), or open bite (o) necessarily results.*

*Diagram B illustrates four horizontal architectural counterparts: the middle cranial fossa (4), the ramus (5), the maxilla (6), and the mandibular corpus (7). The anterior cranial fossa is also involved but is not included in this particular diagram. If all these horizontal counterparts are balanced, as in B, their effective dimensions are very close to an exact match and the bones "fit" relative to each other. Diagram x shows a maxillary arch that is excessive relative to a disproportionately smaller corpus. The other segments (4 and 5) are balanced. The result is maxillary protrusion. Diagram y demonstrates a similar maxillary-mandibular imbalance, but the ramus, although actually out of balance to the cranial base, serves to provide dimensional compensation, so that aggregate balance is achieved in the overall composite of parts. Diagram z illustrates a "long" middle cranial fossa that is not dimensionally in balance with a "short" ramus. However, aggregate balance results because an imbalance also exists between the corpus and the maxilla, so that the sum of all their dimensions has been balanced. It is apparent that many other similar combinations are possible among the composites of these architectural counterparts which will produce either a balanced or imbalanced face. **Note:** The important factor of **alignment** is not included in these figures. (From Enlow, D. H., R. E. Moyers, W. S. Hunter, and J. A. McNamara, Jr.: A procedure for the analysis of intrinsic facial form and growth. Am. J. Orthod., 56:6, 1969.)*

**FIGURE 9–33**

*Headfilm tracing of a Class II patient. Construction lines have been added for the counterpart analysis. (From Enlow, D. H., T. Kuroda, and A. B. Lewis: Angle Orthod., 41:161, 1971.)*

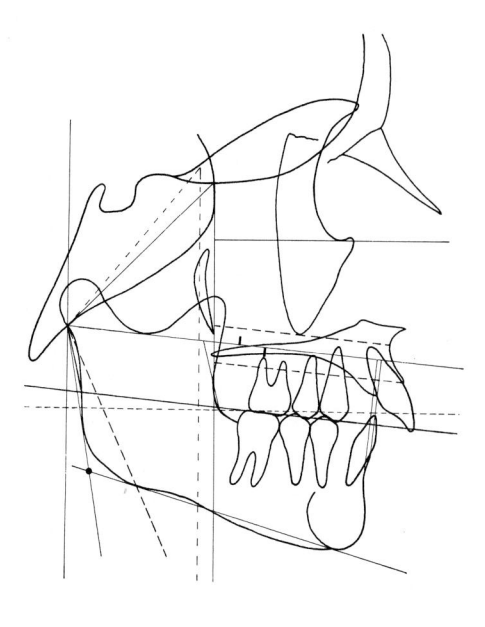

Note that *(1)* the dental **and** skeletal dimensions of the mandibular corpus in this individual exceed maxillary arch length; *(2)* the corpus is aligned (rotated) markedly downward; *(3)* the middle cranial fossa is aligned backward; and *(4)* the ramus is aligned forward. These relationships are all mandibular protrusive or maxillary retrusive, and they combine to produce the prognathic face and Class III malocclusion. The horizontal breadth of the ramus, however, is less than its counterpart, the middle cranial fossa. This is a compensatory feature that, in this particular individual, has partially offset the composite effects of the other relationships and thereby reduced the extent of the malocclusion.

For a more detailed description of the construction lines used, how to deter-

**FIGURE 9–34**

*Headfilm tracing of a Class III patient. Compare with Figure 9–33. (From Enlow, D. H., T. Kuroda, and A. B. Lewis: Angle Orthod., 41:161, 1971.)*

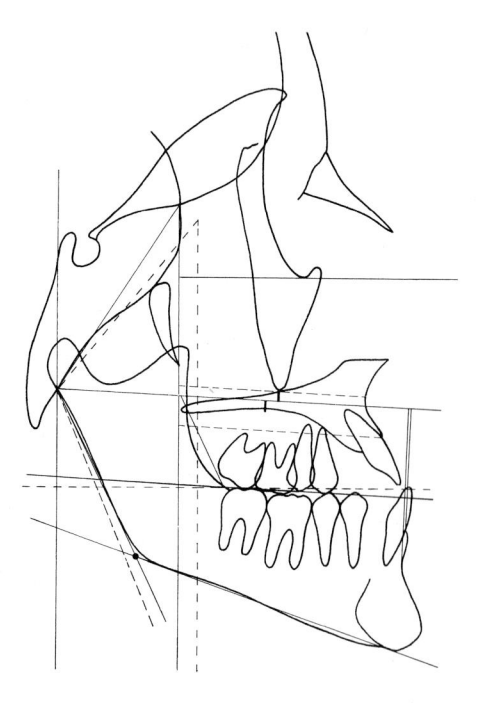

mine the actual dimensions, and how to establish the "neutral" alignment planes for any given individual, see Enlow et al.: Angle Orthod., 41:161, 1971.

The "counterpart analysis" is not intended as a routine clinical tool for everyday office use in diagnosis and treatment planning. It is not needed for this, because the rationale for treatment procedures, at least today, is not usually based on corrections of the actual underlying causes of malocclusions and other kinds of facial and cranial dysplasias. The counterpart analysis is useful, however, in determining what treatment has done in terms of the specific anatomic and developmental changes that have been brought about, more so than most types of analyses because the others deal more with correlative geometry than with morphologic and morphogenic relationships. Actually, the immediate payoff for the counterpart analysis has already been largely achieved. It has pointed out more clearly the multifactorial basis for malocclusions and just what some of the specific anatomic and developmental factors are. It has shown how a number of compensatory features participate. It has explained how and why population groups have either Class II or Class III tendencies. Except for such specific types of research studies, however, the counterpart analysis is inappropriate as a routine clinical method. When the intrinsic control processes of facial growth become better understood, when the control processes themselves can be controlled, and when treatment procedures can **then** become based on the real causative factors that underlie structural imbalances, then cephalometric analyses utilizing genuine anatomic and developmental relationships will become increasingly more relevant. The counterpart analysis itself, of course, is far from complete and is only a beginning, but it is a concept. It is also very useful in understanding the rationale for the basic plan of normal facial construction as well as malocclusions (as in Chapters 3 and 5) and in explaining and teaching this complex subject to students in a way relatively easy to understand.

## GLOSSARY OF TERMS

The terms to be defined are primarily related to landmarks used in roentgenographic cephalometry. The definitions used are those most commonly found in the craniometric and orthodontic literature. Where applicable, the term is preceded by its abbreviation.

Frontotemporale:   a point near the root of the zygomatic process of the frontal bone at the anterior-most point along the curvature of the temporal line

Metopion:   a point in the median line of the forehead between the summits of the frontal eminence

Bregma:   the point on the skull corresponding to the junction of the coronal and sagittal sutures

Vertex:   the most superior point on the cranial vault

Zygion:   the point on the zygoma on either side, at the extremity of the bizygomatic diameter

Dacryon: a point on the inner wall of the orbit at the junction of the frontal and lacrimal bones and maxilla

Gl, Glabella: most anterior point on the frontal bone

N, Nasion:   in craniometry, the junction of the internasal suture with the nasofrontal suture; in roentgenographic cephalometry, the most anterior point of the nasofrontal suture as viewed from norma lateralis

ANS, Anterior Nasal Spine:   the tip of the anterior nasal spine as seen on the lateral film

A, Subspinale:   the deepest midline point on the premaxilla between the anterior nasal spine and prosthion

Pr, Prosthion:   the most anterior point of the alveolar portion of the premaxilla, usually between the upper central incisors

Is, Incisor Superius:   the tip of the crown of the most anterior maxillary central incisor

Ii, Incisor Inferius:   the tip of the crown of the most anterior mandibular central incisor

Id, Infradentale:   most anterior point of the tip of the alveolar process between the mandibular central incisors

B, Supramentale:   the most posterior point on the outer contour of the mandibular alveolar process

Pog, Po, Pogonion:   most anterior point of the bony chin as seen on the lateral film

Gn, Gnathion:   the midpoint between the most anterior and inferior points on the outline of the bony chin

Me, Menton:   the most inferior point on the outline of the symphysis as seen on the lateral film

SE, Sphenoethmoidal Suture:   the most superior point of the suture

Si:   the most inferior point on the lower contour of sella turcica

Sp:   the most posterior point on the posterior contour of sella turcica

S, Sella:   the midpoint of the sella turcica as determined by inspection

SO, Spheno-occipital Synchondrosis:   the most superior point of the junction

Te, Temporale:   intersection of the shadows of the ethmoid and the anterior wall of the infratemporal fossa

Ba, Basion:   the median point of the anterior margin of the foramen magnum

Op, Opisthion:   the most posterior point of the foramen magnum

PNS, Posterior Nasal Spine:   the tip of the posterior spine of the palatine bone in the hard palate

Go, Gonion:   a point midway between the most inferior and most posterior points on the angle of the jaw; in some applications, the intersection of lines tangent to the mandibular base and to the posterior margin of the ascending ramus

Ar, Articulare:   the point of intersection of the contour of the external cranial base and the posterior contour of the condylar processes

Cd, Condylion:   the most superior point on the head of the condyle

Po, Porion:   "anatomic porion" is the outer upper margin of the external auditory canal; "machine porion" is the uppermost point on the outline of the metal rings on the ear rods of the cephalostat

Ptm, Pterygomaxillary Fissure:   the projected contour of the fissure on the lateral film; anterior wall represents maxillary tuberosity outline, and the posterior, the anterior curve of the pterygoid process

Or, Orbitale:   the lowermost point on the inferior margin of the left orbit; actually, when double projection occurs, the midpoint between the two outlines is used

SOr, Supraorbitale:   the uppermost point of the orbital ridge; on the lateral film, it can be located at the junction of the roof of the orbit and the lateral contour of the orbital ridge

Bo, Bolton Point:   the highest point in the concavity behind the occipital condyle

FH, Frankfort Horizontal Plane:   a horizontal plane intersecting right and left porion, and right and left orbitale; in actual practice, midway points between right and left porion and right and left orbitale are used, which results in a line rather than a true plane

# Chapter 10

# Prenatal Facial Growth and Development*

# PART ONE

Webster delimits the face as the front part of the head comprising the nose, cheeks, jaws, mouth, forehead (although not actually part of the true face or viscerocranium), and the eyes. The word face itself means "form" or "shape." A 1 month old embryo has no real face. But the key primordia have already begun to gather, and these slight swellings, depressions, and thickenings are rapidly to undergo a series of mergers, rearrangements, and enlargements that will transform them, as if by slight of hand, from a cluster of separate masses into a **face.** It's a great story.

FIGURES 10–1, 10–2, AND 10–3

**Concept 1.** The "head" of a 4 week old human embryo is mostly just a brain covered by a thin sheet of ectoderm and mesoderm. Where the mouth will be is marked by a tiny depression, the **stomodeum.** The eyes have already begun to form by a thickening of the surface ectoderm (the future lens), which meets an outpouching from the brain (the future retina). The eyes are still located at the sides of the head, however, as in a fish. As the brain continues to grow and expand, the eyes are rotated toward each other and toward the midline of what is soon to become a face. Doesn't this greatly reduce the intervening span between the right and left eyes? Yes, but in a relative sense. **Everything** is increasing in size, including the interorbital dimension. The eyes are actually moving farther apart, but because other parts of the head are enlarging even more, the proportionate size of the interorbital area is becoming decreased. When illustrating the process of facial growth, it has always been traditional to show all of the stages as equal in size. Keep in mind, however, that there is actually considerable overall enlargement involved. These changes proceed very swiftly in the early stages.

---

*Written in collaboration with Candace Mauser, M. A., Department of Anatomy, West Virginia University School of Medicine.

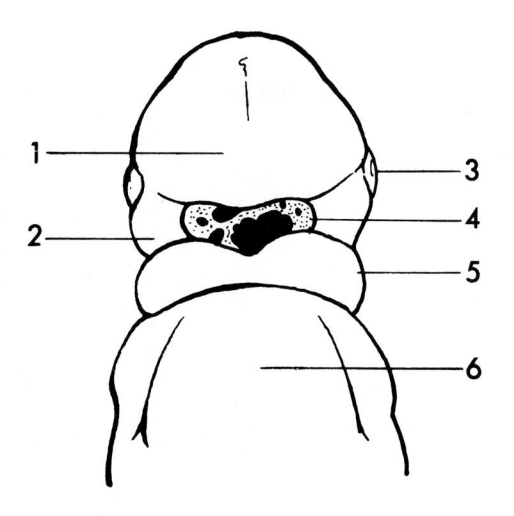

**FIGURE 10–1**

*Human head region at about 4 weeks.* 1, *Forebrain.* 2, *Maxillary region.* 3, *Optic vesicle.* 4, *Stomodeal plate (already rupturing).* 5, *Mandibular (first) arch.* 6, *Cardiac prominence.* (Modified from Patten, B. M.: Human Embryology. 3rd ed. New York, McGraw-Hill, 1968.)

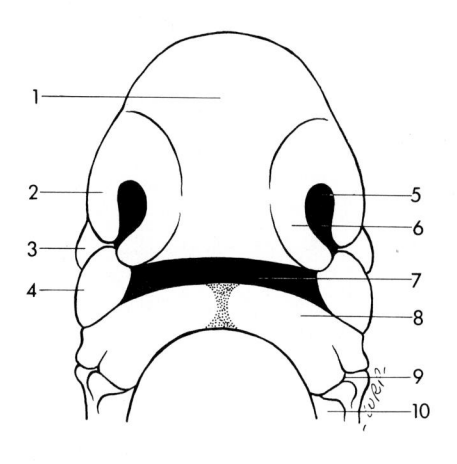

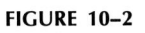

**FIGURE 10–2**

*Face at about 5 weeks.* 1, *Frontal prominence.* 2, *Lateral nasal swelling.* 3, *Eye.* 4, *Maxillary swelling.* 5, *Nasal pit.* 6, *Medial nasal swelling.* 7, *Stomodeum.* 8, *Mandibular swelling.* 9, *Hyomandibular cleft.* 10, *Hyoid arch.*

**FIGURE 10–3**

*Face at about 7 weeks.*

**FIGURE 10–4**

**Concept 2.**   As the whole head markedly expands, the membrane that covers the stomodeum does not keep pace with it. This thin sheet quickly breaks through, and the **pharynx** becomes opened to the outside. Everything in front will become the face, and **this** is what is now going to develop. The stomodeum of the embryo is about where the tonsils of the adult are located, deep within the face, so one can see that considerable growth and development have yet to occur in front of it.

The mammalian pharynx is the homologue of the region that develops into the branchial chamber and the gill system of fishes. The human pharyngeal pouches and clefts, however, did not "evolve from gills." More correctly, the primordia that developed into the fish's branchial system were phylogenetically converted to develop into **other** structures instead of gills. This is where many of the parts of the face come in.

**FIGURE 10–5**

**Concept 3.**   The pharynx is the anterior-most segment of the endodermally lined, embryonic gut. Its lumen is bounded on the right and left sides by the **pharyngeal arches** (also called visceral and, inappropriately, branchial or gill arches). Between the arches are the pharyngeal **clefts** on the outside and the **pouches** (see Fig. 10–4) on the inside. Where each cleft meets its pouch, a mesodermally reinforced contact between ectoderm and endoderm occurs. All these arches and some of the clefts and pouches give rise to **specific** adult structures in the face, in other areas of the head, and in the neck. An important point is that many of the fundamental embryonic relationships among them are retained in adult anatomy. The tissues in each arch develop into specific muscles, bones, and cartilages, and the arrangement in the adult is carried forward from the pattern that exists in the embryo. Remember this: each arch has a specific cranial nerve, and each nerve thereby supplies the other structures that are derived from that particular arch.

**FIGURE 10–6**

**Concept 4.**   The **first** pharyngeal arch gives rise to the tissues that will eventually become the mandible and its muscles. It is thus called the **mandibular arch.** A bud develops from it to become the "maxillary swelling," and this is the anlage (that is, primordium) for part of the maxillary arch that is soon to begin forming. The specific cranial nerve to the first arch is the mandibular (V), and it thus innervates the various **muscles of mastication.** The cartilage of the first arch (Meckel's cartilage) serves as the anlage for two of the ear ossicles (malleus and incus). This cartilage does not develop into the mandible itself. The bone of the lower jaw forms intramembranously **around** Meckel's cartilage, and the cartilaginous condyle develops from a separate secondary cartilage that appears later.

**Concept 5.**   The **second** pharyngeal arch is called, appropriately, the **hyoid arch.** It forms the cartilaginous model from which part of the hyoid apparatus develops and also the third ear ossicle (the stapes). The mesenchyme of this arch gives rise to the hyoid muscles. It also forms all of the various **muscles of facial expression.** These developing, sheetlike muscles spread up and over the face like a superficial sleeve. They are "cutaneous" muscles located in the deep part of the facial skin, and they are much more highly "developed" in man and some primates than in other mammals. This gives our face the capacity for a characteristic repertoire of changing

**FIGURE 10–4**

*Internal view of pharyngeal region. 1, Forebrain. 2, Stomodeum. 3, Cardiac prominence. 4, Maxillary process. 5, Mandibular process. 6, Pouch between second and third arches. (Modified from Langman, J.:* Medical Embryology. *Copyright 1969, The Williams & Wilkins Company, Baltimore.)*

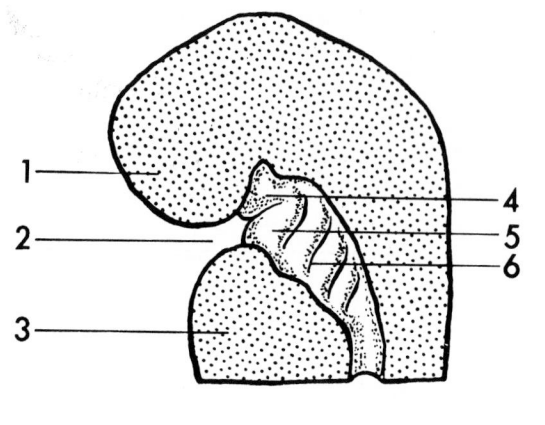

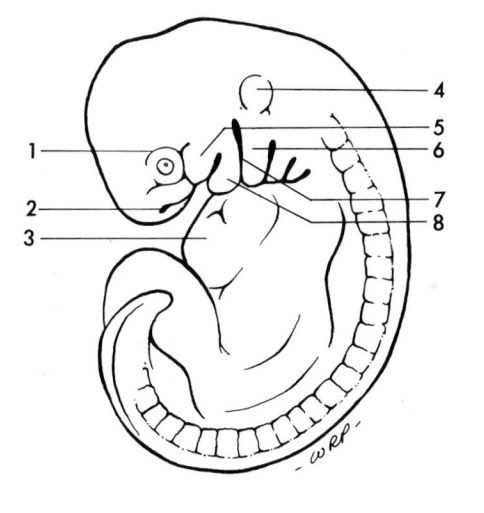

**FIGURE 10–5**

*Human embryo at about 5 weeks. 1, Eye. 2, Nasal pit. 3, Cardiac prominence. 4, Auditory vesicle. 5, Maxillary process. 6, Hyoid arch. 7, Hyomandibular cleft. 8, Mandibular arch. (Modified from Patten,* Human Embryology. *3rd ed. New York, McGraw-Hill, 1968.)*

**FIGURE 10–6**

*Pharyngeal arch derivatives (I to VI). 1, Meckel's cartilage. 2, Intramembranous bone developing around Meckel's cartilage. 3, Superior part of body and lesser horn of hyoid. 4, Sphenomandibular ligament. 5, Malleus. 6, Incus. 7, Stapes. 8, Styloid process. 9, Stylohoid ligament. 10, Greater horn of hyoid bone. 11, Inferior part of hyoid body. 12, Laryngeal cartilages.*

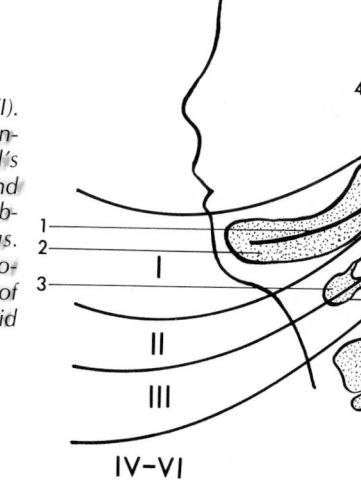

expressions. The specific cranial nerve to the second pharyngeal arch is the **facial** (VII). You can always figure out which nerve goes to which muscle in the face and neck by remembering the simple relationships in Concepts 4, 5, and 6.

**Concept 6.**    The third, fourth, and sixth pharyngeal arches (the fifth drops out) give rise to the remainder of the hyoid apparatus, the laryngeal cartilages, and the muscles of the larynx. The nerves to these arches are the glossopharyngeal (third arch) and the vagus (fourth and sixth arches). In addition, the parathyroids and thymus develop from third and fourth arch tissue.

**FIGURE 10–7**

**Concept 7.**    The main body of the **tongue** develops from the right and left first (mandibular) arches, where they join in the floor of the pharynx, by a fusion of the paired **lingual swellings.** The mucosal covering has sensory innervation, quite naturally, by the fifth cranial nerve (with a branch joining it from the nerve of the adjacent second arch). The root of the tongue develops from third and fourth arch tissue, and its sensory innervation is thereby provided by the glossopharyngeal and vagus nerves. The primordium for the **thyroid gland** forms by a deep diverticulum from the lining, endodermal epithelium into the floor of the pharynx exactly between the first and second arches. Later, it becomes relocated into the neck along with the parathyroids.

**FIGURE 10–8**

**Concept 8.**    Note that the **auditory vesicle,** which was formed by an invagination of the surface ectoderm, is positioned close to the second pharyngeal arch. It is developing into the **inner ear** apparatus (semicircular canals and cochlea). The **external** ear is being formed by the superficial tissue surrounding the first pharyngeal cleft, and the cleft itself is soon to become the external auditory canal. The **middle** ear chamber, which will contain the auditory ossicles, is formed by an expansion of the first pharyngeal **pouch,** and the ossicles develop from the cartilages of the first and second **arches** which are, conveniently, right there.

**FIGURES 10–9 AND 10–10**

**Concept 9.**    Note that the first pharyngeal arch has given rise to the two sizable pairs of **mandibular** and **maxillary** swellings. Below the "forehead" is a pair of U-shaped swellings, the **nasal primordia.** At this stage, the embryo is 5 weeks old, but in only about 2 more weeks a fast-moving sequence of changes takes place, and the result is a recognizable face. On each side, the maxillary swellings fuse with the **medial** limbs of the nasal swellings, and this composite forms the closed maxillary arch.* The middle portion is the "premaxillary" segment that will later house the incisors. It also gives rise to the **philtrum** (cupid's bow) of the upper lip. Above it, the medial limbs merge to form, in addition, the middle part of the nose; the lateral limbs become the nasal wings. Bone is beginning to form in the maxillary and mandibular arches, the eyes are continuing to be displaced into more forward-pointing positions by the enlarging brain, the ear lobes are forming, and there is now a **face.** Within the face, a nasal septum has formed, "shelves" from the right and left maxillae develop and fuse at the midline to form a palate, and the oral and the paired nasal chambers thereby all become partitioned from one another. This will make possible, after birth, breathing and eating at the same time.

---

*In normal development, this merger involves a closure of the **furrow,** not an actual cleft, between these parts.

**FIGURE 10–7**

1, *Body of tongue (lateral lingual swellings and tuberculum impar).* 2, *Thyroid diverticulum.* 3, *Mandibular arch.* 4, *Pouch between first and second arches.* 5, *Root of tongue (copula).* 6, *Arytenoid swellings.* 7, *Trachea.* 8, *Esophagus.*

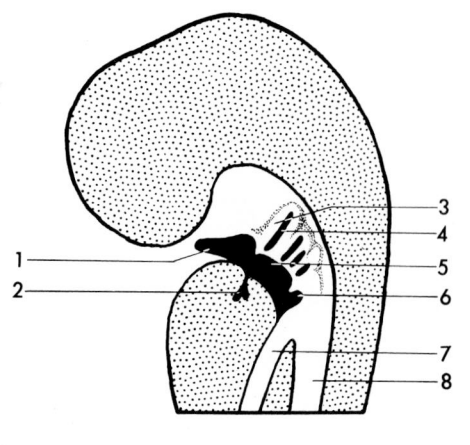

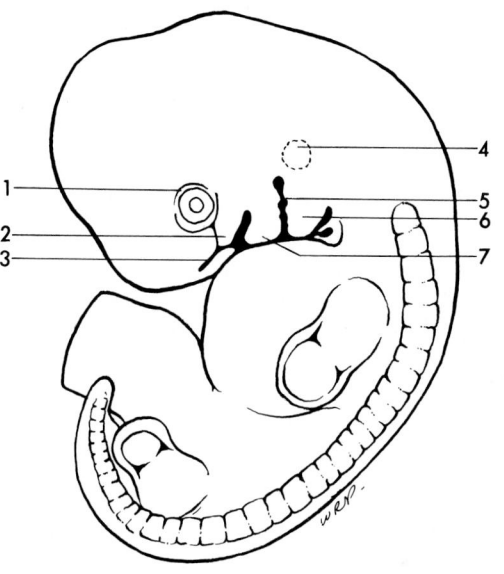

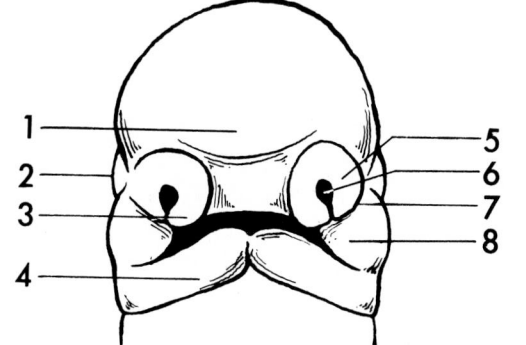

**FIGURE 10–8**

*The face at about 6 weeks.* 1, *Eye.* 2, *Maxillary process.* 3, *Nasal pit.* 4, *Auditory vesicle.* 5, *Hyomandibular cleft.* 6, *Hyoid arch.* 7, *Mandibular process of first arch. (Modified from Patten, B. M.: Human Embryology. 3rd ed. New York, McGraw-Hill, 1968.)*

**FIGURE 10–9**

*The face at about 5 weeks.* 1, *Frontal prominence.* 2, *Eye.* 3, *Medial nasal swelling.* 4, *Mandibular swelling.* 5, *Lateral nasal swelling.* 6, *Nasal pit.* 7, *Nasolacrimal groove.* 8, *Maxillary swelling. (Modified from Langman, J.: Medical Embryology. Copyright 1969, The Williams & Wilkins Company, Baltimore.)*

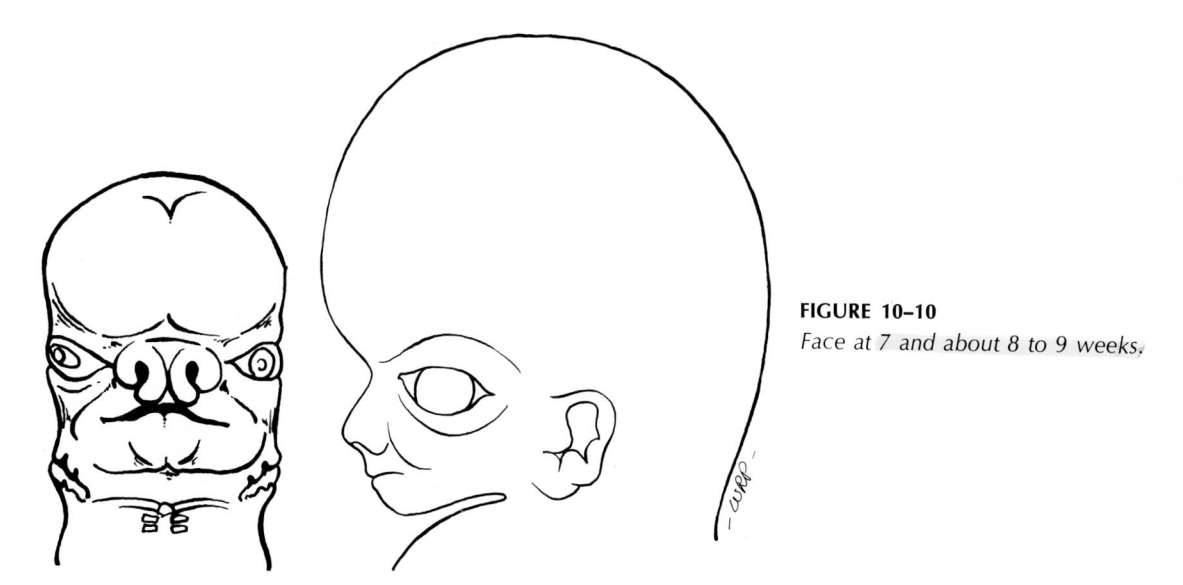

**FIGURE 10–10**

*Face at 7 and about 8 to 9 weeks.*

**FIGURE 10–11**

**Concept 10.**    In short order, centers of ossification appear for most of the other major bony parts of the face and cranium, some intramembranously and some endochondrally. The bone tissue of each center spreads until the definitive shape of that bone is attained. **Then,** the bone begins to "remodel" as it grows. This remodeling process first starts at around 14 weeks for most of the various separate bones and their parts. In Chapter 3, you will recall that the facial growth and remodeling process in the growing child was explained. The various patterns of resorption and deposition on the surfaces of all the bones were described, and the different growth movements of each bone were accounted for. Is the growth and remodeling process for the **fetal** face and cranium the same? In general, yes. The principal differences are in the anterior parts of the upper and lower jaws and the zygoma. In the face of the child after about 6 or so years of age, these surfaces characteristically become **resorptive** (Fig. 3–151). Throughout fetal life and the early part of childhood, however, they remain depository. The reason is that the bony maxillary and mandibular arches must expand anteriorly to accommodate the development of the primary dentition and the buds for permanent teeth. The bony arches thus grow **forward** as well as posteriorly. The maxillary arch, at the same time, is also growing downward as the nasal chambers expand. Sometime before about 5 or 6 years of age, however, the outer surfaces of the forward part of both the maxilla and mandible become resorptive. Subsequent lengthening of the bony arch then proceeds only posteriorly. The characteristic postnatal resorptive fields in the anterior parts of the arches develop in conjunction with the continuing **vertical** growth of the maxilla and mandible, as explained in Chapter 3.

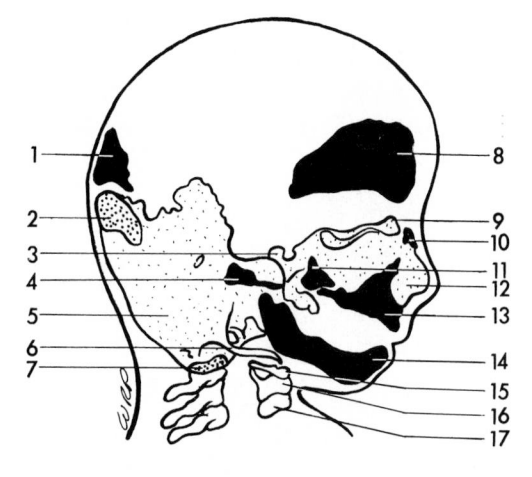

**FIGURE 10–11**

*Developing skull at about 9 weeks. 1, Occipital bone (interparietal part). 2, Supraoccipital. 3, Dorsum sellae (still cartilaginous). 4, Squamous part of temporal. 5, Cartilage. 6, Styloid process. 7, Occipital (basal part). 8, Frontal bone. 9, Crista galli (still cartilaginous). 10, Nasal bone. 11, Malar. 12, Cartilage of nasal capsule. 13, Maxilla. 14, Mandible (surrounding Meckel's cartilage.) 15, Hyoid. 16, Thyroid cartilage. 17, Cricoid cartilage. Endochondral sites of ossification are shown by dark stippling, and intramembranous sites are shown in black. (Modified from Patten, B. M.: Human Embryology. 3rd ed. McGraw-Hill, 1968.)*

# PART TWO ────────────────────────────────

A 1 month old embryo has no "face" as such. The stage, however, is set. The primordia that will now rapidly develop into the jaws, nose, eyes, ears, mouth, and the many deep structures located within these parts have already begun to form.

Below the bulging **frontal eminence,** an ectodermally lined surface depression marks the developing site of the future mouth. This shallow pit, the **stomodeum,** is separated from the foregut by a thin ectoderm-endoderm floor, the **bucco-pharyngeal membrane** (Fig. 10–12). This membrane is already beginning to rupture and disappear. The structures around the stomodeum grow and enlarge at a rapid rate. The membrane itself, however, does not continue to grow, so it becomes broken through as massive expansion and separation of the structural parts around it take place. To appreciate how much facial growth is going to occur, realize that the location of the buccopharyngeal membrane in the 1 month old embryo is at the level of the tonsils in the adult. An enormous amount of facial expansion thus will occur in front of the stomodeum. On the other side of this opening is the endoder-mally lined pharynx. The pharynx is that part of the foregut characterized by the **pharyngeal** (visceral, branchial) **arches** (Fig. 10–13). Within the pharynx, a **pharyngeal pouch** lies between each arch, and on the outside a **pharyngeal cleft** occurs between each arch. The ectoderm-endoderm contact between each cleft and pouch is termed the **branchial membrane.** All these various pharyngeal parts are major participants in the subsequent formation of many component structures in the head and neck.*

Each right and left pharyngeal arch has a specific nerve, a specific artery (aor-tic arch), and mesenchyme which develops into specific muscles and specific em-bryonic cartilages (Fig. 10–14). Certain bones are associated with specific pharyngeal arches. This is a basic and important concept because, if one can un-derstand the simple embryonic relationships involved, it makes an understanding of the exceedingly complex adult anatomy so much easier. The **muscles** that develop in relation with each arch associate directly with the **bones** forming in that arch and are innervated by the resident cranial **nerve** of the same arch. Some of the embry-onic pharyngeal pouches and clefts also establish the characteristic anatomic rela-tionships for their adult derivatives. All of this has a logical, systematic, readily rec-ognizable developmental rationale in the embryo. Remembering these specific prenatal relationships, the far less fathomable plan for the adult morphology makes sense.

In the human embryo, there are five pairs of pharyngeal arches. The first is the right and left **mandibular** arch. A bud develops from each first arch to form the paired **maxillary processes.** Both the mandibular and maxillary primordia are thus of first arch origin. The second pharyngeal arch is the **hyoid** arch (Fig. 10–13). The remaining arches are identified by their respective numbers only.

---

*It is believed that migrating cranial neural crest cells contribute extensively to the early primordia of many tissues developing in the face and the pharyngeal region (see Johnston, 1973).

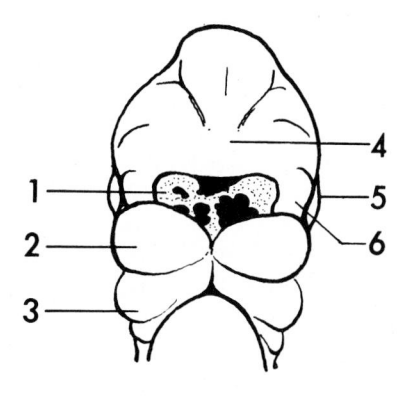

**FIGURE 10–12**

*Human face at about 4 weeks. 1, Stomodeal plate (buccopharyngeal membrane). 2, Mandibular arch (swelling or process). 3, Hyoid arch. 4, Frontal eminence (or prominence). 5, Optic vesicle. 6, Region where the maxillary process (or "swelling") of the first arch is just beginning to form. (Modified from Patten, B. M.: Human Embryology, 3rd ed. McGraw-Hill, 1968.)*

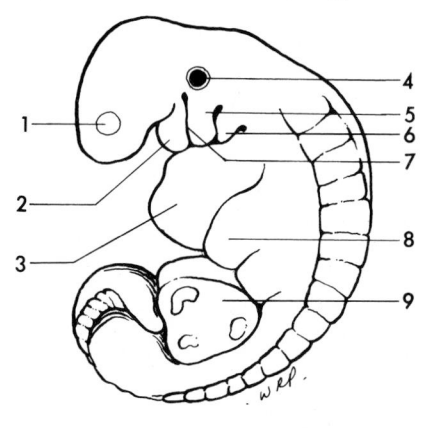

**FIGURE 10–13**

*Human embryo at about 4 weeks. 1, Optic vesicle. 2, Mandibular arch (process or swelling). 3, Cardiac prominence. 4, Auditory (optic) vesicle. 5, Hyoid arch. 6, Third arch. 7, Hyomandibular cleft. 8, Hepatic prominence. 9, Primitive umbilical cord. (Modified from Patten, B. M.: Human Embryology, 3rd ed. New York, McGraw-Hill, 1968.)*

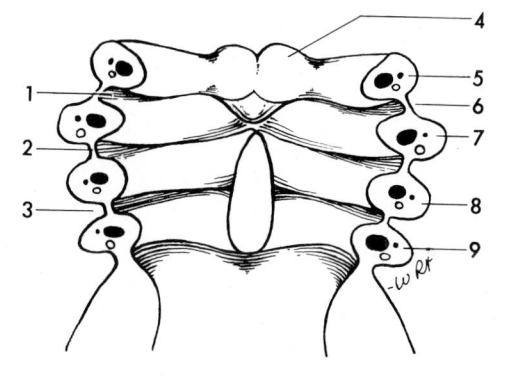

**FIGURE 10–14**

*Internal view of pharyngeal floor and cut arches. 1, First pharyngeal pouch between first and second arches (to become middle ear chamber). 2, Branchial membrane. 3, Pharyngeal cleft. 4, Region that will develop into the anterior two-thirds (body) of tongue. 5, First (mandibular) arch containing its specific cartilage, cranial nerve, and aortic arch. The pharyngeal arch is also filled with branchiomeric mesenchyme. 6, First pharyngeal cleft (hyomandibular) to become external ear canal. 7, Second (hyoid) pharyngeal arch. 8, Third pharyngeal arch with its own cartilage, aortic arch, cranial nerve, and branchiomeric mesenchyme. (Modified from Moore, K. L.: Before We Are Born. Basic Embryology and Birth Defects. Philadelphia, W. B. Saunders Company, 1974.)*

The cartilage of the first pharyngeal arch is **Meckel's** cartilage, right and left (Fig. 10–6). It occupies a location that will later be the core of the mandibular corpus which forms around it. The bony mandible itself develops, independently, directly from the embryonic connective tissue that surrounds Meckel's cartilage. Most of this cartilage actually disappears, but parts of it give rise to the anlagen for two ear ossicles (the malleus and incus), and the perichondrium of Meckel's cartilage forms a ligament, the sphenomandibular ligament.

The cartilage of the hyoid (second) arch is **Reichert's** cartilage. It forms the third of the three ear ossicles on each side, the stapes. The remainder gives rise to the styloid process of the cranium, the stylohyoid ligament, the lesser horn of the hyoid bone, and a portion of the hyoid body (Fig. 10–6).

Muscles form from the mesenchyme of the arches. This mesenchyme is termed **branchiomeric** (Gr., *branchia,* gills; Gr., *meros, segment*) because of its origin, in contrast to mesenchyme of somite origin elsewhere in the body. From the branchiomeric mesenchyme of the first arch, the muscles of mastication, the anterior belly of the digastric, and the tensor tympani muscle all develop. From the branchiomeric mesenchyme of the second arch develop the muscles of facial expression and the muscles of the hyoid arch. The specific cranial nerves (Fig. 10–15) that enter and supply the first arch are the mandibular and maxillary branches of the trigeminal (V). The specific cranial nerve for the second arch is the facial nerve (VII). Thus, the muscles of the first arch (muscles of mastication, etc.) are innervated by the mandibular division of V, regardless of the place in which each muscle finally becomes located later in development. The muscles of facial expression formed by the branchiomeric mesenchyme of the second pharyngeal arch correspondingly are all innervated by the facial nerve, as are the hyoid muscles.

Note the gathering of many structures related to the ear in and around the first and second pharyngeal arches (Fig. 10–13). The **auditory placode** differentiates early as a surface thickening of the ectoderm just above and behind the first pharyngeal cleft. This placode rapidly invaginates to form the auditory (otic) vesicle, which then differentiates into the structures of the inner ear (semicircular canals, cochlea). The first pharyngeal **cleft** (between the first and second arches) forms the external auditory meatus and outer ear canal, and the branchial mem-

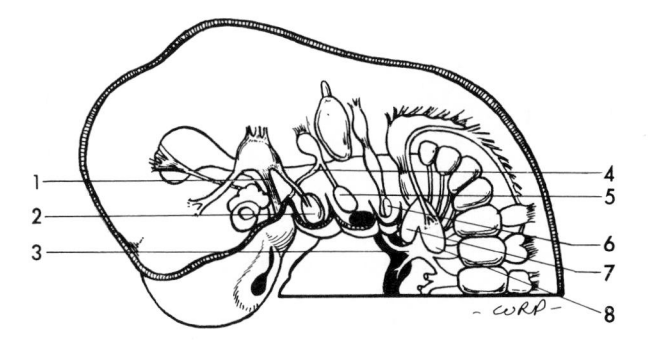

**FIGURE 10–15**

*Five-week embryo showing cranial nerve distribution and developing muscle origins (shown schematically) of the face. 1, Maxillary division of the trigeminal (no skeletal muscle innervation). 2, Muscles of mastication developing from branchiomeric mesenchyme of the mandibular (first) arch supplied by the mandibular division of V. 3, Hypoglossal nerve to the intrinsic muscles of the tongue. 4, Chorda tympani of the facial nerve leaving second pharyngeal arch to enter tongue and provide sensory (gustatory) innervation. 5, Facial muscles developing from hyoid arch mesenchyme, supplied by the facial nerve (VII). 6, Stylopharyngeus muscle of third arch origin, supplied by IX. 7, Pharyngeal muscles, supplied by X. 8, Trapezius and sternocleidomastoid muscles, supplied by the spinal accessory nerve. (Modified from Patten, B. M.: Human Embryology. 3rd ed. New York, McGraw-Hill, 1968.)*

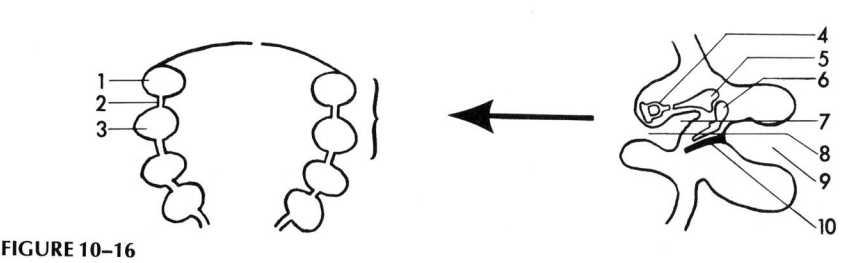

**FIGURE 10–16**

*Developing ear region. 1, Mandibular arch. 2, Branchial membrane between the cleft on the outside and pouch on the inside of the pharynx. 3, Hyoid arch. 4, Stapes. 5, Incus. 6, Malleus. 7, Middle ear chamber to expand as tympanic cavity surrounding auditory ossicles. 8, Auditory (eustachian) tube. 9, External ear canal. 10, Anlage for the tympanic membrane.*

brane between the cleft and pouch undergoes remodeling changes to participate in the formation of the tympanic membrane (Fig. 10–16). The first pharyngeal **pouch** becomes expanded into the middle ear chamber, and it also forms the auditory (eustachian) tube which retains continuity between the middle ear and the pharynx. The ear ossicles, developing from the cartilages of the first and second arches, lie conveniently next to this area and soon become enveloped within the expanding first pharyngeal pouch (middle ear chamber). They function as the bridge between the tympanic membrane and the inner ear. The auricle of each external ear develops from the surface swellings around the first pharyngeal cleft, and the bumps already present on these embryonic primordia form the characteristic hillocks of the adult ear lobe (Fig. 10–10).

The cartilage of the **third** pharyngeal arch produces the greater horn of the hyoid bone and part of the body (Fig. 10–6). The single muscle that develops from third arch branchiomeric mesenchyme is the stylopharyngeus. The specific cranial nerve entering the third arch is the **glossopharyngeal.** It thereby supplies the muscle that develops from this arch (Fig. 10–15). The cartilages in the remainder of the arches form into the thyroid, cricoid, and arytenoid components of the larynx. From fourth arch branchiomeric mesenchyme develops the cricothyroid and pharyngeal constrictor muscles. The specific nerve of the fourth pharyngeal arch is the superior laryngeal branch of the **vagus.** The intrinsic laryngeal muscles develop from the sixth arch and are innervated by that arch's nerve, which is the recurrent laryngeal branch of the vagus.

In each second pharyngeal **pouch,** the lining endoderm and underlying mesenchyme proliferate to form the paired **palatine tonsils** (Fig. 10–17). From the

**FIGURE 10–17**

*Pharyngeal pouch derivatives. 1, Mandibular arch. 2, Hyomandibular cleft. 3, Hyoid arch. 4, Palatine tonsil. 5, Fourth arch. 6, Thyroid diverticulum. 7, Third arch. 8, Parathyroid III. 9, Thymus. 10, Parathyroid IV.*

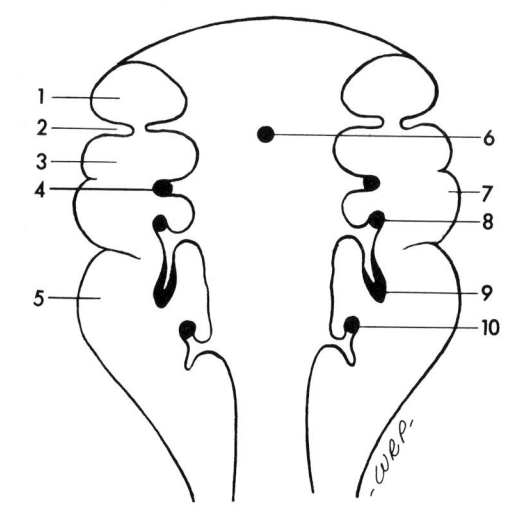

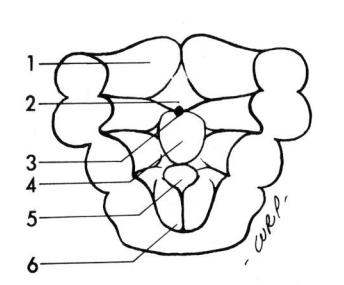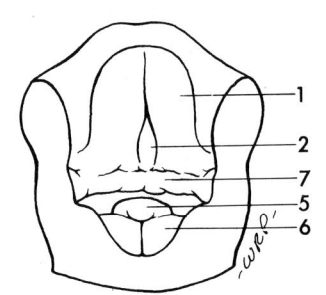

**FIGURE 10–18**

*Developing tongue at 6 and 8 weeks. 1, Lateral lingual swelling. 2, Tuberculum impar. 3, Foramen caecum. 4, Copula. 5, Epiglottis. 6, Arytenoid swellings. 7, Root of tongue.*

lining of the third pouch develops **parathyroid III** (so called because of its third arch origin). This will form the "inferior" parathyroid because it later descends to a level below parathyroid IV. The thymus also develops from the lining of the third pharyngeal arch. Parathyroid IV (the "superior" parathyroid) develops from the fourth pouch.

In the floor of the pharynx, the first (mandibular) arches form rapidly growing **lingual swellings** (Fig. 10–18). A smaller midline swelling, the **tuberculum impar**, is also present, and these three structures develop into the mucosal covering for the anterior two-thirds, or body, of the tongue. Since the mandibular nerve supplies first arch tissue, it therefore provides the sensory (tactile) innervation for the mucosa of the body of the tongue. The chorda tympani, which is a branch of VII that jumps from the second to the first arch by crossing through the branchial (tympanic) membrane to join the mandibular nerve (lingual branch), provides gustatory innervation for the tongue's mucosa.

At the root of the midventral parts of the second, third, and fourth pharyngeal arches, another prominent swelling occurs, the **copula.** This general region develops into the posterior one-third (root) of the tongue. The cranial nerves supplying the third and fourth arches are the glossopharyngeal and vagus, and these are thus the sensory nerves that innervate the mucosa of the root of the tongue. The core of the tongue is occupied by its "intrinsic" muscles. These originate from a more caudal region (probably from occipital somatic mesoderm) and grow into the expanding mucosal covering for the tongue being formed by the floor of the pharynx (described above). The motor innervation to these muscles is provided by the paired hypoglossal (XII) nerves. They are carried along with the intrinsic muscles as they migrate anteriorly into the body of the tongue.

Anatomically, the body of the tongue is separated from the root by a V-shaped sulcus (the **sulcus terminalis,** Fig. 10–19). This marks the approximate line between the derivatives of first arch origin from the arches behind it. At the midline in this developing groove, between the tuberculum impar and the copula, the thyroid primordium develops as an epithelial diverticulum into the pharyngeal floor (Figs. 10–7 and 10–17). It then separates from the mucosal lining and migrates caudally. The point of invagination, however, remains as a permanent pit, the **foramen caecum** (Fig. 10–18). It is located at the apex of the "V" and is a landmark that identifies the adult position of the embryonic boundary between the first and second arches. As for most glandular tissues, the thyroid is thus of epithelial origin, and because the primordium develops from the pharyngeal lining, it is of endodermal derivation.

By the time an embryo is about 5 weeks old, the first pharyngeal arch has formed recognizable **maxillary** and **mandibular swellings** (Figs. 10–20 to 10–23). Just above the stomodeum, the paired, laterally located **nasal placodes** have already formed by thickenings of the surface ectoderm, and horseshoe-like ridges

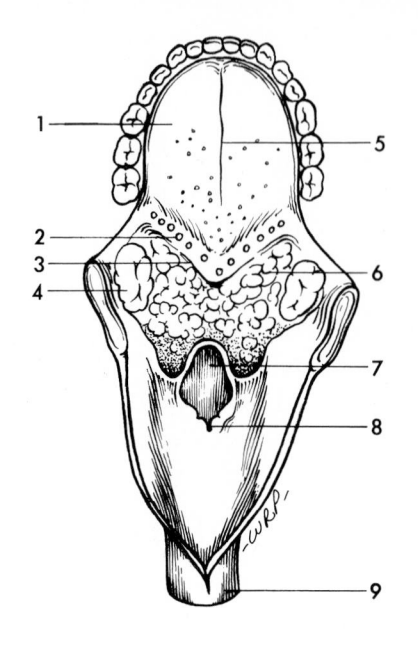

**FIGURE 10–19**

Adult tongue. 1, Body. 2, Circumvallate papilla. 3, Sulcus terminalis. 4, Palatine tonsil. 5, Median sulcus. 6, Lingual tonsil in root of tongue. 7, Epiglottis. 8, Interarytenoid notch. 9, Esophagus.

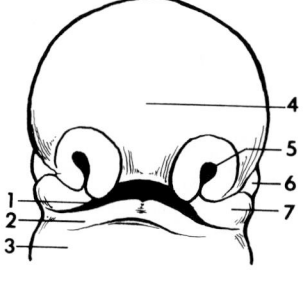

**FIGURE 10–20**

Facial region at about 5 weeks. 1, Stomodeum. 2, Mandibular swelling. 3, Hyoid arch. 4, Frontal prominence. 5, Nasal pit. 6, Optic vesicle. 7, Maxillary swelling. (Adpated from Langmen, J.: Medical Embryology. Copyright 1969, The Williams & Wilkins Company, Baltimore.)

**FIGURE 10–21**

Facial region at about 5½ weeks. 1, Forebrain. 2, Optic vesicle. 3, Lateral nasal swelling. 4, Mandibular process. 5, Medial nasal swelling. 6, Nasolacrimal groove. 7, Maxillary process. 8, Hyomandibular cleft. 9, Hyoid arch. (Modified from Patten, B. M.: Human Embryology. 3rd ed. New York, McGraw-Hill, 1968.)

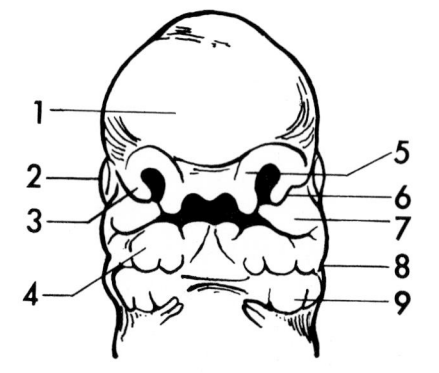

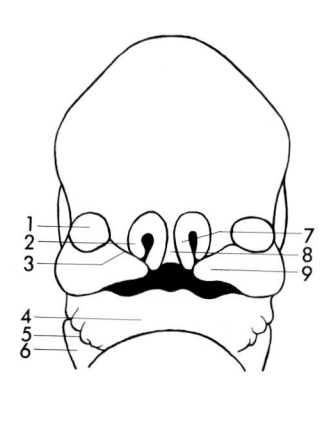

**FIGURE 10–22**

The face between 6 and 7 weeks. 1, Eye. 2, Lateral nasal swelling. 3, Nasolacrimal groove. 4, Mandibular swelling. 5, Hyomandibular cleft. 6, Hyoid arch. 7, Medial nasal swelling. 8, Midline nasal region where nasal septum is forming. 9, Maxillary swelling.

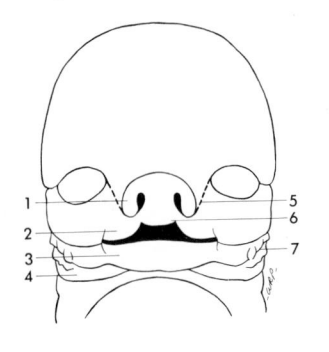

**FIGURE 10–23**

*The face between 7 and 8 weeks. 1, Nasolateral process. 2, Maxillary process. 3, Mandible. 4, Hyoid arch (note formation of external ear lobes). 5, Merger line of nasolacrimal groove. 6, Philtrum. 7, Hyomandibular cleft.*

**(nasal swellings)** have developed around them to form deepening **nasal pits.** The floor of each pit is termed the oronasal membrane, but it is a transient structure that soon breaks through, thus opening the nasal pits directly into the oral cavity. At the same time, the semicircular nasal swellings continue to enlarge. Each swelling is composed of a lateral and medial limb. The expanding **medial** limbs merge at the midline to form the primordium that will differentiate into the middle part of the nose, the philtrum of the lip, the "incisor" part of the maxilla (premaxilla), and the small primary palate.

The rapidly growing **lateral** limbs of each nasal swelling form the alae of the nose. While these changes occur, the maxillary swellings are also enlarging, and they subsequently merge with the medial limbs of the nasal swellings. The furrow between them (not a complete cleft in normal development) disappears, and a closed, U-shaped arch is thereby formed. The medial limbs form the middle span of the maxillary arch and the upper lip. The canine, premolar, molar, and lateral lip parts of the upper arch develop from the maxillary processes, and the incisor and medial lip (philtrum) parts develop from the medial nasal swellings. **These** are some of the lines of merger that can be involved in cleft lip and jaw. Sometimes, developmental variations are encountered in which the blastema of a tooth is caught on the "wrong" side of a cleft, and this always causes much excitement because that's not the way it's supposed to be.

An oblique groove is present between the maxillary swelling and the lateral limb of the nasal swelling. This is the **nasolacrimal groove** which will soon close, but the line of merger establishes a developmental pathway for the later formation of the nasolacrimal duct. If this merger fails, a permanent facial cleft or fissure results. The superficial tissues in lateral areas of the maxillary process fuse with the mandibular process to form the cheek. **Epithelial pearls** often occur along such lines of mucosal and cutaneous fusion. These are small islands of epithelial cells that were "programmed" to form but which were caught up in the fusion process. **Fordyce spots,** which are remnants of cutaneous sebaceous glands, can similarly be found in the adult buccal mucosa, for the same reason, along lines of fusion.

The growing right and left mandibular swellings join at the midline to form the lower jaw and lip. A cartilaginous interface forms at this junction.*

---

*While a limited amount of endochondral ossification will later occur here, the two halves of the mandible fuse completely after birth, unlike the permanently separate lower jaw halves of nonprimates. Except in the "secondary cartilage" of the mandibular condyle (and to a much lesser extent in the mandibular symphysis and a small cartilage on the coronoid process), the greater part of mandibular ossification is by the intramembranous mechanism. Meckel's cartilage does not participate in the endochondral ossification process (except for spots here and there) and disappears except for its contribution to the ear ossicles and ligaments. The maxilla is entirely of intramembranous origin.

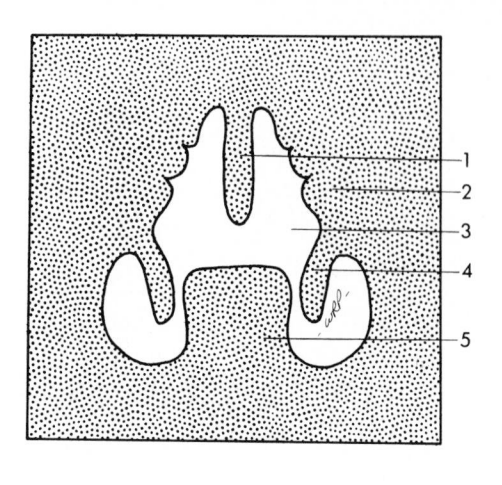

**FIGURE 10–24**

*Frontal section through the oronasal region in a 6¹/₂-week embryo. 1, Nasal septum. 2, Lateral nasal wall. 3, Nasal chamber. 4, Palatal shelf. 5, Tongue. (Adapted from Langman, J.: Medical Embryology. Copyright 1969, The Williams & Wilkins Company, Baltimore.)*

The frontal prominence forms the forehead and a vertical zone of tissue between the merging medial nasal swellings. Here, the midline **nasal septum** is formed, which is believed by some to function as a pacemaker in later fetal development when its core becomes cartilaginous.

To date, all these facial changes are occurring at about the same time and have proceeded **rapidly** from about the fourth to the sixth week of embryonic development. The paired **palatal shelves** are now forming from each side of the maxillary arch (Figs. 10–24 to 10–29). The oral cavity is still relatively small, however, and the sizable tongue remains interposed between the right and left shelves. The early shelves necessarily enlarge downward in an obliquely vertical manner because of this. However, the inferior expansion of the whole lower part of the face carries the tongue downward. The oral cavity increases greatly in size. The paired nasal chambers are still continuous with the oral cavity (right and left nasal cavities are

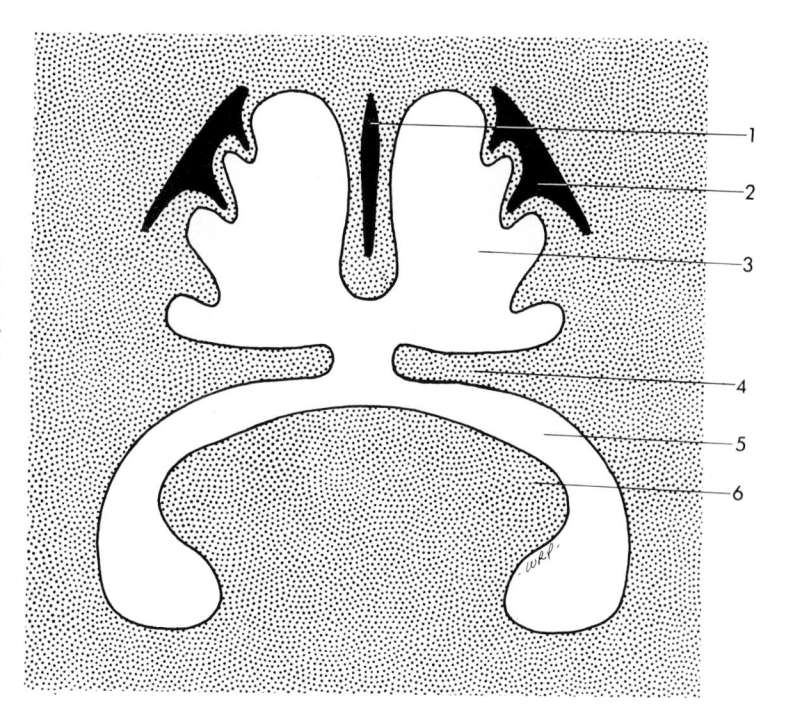

**FIGURE 10–25**

*Frontal section through the oronasal region of a 7¹/₂-week embryo. 1, Cartilage of the nasal septum. 2, Cartilage of the nasal conchae. 3, Nasal chamber. 4, Palatal shelf. 5, Oral cavity. 6, Tongue. (Adapted from Langman, J.: Medical Embryology. Copyright 1969, The Williams & Wilkins Company, Baltimore.)*

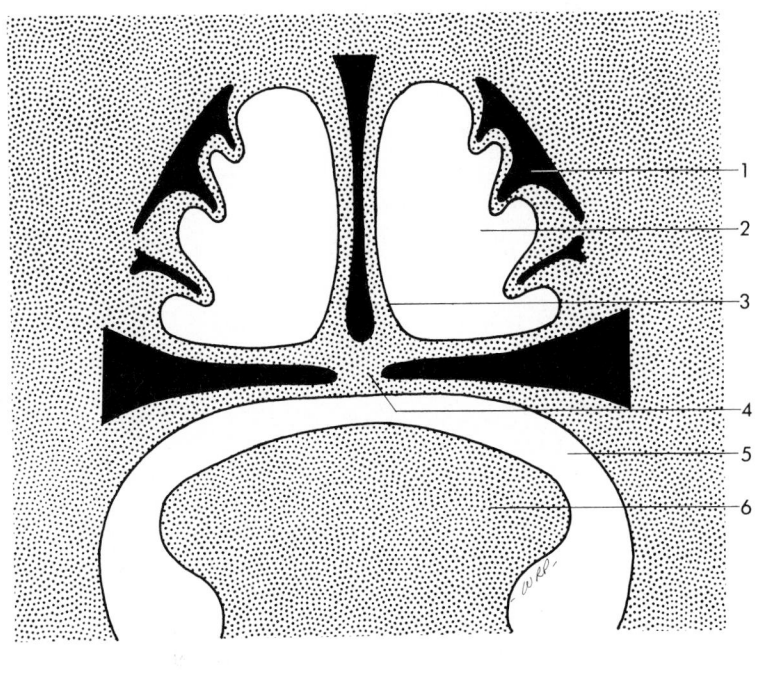

**FIGURE 10–26**

*Frontal section through the oronasal region of a 10-week embryo. 1, Nasal conchae. 2, Nasal chamber. 3, Nasal septum. 4, Palatal shelves, fused at midline and fused with nasal septum. The intramembranous bone of the palatal shelves (from the maxilla) is beginning to form. 5, Oral cavity. 6, Tongue. (Adapted from Langman, J.: Medical Embryology. Copyright 1969, The Williams & Wilkins Company, Baltimore.)*

present because the nasal septum developing downward from the frontal prominence has sustained the original paired nature of the nasal primordia). The oral and nasal cavities at this stage are separated from each other only in the anterior-most region by the tiny, unpaired **primary palate** (median palatine process). The latter was formed by the fusion of the nasomedial ("premaxillary") processes. The whole lower part of the developing face, including the tongue and the floor of the oral cavity, now becomes displaced inferiorly to a greater extent than the palatal shelves are descending, so that the newly formed shelves of the maxilla are free to expand **medially** as well. They come together and soon fuse along the midline (the palatal raphe). The shelves swing upward in order to contact each other, but some of this apparent upward rotation is also produced by differential growth. The shelves are growing and, especially, are becoming displaced inferiorly. By differential growth, the shelves expand downward as well as toward each other, but the **entire** nasal chamber on each side also expands laterally and inferiorly, and some of this apparent "upward swing" is relative and is a consequence of actual downward growth.

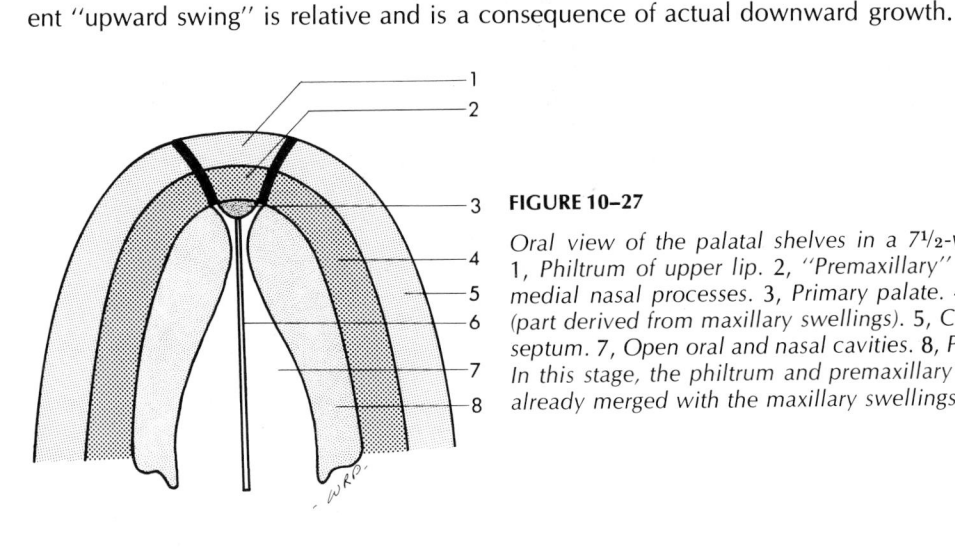

**FIGURE 10–27**

*Oral view of the palatal shelves in a 7½-week embryo. 1, Philtrum of upper lip. 2, "Premaxillary" segment from medial nasal processes. 3, Primary palate. 4, Upper arch (part derived from maxillary swellings). 5, Cheek. 6, Nasal septum. 7, Open oral and nasal cavities. 8, Palatal shelves. In this stage, the philtrum and premaxillary segment have already merged with the maxillary swellings.*

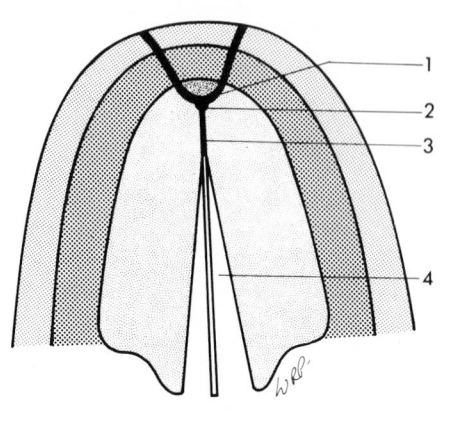

**FIGURE 10–28**

*Oral view of palate showing beginning of fusion. 1, Merger of midline primary palate with bilateral secondary palatal shelves. 2, Incisive foramen. 3, Palatal raphe (midline fusion). 4, Open nasal and oral chambers.*

This is a process in which the different parts in the two nasal chambers and the oral cavity all grow at differential rates and to different extents as the whole midfacial region rapidly increases in size.

The merger of the right and left palatal shelves forms the **secondary palate.** Bone tissue soon appears within it, and this part of the palate is a direct extension of the maxilla from which it develops. The original primary palate, formed from the nasomedial (premaxillary) processes, is retained as a small median, unpaired, triangular-shaped segment of the palatal complex in the anterior region just ahead of the incisive foramen, a landmark which identifies the midline boundary between the primary and secondary parts of the palate (Fig. 10–29). The separate palatine bone itself and its posterior contribution to the palatal complex does not develop until somewhat later. In the meantime, the nasal septum has merged with the superior surface of the palate. The two nasal chambers are now completely separated, and both have been closed off from the oral cavity along the length of the palate. While these various changes take place, the nasal conchae are already developing as medial and inferior-growing processes from the lateral walls of each nasal chamber.

## CLEFT LIP AND PALATE

There are over a hundred craniofacial **syndromes** that are known to involve clefts of the lip and palate. Some are caused by mutant genes or chromosomal aberrations, but the underlying, multifactorial etiologies for most are poorly under-

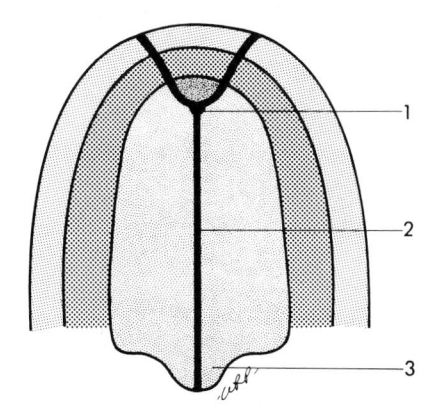

**FIGURE 10–29**

*Full length palatal fusion. 1, Incisive foramen. 2, Palatal raphe. 3, Uvula.*

stood. An **enormous** amount of research has been carried out on the problems of clefting, much more so than on the other congenital facial dysplasias. One reason is that cleft lip is relatively common, about one in 700 to 800 or so births among Caucasians (higher for cleft palate, about 1:2500). The incidence is much lower among American Blacks and somewhat higher among the Japanese. Many variations in the extent, severity, and combinations of clefts can occur in the different parts of the jaws and face. Most, of course, appear along the various principal lines where embryonic fusions normally occur. If the two or more parts to become merged are carried apart by other developing organs (such as a separation of the maxillary halves by a wide-growing brain and cranial floor), if they are held apart (as by the tongue, a controversial theory), if a biochemical or tissue barrier intercedes between the parts, or if the fusion process fails, a **cleft** results. This may cause only a scarcely noticeable groove in the red part of the lip, or the entire upper lip may be totally cleft, with the defect extending vertically beside the nose following the nasolacrimal plane. Cleft lips usually occur at the junction between the philtrum (nasomedial process) and the lateral part (maxillary process) of the upper lip. Midline clefts between the right and left nasomedial processes are much more infrequent; this would be a "hare" (rabbit) lip, although this term is not to be favored and thankfully is rarely heard today. Such a midline cleft may also extend into the overlying nasal region to form a groove or depression along the cutaneous surface overlying the midline nasal septum. This produces a **bifid nose.**

A cleft may be restricted to the upper lip, or it can extend much deeper along the lines of the maxillary and palatal fusion planes. Thus, the upper jaw may show a cleft between the lateral incisor and the canine, that is, between the nasomedial (premaxillary) segment and the maxillary process. This can produce a **unilateral** cleft lip and maxilla, or it may occur on both sides as a **bilateral** cleft. The cleft may stop here, or it may continue along the primary palate on one or both sides to form either a unilateral or bilateral cleft in the anterior part of the palate. The Y-shaped line of clefting then continues back to the midline where the right and left halves of the maxillary palatal shelves ordinarily fuse. Failure of merger here causes a midline palatal cleft. Or, the entire lip-jaw-palatal complex can fuse normally with only the uvula becoming involved—a **bifid uvula.** The tongue may also be bifid (incomplete fusion of the paired lingual swellings). A median cleft of the whole lower lip (not just chin point) and/or the bony mandible may occur at the midline fusion plane between the mandibular swellings, but this is infrequent (see page 6).

Clefts of the lip and palate are often only one of many expressions that characterize different kinds of **syndromes** involving combinations of regional craniofacial malformations. In most of the various syndromes, diseases, and malformations briefly outlined below, keep in mind that there is often overlap, and many of the same regional deformities characterize a number of the different syndromes. Hypertelorism (excessively wide-set eyes), orbital distortions, mandibular underdevelopment (micrognathia), nasal deformities, underdeveloped upper jaws, ear malformations, lip and palate clefts, facial clefts, meningoceles, forehead and other calvarial deformities, brain distortions, and other such conditions may be involved in many of the more severe composite types of cases. There are a great many opportunities for genetic disturbances, congenital insults, diseases, and traumatic injuries to affect the normal and fragile course of facial and cranial development. The effects of **trisomy 21** (Down's syndrome, "mongolism"), for example, are well known. Much research is presently being carried out with regard to the genetics and the etiology of this condition. It involves maxillary hypoplasia to varying extents, cranial base distortions (and hence characteristic facial changes such as mandibular pro-

trusion), a fissured tongue, delayed eruptions of teeth, missing or malformed teeth, and malocclusions. Other types of trisomies, triploidy, and different combinations of the X and Y chromosomes underlie a wide range of deformities, including clefts, prognathism, micrognathia, hypoplastic salivary glands, and bifid tongue and uvula.

The basic, underlying causes of most of the severe **congenital** types of dysplasias are virtually unknown at present. **Craniostenosis** is such a condition. It is the premature closure of cranial vault sutures; distortions of the brain, calvaria, and orbits necessarily follow. **Scaphocephaly** results from early closure of the sagittal suture of the skull roof, and a long, narrow calvaria is produced with a prominence of the frontal and occipital areas. The cranial base may also be affected, and this then can be passed on to cause deformities of the face during its development. The coronal sutures may also close **(brachycephaly)** prior to the completion of brain growth, and the distorted lateral expansion of the brain can result in a widening of the skull. This may decrease cranial base flexure to cause, in turn, a greater or lesser amount of mandibular protrusion. In oxycephaly, all the sutures close prematurely; an abnormally small brain occurs, and marked facial distortions develop. **Achondroplasias** may occur in the cranial base cartilages, although it is not clear if this is a cause or an effect. Facial growth, however, is directly involved, since the cranial base is the template for facial development. Prognathism and a Class III malocclusion are consequences. The **Pierre Robin** syndrome is a relatively rare condition that involves severe mandibular hypoplasia and a cleft palate (not lip). Ear and eye defects are also common. The **Treacher Collins** (mandibulofacial dysostosis) syndrome is characterized by slanting eyes, ear defects, malocclusion, coloboma, a characteristic flattened, retrusive malar region, and sometimes a cleft palate. The frontonasal area is unaffected and tends to dominate the face. Conditions such as the Pierre Robin and the Treacher Collins syndromes, cleft lip and jaw, and deformities of the external and middle ear are sometimes categorized as **first and second arch syndromes** because of their embryologic relationships with these pharyngeal arch primordia. In **Crouzon's** disease (craniofacial dysostosis), the eyes bulge (exophthalmia) as a consequence of premature synostosis, the nose is beaked, an antimongoloid obliquity of the palpebral fissure occurs, and external strabismus, a hypoplastic maxilla, a characteristic protruding lower lip and jaw, and a half-open mouth are all present. Cleft palate and uvula and ear anomalies may also occur. **Hurler's** syndrome involves mandibular protrusion, large chin, wide-spaced eyes, saddle-shaped nose, wide nostrils, high palate, low-set ears, thick lips and tongue, anodontia, and a tendency for frontal bossing. **Cebocephaly** (Gr., *kebos,* monkey) is a condition in which the nasal area is flattened or deformed and the eyes are quite close set. **Apert's** syndrome is a congenital malformation in which the top of the head is pointed (acrocephaly) in conjunction with early fusion of the coronal and sagittal sutures, and syndactyly is characteristically present in the hands and feet. The forehead is broad, and hypertelorism may occur. Palate malformations are sometimes present, as well as spina bifida. The orbits are shallow, a consequence of cranial synostosis, and the eyes may be protrusive. In **cyclops,** the paired eye primordia merge, and variable extents of eyeball fusion occur within a single, midline orbit. This then blocks the downward expansion of the frontal prominence, and some of the tissue programmed to form the nose remains isolated in the forehead region. A protruding, fleshy proboscis thus forms above the median, single eye, and the suborbital nasal region itself develops incompletely or is largely absent.

## COMPARISON OF PRENATAL AND POSTNATAL GROWTH PROCESSES IN THE FACE AND CRANIUM

The process of "remodeling," involving periosteal resorptive surfaces, first begins in the fetus at about 10 weeks in two principal locations: on lining surfaces of the bone around tooth buds and on the endocranial surface of the frontal bone. The major remodeling throughout the remainder of the early facial skeleton begins at about 14 weeks. Before this time the bones enlarge in all directions from their respective ossification centers. Remodeling, as a process that accompanies growth, starts when the definite form of each of the individual bones of the face and cranium is attained (see Fig. 10–30).

### Nasomaxillary Complex

The anterior part of the maxilla in both the fetus and the child is depository on lingual surfaces and resorptive on nasal lining surfaces. A major difference exists, however, on the anterior-most (labial) surface. Here it is depository in the fetus but characteristically becomes resorptive in the postnatal face after the first few years following birth. During the fetal period, the exterior surface of the entire maxilla, including its anterior part,* remains depository to provide for increasing arch length in conjunction with the development of the tooth buds and their subsequent enlargement. Resorption occurs on all alveolar lining surfaces surrounding each of the tooth buds. The fetal maxillary arch thus lengthens horizontally in both posterior **and** anterior directions, in contrast to the largely posterior mode of elongation in the later periods of childhood growth.

Beyond about 5 or 6 years postnatally, anterior expansion of the maxillary arch no longer takes place except for small amounts at the margins along the alveolar crest. Not only have all the tooth buds been formed, but also the deciduous teeth have erupted and are beginning to be shed, making way for the permanent teeth. The anterior surfaces subsequently become resorptive as a part of the growth and remodeling process that continues to produce the **downward** growth movement of the premaxillary region.†

The posterior and infraorbital surfaces of the maxilla proper are depository in both prenatal and postnatal life. The process of posterior deposition on the maxillary tuberosity progressively increases the maxilla in horizontal length. Deposition on the orbital floor in the fetal skull keeps it in a constant positional relationship with the eyeball, just as in the growing child. The eyeball enlarges in volume at a decreasing rate after the fourth to fifth fetal months. Its volume increases by over 100 per cent before the fifth month, by 50 per cent during the sixth and seventh months, and only 23 to 30 per cent in the eighth and ninth months. Remodeling of the orbital floor takes place because the entire maxilla, including the orbit, is displaced in a progressively inferior direction in relation to continuing new bone growth at the frontomaxillary suture. At the same time, deposition on the orbital

---

*An old game among facial researchers is arguing whether or not a premaxilla exists as a separate bone in man and how many ossification centers are involved.

†During postnatal growth, the vertical length of the maxilla increases more than its width. Upper jaw alveolar width and bigonial width increase about twofold beyond the neonatal size. Facial height increases threefold relative to the orbits. Mandibular corpus height increases about 2.5-fold, ramus height 3.5-fold, and ramus depth about 1.5-fold. The maxillary sinus increases in diameter from about 5 to 6 mm. at birth to 12 to 14 mm. at 5 to 6 years to 20 to 26 mm. in the adult.

**FIGURE 10–30**

*Human skull at about 3 months. Intramembranous bones are shown in black. Cartilage is represented by light stippling, and bones developing by endochondral ossification are indicated by darker stippling. Approximate time of appearance for each bone is indicated in parentheses. 1, Parietal bone (10 weeks). 2, Intraparietal bone (8 weeks). 3, Supraoccipital (8 weeks). 4, Dorsum sellae (still cartilaginous). 5, Temporal wing of sphenoid (2 to 3 months; the basisphenoid appears at 12 to 13 weeks, orbitosphenoid at 12 weeks, and presphenoid at 5 months). 6, Squamous part of temporal bone (2 to 3 months). 7, Basioccipital (2 to 3 months). 8, Hyoid (still cartilaginous). 9, Thyroid (still cartilaginous). 10, Cricoid (still cartilaginous). 11, Frontal bone (7½ weeks). 12, Crista galli, still cartilaginous (inferiorly, the middle concha begins ossification at 16 weeks, the superior and inferior conchae at 18 weeks; the perpendicular plate of ethmoid begins ossification during the first postnatal year, the cribriform plate during the second postnatal year, the vomer at 8 fetal weeks). 13, Nasal bone (8 weeks). 14, Lacrimal bone (8½ weeks). 15, Malar (8 weeks). 16, Maxilla (end of 6th week; premaxilla, 7 weeks). 17, Mandible (6 to 8 weeks). 18, Tympanic ring (begins at 9 weeks, with complete ring at 12 weeks; petrous bone, 5 to 6 months). 19, Styloid process, still cartilaginous. (Modified from Patten, B. M.: Human Embryology, 3rd ed. New York, McGraw-Hill, 1968.)*

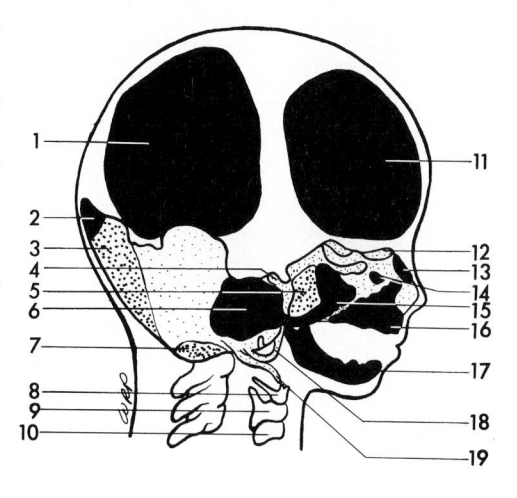

floor serves to carry it superiorly, thereby maintaining it in a constant position relative to the eyeball. The infraorbital canal is also moving upward during maxillary growth by resorption superior to and deposition inferior to the infraorbital nerve. These processes maintain a constant relationship between the nerve and the orbital floor, along which the infraorbital nerve passes before entering the infraorbital canal.

The external surface of the frontal process of the maxilla is depository during both prenatal and postnatal facial development. The contralateral nasal side is mostly depository, with some resorption at older fetal ages, but it is entirely resorptive postnatally. This area in the rapidly growing young child is characterized by a massive lateral expansion of the lateral nasal walls, including the ethmoidal plates

**FIGURE 10–31**

*Resorptive (darkly stippled) and depository (lightly stippled) patterns of growth in a 20-week fetus. See text for descriptions. (From Mauser, C., D. H. Enlow, D. O. Overman, and R. McCafferty: Growth and remodeling of the human fetal face and cranium. In McNamara, J. A. (Ed.): Determinants of Mandibular Form and Growth. Center for Human Growth and Development, University of Michigan, Monograph 5, Craniofacial Growth Series, 1975.)*

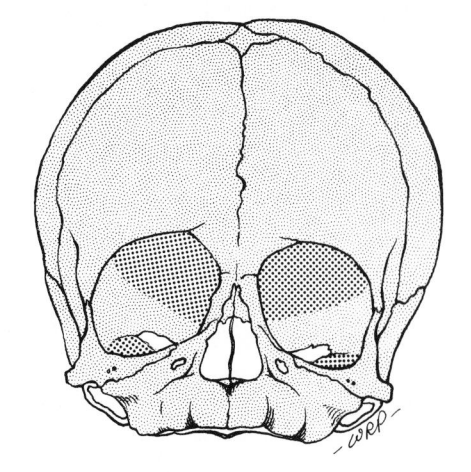

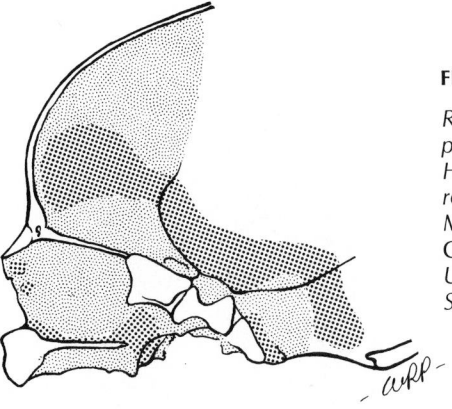

**FIGURE 10–32**

*Resorptive (darkly stippled) and depository (lightly stippled) patterns of growth in a 20-week fetus. (From Mauser, C., D. H. Enlow, D. O. Overman, and R. McCafferty: Growth and remodeling of the human fetal face and cranium. In McNamara, J. A. (Ed.):* Determinants of Mandibular Form and Growth. Center for Human Growth and Development, University of Michigan, Monograph 5, Craniofacial Growth Series, 1975.)

and sinuses. The latter part of fetal life appears to be a transitory period in which these surfaces are just beginning a major lateral expansive movement.

In both the fetal and postnatal periods, the nasal side of the palate (including the palatine bone) is resorptive except along the midline, and the oral surface is depository. This provides for an inferior growth movement of the palate and a vertical enlargement of the nasal chambers.

The mucosal surface of the vertical plate of the palatine bone is resorptive, and the opposite lateral surface is depository in both the prenatal and postnatal periods. This provides for the expansion of this part of the nasal chamber in width.

At 18½ weeks, the **vomer** appears as a U-shaped bone anteriorly and Y-shaped posteriorly. Its sloping anterior margin forms a trough for the inferior edge of the proportionately sizable cartilaginous nasal septum. Remodeling of the vomer has already begun at this age. The anterior part is depository inferiorly in the suture between it and the palate. It is also depository laterally and superiorly in the area supporting the cartilaginous nasal septum but is resorptive within the trough. The Y-shaped posterior part of the vomer is depository inferiorly on the surface adjacent to the suture and also superiorly where it abuts the cartilaginous nasal septum. The vomer in this posterior region is resorptive on the inferior-lateral surface, depository on the superior-lateral surface, and resorptive on the medial surface adjacent to the area of lateral deposition.

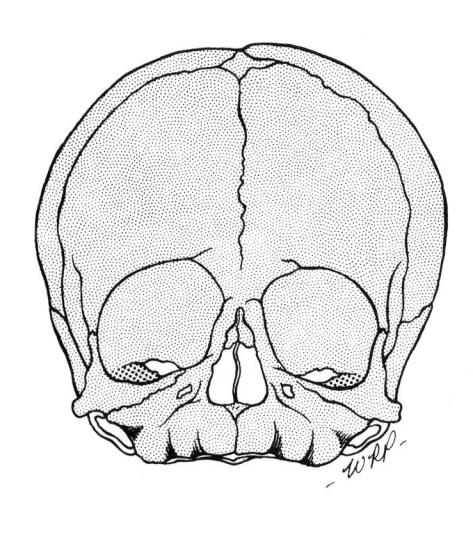

**FIGURE 10–33**

*Resorptive (darkly stippled) and depository (lightly stippled) growth patterns in a term fetus. (From Mauser, C., D. H. Enlow, D. O. Overman, and R. McCafferty: Growth and remodeling of the human fetal face and cranium. In McNamara, J. A. (Ed.):* Determinants of Mandibular Form and Growth. Center for Human Growth and Development, University of Michigan, Monograph 5, Craniofacial Growth Series, 1975.)

**FIGURE 10–34**

*Resorptive (darkly stippled) and depository (lightly stippled) growth patterns in a term fetus. (From Mauser, C., D. H. Enlow, D. O. Overman, and R. McCafferty: Growth and remodeling of the human fetal face and cranium. In McNamara, J. A. (Ed.): Determinants of Mandibular Form and Growth. Center for Human Growth and Development, University of Michigan, Monograph 5, Craniofacial Growth Series, 1975.)*

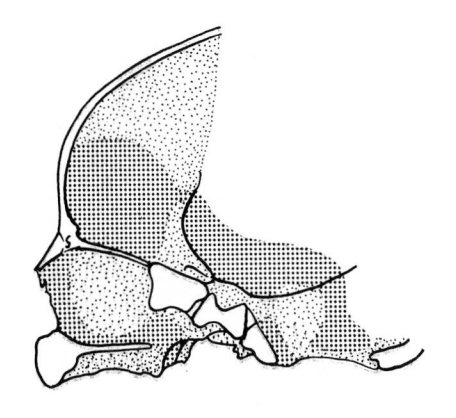

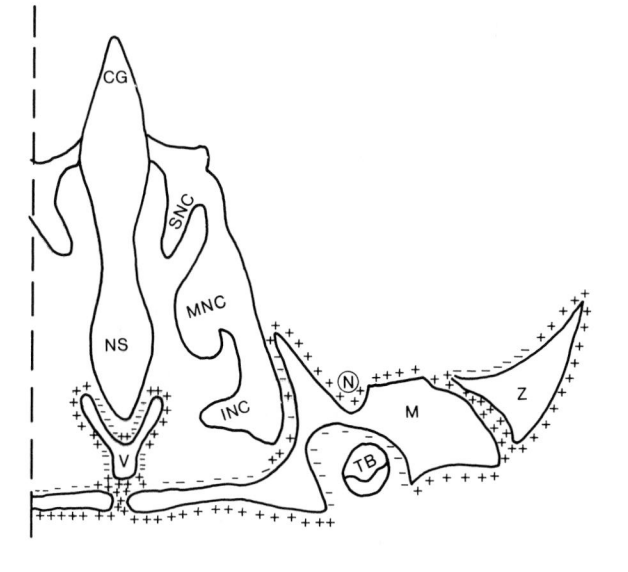

**FIGURE 10–35**

*Resorptive (−) and depository (+) fields in a frontal section through the face of a 26-week fetus. M, Maxilla; Z, zygomatic bone; TB, tooth bud; V, vomer; CG, crista galli; SNC, MNC, INC, superior, middle, and inferior nasal conchae; N, inferior orbital nerve; NS, nasal septum. (From Mauser, C., D. H. Enlow, D. O. Overman, and R. McCafferty: Growth and remodeling of the human fetal face and cranium. In McNamara, J. A. (Ed.): Determinants of Mandibular Form and Growth. Center for Human Growth and Development, University of Michigan, Monograph 5, Craniofacial Growth Series, 1975.)*

**FIGURE 10–36**

*Resorptive (−) and depository (+) fields in a parasagittal section through the 26-week fetal skull. M, Maxilla (with tooth buds); S, sphenoid; FR, foramen rotundum; T, temporal bone; LWS, lesser wing of sphenoid; F, frontal bone, showing the pattern found until the last trimester. (After Mauser, C., D. H. Enlow, D. O. Overman, and R. McCafferty: Growth and remodeling of the human fetal face and cranium. In McNamara, J. A. (Ed.): Determinants of Mandibular Form and Growth. Center for Human Growth and Development, University of Michigan, Monograph 5, Craniofacial Growth Series, 1975.)*

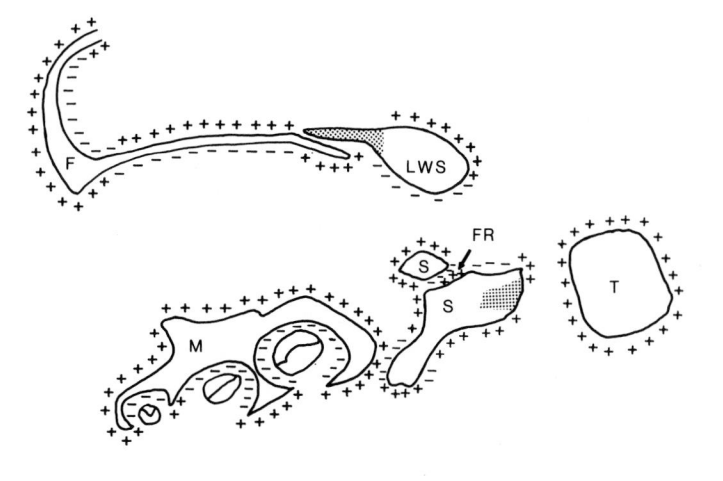

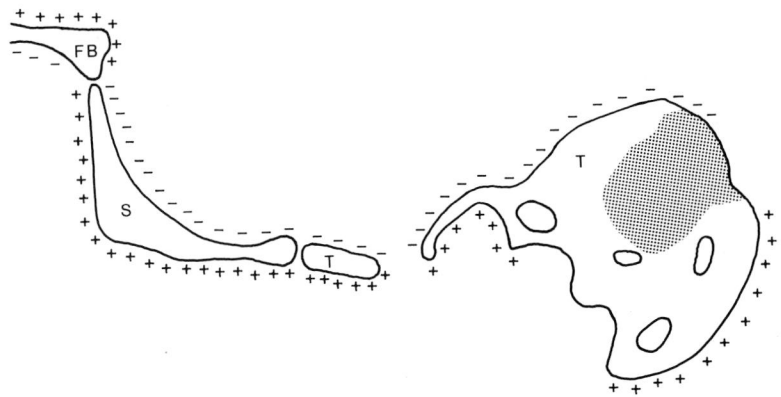

**FIGURE 10–37**

*Resorptive (−) and depository (+) fields in a parasagittal section (lateral to maxillary arch) in a 26-week fetal skull. FB, Frontal bone, showing pattern found until reversals occur in last trimester; S, sphenoid; T, temporal. Stippled area is still cartilaginous. (After Mauser, C., D. H. Enlow, C. O. Overman, and R. McCafferty: Growth and remodeling of the human fetal face and cranium. In McNamara, J. A. (Ed.): Determinants of Mandibular Form and Growth. Center for Human Growth and Development, University of Michigan, Monograph 5, Craniofacial Growth Series, 1975.)*

The cartilages in the **inferior nasal conchae** begin endochondral ossification between 15 and 17 weeks. At this age, the cartilage is hypertrophic, and endochondral ossification is in an early stage. By 22 weeks ossification is well under way, and by 26 weeks remodeling has become established, the bone surfaces being resorptive inferiorly and depository superiorly. Later stages show the opposite, with resorption superiorly and deposition inferiorly.

The **superior and middle nasal conchae** ossify somewhat later than the inferior conchae, not beginning until 17 weeks. By 22 weeks ossification is established, and by 26 weeks remodeling is taking place in both of these ethmoidal labyrinths. The middle concha is depository superiorly and resorptive inferiorly. The superior concha is depository inferiorly and resorptive superiorly on the anterior and posterior thirds but depository in the middle third.

**FIGURE 10–38**

*Resorptive (−) and depository (+) fields in a midsagittal section of a 26-week fetus. F, Frontal bone (the only area in which the endocranial surface of the forehead is depository is along the midline crest, as seen here); FPM, frontal process of maxilla; SNC and INC, superior and inferior nasal conchae (endochondral ossification has started in these areas but is not shown here); OCH, optic chiasma; Pr S, presphenoid; Po S, postsphenoid; Pi, pituitary; O, occipital. Stippled areas are still cartilaginous. (After Mauser, C., D. H. Enlow, D. O. Overman, and R. McCafferty: Growth and remodeling of the human fetal face and cranium. In McNamara, J. A. (Ed.): Determinants of Mandibular Form and Growth. Center for Human Growth and Development, University of Michigan, Monograph 5, Craniofacial Growth Series, 1975.)*

The **nasal bones** in both the fetal and postnatal periods are resorptive on the mucosal surface and depository on the external side. This moves the bony bridge and roof of the external nasal protuberance anteriorly.

The **lacrimal bone** in the fetus is depository only on the surface adjacent to the nasolacrimal canal. In the adult, however, it is depository on its superolateral and inferomedial surfaces, with the contralateral surfaces showing resorption. In postnatal development, the sutures surrounding this small bone function to adjust to the changing relative positions of the other bones with which it articulates (frontal, maxilla, and ethmoid) by providing sutural "slippage" as they all grow at different rates around it. These facial and cranial changes are not nearly as marked in the fetal skull, however, and the remodeling pattern that characterizes the lacrimal bone in the postnatal skull does not appear until later in childhood when marked vertical midfacial growth and ethmoidal expansion begin to occur relative to a more stable orbit.

The external surfaces of the **zygomatic bone** are primarily depository in fetal life. Resorption does occur, however, on the infraorbital margin where the zygoma overlaps the maxilla. In postnatal development, both the anterior and orbital surfaces are resorptive. Transition from the fetal to the postnatal pattern occurs at some time before the fifth to sixth years when the premaxillary region ceases to grow in a direct anterior direction. Some resorption is already present on the anterior surface of the zygoma and always on the orbital surface, even during the fetal period. The postnatal change in pattern relates to the posterior relocation of the zygoma in conjunction with the posterior mode of maxillary lengthening in later childhood.

### Calvaria and Cranial Base

The squama of the frontal bone in the forehead region has the same pattern of remodeling and growth in both the prenatal and postnatal periods. The ectocranial surface is depository, and the endocranial side is resorptive (except along the midline crest). In the region above the frontal eminence, a reversal line occurs, and superiorly the squama becomes depository on both the intracranial and ectocranial sides. This same pattern continues throughout the postnatal growth period.

Deep to the supraorbital rim in the postnatal skull, the entire dural surface is resorptive. On the orbital side beneath the rim, the surface is depository medially and resorptive laterally. The roof of the orbit is entirely resorptive on the dural side and depository on the orbital side. During prenatal growth the medial half of the rim of the orbit, however, is characteristically resorptive, and the lateral half is depository on the intraorbital side. On the dural surface the pattern is reversed, with deposition on the medial half and resorption on the lateral half. The roof of the orbit is resorptive on the intraorbital side and depository on the dural side (floor of the anterior cranial fossa) in early fetal life. In the last trimester, complete reversals of the fetal pattern, however, occur. By 39 weeks the roof of the orbit exhibits the characteristic postnatal pattern.

The difference between the prenatal and postnatal patterns of growth and remodeling in the thin-layered roof of the orbit and floor of the anterior cranial fossa appears to be based on differential growth rates by the eyeball and the cerebral hemispheres. As previously mentioned, the relative rate of eyeball growth is much faster before 24 weeks' gestation than after. The frontal lobes of the brain "lag" in their growth until 25 or 26 weeks. The frontal lobes are only 4.5 per cent of their adult size at 6 months fetal life and about 11 per cent at birth. After birth a

rapid spurt of growth takes place, until about 47 per cent of the adult size is attained at 11½ months and 93 per cent at 7 years.

As the eyeballs pass their stage of maximum expansion and the frontal lobes begin theirs, a reversal in the growth pattern of these parts then occurs to accommodate the changes. Slight variations in the reversal times during this period probably relate to individual variations in the differential growth rates of the different soft tissues involved. The change in the remodeling pattern coincides with the period when the growth of the eyeballs is just slowing and the subsequent spurt of the frontal lobe growth is just beginning.

It is usually stated that the medial part of the anterior cranial base (the mesethmoid) does not begin to ossify until after birth. However, ossification can be noted in some 33 to 39 week old specimens. When seen, the superior surface of the thin bony plate is resorptive, and the inferior surface is depository. This is the same pattern characterizing the bone in the postnatal skull, and it functions to lower the mesethmoid in conjunction with the inferiorly directed growth and displacement of the remainder of the anterior cranial floor.

The prenatal and postnatal growth and remodeling patterns of the greater wing of the sphenoid are essentially the same. The bone is characteristically resorptive endocranially and depository ectocranially at all fetal ages after its early form has been attained (that is, after 14 weeks). This accommodates the expansion of the temporal lobes; these lobes develop earlier in prenatal life than do the frontal lobes.

In fetal growth, the maxillary nerve and the foramen rotundum through which it passes both move forward. The foramen undergoes this remodeling movement by resorption anterior and deposition posterior to the nerve. The nerve and its canal thereby keep pace with the anterior growth expansion of the entire facial complex.

In the postnatal period the pterygoid plates are depository on all sides at their base where they connect with the rest of the sphenoid. Inferiorly, however, most of the periosteal surfaces are resorptive except posteriorly, in the trough, where deposition occurs. The fetal pattern is similar to the postnatal pattern in the last trimester, most surfaces being depository but with some resorption on the anteroinferior surfaces. Variable amounts of resorption are also noted on the posterior surface. This functions to enlarge the pterygoid fossa; it is a common variation in the postnatal skull until the forward-growing plates make contact with the maxillary tuberosity, at which time downward growth and fossa expansion can only take place by deposition on the lining surface of the downward-facing fossa itself, with resorption on the anterior side.

The postnatal presphenoid is largely resorptive on the endocranial surface. However, in the fetus the lesser wing is resorptive only on its posterior and inferior edges where it borders the superior orbital fissure. It is depository superiorly where this bone forms the posterior part of the anterior cranial floor. Resorption also occurs anterior to the optic nerve and in the chiasmatic sulcus. These patterns coincide with the early remodeling pattern of the orbital roof at the time when the eyeball is expanding more rapidly than the frontal lobes. The optic nerve is moving forward to keep pace with the orbit, which also moves forward with the growth of the rest of the face and anterior cranial floor. The dural surface of the presphenoid becomes increasingly resorptive endocranially during a later period, coinciding with the massive growth of the frontal lobes. In the young child, this surface area of the bone usually becomes completely resorptive.

The cortical lining for most of the sella turcica is usually depository during both the prenatal and postnatal growth periods. The superior part of the posterior

**FIGURE 10–39**

*Fetal skull enlarged to same size as an adult skull to dramatize the marked differences in regional proportions between the two.*

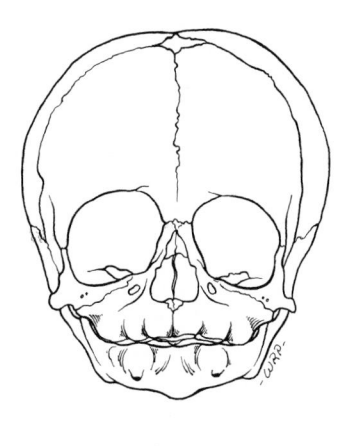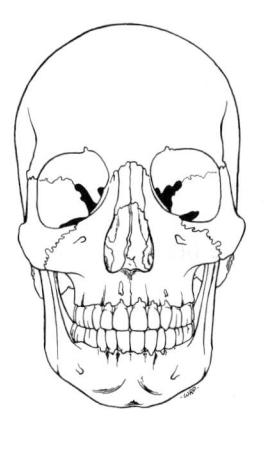

wall lining the fossa, however, is typically resorptive prenatally as well as post-natally. Variations in reversal line placement exist in the fetus just as they do in the growing child. The floor as well as the anterior and posterior lining walls are variable in growth pattern because of the markedly differential rates of growth of the many separate soft tissue parts in this general region. (See Chapter 3.)

The petrous part of the temporal bone follows essentially the same pattern of growth during both the prenatal and postnatal periods. This involves deposition on the medial surfaces of the petrous ridge, with resorption on the far lateral portion where it grades into the side of the calvaria. The petrous bone is characteristically depository ectocranially; resorption occurs within the jugular fossa in both age groups.

As in the postnatal period, the fetal basioccipital is resorptive endocranially and depository ectocranially. The lateral part of the occipital complex also follows this pattern. Resorption is common on the surface facing the jugular fossa. Cartilage remains in the occipital condyle at birth.

The overall **length** of the whole braincase at birth is about 63 per cent of its total growth. By the end of the first year, 82 per cent of growth is complete; by 3 years 89 per cent, and by 5 years 91 per cent of braincase length is attained. At about 15 years, 98 per cent is reached. The anterior part of the cranial floor (basion-nasion) shows about 56 per cent of adult growth by birth and 70 per cent at 2 years of age. In the newborn, the **width** of the skull base has grown to about 100 mm. By the sixth postnatal month, 50 mm. have been added, and by the first year, 20 mm. more growth occurs. Thereafter, the rate declines, until only about 0.5 mm. per year is added from 3 to 14 years. At birth, the **weight** of the brain is about half that of the adult. By the third year, the brain's weight is 80 per cent of the adult weight, and at 5 to 8 years 90 per cent.

In the newborn, the four parts of the occipital bone are still separated by synchondroses but are fused by about the fifth year (with a range of about ± 2 years). At birth, the growth activity of the intersphenoidal synchondrosis has largely ceased. The sutures among the three parts of the temporal bone have undergone partial fusion by about 2 to 4 years. The closure time (or more importantly, the time of growth cessation) of the sphenoethmoidal synchondrosis is not known with certainty, and estimates range from 5 to over 20 years. The early cartilaginous junction can later become membranous, and any subsequent growth proceeds intramembranously. The growth of the spheno-occipital synchondrosis ceases around 15 years, and closure has occurred by about 20 years. The metopic suture usually

fuses at about the second year, and the maxillary-premaxillary sutures close during the first or second year. The right and left halves of the mandible fuse during the first postnatal year. The sphenopetrosal and petro-occipital synchondroses may persist in the adult.

At birth, the bones of the calvaria are not yet fully united, and the surfaces of the bones are smooth. The bony cortices are thin and single-layered (lacking a diploë). Six fontanelles are present in the newborn; the posterior closes with bone around the time of birth, the anterior at about the first year, the anterolaterals at about 15 months, and the posterolaterals by 18 months. In the adult, the suture lines are much more jagged, and by the sixth year the bones have become three-layered with a diploë.

In the newborn, the external acoustic meatus characteristically faces downward, but it is already nearly adult-sized. The tympanic bone is still a ring. The mastoid process and pneumatic cells are not present at birth.

## The Auditory Ossicles (Fig. 10–40)

All three pairs of the tiny ossicles are fully formed in cartilage by about $8^{1}/_{2}$ weeks. The intramembranous bone of the mandible has already begun to form around Meckel's cartilage, which is still connected with the formative malleus. The external canal and the auditory tube are approaching the developing ossicles from the lateral (auricular) and medial (pharyngeal) sides, respectively. Ossification begins first in the malleus and incus at 16 weeks and follows within 2 weeks in the stapes.

There is one notable difference between the growth and development of the ear ossicles and that of all other bones. They do not undergo extensive remodeling. Their linear dimensions are, for the most part, already established by the cartilaginous anlagen that precede them. Because these tiny bones do not grow in length, the factor of "relocation" is not involved, and the remodeling processes of reshaping and resizing thus do not take place.

The **malleus** is identifiable as a mesenchymal mass in the 10-mm. embryo, and this has developed into cartilage in the 28-mm. embryo. As Meckel's cartilage begins to undergo degenerative changes anticipating replacement by ligamentous tissue, a center of ossification appears on the cartilaginous manubrium of the malleus at about 15 weeks. This is "perichondral" bone. That is, it develops intramembranously within the connective tissue membrane (the former perichondrium) enclosing the mass of cartilage. Ossification spreads until the entire surface of the element becomes covered with a thin, bony shell (except at articular facets and points of ligament attachment). The core is still cartilaginous, but the chondrocytes now undergo hypertrophy, and the matrix calcifies. Vascular buds invade the cartilage, and partial endochondral bone replacement then takes place. Small medullary spaces form by resorption and become filled with undifferentiated connective tissue, and endosteal bone is laid down in these spaces. Within the perichondral shell of bone, the medullary region is thus composed of endochondral bone, compacted cancellous bone, and islands of calcified cartilage matrix remnants. At the tip of the manubrium, perichondral bone does not develop, but the core of the cartilage receives partial endochondral replacement. It remains surrounded by a covering of the original cartilage.

In the medulla of the malleus, trabecular reconstruction takes place and is said to continue, albeit very slowly, through old age. Major changes, however, do not

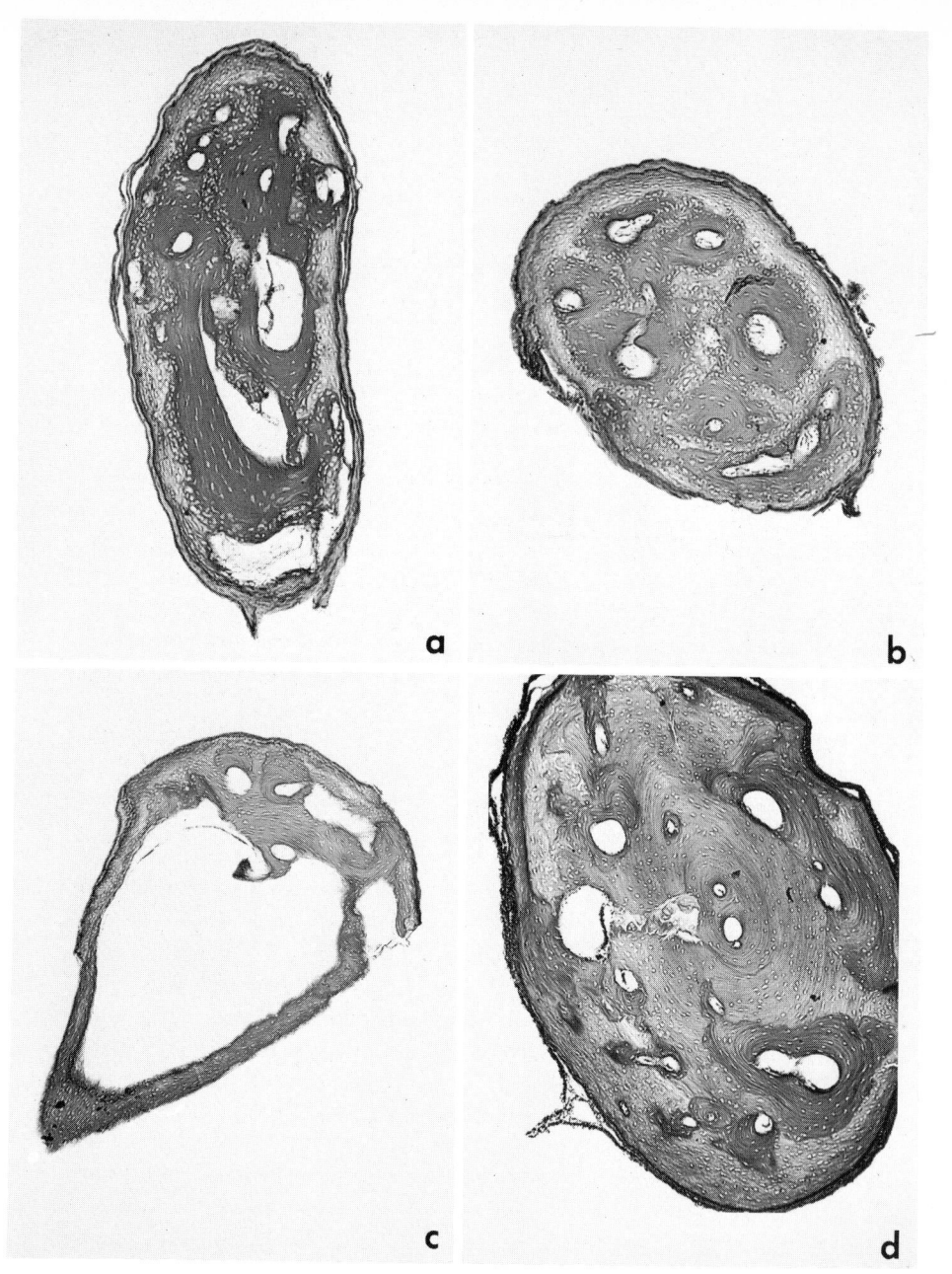

**FIGURE 10–40**

a, *Malleus, newborn.* This section, from near the head, shows a thin periphery of cartilage enclosing the core of trabecular bone. The latter has completely replaced the endochondral bone which originally occupied the medulla.

b, *Malleus, newborn.* Near the manubrium, a very thin crust of the original cartilage remains. The core of endochondral bone has been largely replaced by several irregular, haversian-like structures that have been deposited within the medullary resorption spaces. Extensive haversian reconstruction does not occur, however.

c, *Stapes, newborn.* The obturator foramen in this section is bounded by two crura and the somewhat thicker base (at top). Each crus is composed of perichondral (intramembranous) bone. Although only a portion of the cortex is seen in this section, the crura are hollowed along their medial obturator face, and the cartilaginous core has become completely removed and replaced by mucosal tissue. Note that the external surface of the base retains a thin cartilaginous cover. The perichondral cortex adjacent to the obturator foramen has already been resorbed, and the original endochondral bone in this particular section has been entirely removed.

d, *Incus, 3-month infant.* The bone is enclosed by a thin peripheral layer of perichondral (intramembranous) bone tissue. The original endochondral bone of the core has been largely replaced by irregular trabecular bone that has undergone compaction. A few haversian-like structures were produced by deposition of lamellar bone within medullary spaces. Scattered calcified cartilage matrix spicules remain in the medulla. *(From Enlow, D. H.:* The Human Face. *New York, Harper & Row, 1968, p. 224.)*

occur, and extensive haversian reconstruction does not take place. Unlike the stapes, the medulla of the malleus (and also the incus) remains occupied by bone and is not hollowed to become replaced by pharyngeal mucosal tissue. The anterior process of the malleus is the only part that does not develop in direct association with cartilage. It forms in the adjacent connective tissue.

As in the malleus, the **incus** is formed in and around a cartilaginous prototype that approximates adult linear dimensions. A perichondral bony cortex develops around this cartilage, and subsequent replacement by endochondral trabeculae then occurs. Trabecular reconstruction and compaction of some medullary spaces take place very slowly through the years. Small remnants of calcified cartilage, however, may still survive in the aged adult. These various changes involve internal **reconstruction** and not the "remodeling" that ordinarily accompanies the growth process in all other bones.

The cartilage primordium of the **stapes** also has a general configuration similar to the adult form. The stirrup-shaped stapes has a broad, platelike base with two projecting crura encircling the central obturator foramen. The crura converge at the other end to form the head of the stapes, and this is in articular contact with the formative incus. A primary ossification center appears on the surface of the basal plate between the crura. Perichondral bone then forms on the surface of the cartilage within the covering membrane and spreads over the base and onto the crura to form a cortical crust of intramembranous bone. Cartilage is retained, however, on the articular contact surface of the head and also on the inferior side of the base. In both the base and head, the core of cartilage undergoes hypertrophy and calcification, and endochondral bone replacement follows.

Major reconstruction changes now take place on all the medial bone surfaces lining the periphery of the obturator foramen (that is, the two lateral crura and the base and head on the bottom and top). The thin shell of perichondral bone undergoes resorptive removal on these particular surfaces. In the head and base, this exposes the underlying core of endochondral and trabecular bone, which is undergoing compaction to form the irregular, "convoluted" type of bone tissue. Scattered spicules of calcified cartilage survive. The basal plate and the head become thinned as a consequence of the removal of the perichondral bone shell.

In the two crura, unlike the base and head of the stapes, endochondral bone replacement of the cartilaginous core is negligible. As the crust of perichondral bone on the medial surface of each crus is removed, the resorptive process continues on into the cartilage and entirely removes it. The result is a hollowing of the paired crura on their obturator (medial) sides to form trough-shaped processes connecting the base with the head. The hollowed basin of each crus becomes lined by a mucosa of pharyngeal origin. Thus, the crura are composed only of a thin, U-shaped cortex made up of perichondral bone tissue. Original cartilage is retained on the articulating surfaces of the base and head. As in the other ossicles, these cartilage plates do not function as epiphyseal growth centers, and linear growth increases do not occur. Remodeling changes associated with growth are thus not involved. Haversian reconstruction is also negligible in the stapes, and, unlike in the malleus and incus, trabecular reconstruction is all but lacking. The histologic structure of the stapes remains essentially unchanged throughout life.

## The Mandible

The beginning fetal mandible, as in the earliest growth stages of the other bones of the skull, initially has outside surfaces that are entirely depository in char-

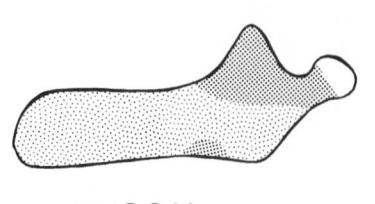

**BUCCAL**

**FIGURE 10–41**

*Mandible in the last trimester of fetal development. Dark stippling represents resorptive fields, and light stippling indicates depository fields.*

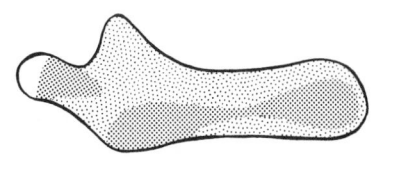

**LINGUAL**

acter. At about 10 weeks, however, resorption begins around the rapidly expanding tooth buds and is present thereafter. By 13 weeks, distinct resorptive fields are becoming established on the buccal side of the coronoid process, on the lingual side of the ramus, and on the lingual side of the posterior part of the corpus. The anterior edge of the ramus is already resorptive, and the posterior border is depository. In some specimens, however, the anterior margin along the tip of the coronoid process shows deposition, suggesting a "rotation" to a more upright position (see Chapter 3). By 26 weeks, the basic growth and remodeling pattern that continues on into postnatal development is seen except, notably, in the incisor region (Fig. 10–41). In the fetal and early postnatal mandible, the entire labial side of the anterior part of the corpus is depository. As in the fetal and young postnatal maxilla, the fetal mandibular corpus grows and lengthens **mesially** as well as distally in conjunction with the establishment of the primary dentition. The lingual side of the fetal corpus in the incisor region is resorptive after about the 15th week in most (but not all) mandibles. This contributes to a forward growth movement of the entire incisor region of the corpus. Some time before about the fifth or sixth year of childhood growth, however, the alveolar bone on the labial side in the forward part of the arch undergoes a reversal to become **resorptive,** and the lingual side becomes uniformly depository. This change occurs in conjunction with the unique lingual direction of incisor movement in the child's mandible. From this time, the chin begins to take on progressively more prominent form; the mental protuberance continues to grow anteriorly, while the alveolar bone above it moves posteriorly until the lower permanent incisors reach their definitive positions.

# Chapter 11

# Bone and Cartilage

# PART ONE

**Concept 1.** Because of the unique nature of its intercellular matrix, cartilage is a rigid and firm tissue, but it is not hard. Cartilage provides three basic functions. It gives **flexible** support in appropriate anatomic places (the nasal tip, ear lobe, thoracic cage, tracheal rings); it is a **pressure-tolerant** tissue located in specific skeletal areas where direct compression occurs (such as the articular cartilage); and it functions as a **"growth cartilage"** in conjunction with certain enlarging bones (e.g., a synchondrosis, condylar cartilage, epiphyseal plate). Cartilage is a non-vascular connective tissue that is ordinarily noncalcified (both vascularization and calcification, however, are involved as steps in the replacement process by bone tissue).

**Concept 2.** Cartilage usually has a perichondrium, but, significantly, it can exist without this covering membrane. Cartilage grows **appositionally** by the activity of its chondrogenic membrane, and it also grows **interstitially** by cell divisions of chondrocytes and by additions to its intercellular matrix. Together with the non-calcified nature of the matrix and the absence of vessels, these various features combine to allow the cartilage to function and to grow in areas of direct pressure. Because it can exist without a covering membrane, it is suited for articular surfaces, synchondroses, and epiphyseal plates. Because it can expand interstitially, cartilage can thereby grow even without a membrane. Because its matrix is noncalcified, unlike bone, cell divisions are possible, and diffusion of nutrients and wastes can take place through the permeable matrix. Because there are no vessels to press closed, either in a covering membrane or within the matrix, cartilage is pressure-adapted and can **grow** where compression exists because of its noncalcified, inter-stitial expansion features.

**Concept 3.** Unlike cartilage, which is pressure-related in many of its ana-tomic locations, bone is **tension-adapted.** Bone **must** have a vascular, osteogenic covering membrane (or other soft tissue), and it can only grow appositionally. Bone cannot grow directly in heavy pressure areas because its growth is dependent upon a sensitive, vascular membrane. Bone needs a membrane in order to support its in-ternal vascular system, which, in turn, is essential because the matrix is calcified and will not allow diffusion of oxygen, nutrients, and metabolic wastes to and from cells. Moreover, bone cannot grow interstitially because of its calcified matrix.

**Concept 4.** In all areas of skeletal growth, bone grows **intramembranously** in tension areas and **endochondrally** in pressure areas. "Growth cartilages" take part in the latter ossification process. They provide for the **linear** growth of a bone

**toward** the direction of pressure. As interstitial cartilage expansion provides pressure-adapted growth on the pressure side of the cartilage plate, an equal amount of cartilage is removed and replaced by bone on the other side. This allows the bone to lengthen toward its force and weight-bearing articular contacts. The remainder of the bone, including all its cortical plates, grows by membranous ossification in conjunction with the periosteal and endosteal membranes.

**Concept 5.** The membranes associated with bone (periosteum, sutures, periodontium) have their **own** internal growth and remodeling process. The whole membrane, as a sheet, does not merely move away as new bone is deposited by it. Rather, it undergoes extensive fibrous changes in order to sustain constant connections with the bone by means of collagenous fiber continuity from the membrane into the matrix of the bone. As fibers in the membrane become enclosed within the new bone deposits, the membrane-produced fibers become incorporated as bone fibers. This is accompanied by fibrous remodeling within the membrane to provide continuity between membrane and bone fibers. The membrane **grows** outward rather than just backing off as bone is laid down by it. The process involves complex linkages among the periosteal collagenous fibers (Fig. 11–15). The movements of muscle attachments along remodeling surfaces of bones and the insertion of muscles on **resorptive** bone surfaces are also carried out by this membrane remodeling and relinkage process.

**Concept 6.** The periodontal membrane converts the **pressure** exerted by a tooth during chewing into **tension** on the collagenous fibers attaching the tooth to the pressure-sensitive alveolar bone. This is accomplished by placing the tooth in a sling of membrane fibers (that is, the periodontium), so that inward pressure by the tooth caused by biting is thereby translated into a pull on the fibers. The periodontal membrane **also** provides for the mesial, lateral, vertical, and distal drifting and for the eruption, tipping, and rotations of each individual tooth. As the face enlarges, the teeth are **moved** along with the various parts of the growing bones. There are significant amounts of such tooth movements associated with progressive facial growth. The maxillary teeth, for example, move inferiorly for a considerable distance in conjunction with vertical nasal expansion. Everybody is familiar with "mesial" drift, but the movements of teeth during the facial growth process involve **much** more than just this. All the various tooth movements are carried out by the periodontal membrane. It is customary to describe drift as a process of resorption on the lead or "pressure" side and deposition on the trail or "tension" side of the alveolar socket. This is not, however, simply a two-dimensional process (Fig. 11–30). This **same** deposition and resorption of bone **also** brings about vertical movements of the teeth as the jaws and other parts of the face grow vertically, and it provides for all of the many other growth movements of the teeth as well (Fig. 3–118).

**Concept 7.** Tooth movements are accomplished by relinkage changes of the collagenous fibers within the periodontal membrane (Fig. 11–31). The membrane does not simply "move" in the direction of the tooth movement; it **grows** in this direction. As parts of the periodontal fibers become enclosed by bone on the depository side of the moving alveolar socket, the fibers within the periodontium are lengthened by the fibroblasts in the membrane, and new fibers are produced as well. This involves continued fibrous remodeling to provide direct continuity between the fibers in the membrane and the attachment fibers embedded in the bone. The remodeling process is one in which slender "precollagenous" fibrils link, relink, lengthen, or shorten to provide continuous, uninterrupted connections between the coarse "mature" collagenous fibers attached to the tooth on one end and

the alveolar bone on the other (Figs. 11–32 to 11–35). This is presumably done in some parts of the membrane by enzymatic release of ground substance to free the precollagenous fibrils (which are bundled together to form the larger mature fibers, using ground substance as the binding agent), by fibroblast activity to lengthen or shorten them in the intermediate layer, and by additions of ground substance by fibroblasts to bundle them up in some other parts of the membrane. The fibrous membrane thereby grows to new locations along with the drifting teeth and bone. In some instances an **intermediate plexus** of precollagenous fibrils can be distinguished as a distinct middle layer in the periodontal membrane during active periods of tooth movement. Or linkage changes may be more diffuse throughout most of the membrane without producing a distinguishable, separate intermediate plexus. The remodeling process in the membrane, in either case, continues to occur as long as the tooth actively moves. When bone growth and tooth movements largely cease, this active, dynamic membrane then functions essentially as a ligament. The term periodontal "ligament," rather than periodontal membrane, is not appropriate during the growth period because of the many different functional activities involved. This remarkable membrane is much more than a stable "ligament."

**Concept 8.** Histology textbooks teach (at least at present) that the haversian system (secondary osteon) is the structural unit of bone. This is incorrect. In the bone of the young, growing child, the haversian system is **not** a major structural feature. This old haversian system notion not only has misled students of dentistry and medicine, but also has concealed an important concept. There are other, much more widespread kinds of bone in the child's growing skeleton. The concept is this: different functional circumstances exist, and there is a specific type of bone tissue for each circumstance. Some bone types are fast-growing, others slow-growing. Some bone types grow inward, others outward. Some are associated with muscle, tendon, or periodontal attachments; others are not. Some bone types form a thick cortex, others a thin cortex. Some relate to a dense vascular supply, others to scant vascularization, and so on. "Haversian systems" could not do all this. These are important points because a basic feature of bone is its developmental versatility and adaptability as a tissue.

**Concept 9.** **Primary vascular** bone tissue (Fig. 11–1) is the principal type of periosteal cortical bone in the growing skeleton of the child. The vessels are enclosed within canals as new bone is laid down around each vessel in the osteogenic part of the periosteum. These canals are not surrounded by concentric "haversian" lamellae. If the bone is fast-growing, many vessels and their canals characteristically become enclosed. If it is slower growing, fewer or even no canals are incorporated within the compact bone substance. **Compacted coarse cancellous** bone (Fig. 11–2) is the principal cortical type formed by the endosteum. Half to two-thirds of all the cortical bone in the body is composed of this important, distinctive structural type. It is formed by the inward growth of the cortex into the medulla (that is, periosteal resorption and endosteal deposition); medullary cancellous bone is converted into cortical compact bone by filling the spaces until they are reduced to vascular canal size. **Fine cancellous** bone (Fig. 11–3) is one of the fastest growing types. It is formed throughout the fetal skeleton and also occurs in rapidly enlarging parts of a postnatal bone. This type of cortical bone tissue is characterized by spaces that are larger than ordinary vascular canals but smaller than the coarse cancelli of the medulla. **Nonlamellar** bone is also a rapid-growing type, and it occurs in conjunction with fine cancellous bone formation, although many "compact" areas of the cortex may also be nonlamellar. **Lamellar** bone is a slower

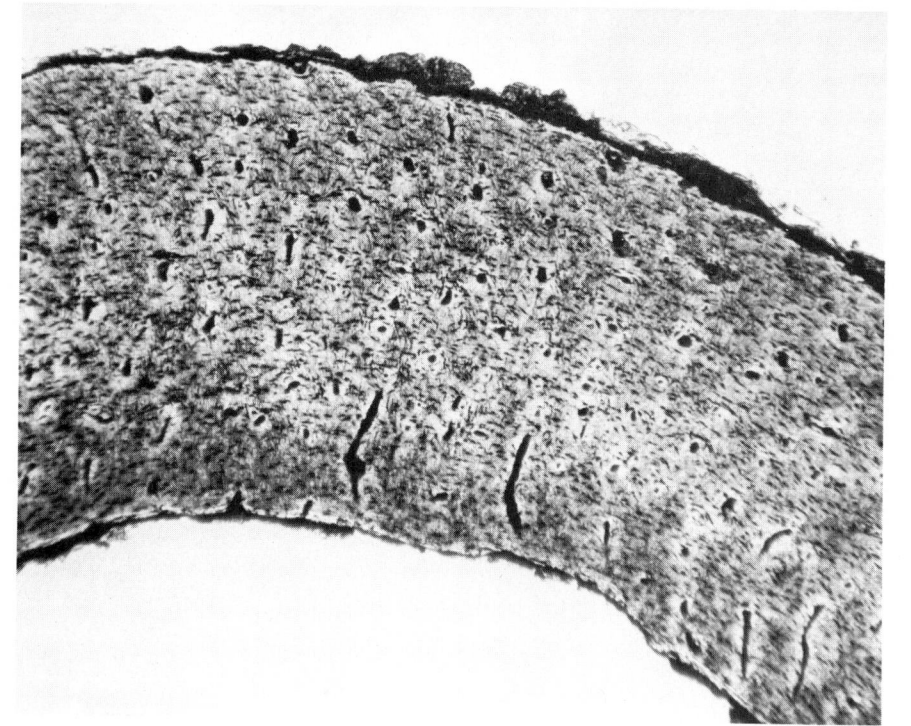

**FIGURE 11–1**

*Primary vascular bone tissue. This is the "standard" type of bone found in the growing human child as well as in most other vertebrates. Note that haversian systems are not present in this transverse section of cortical bone. (From Enlow, Donald H.:* Principles of Bone Remodeling, *Courtesy of Charles C Thomas, Publisher, Springfield, Illinois.)*

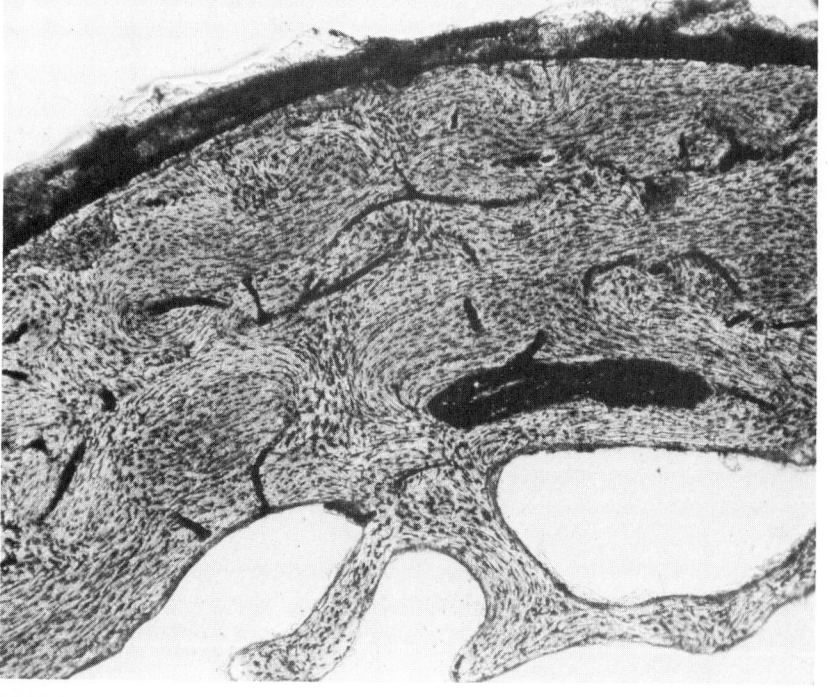

**FIGURE 11–2**

*This section was taken from an inward-growing region of the cortex. It is compacted cancellous bone produced by endosteal deposition and periosteal resorption. The large spaces between the coarse cancellous trabeculae have been filled with lamellar bone. (From Enlow, D. H.:* Am. J. Anat., *110:79, 1962.)*

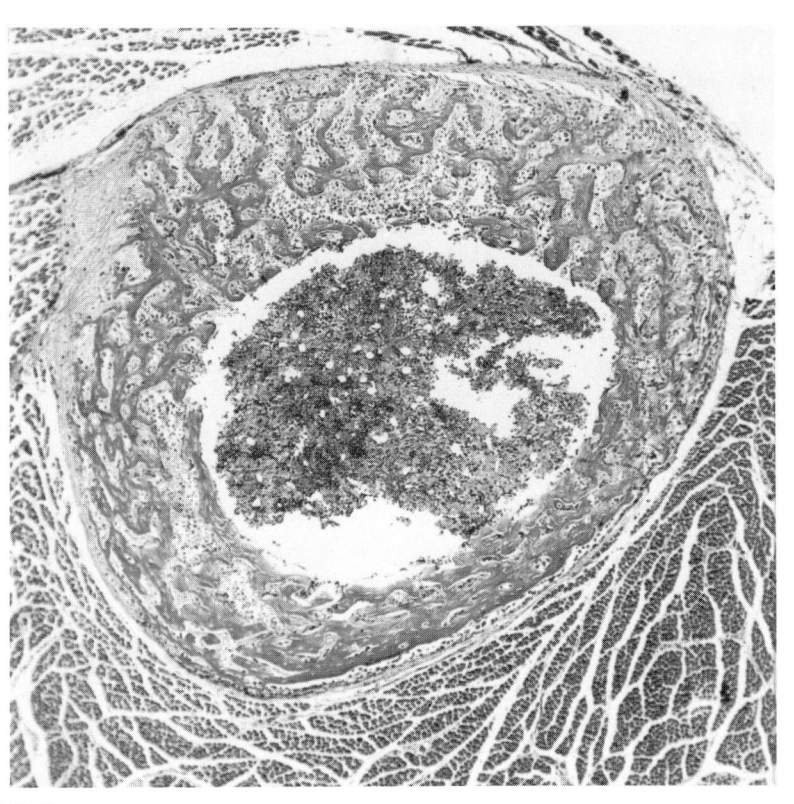

**FIGURE 11–3**

*The cortices in fetal bones are composed of fine cancellous, nonlamellar bone tissue. Note the relatively small connective tissue–filled spaces. Areas of very fast growth in postnatal bones can also be fine cancellous in structure. (From Enlow, D. H.: Am. J. Anat., 110:79, 1962.)*

growing type found throughout most parts of the adult's skeleton and the slower forming areas of the child's skeleton. The various sections of cortical bone seen in Figures 11–41, 11–42, 11–43, and 11–44, for example, are all composed of lamellar bone tissues. A limited distribution of **haversian** bone is formed in some areas of muscle attachment on resorptive or remodeling surfaces of a bone in the child (Fig. 11–4), but it is not a principal feature of the young skeleton. Most haversian systems develop much later in life and are concerned with the secondary reconstruction of the original primary cortical bone (for various reasons described in Part Two). A great many species, however, do not have **any** haversian systems at any age. **Bundle bone** is characterized by dense inclusions of attachment fibers from the periodontal membrane (Fig. 11–33). This bone type is formed only on the depository sides of the alveolar socket. The resorptive side is usually composed of compacted coarse cancellous (endosteal) bone or, if the alveolar plate is very thin, of bundle bone formed on the depository side but translated over to the resorptive side as alveolar drift proceeds. **Chondroid** bone (Fig. 11–39) is found at the apex of the alveolar rim and other rapidly forming areas throughout the skeleton (such as the apex of growing tuberosities where tendons attach). This bone tissue type resembles cartilage because of the large, rounded appearance of its crowded osteocytes surrounded by a nonlamellar, basophilic matrix. Because it undergoes internal metaplasia into **other** bone tissue types, chondroid bone is perhaps the only kind of bone tissue that actually has what might be regarded as an interstitial mode of growth.

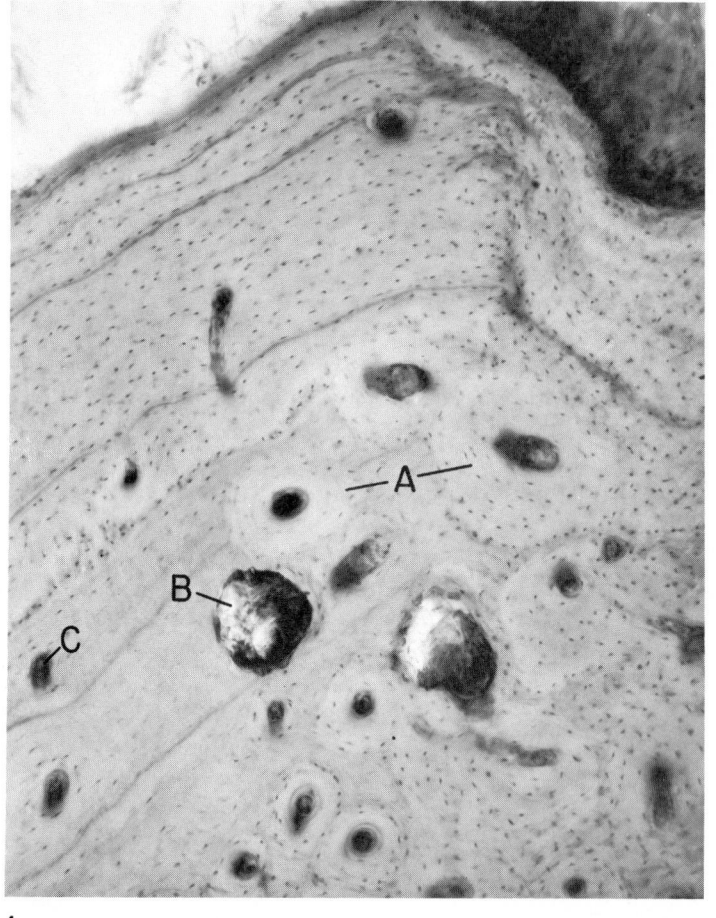

**FIGURE 11–4**

*These secondary osteons are located in a tuberosity that is undergoing a remodeling movement. The shifting of the attached muscle involves the continued formation of haversian systems. A primary vascular canal (C) is enlarged into a resorption canal (B), and concentric lamellae subsequently deposited within the resorption spaces result in fully formed secondary osteons (A). (From Enlow, D. H.: Am. J. Anat., 110:269, 1962).*

# PART TWO

Haversian systems have long been popularized as the "unit" of bone tissue structure, but this is not really the case at all.* It is important to know that **other**, quite basic kinds of bone tissue exist and that they are directly involved in the growth process. Elementary texts adequately describe the two basic modes of bone growth, endochondral and intramembranous, but they do not explain **why** both occur, and this is important. The present chapter provides a short supplement to standard histology texts in order to provide this kind of essential and useful information.

## CARTILAGE

This tissue, historically, was one of the first to be described histologically (bone is another) because of its stiff, turgid matrix which can be readily sectioned and studied microscopically. It was a basic tissue utilized by Schwann in his landmark development of the animal cell doctrine. Curiously, it is a relatively poorly known tissue today compared with most of the others. Many monographs and textbooks have been written on bone, but (at this writing) few such encyclopedic references exist for cartilage. Numerous symposia have been conducted on the subjects of bone, the lymphocyte, muscle, nervous tissue, and so on, but very few have dealt especially with cartilage. There are no cartilage "societies," as there are for many other tissues and organs. Many internationally recognized authorities are identified with most kinds of tissues, but only a relative handful are concerned with the normal structure, physiology, and growth of cartilage. Considering that the controversial cartilaginous growth sites are presumed by many to represent **the** key growth pacemakers for so many parts of the body, including the face and braincase, this is

---

*Another basic point is **why** haversian systems develop. This is explained later.

**FIGURE 11–5**

*The dual nature of cartilage growth is illustrated in this series of growth changes. In a center of chondrogenesis (a) some of the mesenchymal cells differentiate into a mass of precartilage (2), and other stem cells become involved in the formation of the enclosing perichondrium (1). Note that the perichondrium has blood vessels, but the cartilage itself is avascular. Then, in stage c, some of the chondroblasts present in the inner part of the perichondrium of stage b secrete matrix. When fully enclosed within its own secretory material (fibers and ground substance), each cell becomes a chondrocyte (4). Note that the new, young chondrocytes are still elongate; the biomechanical forces acting on the perichondrium have caused its cells, and the other cells in the cartilage derived from them, to be elliptical in shape. The laying down of new cartilage by the perichondrium is* **appositional** *growth. The cartilage cells already existing also enlarge (lipids and glycogen accumulate in the cytoplasm in large amounts), and the cells become rounded (3). The chondrocytes become separated from each other by the increasing amounts of intercellular matrix. These changes, combined with repeated divisions of the chondrocytes, constitute* **interstitial** *growth (5). The inset in stage c shows the fibrous matrix as it would appear if the hyalinizing ground substance were removed. It is a feltwork of fine, densely arranged collagenous fibers. (From Enlow, D. H.:* The Human Face. *New York, Harper & Row, 1968, p. 4.)*

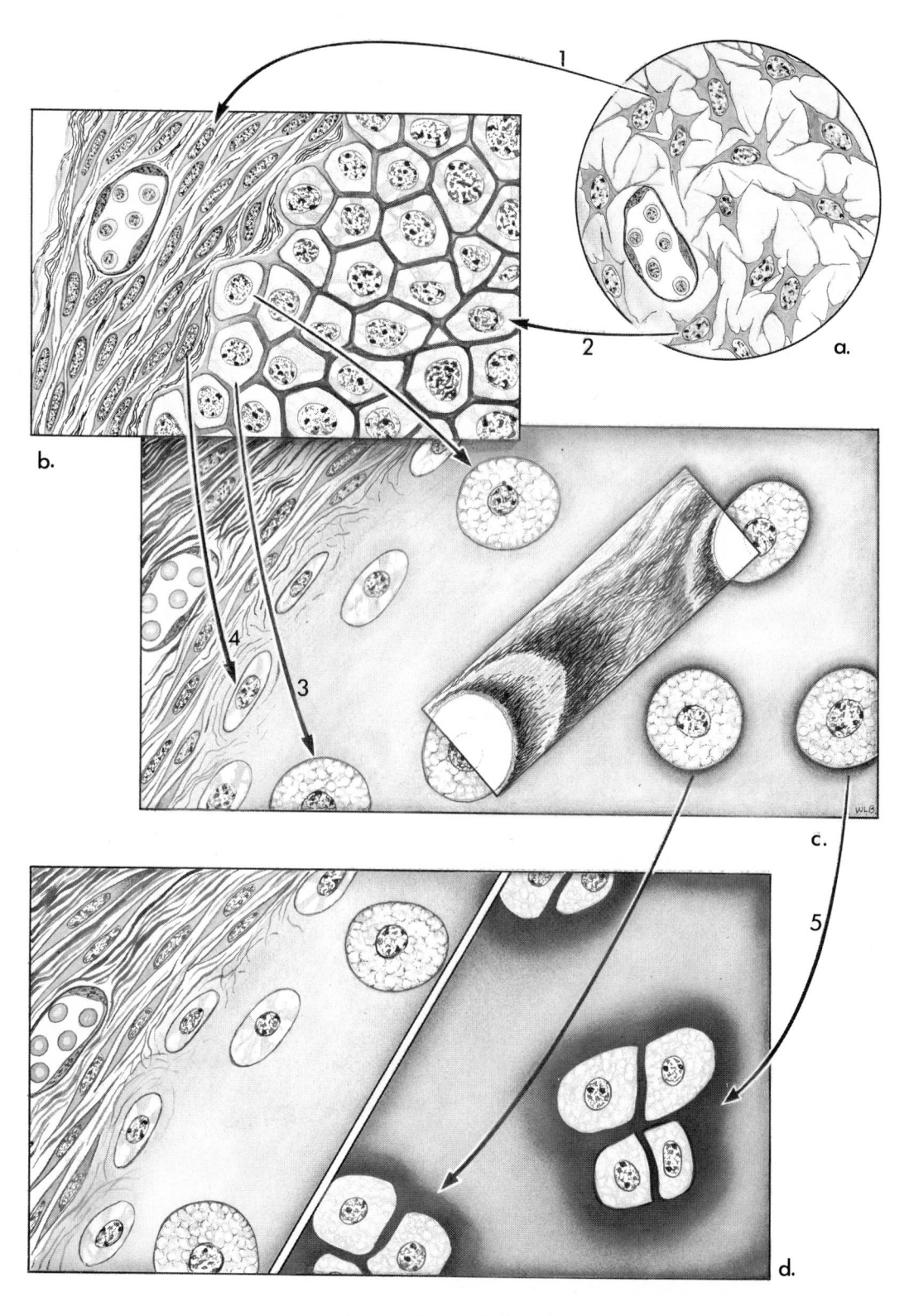

See opposite page for legend.

not as it should be. There are great opportunities here for researchers in a relatively uncrowded but very important area.

Several distinctive structural features relate to cartilage (Figs. 11–5 and 11–6). First these are listed, and then the nature of their interrelationships is explained in terms of the various fundamental functions of cartilage.

1. Cartilage has a stiff but **not hard** intercellular matrix. It provides rigid support, but it is so soft that it can be cut with a fingernail. This feature is based on the exceptionally high content of water-bound ground substance. The rich amount of chondroitin sulfate (Gr., *chondros*, cartilage) in the cartilage matrix is associated with a noncollagenous protein, and this combination has the special property of marked hydrophilia. This gives the turgid, firm character to the matrix. Cartilage develops in locations around the body where **flexible** (not brittle, unyielding) support is appropriate.

2. The matrix of ordinary cartilage is noncalcified.

3. The matrix is nonvascular.

4. Cartilage can grow **both** interstitially and appositionally.

5. Cartilage has an enclosing, vascular membrane, but it can exist without one; it does this in certain specific locations.

6. Cartilage is uniquely pressure-tolerant.

Now, let us combine these various structural and physiologic features in order to relate cartilage to its special roles in the body and particularly in the face.

1. Because the matrix is noncalcified, the matrix is also able to be nonvascular. Nutrients and metabolic wastes diffuse directly through the soft matrix to and from cells. Thus, blood vessels are not required in cartilage as they are in bone, with its hard, impervious matrix.

2. Because the matrix is nonvascular, cartilage is pressure-tolerant (one of several reasons). There are no vessels near the surface to mash closed by compression, in contrast to other kinds of soft tissues, thereby allowing cartilage to operate metabolically because there are no lines of supply to occlude. The water-bound, noncompressible matrix is not badly distorted by the force, and its turgid nature protects the cells within it.

3. Unlike bone, cartilage can function without a covering membrane. This is possible because the matrix is noncalcified (allowing diffusion) and because the matrix is nonvascular. It is thus not dependent upon **surface** blood vessels in an enclosing **surface** membrane, since it does not have vessels within the matrix.

4. Because cartilage can function without a covering membrane, it is especially adaptable to sites involving pressure, as on articular surfaces and the surfaces of epiphyseal plates. If a soft connective tissue membrane were present, its vessels would be closed off by the pressure, and the cells would be subject to anoxia as well as the direct pressure itself. Further, a delicate perichondrial membrane could not withstand abrasive articular movements. In conjunction with synovial secretions, however, the naked surface of the articular cartilage provides relatively frictionless movements while bearing great weight under severe pressure.

5. Because cartilage has an interstitial as well as a membrane-dependent ap-

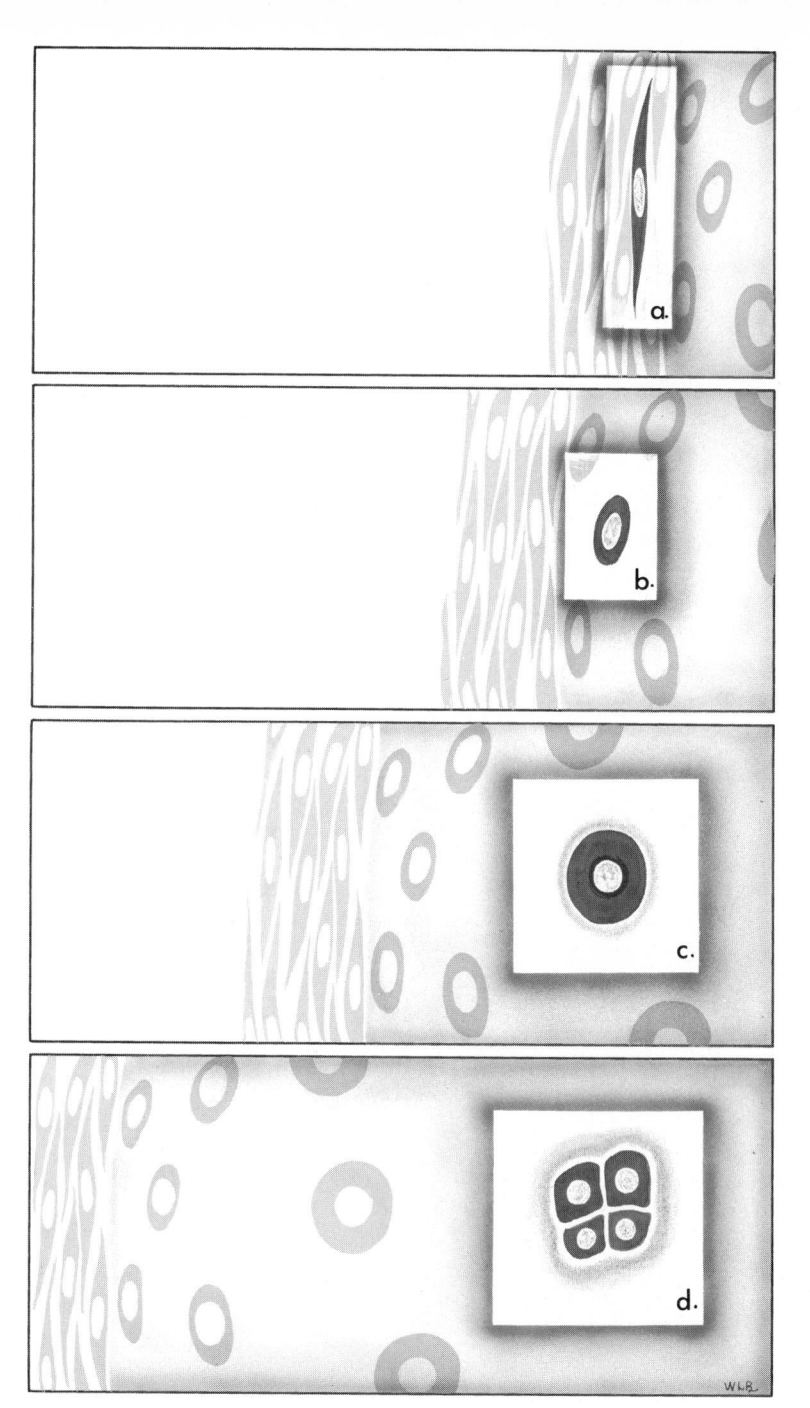

**FIGURE 11–6**

The sequence of changes involved in cartilage growth is schematized in this illustration. An undifferentiated connective tissue cell in stage a becomes activated into a chondroblast. When its own matrix secretion completely encloses it, this cell is then a chondrocyte (b). This is **appositional** growth, a process that requires a connective tissue membrane. As new cartilage continues to be formed, the chondrocyte becomes buried more and more deeply within the enlarging cartilage mass. The cell increases in size and becomes rounded. Intercellular matrix is produced by all the cells, and this pushes them apart (c). The chondrocytes, soon after inclusion, undergo mitotic cell division. These are **interstitial** growth changes. When growth slows and finally ceases, matrix production is curtailed, and the cells are not thereby further separated. They remain clustered in isogenous groups (d). In many **growth cartilages,** these isogenous cell groups have a special linear arrangement that coincides with the longitudinal direction of growth of the bone. The mandibular condyle is an exception. (From Enlow, D. H.: The Human Face. New York, Harper & Row, 1968, p. 6.)

positional mode of growth, it can thereby still grow in those pressure areas where an enclosing connective tissue membrane is absent. This includes articular surfaces, synchondroses, and epiphyseal plates.

6. Because the ordinary cartilage matrix is not calcified, cell divisions can take place, unlike in bone, thus providing interstitial growth.

One can readily see how each and all of the above features interrelate and are directly interdependent. To perform the functions of a **growth cartilage,** no one of the features could work without all the others. Cartilage exists as a separate, special tissue type because of these special features, and its functions could not be carried out by **any** of the other soft or hard tissues.

## BONE

Bone provides the specialized feature of **hardness.** Because of this, it has several unique developmental characteristics. A bone, of course, cannot enlarge by interstitial growth because its cells are locked into a nonexpandable matrix. It is thus dependent upon a covering vascular membrane to provide the osteogenic capacity for an appositional system of growth. This is also why bone is necessarily a **traction** (tension)-adapted kind of tissue. The covering soft tissue membrane is sensitive to direct compression, because any undue amount would occlude blood vessels and interfere with osteoblastic deposition of new bone. Actually, it is the **membrane** and not the hard part of the bone itself that is pressure-sensitive. In fact, most **concave** surfaces of bone **are** in a state of "compression," whereas convex surfaces are under "tension" with regard to weight-bearing forces or the action of muscles. Many concave surfaces involving compression are osteogenic, not osteoclastic. These factors may relate to the differential piezo responses of bone to different mechanical forces. When it is stated, however, that bone is a "tension-adapted tissue" and that it is "pressure-sensitive," the reference is to its soft tissue membrane rather than to the actual hard part.

The degree of vascular flow is affected by the amount and type of mechanical force acting on a soft tissue, and this is directly involved in initiating either chondrogenesis or osteogenesis. More extreme levels of anoxia caused by higher levels of pressure are known to stimulate formation of chondroblasts rather than osteoblasts from undifferentiated connective tissue cells. For all of these various reasons, two basic modes of bone growth exist, one adapted to a localized environment of tension (or at least moderate levels of pressure) and the other to more extreme compression.

In osteogenic areas where **membrane** tension* exists or where the level of any

*Text continued on page 336.*

---

*The whole subject of "pressure and tensions" is greatly oversimplified in most routine descriptions, as it is also in the present discussion. The actual nature of the forces that act on a bone is quite complex and can seldom be designated purely as either tension or pressure. A bundle of periosteal fibers, for example, can exert tension on a bone surface, but that same surface can also be under a compressive influence from other sources, such as a flexure pressure effect on a concave curvature. Pressure effects can also be exerted by intercellular fluids in a region otherwise classified as under tension. An osteogenic cell located between collagenous fibers under tension actually receives a direct pressure effect. Moreover, compressive effects on the hard part of the bone itself can have an osteoblastic trigger effect, while compressive effects on blood vessels can have an osteoclastic effect. The old, greatly oversimplified, and inaccurate concept that tension-deposition and pressure-resorption relationships exist is no longer acceptable, as described in Chapter 7. The overall control system is much more complex, and it is incompletely understood at present.

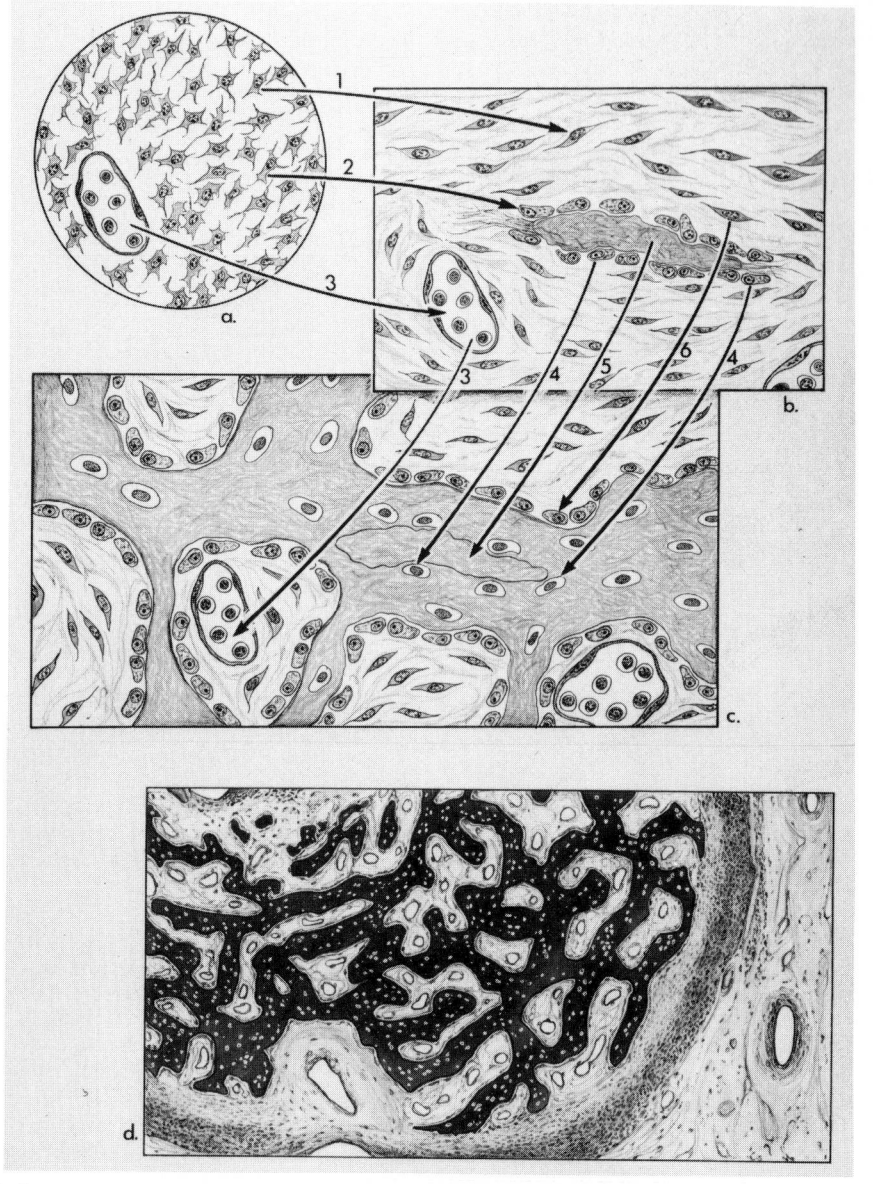

**FIGURE 11–7**

*Intramembranous bone formation. In a center of ossification (a), the cells and matrix of the undifferentiated connective tissue (late mesenchyme) undergo a series of changes that produce small spicules of bone. Some cells (1) remain relatively undifferentiated, but others (2) develop into osteoblasts that lay down the first fibrous bone matrix (osteoid), which subsequently becomes mineralized (stage b). Blood vessels are retained within the spaces among the formative bony trabeculae (3). As bone deposition by osteoblasts continues, some of these cells are enclosed by their own deposits and thus become osteocytes (4). Some undifferentiated cells develop into new osteoblasts (6), and other remaining osteoblasts undergo cell division to accommodate enlargement of the trabeculae. The outline of an early bone spicule (5) is shown in the enlarged trabeculae for reference. The spaces contain a scattering of fibers, undifferentiated connective tissue cells, and osteoblasts. At lower magnification, the characteristic fine cancellous nature of the developing cortex is seen. This bone tissue type is widely distributed in the prenatal as well as young postnatal skeleton. It is a particularly fast-growing variety of bone tissue. Note that the periosteum (also formed from undifferentiated cells in the ossification center) has become arranged into inner (cellular) and outer (fibrous) layers. (From Enlow, D. H.:* The Human Face. *New York, Harper & Row, 1968, p. 44.)*

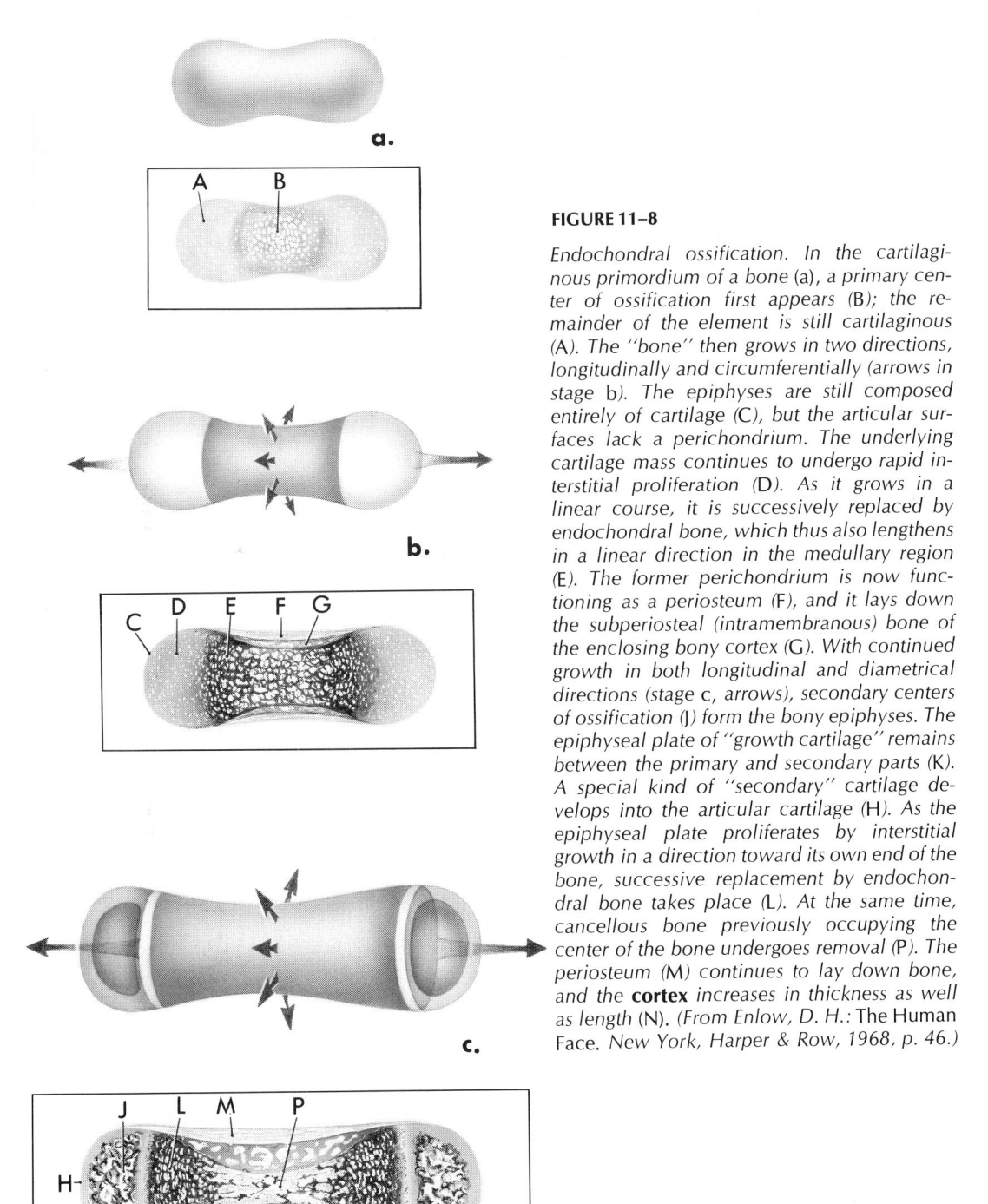

**FIGURE 11–8**

*Endochondral ossification. In the cartilaginous primordium of a bone (a), a primary center of ossification first appears (B); the remainder of the element is still cartilaginous (A). The "bone" then grows in two directions, longitudinally and circumferentially (arrows in stage b). The epiphyses are still composed entirely of cartilage (C), but the articular surfaces lack a perichondrium. The underlying cartilage mass continues to undergo rapid interstitial proliferation (D). As it grows in a linear course, it is successively replaced by endochondral bone, which thus also lengthens in a linear direction in the medullary region (E). The former perichondrium is now functioning as a periosteum (F), and it lays down the subperiosteal (intramembranous) bone of the enclosing bony cortex (G). With continued growth in both longitudinal and diametrical directions (stage c, arrows), secondary centers of ossification (J) form the bony epiphyses. The epiphyseal plate of "growth cartilage" remains between the primary and secondary parts (K). A special kind of "secondary" cartilage develops into the articular cartilage (H). As the epiphyseal plate proliferates by interstitial growth in a direction toward its own end of the bone, successive replacement by endochondral bone takes place (L). At the same time, cancellous bone previously occupying the center of the bone undergoes removal (P). The periosteum (M) continues to lay down bone, and the* **cortex** *increases in thickness as well as length (N). (From Enlow, D. H.: The Human Face. New York, Harper & Row, 1968, p. 46.)*

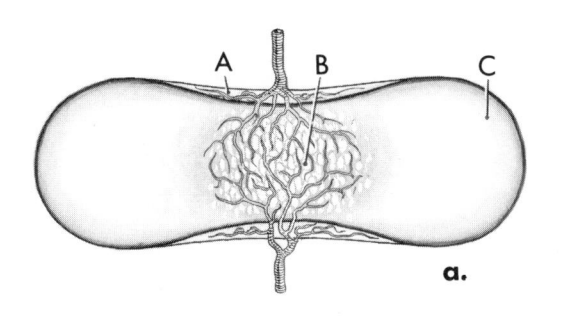

**FIGURE 11–9**

*Endochondral ossification. The cartilaginous prototype of the bone (C) has a primary center of ossification (B) that involves chondrocyte hypertrophy, matrix calcification, and invasion by vascular buds from the periosteum (A) bringing in undifferentiated connective tissue cells. In stage b, the calcified matrix (E) has become permeated by anastomosing erosion tunnels, and each space (D) contains vessels and undifferentiated cells (H). Osteoblasts develop from these cells, and in stage c, a thin shell of bone (J) has been laid down on the remnants of the calcified cartilage matrix (K). Endochondral bone, as a type, can easily be recognized in sections by the presence of these identifying spicules. Some of the osteoblasts (M) have been incorporated in the fine cancellous trabeculae as osteocytes (L). (From Enlow, D. H.: The Human Face. New York, Harper & Row, 1968, p. 48.)*

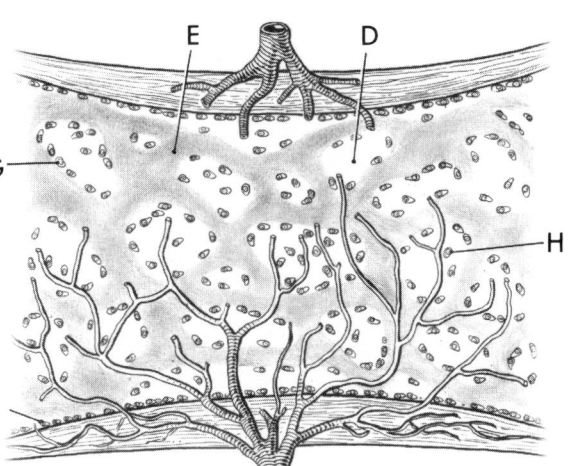

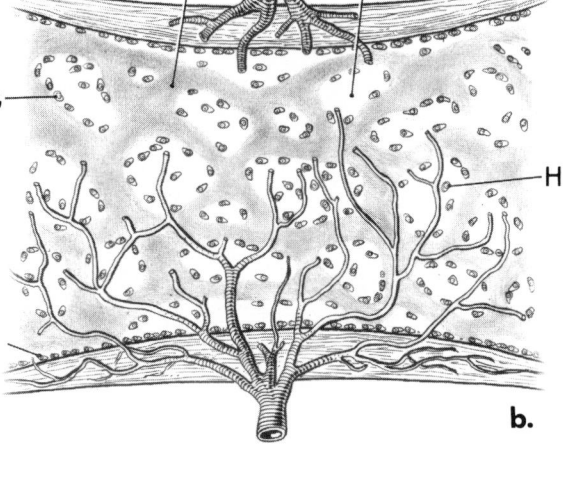

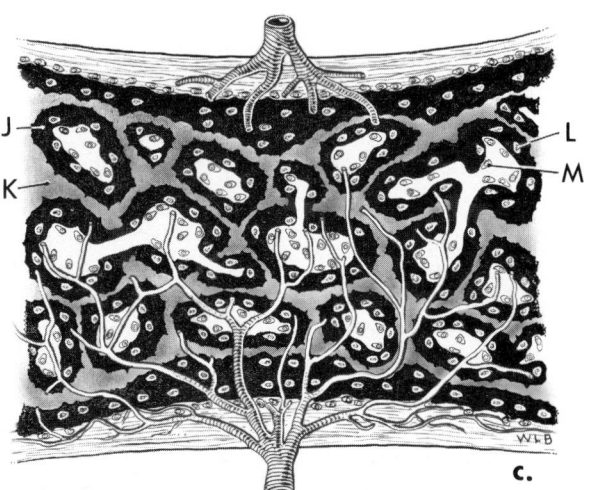

pressure present is minimal, the intramembranous mode of bone growth and development takes place (Fig. 11–7). Thus, bone formed by the periosteum and by sutures occurs in locations where severe membrane compression is not involved. In specific places where such compressive forces do occur, however, the endochondral mode of osteogenesis occurs (Figs. 11–8 and 11–9). Chondroblasts rather than osteoblasts develop from stem cells, since, presumably, the level of hypoxia is increased because of decreased vascular capacity.

Soft tissues, in general, grow (1) by increasing the **number** of cells (as in epithelia); (2) by increasing the **size** of cells (as in skeletal muscle); or (3) by increasing the amount of **matrix** between cells (as in loose connective tissue). Many tissue types combine two or all three of these different modes of growth (as in cartilage). All are interstitial systems of growth because they involve expansive changes of tissue components already present. Bone, because it is hard, must necessarily grow by a process of adding new cells and new matrix onto the existing **surfaces** of previously formed generations of bone tissue. It cannot, of course, expand interstitially by division and proliferation of osteocytes because the cells have no place to divide to; they and their genes are locked into their calcified, nonexpanding matrix. Bone **must**, therefore, grow in relationship to a covering or lining membrane, or some other soft tissue such as cartilage or tendon. It is the bone surface, either periosteal or endosteal, that is the site of growth activity. This type of growth is termed "appositional," in contrast to interstitial expansion. All bone that is fully calcified grows in this manner regardless of its mode of osteogenesis (endochondral or intramembranous).

Where **compression** is involved, as mentioned above, the intramembranous growth mechanism (which is dependent upon vascular membranes, as the name indicates) does not have the capacity to function. Thus, the articular ends of a bone and the epiphyseal plates are composed of cartilage, which **can** grow and function in a pressure environment. The cartilage grows **toward** the site of compression. An epiphyseal plate, synchondrosis, and other "growth cartilages" provide for the linear enlargement of bones that have pressure contacts at their ends. The cartilage grows interstitially on one side, and as it does so, the older part of the cartilage on the other side is removed and replaced by bone. The cartilage functions essentially as a kind of advance ram that protects the sensitive endosteal bone membrane beneath it, and, importantly, **also provides for the growth elongation of the bone at the same time.** The **other** areas of the bone grow intramembranously.

**FIGURE 11–10**

During active growth, a bone surface is constantly changing because of new additions; any given point within the compact substance of bone **used** to be an actual exposed surface, either periosteal or endosteal. The point in stage *a* is on the periosteal surface, but as new deposits are laid down by the osteogenic layer of the periosteum, it is "relocated" in stage *a'*. The point itself has not moved, but its relative position has become shifted because of the surface additions which cover it. If a **metallic implant marker** or **vital dye** (such as alizarin or the procion dyes) is used on a living bone, growth changes that occur after the marker is implanted or the dye is administered can be determined. Such markers and the lines formed by the vital dyes become covered over wherever subsequent surface growth occurs, just as point *a* is covered. (Note: vital dyes stain only that bone actively being laid down during the period in which the dye is in the bloodstream. Thus, one injection forms a thin, colored line on the surface of the bone. Subsequently formed bone deposits are not colored.)

a

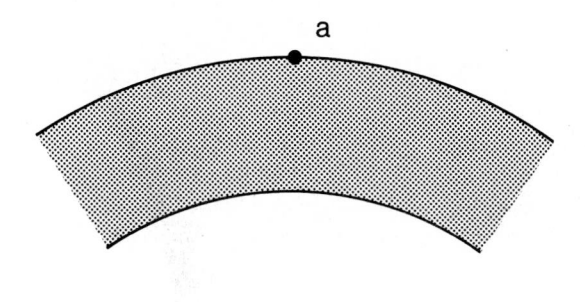

a'

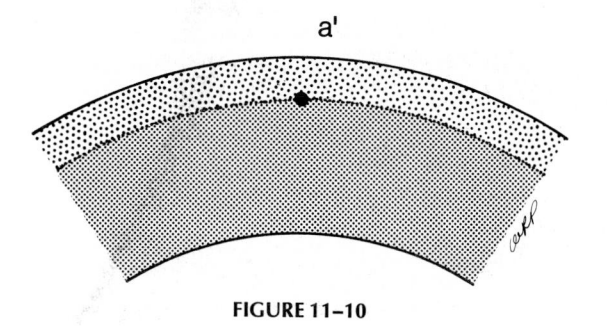

**FIGURE 11–10**

**FIGURE 11–11**

Bone **deposition** is only part of the overall process of bone enlargement; it is one phase of a multiphase growth system. **Resorption** is another part, and this is just as important and necessary as deposition. Ordinary resorption associated with growth is not "pathologic," although beginning students sometimes mistakenly view it as evil because resorption is a destructive process or because it can also be involved in some diseases. Resorption **must** accompany deposition, and deposition on one side of a cortical plate (or cancellous trabecula) with resorption from the other brings about the growth movement of the part. Thus, deposition on the periosteal side with resorption on the endosteal surface of a cortex moves the whole cortex outward and at the same time proportionately increases its thickness. As pointed out in Chapter 2, however, **remodeling** also takes place throughout the entire bone because the bone does not simply expand by generalized external deposition and internal resorption.

**FIGURE 11–12**

Remodeling involves various combinations of deposition (*1, 3*) and resorption (*2, 4*) on different periosteal and endosteal surfaces in order to **move** a part of a growing bone into a new location (see Chapter 2). The process of displacement also occurs as a basic part of the growth mechanism.

Except at contact surfaces involving direct, high-level pressure, the bone grows by the surface depository activity of the osteoblasts present as the innermost cellular layer of the thick periosteum or the very thin endosteum (the latter is only about one cell thick and lacks a thick, fibrous layer because muscles, tendons, and other such force-adapted tissues are not attached to it). Vessels enclosed within endosteal circumferential layers of bone characteristically enter at a right angle because the endosteum is not under tension and its vessels therefore are not drawn out toward the long axis of the bone. Also, periosteal "slippage" on the bone is involved in the periosteum in relation to the forces acting on it as it grows, and this causes the periosteal vessels to become enclosed at much more acute angles. The old concept of perpendicular "Volkmann's" canals entering the bone from the periosteum needs to be forgotten.

**FIGURE 11–13**

As new bone is progressively laid down, the covering periosteal membrane **moves** outward. The membrane moves inward if the periosteal surface is resorptive. Thus, the periosteum *a* moves in an outward direction to *a'* as its osteoblasts lay down the new bone in "space" *x*. The endosteum *b* moves to *b'* as its osteoclasts remove bone. (Reverse the sequence if the bone is growing inward by endosteal deposition and periosteal resorption.) However, these membranes do not simply "back off" as they bring about the cortical movement. They are not merely pushed or pulled into their new positions. Rather, the membranes each **grow** from the one location to the other. It is the growth movement of the membranes that leads the growth movement of the bony cortex located between the membranes. The membrane has its own internal, interstitial growth process. Just as the bone **remodels** during growth, the periosteum also undergoes its own internal remodeling process. Remember that the membrane itself paces the bone changes and that the "fields" of growth activity (described on page 10) reside in this membrane and the other soft tissues rather than in the bone.

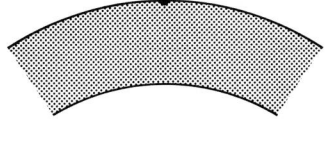

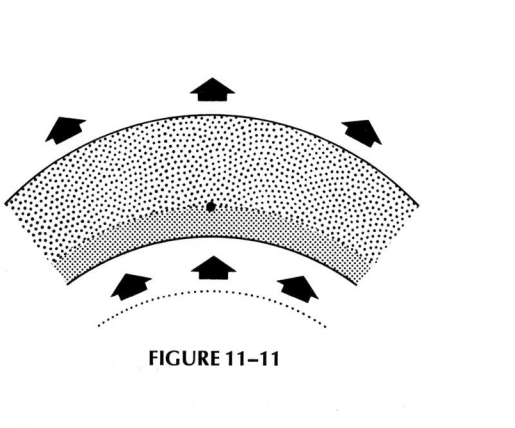

**FIGURE 11–11**

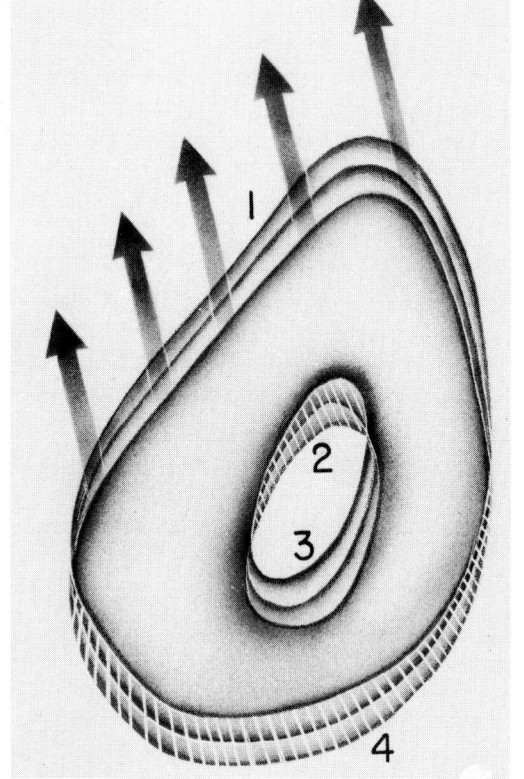

**FIGURE 11–12**

*(From Enlow, Donald H.: Principles of Bone Remodeling, 1963. Courtesy of Charles C Thomas, Publisher, Springfield, Illinois.)*

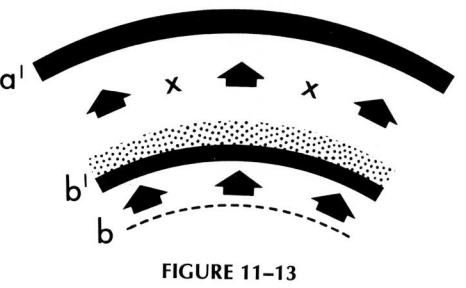

**FIGURE 11–13**

**FIGURES 11-14 AND 11-15**

As collagenous fibers and ground substance are laid down by the osteoblasts (g), this layer of **osteoid** almost simultaneously undergoes calcification to become a new layer of bone tissue (x). Some of the osteoblasts are enclosed to become new osteocytes, and some of the periosteal blood vessels (f) contiguous with the bone surface are also incorporated. These anastomosing vessels then lie within a network of **vascular canals** as bone is formed around them. Note that the fibers of attachment (Sharpey's fibers) become more deeply embedded as new bone is formed around the collagenous fibers in the innermost layer of the periosteum (d). The periosteal fibers now **grow** outward. Fiber segment d lengthens on the outside while it is being enclosed by new bone on the inside (e'). It does this by a **conversion** of segment c into a new addition onto d, which thereby elongates by this process. Segment c is a special **precollagenous** fibril. It is a very thin fibril that requires special staining methods (see Kraw and Enlow, 1967). Many such fibrils (that is, "linkage" fibrils) form a distinctive zone within the intermediate part of the periosteal membrane. Under the control of the rich population of fibroblasts, these slender fibrils become remodeled into the thick collagenous fibers that form the lengthening segments of d. This is done by binding many of the fibrils together with ground substance (the various mucopolysaccharides such as chondroitin sulfate).

Segment c now lengthens in a direction away from the bone surface. It is not presently known whether this is done by a remodeling conversion from segment b through enzymatic removal of the binding ground substance to release its numerous slender precollagenous fibrils, or whether new precollagenous lengths are added directly to c by the fibroblasts in this zone. As these changes occur, however, new segments b are being formed by fibroblastic activity as they join the expanding outer "fibrous" layer of the periosteum (a). The entire periosteum thus **drifts** outward as the bone surface correspondingly drifts in the same direction. If the periosteal surface is **resorptive** rather than depository, the sequence of operations is the same but the direction is reversed. That is, the periosteum and its fibers grow **toward**, rather than away from, the bone surface which is moving inward rather than outward.

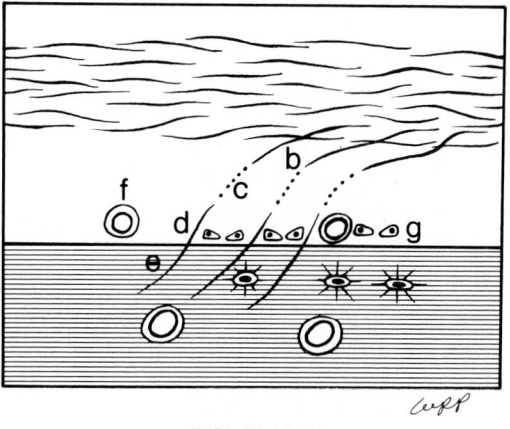

**FIGURE 11–14**

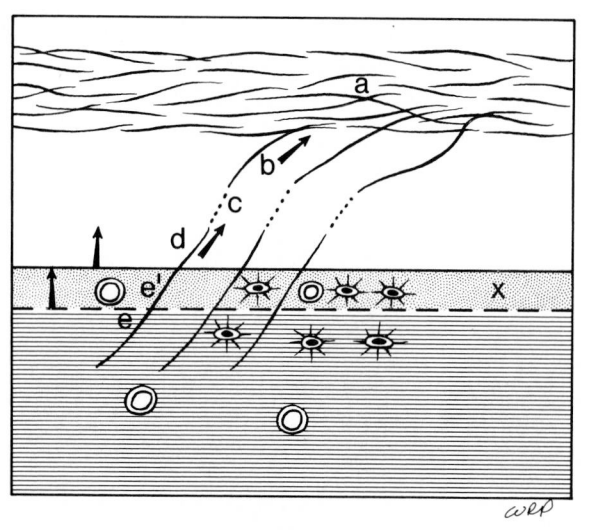

**FIGURE 11–15**

## FIGURES 11–16 TO 11–21

How does a muscle or tendon maintain continuous attachment onto a bone surface that is **resorptive**? Moreover, how does a muscle **migrate** over a bone surface (whether resorptive or depository) as the bone lengthens? For example, the muscle shown in Figures 11–16 to 11–18 **moves** its insertion. It must also sustain constant attachment on a surface, part of which is undergoing progressive resorption, thus seemingly destroying its fibrous anchorage. The other muscle shown in Figure 11–18 attaches entirely onto a resorptive surface. Bone resorption is customarily regarded as a process that results in total destruction of the bone tissue, including its fibers of attachment. In many non-stress-bearing locations, this is true. In some (not all) areas involving muscle and tendon attachments, however, the process of fiber destruction is **not** complete. Importantly, some of the fibers in the ordinary bone matrix are not removed by the resorptive process (Figs. 11–19 and 11–20). These fibers become uncovered as the remainder of the bone matrix around them is resorbed, and they are then freed to function as fibers of the **periosteum** while retaining continuity and attachment with the fibers in the bone of which they were once a part. Thus, surface layer m is resorbed (Fig. 11–21). This releases fibers b, which are continuous with the periosteal fibers a on one side and the bone fibers c on the other side. Fibers b thereby become transferred from bone fibers into periosteal fibers.

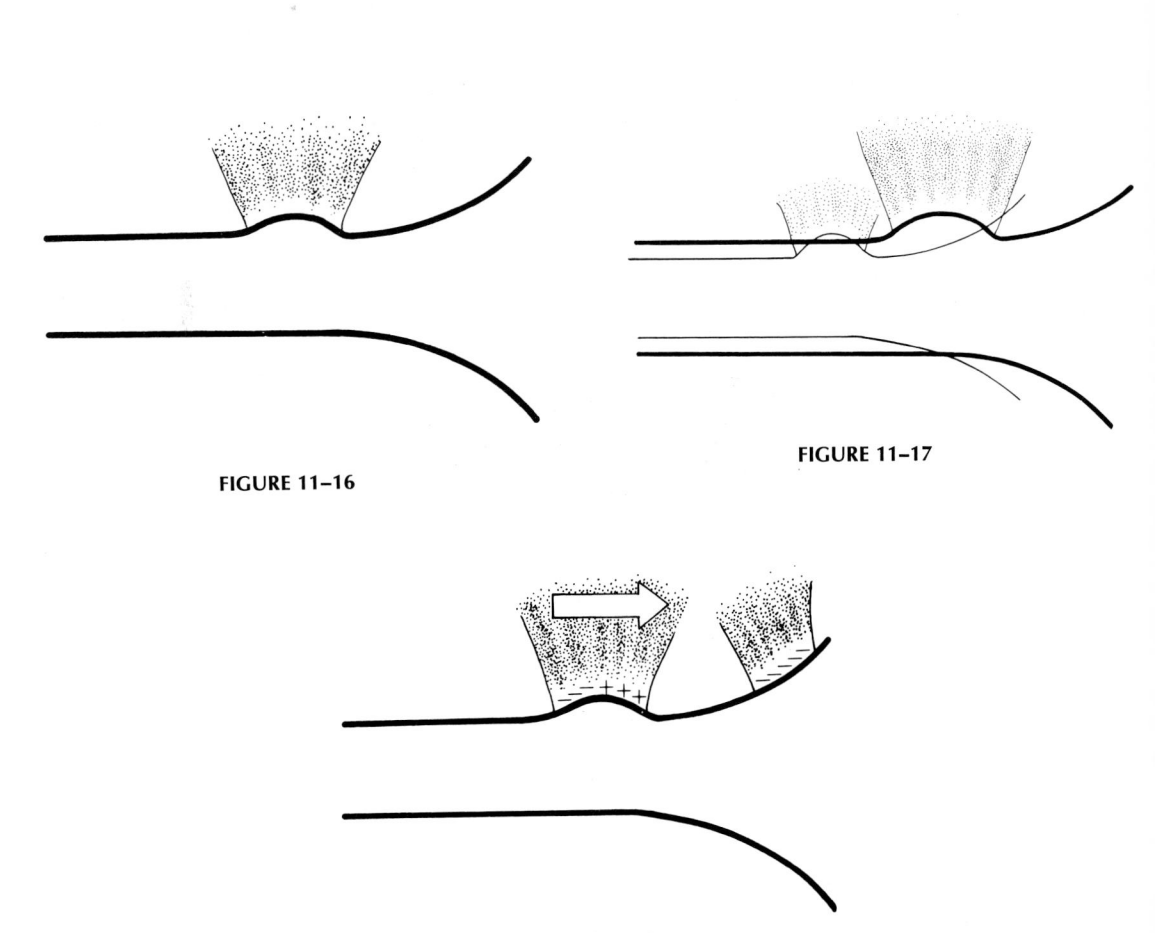

**FIGURE 11–16**

**FIGURE 11–17**

**FIGURE 11–18**

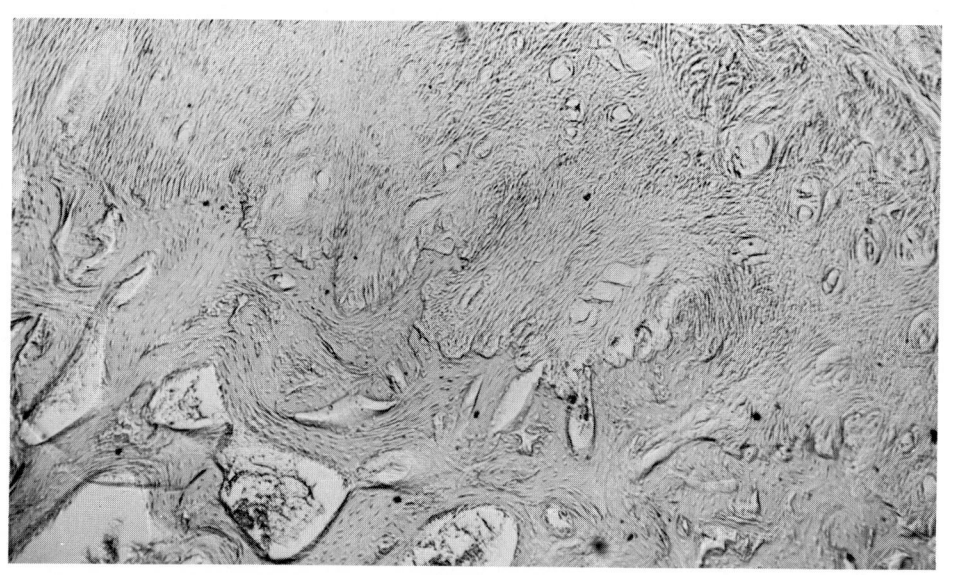

**FIGURE 11–19**

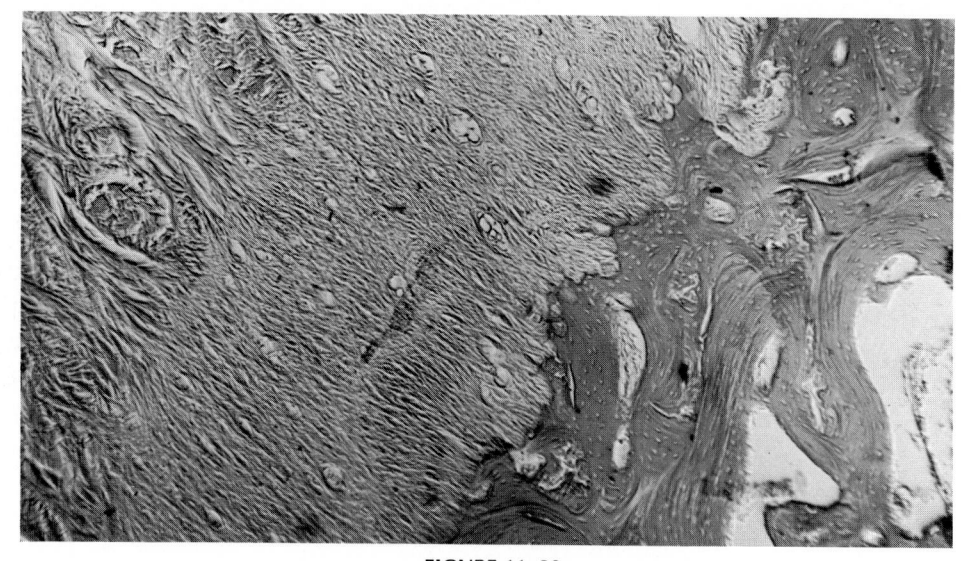

**FIGURE 11–20**

**FIGURE 11–21**

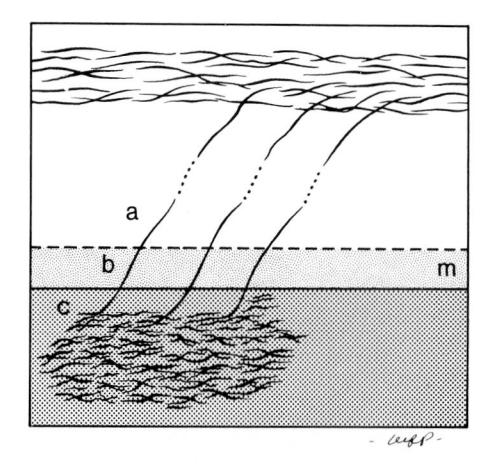

**FIGURES 11–22 TO 11–25**

In other stress-bearing locations, another separate mechanism is seen that provides for continuous fibrous attachments on resorptive and remodeling bone surfaces. This involves a reconstruction of the bone **deep** to the resorptive surface in order to provide fibrous anchorage. "Undermining" resorption takes place in which resorption canals are formed well below the surface of the inward-advancing resorptive periosteal front. New bone is then laid down in these spaces, and it is **this** attachment that provides additional fibrous anchorage while the outside bone surface itself is undergoing resorptive removal. Thus, as periosteal surface *a* is resorbed, a large number of resorption spaces are formed at *b* (in thin sections, many appear as cut-off canals). These spaces anastomose with each other. The fibers in the new bone subsequently deposited in them are continuous with the fibers of the periosteum, and anchorage is thereby maintained. The structural result is a generation of haversian systems (secondary osteons) **deep** to the surface. The fibrous matrix of each osteon and its connection by labile linkage fibrils to the inward-moving periosteum are protected from resorption until the resorptive front reaches them. However, new waves of haversian systems are constantly being formed in advance of the resorptive front, so that new deeper osteons replace, in turn, those that become exposed as they reach the resorptive periosteal surface. It is believed that the combination of this process in some areas with the process of bone matrix fiber release in other areas maintains continuous muscle attachment during the period of active skeletal remodeling. Moreover, a muscle is believed to move and migrate along a bone surface by this same process of haversian formation, as well as by lateral reconnections of the labile linkage fibrils (x) in the intermediate part of the periosteum (or equivalent areas for direct tendon insertions). Thus, the precollagenous linkage fibrils connecting with fiber *a* in the outer part of the periosteum (Figs. 11–24 and 11–25) become recombined with the precollagenous fibrils of fiber *b'* in the inner part of the periosteum, and so one. This progressively moves the entire muscle across the bone surface to keep pace with the elongation of the entire bone. Separations of fiber bundles by enzymatic release of ground substance binding, together with fibril regroupings by new ground substance formation, are believed to carry this out.

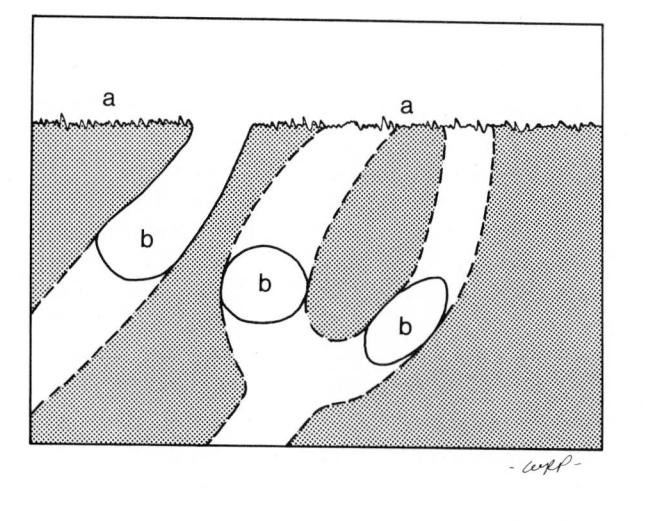

**FIGURE 11–22**

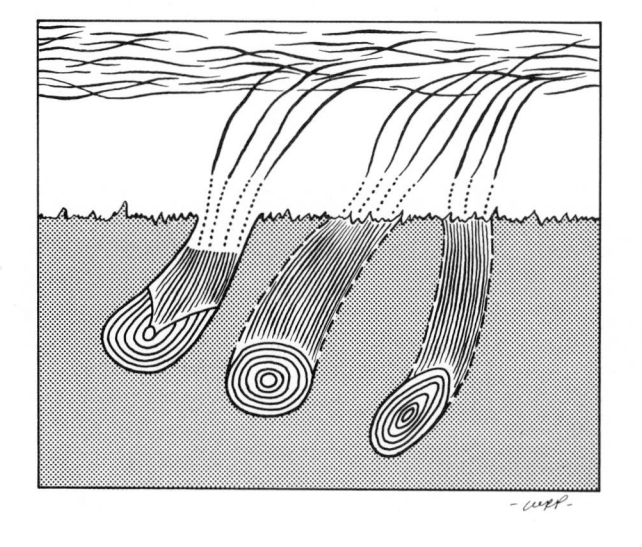

**FIGURE 11–23**

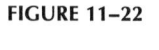

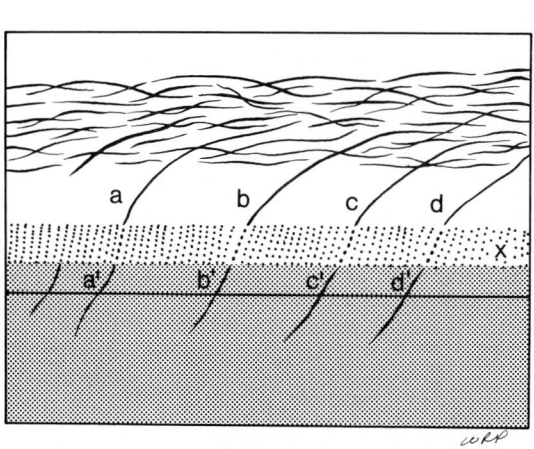

**FIGURE 11–24**

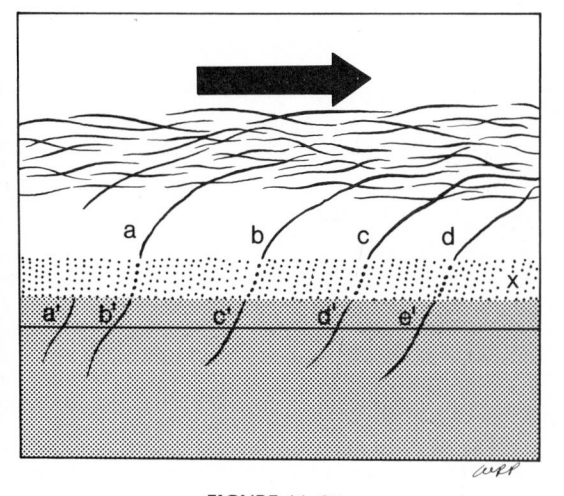

**FIGURE 11–25**

FIGURES 11–26, 11–27, AND 11–28

**Sutures** have an osteogenic process comparable to periosteal bone growth. The suture is an inward reflection of the periosteal membrane, and the various fibrous, linkage, and osteoblastic zones are directly continuous from one to the other. As a new layer of bone is added (x), inner collagenous fibers d become embedded to form new attachment fibers (e') in the bone matrix. Fibers d, however, lengthen by conversion from the labile linkage fibers c (just as previously described for the periosteum), and fibers b become converted, in turn, into lengthening fibers c (or, fibroblastic activity may bring about direct lengthening of c, as in the periosteum). As new bone is added onto the sutural contact surfaces, the bones are simultaneously displaced away from one another (see Chapter 3 for a discussion of the physical forces that cause this displacement). Many sutures have three basic layers (on each side), as indicated here. Some sutural types, however, have another layer of loosely arranged fibers located within the center of the dense-fibered capsular layer dividing the two sides. The basic plan of growth and remodeling, however, is the same. When the process of growth ceases, the suture becomes essentially a mature ligament, and the precollagenous linkage fibrils are no longer present.

Importantly, the source of the propulsive force that produces the "downward and forward" displacement movement of the nasomaxillary complex at its various sutures has long been a subject of controversy. It has recently been suggested that the abundant population of actively contractile fibroblasts ("myofibroblasts," cells m in Figure 11–27B) within the linkage zone of the sutural membrane provide, at least in part, a contractile force that exerts tension on the fibrous framework. This in turn pulls one bone along its sutural interface with another bone. The bone thus "slides" along the suture as new bone tissue, at the same time, is laid down on the sutural edges. The midface, thus, is **pulled** forward and downward along its zygomaxillary, among others, sutural surfaces. Special collagen-degrading and collagen-producing fibroblasts (x and y) provide the fiber remodeling and ground substance relinkage changes also involved. Fibers at level 1, which were formerly linked with fibers at level 1', have become relinked with fibers at 2', and so on. (See Azuma, M., D. H. Enlow, R. G. Frederickson, and L. G. Gaston: A myofibroblastic basis for the physical forces that produce tooth drift and eruption, skeletal displacement at sutures, and periosteal migration. In McNamara, J. A., Jr. (Ed.): Determinants of Mandibular Form and Growth. Center for Human Growth and Development, Craniofacial Growth Monograph Series, The University of Michigan, Ann Arbor, 1975.)

The **periodontal membrane*** is comparable to both the sutural and periosteal membranes. It is a membrane that, phylogenetically, is the adaptive answer to a basic functional problem. If the teeth were attached to the jaw similar to a denture riding on the surface of the bone, **pressure** would result on the periosteum when chewing. The bone's membrane is known to be quite pressure-sensitive, however,

---

*Also commonly called the **periodontal ligament.** It is indeed a mature ligament in terms of its histologic structure in the more stable, adult form. However, the term "membrane" is much more appropriate for the childhood growth period. The periodontium is a connective tissue membrane that is quite active and dynamic, not one which merely physically supports a tooth (that is, a ligament). It (1) contributes to the growth and development of the tooth; (2) is involved directly in the eruption of the tooth; (3) is involved directly in the drifting, tipping, and rotation movements of the tooth; (4) provides for the formation of the bone tissue lining the alveolar socket; and (5) is involved directly in the extensive remodeling of the bone associated with the movements of the teeth. For these reasons, the term periodontal "membrane" is more closely associated with the truly dynamic functions of this connective tissue layer. "Ligament," on the other hand, connotes a more stable, inactive, nonchanging type of tissue that has a single function—fibrous attachment.

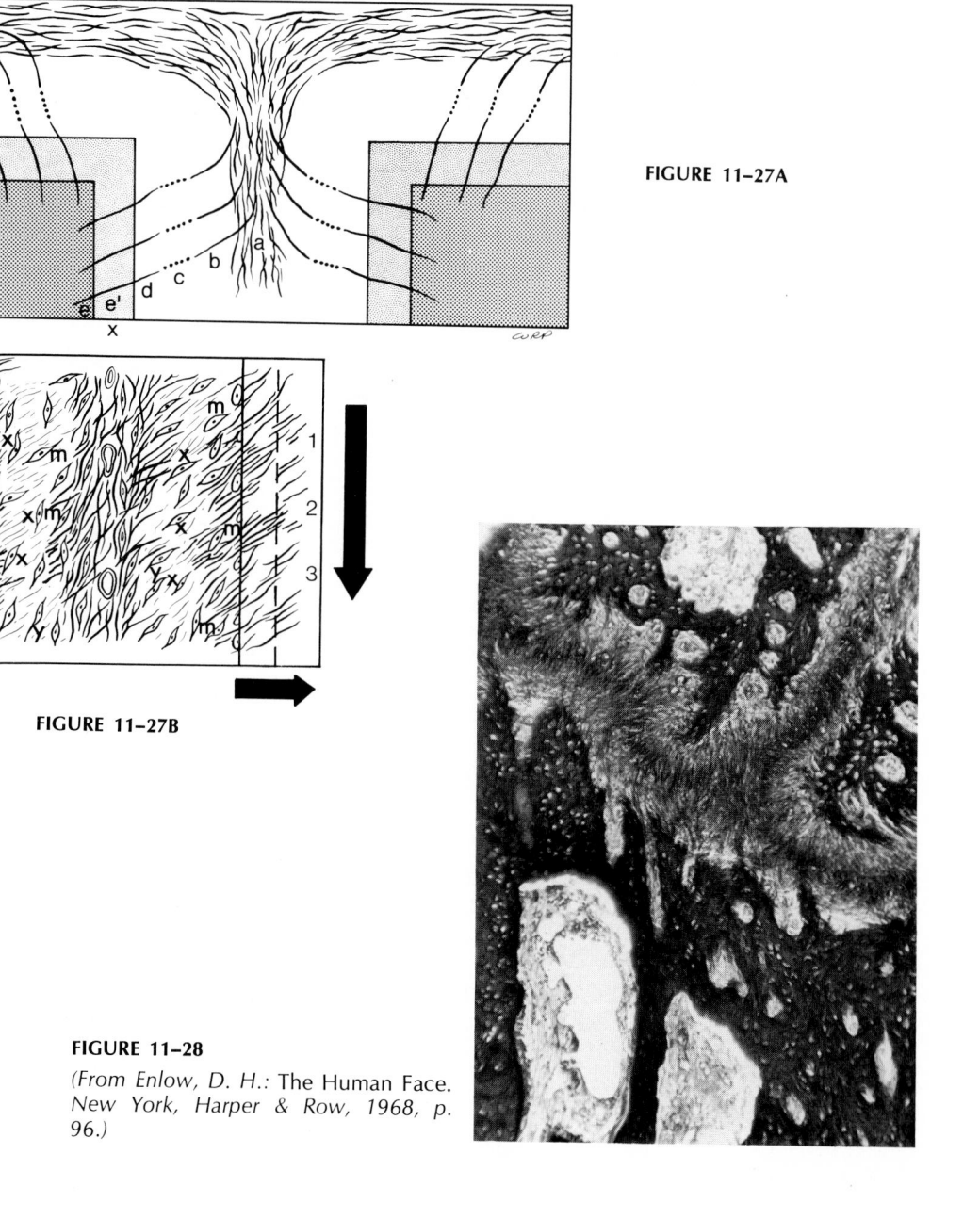

FIGURE 11–26

FIGURE 11–27A

FIGURE 11–27B

**FIGURE 11–28**

*(From Enlow, D. H.: The Human Face.
New York, Harper & Row, 1968, p.
96.)*

and resorption would result because of masticatory compression by the teeth. Would cartilage, a pressure-resistant tissue, function satisfactorily as a buffer tissue between the tooth root and the bony surface? No, because cartilage is severely limited in its capacity for remodeling and could not accommodate the dynamic changes required for tooth development, eruption, and drift.

### FIGURE 11–29

The phylogenetic problem of compression on the bone surface beneath a tooth has been solved in a simple but effective way. Pressure is converted directly into **tension** (which the membranes can handle) merely by suspending each tooth in a connective tissue **sling** within a socket.* By this means, the inward direction of compression by a tooth being pushed into its socket is translated, not as pressure, but as direct tension on the alveolar bone. Thus, the periodontal membrane is not exposed to the killing effects of compression as the tooth is depressed into the socket or as it is tipped or rotated in one direction or another by masticatory forces. This relatively simple plan accomplishes several needed functions. It provides effective support for the tooth, gives resilient yet nonbrittle stability, provides a system for eruption, enables each individual tooth to acquire a functional occlusal position, provides for the growth and maintenance of the alveolar bone, and provides for the vertical and horizontal drifting of the tooth and the corresponding remodeling of the alveolar bone.

The teeth drift for two basic, functional reasons. One, as described in all basic oral histology texts, is to close up the dental arch during growth and keep it closed as the contact edges along the sides of the teeth progressively wear. This braces the arch to better withstand masticatory forces. **The second reason, much less known but of great importance, is to anatomically locate the teeth as the whole mandible and maxilla grow and remodel.** Each tooth (and even the unerupted tooth buds) must drift vertically, laterally, and either mesially or distally in order to retain proper anatomic position. The "molar" region at an early age level, for example, becomes the "premolar" region of the jaw at a later age. The maxillary teeth, for a second example, must drift inferiorly for a considerable distance as the whole bony maxillary arch grows downward to provide for the vertical enlargement of the nasal chambers. Thus, the function of **drift** is far more significant than merely "closing up" the dentition. It is one of the basic processes involved in facial growth.

### FIGURE 11–30

The customary diagram used to illustrate "mesial drift" shows deposition and resorption on the "tension" and "pressure" sides of the alveolar socket, respectively. A *PA* section through the jaw showing several tooth roots gives the familiar histologic picture shown here. However, only the **mesial** direction of drift movement is pointed out in the standard textbooks; the important vertical drift movements and the rotations and tipping that **also** take place are not always explained. These other movements are carried out by the **same** alveolar bone deposits and resorption usually associated only with mesial drift. Drift is a three-dimensional growth process, and the oversimplified, two-dimensional, diagrammatic picture illustrated here does not adequately represent it (see p. 124).

---

*A violation of this anatomic relationship is the basis for many of the problems encountered by the prosthodontist. Dentures are **pressure**-causing appliances fitted onto bone without a tension-converting sling of periodontal fibers. Uncontrolled resorption is a common consequence.

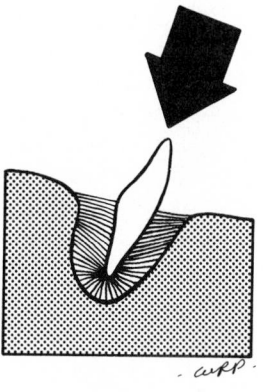

FIGURE 11–29

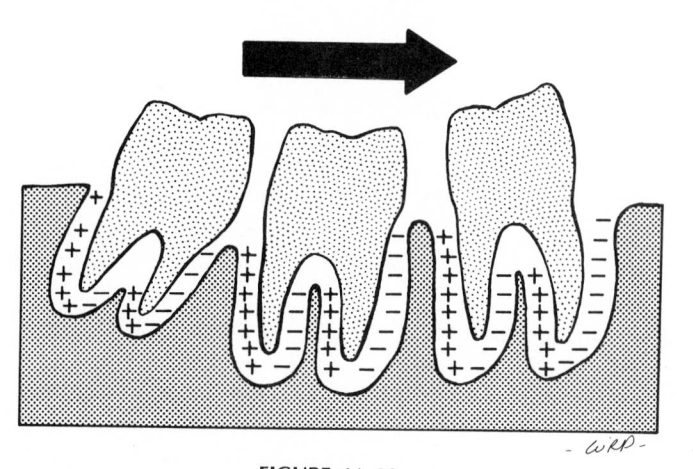

FIGURE 11–30

The "pressure" and "tension" concept of the mesial and distal sides of the alveolar socket is also an oversimplification. The collagenous fibers of the periodontium on the pressure side are often actually under tension (Fig. 11–36). While heavy tooth-to-periodontal membrane-to-alveolar bone surface compression can indeed be involved in extreme tooth movements (as by an orthodontist using heavy forces), such severe levels of force are not usually involved in ordinary physiologic circumstances. On the "tension" side, further, the cells of the periodontium are actually under compression between the taut collagenous fibers.

It has been a major point of controversy for many years whether the pressure that is presumed to trigger alveolar bone resorption acts first on the membrane or directly on the bone, which in turn causes the membrane to respond (see below). It is believed that **pressure** is indeed the actual cause of resorption on the "pressure" side of the alveolar socket. The nature of the cause and effect relationships, however, is argued. One long-standing concept is that **distortion** of the alveolar bony plate by the tooth's root is required in order to trigger alveolar remodeling. The piezo effect is now held by many investigators to be the response to this stress trigger, and it is believed that this bioelectric stimulus serves as a "first messenger" which fires the receptor sites on osteoblastic and/or osteoclastic cell membranes within the periodontium. The source of the pressure that causes the active distortion (bending) of the alveolar plate is still not clear. Some investigators suggest that the vascular system functions as the biomechanical intermediary. It presumably acts as a "hydraulic system" which can transmit variable amounts of pressure to the alveolar bone surface by the degree of distention of the vessels. If the extent of pressure exerted by a **tooth** on the periodontal membrane, caused by heavy forces acting on the tooth, causes an actual compression of the membrane and a closing off of the blood vessels, however, the growth capacity of the membrane is destroyed and remodeling changes on the lining alveolar surface are precluded. This is presumed to trigger **undermining resorption.** In this process, the resorptive changes proceed from the cancellous spaces **deep** to the alveolar lining, since the alveolar surface itself is closed off due to vascular occlusion. While this concept has been well established, some workers now feel that undermining resorption is not necessarily associated only with such heavy forces and resultant vascular occlusion. It is believed that this kind of resorptive process is simply a part of the ordinary growth and remodeling sequence that occurs as the cancellous trabeculae "drift" together with the drifting movements of the alveolar bone and the teeth. (Undermining resorption is also involved in periodontal fibrous attachments, as described later.)

In dentistry, the concepts of "heavy force" and "light force" are regularly used. Heavy forces are presumed to have two results. First, they can cause undermining resorption because the level of pressure causes vascular occlusion, and movements of teeth can thus only proceed, presumably, by resorptive remodeling originating from the cancellous area deep to the alveolar surface. Second, heavy forces (sometimes called "orthopedic" forces) not only cause deformations of the alveolar bone sufficient to activate major changes in tooth positions, but also are presumed to extend **beyond** alveolar bone to the basal bone and on to other separate bones as well. In the "light force" concept, only the alveolar bone is believed to be responsive. Whether or not "distortions" of the alveolar plate are needed is still controversial. However, whatever levels of forces are involved, a number of workers, such as Ackerman, have shown that the **duration** of applied stress is also of basic importance.

The periodontal membrane is the equivalent of both the periosteum and sutures. It has sometimes been regarded as an essentially different membrane, but its general structure is similar to that of others, and its mode of growth is the same. The notable difference, of course, is that one side attaches to a tooth rather than to a muscle or another bone. The periodontal membrane is a reflection of the periosteum into the alveolar socket, and these two membranes are directly continuous.

In its "stable," nonremodeling form, the periodontium is essentially a mature ligament composed of dense bundles of thick collagenous fibers with correspondingly few fibroblasts and little ground substance. During the active period of facial growth, dental development, and the establishment of occlusion, however, this membrane has a much more dynamic function, and its histologic structure is adapted to the complex role it plays. During the growth period, the periodontal membrane is much more highly cellular, and more than just ligament fibers are present. Like the actively growing periosteum and sutures, the periodontal membrane has three basic layers. The middle layer, called the "intermediate plexus," is composed of the same slender, precollagenous **linkage fibrils** that are present in the intermediate layers of the periosteum and sutures. Linkage fibrils provide connections and sequential reconnections between the innermost and outermost dense fibrous layers. Their key function is for **adjustments** involved in tooth drift, eruption, rotations, and alveolar bone remodeling. This layer may be poorly differentiated or absent in nonremodeling periods and in regional locations (and species) where tooth movements are relatively slow. Or, the distribution of the linkage fibrils may be more diffuse rather than forming a single, recognizable zone. During active tooth movements, nonetheless, they are necessarily **always** present.

It has recently been proposed that the actual source of the propulsive mechanical force that brings about eruption, vertical and horizontal drift, and other tooth movements is provided specifically by an abundant population of actively contractile fibroblasts ("myofibroblasts") on the **resorptive** sides of the socket. The contraction of these special cells (m in Figure 11–35B) **pulls** the collagenous framework within the periodontal membrane, and thereby the tooth, in the direction of the resorptive bone front. Simultaneously, special collagen-degrading and collagen-producing cells (x and y) within the linkage zone provide the fiber remodeling and relinkages described below. This occurs in conjunction with ground substance degradation and synthesis, and the tooth is thus propelled in horizontal and vertical drift movements (arrows). The same process also provides for eruption. The fibers at level 1, formerly linked with 1', thus become relinked with fiber level 2', and so on. It is suggested, importantly, that these various cells are the specific targets of the clinical forces utilized by the orthodontist to move teeth. (See Azuma, M., D. H. Enlow, R. G. Frederickson, and L. G. Gaston: A myofibroblastic basis for the physical forces that produce tooth drift and eruption, skeletal displacement at sutures, and periosteal migration. In McNamara, J. A., Jr. (Ed.): *Determinants of Mandibular Form and Growth.* Center for Human Growth and Development, Craniofacial Growth Monograph Series, The University of Michigan, Ann Arbor, 1975.)

### FIGURES 11–31, 11–32, AND 11–33

On either side of the zone of linkage fibrils (*b*) are a layer of coarse collagenous fibers that attach to the alveolar bone (*a*) and a layer of coarse fibers attaching to the cementum of the tooth (*c*).

The activity on the tension side is schematized here ("tension" because the pull of the tooth to the right presumably sets up tension on the bone surface by the periodontal fibers). A new layer of bone is deposited on the alveolar surface. This embeds the periodontal fibers of layer *a*. Note that the attachment fibers are not driven into the bone as with a nail; they are progressively enclosed as new bone deposits form around them. It is apparent that the fibers of zone *a* would soon be used up and become completely enclosed. However, the **linkage** fibrils of the intermediate zone *b* become converted into *a*, thereby lengthening *a* in advance of the drifting alveolar wall. The fibers of layer *a* are thus enclosed by new bone on one side while being lengthened by an equal amount on the other. The conversion from *b* to *a* is accomplished by a bundling together of the thin, precollagenous linkage fibrils into the thick, "mature" fibers of layer *a*. Ground substance is believed to be the binding agent, and the process is carried out by the abundant resident population of periodontal fibroblasts. Layer *b* retains its breadth by elongation of the precollagenous linkage fibrils. It is not presently known if this lengthening process occurs within zone *b* or at the interface between *b* and *c*. New unit fibrils are also constantly added as the tooth grows and as the membrane drifts in conjunction with tooth drift. The fibers of layer *c* are carried in the direction of the tooth's movement. Throughout this membrane remodeling process, continuous attachment between tooth and alveolar bone is thereby maintained. Note that the periodontal membrane as a whole is not simply pushed or pulled along as the tooth moves. It **grows** from one location to the next.

### FIGURE 11–34

The activity on the pressure side of the tooth root ("pressure" because the tooth root, according to long-standing but inadequate theory, exerts direct compression on the periodontal membrane and bony alveolar wall) is the reverse of the remodeling sequence on the opposite ("tension") side. A layer of bone is **resorbed** (*x'*) from the alveolar surface by an abundant sheet of osteoclasts. The resorptive side of the alveolar socket can easily be distinguished microscopically from the depository side by the characteristic chiseled, pitted, eroded appearance of the bone's surface. If the resorptive process was active at the time, numerous osteoclasts are seen in the erosion pits.

### FIGURE 11–35

Note that fibers which were **formerly** a part of the collagenous bone matrix (*a'*) are uncovered by the resorption of bone layer *x'* to become converted into an extension of zone *a*. Zone *b* (the linkage fibers of the intermediate zone) provides a take-up of what would otherwise be an excessive length of fibers *a*. A remodeling takes place by converting the thick fibers of *a* into the thin, precollagenous fibrils of *b*. This involves the separation of the thick fibers into precollagenous fibril subunits. The change is presently believed to be accomplished by an enzymatic breakdown of the ground substance binding the fibrils into the thick bundles. At the same time, the fibrils of layer *b* shorten by an equal amount at the junction between *b* and *c*.

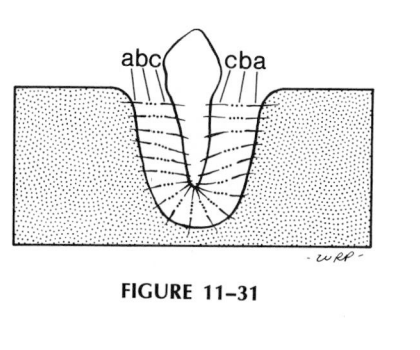

**FIGURE 11–31**

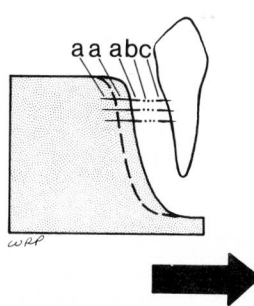

**FIGURE 11–32**

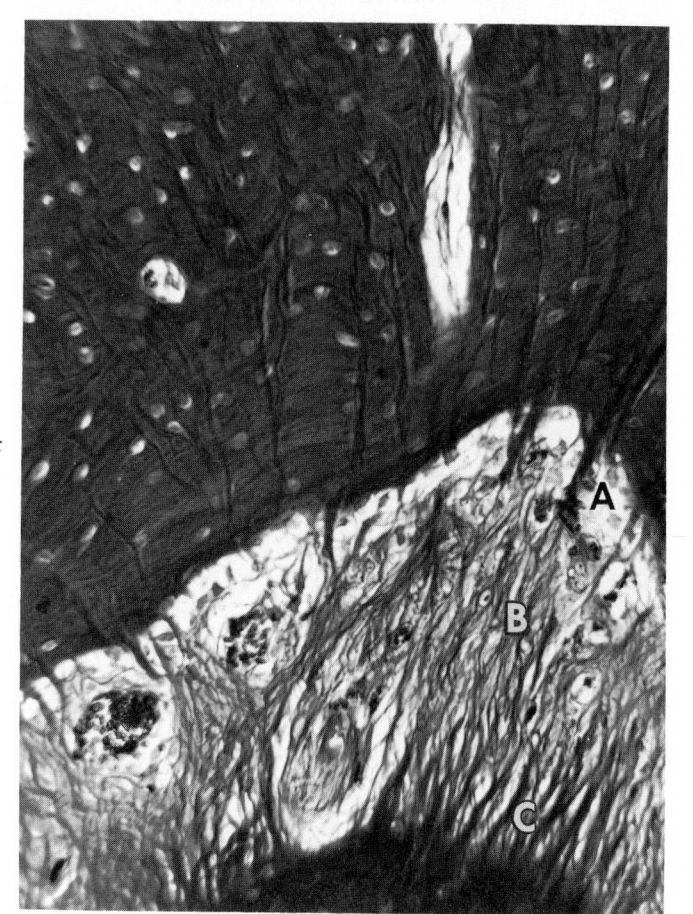

**FIGURE 11–33**

*(From Kraw, A. G., and D. H. Enlow: Am. J. Anat., 120:133, 1967.)*

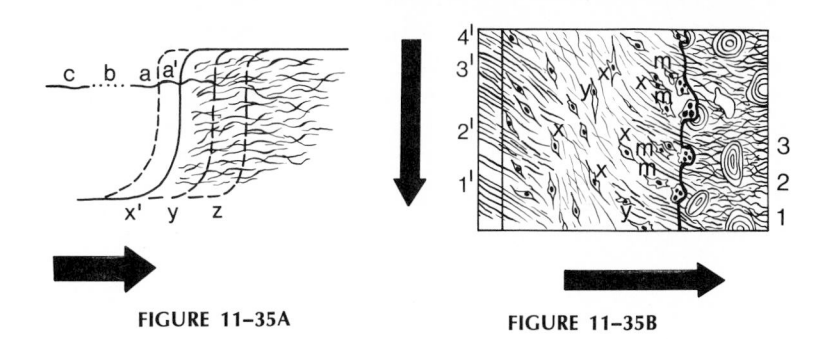

**FIGURE 11–34**                    **FIGURE 11–35A**                    **FIGURE 11–35B**

The whole membrane thus grows in a direction toward the bone surface and away from the tooth as the tooth simultaneously moves in a like direction. The process is continuously repetitive, and the resorptive front advances on to *y* and *z*. The tooth movement itself is provided by myofibroblasts, as described on page 351 and schematized in Figure 11–35*B*.

Even though the bone on the resorptive side of the alveolar socket undergoes progressive removal, periodontal connection between the bone and the tooth is thus still maintained by the release and conversion of bone matrix fibers into periodontal fibers. Wherever such connection is present, the fibers of the membrane are under direct tension (not pressure); they are taut between bone and tooth (Fig. 11–36). However, this process in which the collagenous fibers of the bone become uncovered and then utilized directly as periodontal fibers does not occur in all locations. In some areas, these bone matrix fibers **also** undergo resorptive removal with the rest of the bone. In such locations, one of two situations exists. First, all periodontal fibrous attachment between bone and tooth is **locally** severed by the resorptive process. The tooth in these regional areas of the socket is unconnected; the fibers here are slack. Second, fibrous attachment can be maintained by new bone deposits within the deep resorption spaces produced by "undermining resorption." That is, as resorption proceeds on the alveolar surface, resorption also occurs **deep** to the surface (this is a normal physiologic process that has nothing to do with "heavy forces"). Bone deposits are then formed in these spaces, and new fibers provide local attachment. As the surface resorptive front progressively reaches such subsurface attachment areas, new deep spaces are continuously formed with a new generation of bone and attachment fibers formed within them. It will be recalled that a comparable process also occurs in relation to the attachment of the periosteal membrane.

## BONE TISSUE TYPES

Many different varieties of bone tissue exist, and each of these relates to **specific** circumstances of growth, physiology, and pathology. At present, at least, most histology texts imply that bone generally is composed of haversian systems. The **secondary osteon** is described as the universal "unit" of bone tissue structure. This is not true. Like the 14 clubs in the golf bag, there are special kinds of bone for different circumstances. Very fast-growing types exist wherever rapid bone growth must keep pace with a correspondingly fast rate of soft tissue expansion. There are slow-growing bone types as well. Some kinds of bone tissue are adapted to large masses of deposition, others to thin cortical deposits. Some bone types associate with muscle and tendon attachments, others to processes of cortical reconstruction, and still others to alveolar tooth attachments. Some bone types grow outward, others inward. Some are densely vascularized, others sparsely vascular. The processes of normal remodeling produce almost limitless variations in the localized arrangement of the histologic components of bone. The cortex becomes layered during growth and remodeling, and the pattern of stratification and the bone tissue types in each layer differ in the different parts of a bone (Fig. 11–37). A section made at one level of a bone is always different from sections made elsewhere in that same bone because of the characteristic nature of localized remodeling changes in each of the different parts of any individual bone (Fig. 11–38). Just as a

**FIGURE 11–36**

*The alveolar surface for the socket at top is depository (A), and the lower socket has a resorptive alveolar surface (B). The entire plate of bone was produced by the periodontal membrane for the upper socket. Note that some bone fibers from the bone matrix on the resorptive side are released to function as periodontal fibers (D). Note also that the fibers on this side are taut, not slack. In some areas, resorption of the bone matrix is total, and attachment fibers are severed (C). (From Kraw, A. G., and D. H. Enlow: Am. J. Anat., 120:133, 1967.)*

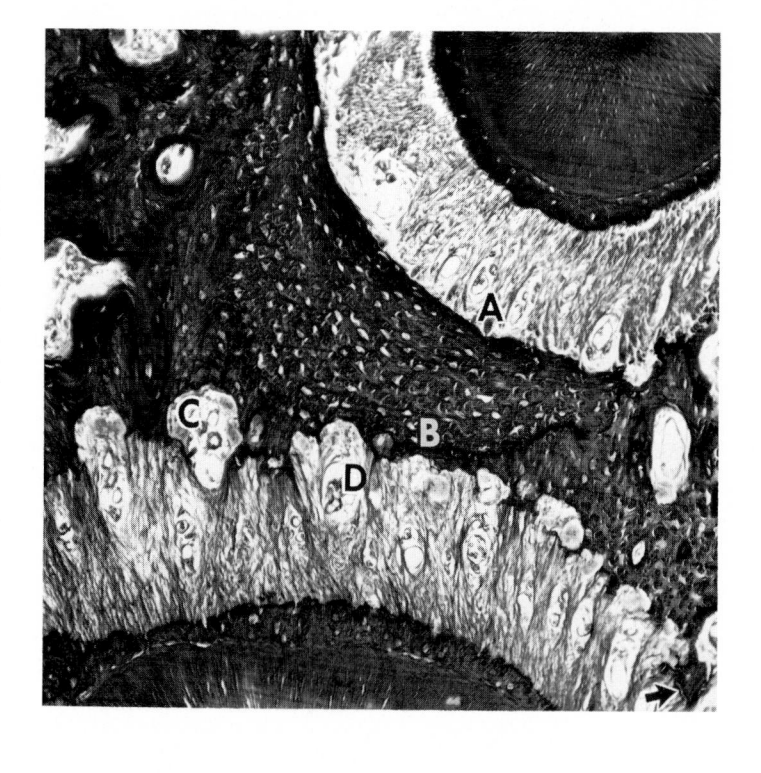

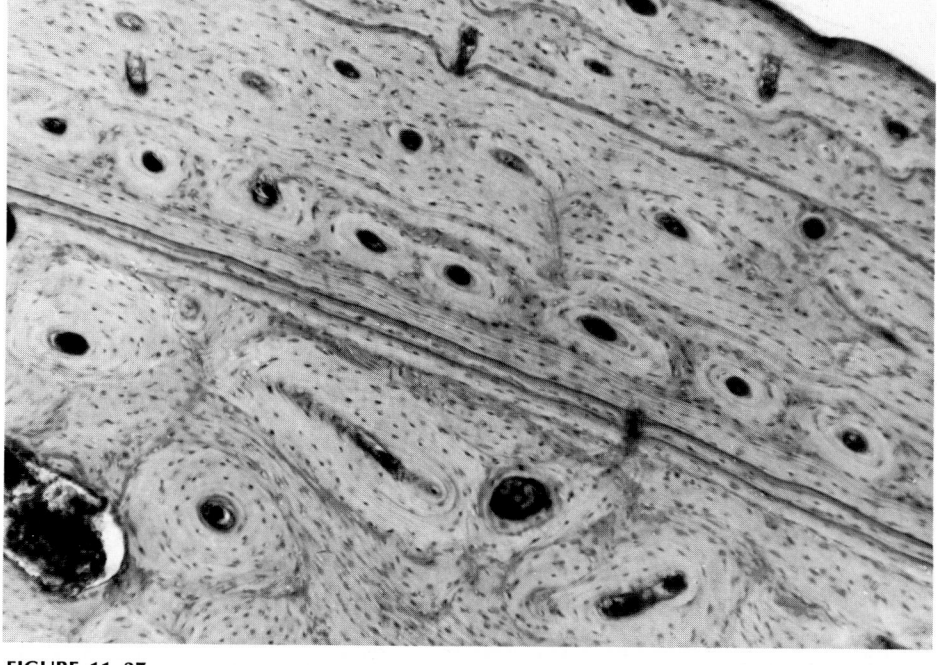

**FIGURE 11–37**

*This cortical section illustrates the stratified nature of the cortex and also shows the difference in structure between secondary osteons (haversian systems) located in the lower region and the smaller primary osteons in the center of the field. See text for details. (From Enlow, Donald H.: Principles of Bone Remodeling, 1963. Courtesy of Charles C Thomas, Publisher, Springfield, Illinois.)*

geologist can reconstruct the history of sedimentation and erosion in large rock outcroppings, the histologist can reconstruct the depository and resorptive remodeling history of a bone recorded in the rocklike substance of the cortex.

The bone tissue of a newborn is quite different histologically from the bone of an older child. This is different from the bone tissue of a mature adult which, in turn, differs from aged bone. The reason is that the **circumstances** (rate of growth, amount, and so forth) are different for each age. The pattern of bone structure in a human being differs from that in a rat. The rat is tiny, the cortex of any given bone is much thinner, the metabolic rate is different, the rates of growth differ, the mechanical forces are different, and the rat lives only a year or two. Research investigators are continually "discovering" that the bone of a 6 month old rat weighing a few ounces is considerably different histologically from a 50 **year** old, 200 **pound** man. Surprise is always expressed because no haversian systems (the legendary universal units of bone) are observed in the rat. The secondary osteon in no way is a "unit" of bone construction. It is not present at all in most vertebrates. In those species in which it indeed occurs, haversian systems are mainly a feature of larger animals having a longer life span and are found in abundance only at older age levels. When present, haversian systems form in specific locations of a bone. The orthodontist, when manipulating the bone tissue of a child, is **not** dealing with the haversian type of bone. What, then, is the structure of bone in the human child?

Outlined below is a brief summary of some major varieties of bone tissue. There are many more. However, these are the types most often encountered by the clinical researcher concerned with human bone and the bone tissues of common laboratory animals. (See Enlow: *The Human Face,* Harper & Row, 1968, for descriptions of other types.)

**Fine cancellous bone tissue** is the fastest-growing bone type of all (Fig. 11–39). It is characteristically found throughout all bones in the fetal skeleton and in some parts of the postnatal skeleton as well. It is also involved in the callus formation of a healing fracture. Osteophytes and exostoses, commonly present in many pathologic conditions, are layers of fine cancellous bone added to the outer or inner surfaces of the compact bone of the cortex.

Fine cancellous bone forms almost the entire cortex of fetal bones, and it is found in some cortical areas of the **fast**-growing parts of a child's bone (as in the posterior border of the growing ramus). Even though this bone type is located in the cortex, it is not "compact" bone; it is porous. The spaces, however, are not nearly as large as those in "coarse" cancellous bone (another major type, described below). Immature connective tissue, not marrow, is located within the spaces. Fine cancellous bone can be formed by both the periosteum and the endosteum.

**Coarse cancellous bone** has irregularly large spaces containing red or yellow marrow. This is the familiar "spongy bone" composed of trabeculae or thin bony plates. It is a feature of the medulla and is particularly abundant in epiphyses and between the cortices of the flat bones of the skull (where it is termed "diploë"). Coarse cancellous bone is always produced by the endosteum; it never normally forms on the outside of a bone.

**Nonlamellar bone** is the type that makes up fine cancellous bone tissue (Fig. 11–39). It has a matrix in which fiber orientation does not produce the distinct stratification that characterizes the more familiar lamellar bone tissue. One common type of nonlamellar bone is sometimes termed "woven" bone tissue because of the interwoven nature of the fibers throughout the matrix. The osteocytes have a characteristically irregular, jumbled arrangement with no alignment into rows. This is in

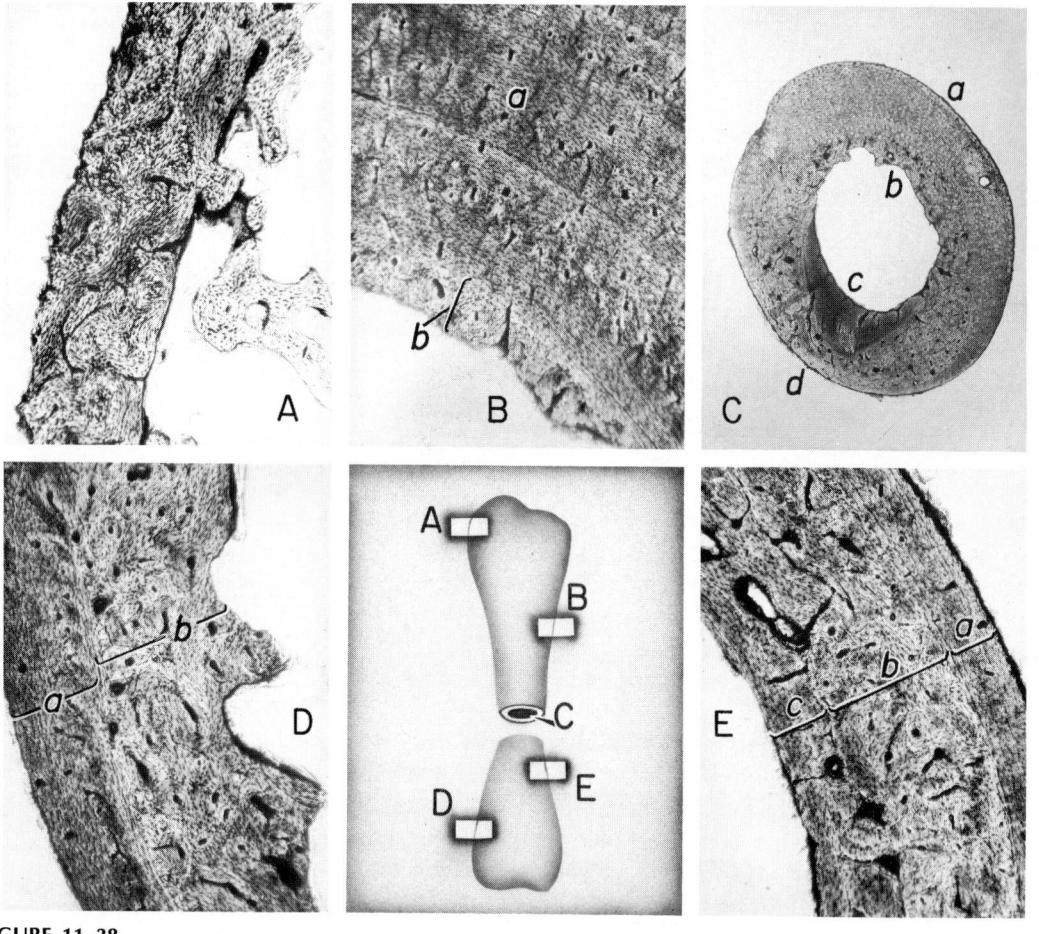

**FIGURE 11–38**

Sections taken from different parts of a bone always show marked differences in histologic pattern due to the differences in **regional** growth and remodeling. The cortex taken from location A is composed entirely of endosteal bone. The periosteal surface is resorptive, and the endosteal surface is depository. The bone tissue is of the "compacted cancellous" type. Section B shows a wide periosteal zone (a) composed of primary vascular bone and a remnant of an endosteal zone (b) that now has a resorptive inner surface. In section c, the cortex is drifting in a northeast direction to form the lateral curvature of the bone. Surface a is depository, b is resorptive, c is depository, and d is resorptive. Section D shows a zone of periosteal bone (a) separated from a layer of endosteal bone (b) by a reversal line. The endosteal surface, formerly depository, is now resorptive following outward reversal. Section E is composed of a layer of periosteal bone (a), a middle zone of compacted cancellous bone (b), and an inner layer of circumferential bone (c). Compare with Figure 11–50.

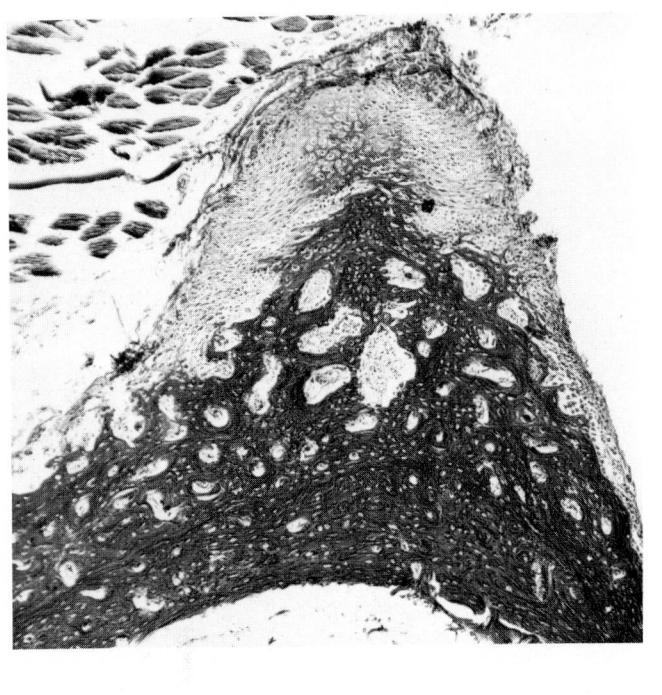

**FIGURE 11-39**

*At the apex of many tubercles and tuberosities, as seen in this section, chondroid bone is formed. This unique bone type is also found on alveolar crests in the mandible and maxilla during the active growth period. Note the cartilage-like appearance of the large cells. Fine cancellous, nonlamellar bone occurs deep to the chondroid apex of the tubercle. (From Enlow, D. H.: Am. J. Anat., 110:79, 1962.)*

contrast to **lamellar** bone in which the collagenous fibers all have a common alignment in each layer (lamella), and the cells are arranged so that their long axes are parallel. Because the fiber directions differ in adjacent layers, a stratified appearance is produced. This fiber arrangement gives a plywood-like strength to the bone substance.

While composing the fine cancellous type of bone, nonlamellar bone tissue may also be formed as compact masses, and this occurs when the rate of deposition is somewhat less rapid. In the child, thus, nonlamellar bone is present wherever growth involves a moderate to rapid rate of bone deposition to accompany surrounding soft tissue expansion. Lamellar bone is slower growing. In any given bone of a young child, mixed layers and areas of lamellar and nonlamellar bone always occur. This is because all parts of a bone do not grow at the same rate, and because all the various areas of a bone become shifted (as a result of "relocation") into new regions that have different growth rates. Layers of these and other different bone types thus accumulate in the cortex.

**Primary vascular bone** is the most common type of bone tissue of periosteal origin in the child's skeleton, including all the facial and cranial bones. This is **the** standard periosteal type of bone in the growing skeleton, and it may also be formed in some endosteal areas as well (Figs. 11–41, 11–42, and 11–53). Curiously, it has never been included in elementary histology texts. There are two general classes of bone tissue—primary and secondary. All bone deposited directly on the outside and inside surfaces of the cortex is "primary," and blood vessels enclosed in these deposits (as the bone matrix is formed around them) are correspondingly designated as primary in type. In certain circumstances, primary bone can undergo a process of cortical reconstruction. The result is "secondary" bone, which includes the haversian system. Primary vascular canals do not have concentric rings of lamellae surrounding them; they are simply canals located within a matrix that is composed largely of circumferential lamellae. Primary vascular bone is also the

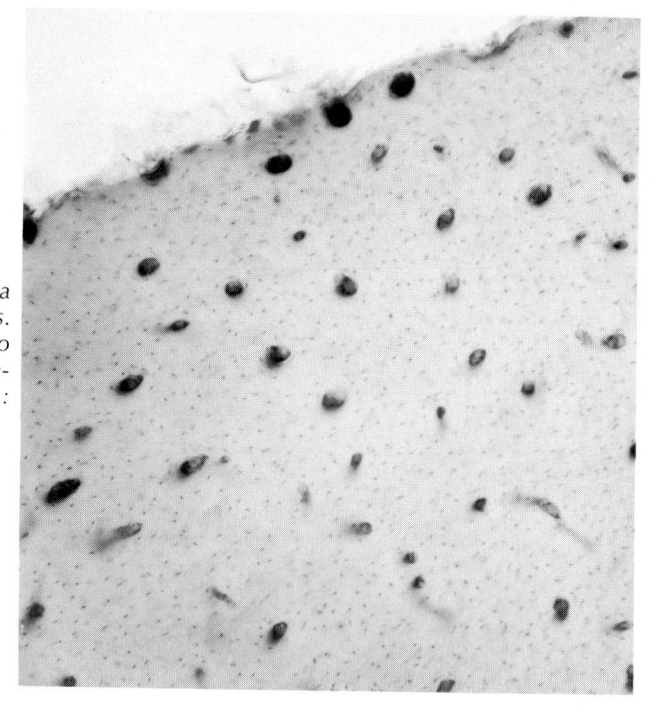

**FIGURE 11–40**

*Areas of fast-forming bone tend to have a richer concentration of vascular canals. This kind of bone is also more resistant to the normal process of necrocytosis. Compare with Figure 11–43. (After Enlow, D. H.: Am. J. Anat., 110:269, 1962.)*

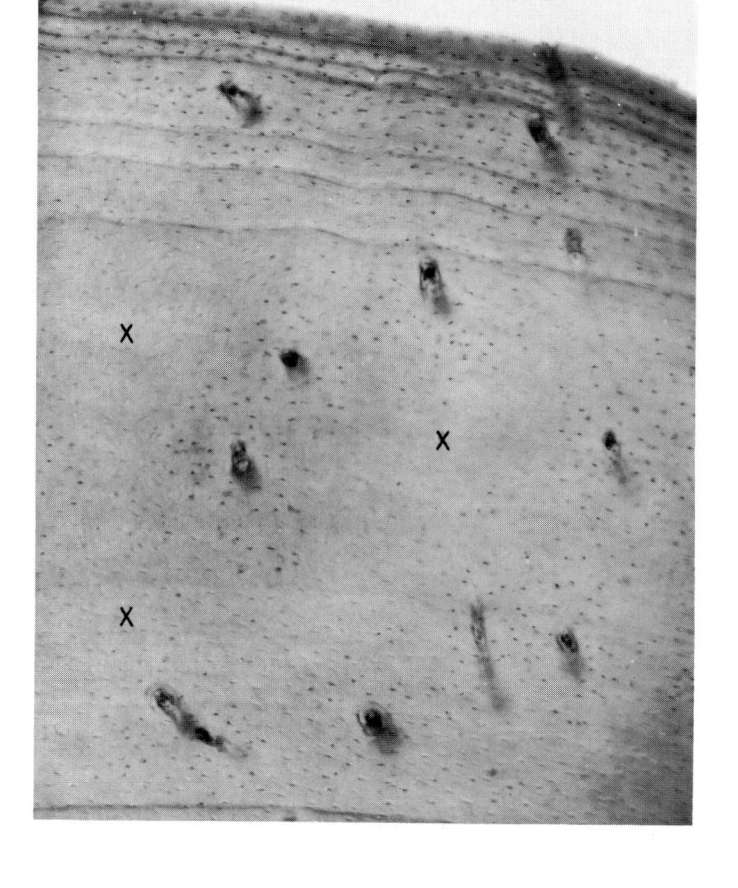

**FIGURE 11–41**

*This section of cortical bone illustrates the concept of the "tissue cylinder." The capillary in each canal supplies a cylindrical region of cells surrounding it. When the spread of capillaries is relatively scant, the intervascular areas outside these physiologic cylinders of effective capillary supply eventually undergo necrosis (note the empty lacunae in area X). In time, such necrotic areas spread as the canaliculi and canals become filled with minerals, and the bone tissue is replaced by the process of haversian reconstruction. The resorption canals and secondary osteons that then develop represent the radius of effective supply from the blood vessel to the osteocytes. The haversian system is a graphic example of the "tissue cylinder"; such physiologic cylinders exist for all vascular soft tissues as well. Compare with Figure 11–46. (After Enlow, D. H.: Am. J. Anat., 110:269, 1962.)*

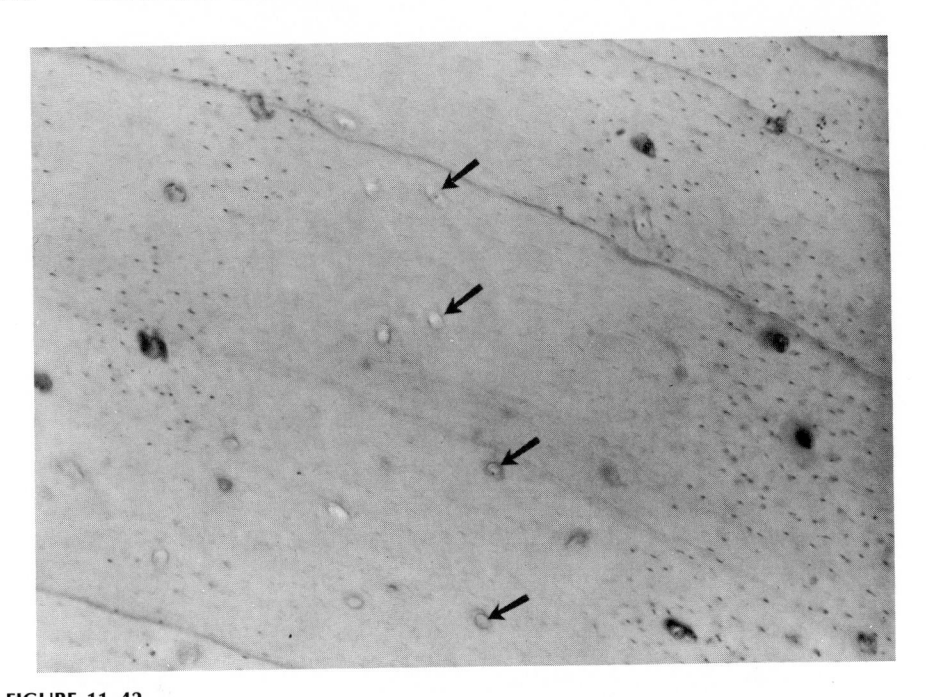

**FIGURE 11–42**

*Empty vascular canals or canals that have become plugged with mineral (arrows) lead to the early necrosis (note empty lacunae) of the "tissue cylinders" surrounding them.*

standard form of periosteal bone tissue among most other animals. Only in relatively a few groups is this basic kind of bone replaced by secondary bone tissue. While primary vascular bone is the dominant type in the human child and young adult, many primary vascular canals are still found in the aged skeleton.

**Nonvascular bone** is found as the sole type of bone in some species. It is a very **slow**-growing variety. In the human skeleton, it forms in specific locations where the cortex forms leisurely over a period of time. In **all** bone tissue types, in general, the extent of vascularity is an index of the rate of bone deposition. Abundant vascular canals are characteristic in any cortical region that is rapidly formed (Fig. 11–40). The spectrum of canal density then grades down to totally nonvascular areas that are the slowest forming of all (as in the zone seen in Fig. 11–43).

Bone undergoes a normal process of **necrosis.** Osteocytes, like most connective tissue cells, have a limited life span. In bone, the cells live for about 7 years or so, but the length of time is quite variable; the degree of vascularization is a basic factor determining this life span. Less abundant vascular canals which are farther apart favor earlier bone cell death. There is a cylinder-like area of physiologic tissue dependence surrounding each vascular canal (this is a general rule, also, for the various kinds of soft tissues and the capillaries within them). The vessel in each canal supplies the particular cylindrical field around it. When vascular canals are abundantly distributed, all the osteocytes have adequate vascular supply. However, when the canals are widely separated, as in the slower-growing areas, the outlying regions between the canals have a much lower level of vascular supply because the transport distance from the capillary through the canalicular system of the lacunae is greater. In time, these areas farther removed from the blood vessels begin to undergo necrocytosis (Figs. 11–41 and 11–43). It is a common and, indeed, normal

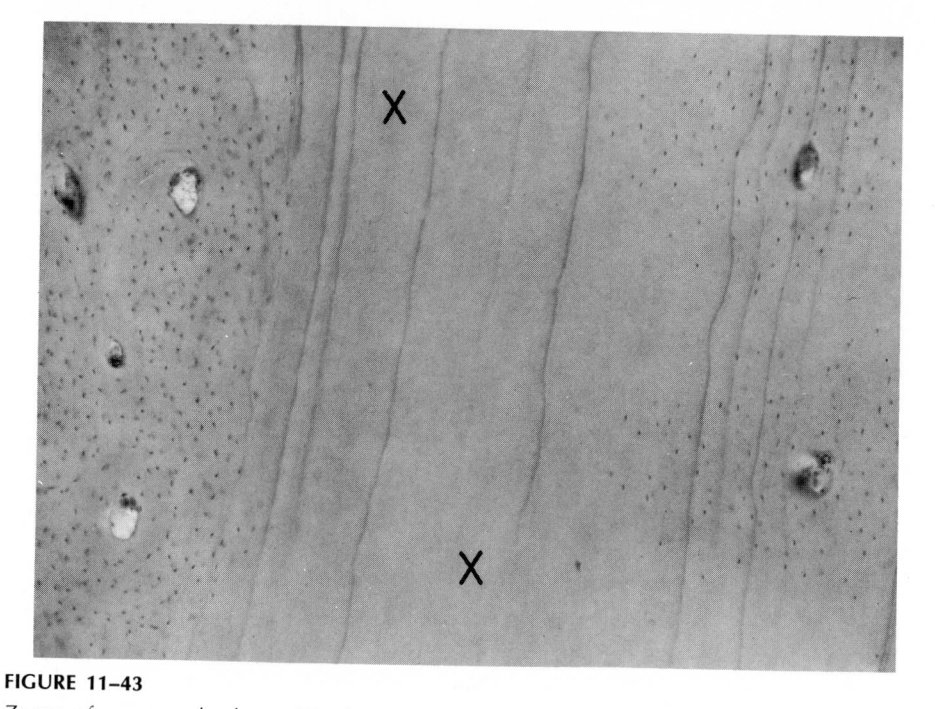

**FIGURE 11–43**

Zones of nonvascular bone (X) often occur in thick, slow-forming cortical areas. Note that this layer has become necrotic (empty lacunae). The bone cells surrounding the vascular canals, however, still survive. (From Enlow, D. H.: Am. J. Anat., 110:269, 1962.)

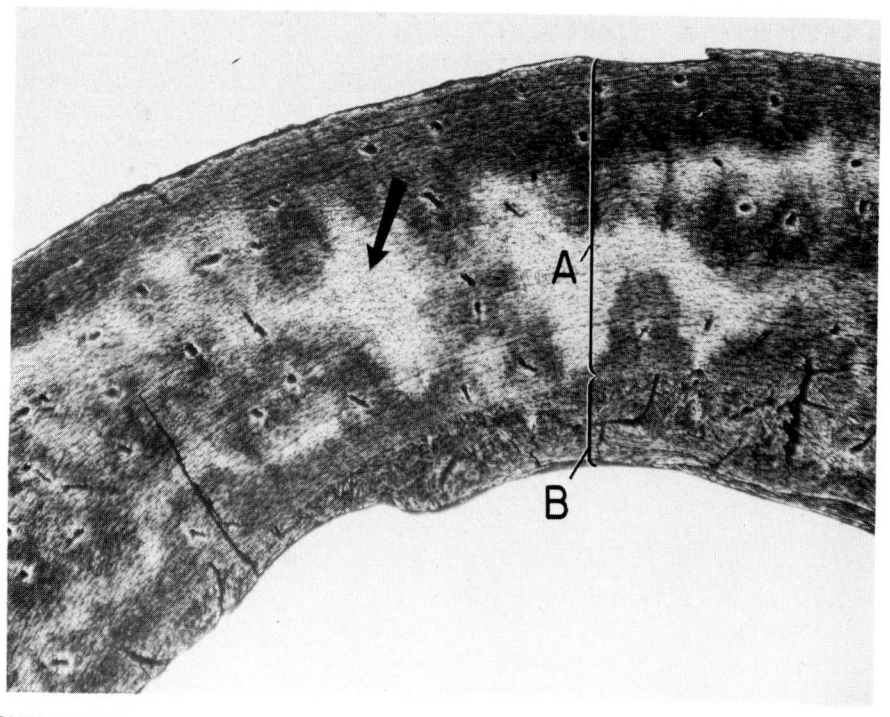

**FIGURE 11–44**

The outer zone of this cortex (A) is composed of primary vascular bone tissue. Zone B is composed of irregular endosteal bone. Note the intervascular areas of necrosis (arrow). In a ground section, such as this, necrosis can be recognized by the cleared appearance of the matrix (as seen also in sclerotic dentin). (From Enlow, D. H.: Am. J. Anat., 110:269, 1962.)

circumstance. Moreover, the canaliculi and even the lacunae and vascular canals often become plugged with calcium deposits. This causes the death of all the bone cells located downstream (Fig. 11–42). In Figure 11–40, the close proximity of the vascular canals is such that the effective physiologic **tissue cylinder** of each is sufficient to reach all osteocytes. In Figure 11–41, however, the spread of canals is such that the intervascular areas lie beyond the physiologic limits that the canals can effectively supply over a long period. The osteocytes in these areas undergo necrosis in time. Localized necrotic regions can be recognized because the lacunae are empty, and they are often filled with calcium deposits (the clear areas in Figure 11–44; this circumstance is comparable to that found in **sclerotic dentin**). Such necrotic spots are more widespread in adult bone because nonvascular areas or regions having less dense vascularization are correspondingly more widespread, as noted below.

Haversian bone tissue is "secondary" because it replaces the original primary vascular bone. This involves a process of cortical reconstruction. When it is said that "bone remodels throughout life," this is what is meant. There are different kinds of remodeling, as noted in Chapter 2. **Biochemical remodeling** involves the exchanges of minerals and other ions between the bone substance and blood in order to maintain calcium and other levels. **Growth remodeling** carries out the relocation of a bone's various parts as the bone increases in size, and the remodeling changes are produced by various combinations of resorption and deposition on the periosteal and endosteal surfaces. **Haversian remodeling** is a process of internal reconstruction **within** the cortex, not changes on the outer and inner cortical surfaces. There are several functional reasons why it occurs.

During the period of childhood growth, the constant replacement of bone takes place because of **growth** remodeling. That is, new bone is constantly being formed, and previously deposited bone is removed. The bone is not present long enough for any marked extent of osteocyte necrosis to develop. Moreover, the childhood types of bone tissue are more highly vascular because of their more rapid rates of formation. This favors a longer life span for the bone cells. As the child matures, however, slower-growing types of bone are then laid down, and these tend to be less vascular or even nonvascular in many areas. The decreasing abundance of vascular canals in the adult skeleton leads to earlier cell death. Because the bone is no longer being replaced, owing to the turnover by the process of growth remodeling, it remains in the skeleton until much more widespread cell necrosis has time to occur. **Haversian** remodeling is a process that replaces this older dead or dying bone with progressively new generations of living bone tissue. Secondary osteons thus develop by a process of reconstructing the primary bone. This occurs by the same resorptive and depository mechanism utilized in growth remodeling. **Resorption canals** develop by enlarging the original primary canals through osteoclastic activity (Fig. 11–45). New resorption canals can be formed in previously nonvascular regions by vascular invasion. The resorption canal represents the removal of the primary physiologic "tissue cylinder," which is that specific territory supplied by the vessels of each canal. The radius of this tissue cylinder and the resorption canal that replaces it is determined by the effective distance material can diffuse to and from the centrally located capillary. Lamellar bone is then laid down from outside to inside within each resorption space. The result is a secondary osteon, and it is structurally (as well as functionally) a true cylinder. **All** vascular tissues, soft as well as hard, have "tissue cylinders"; the Haversian system in bone is a graphic example that can actually be seen (Fig. 11–46). The haversian reconstruction process also contributes to **mineral homeostasis** in the older individ-

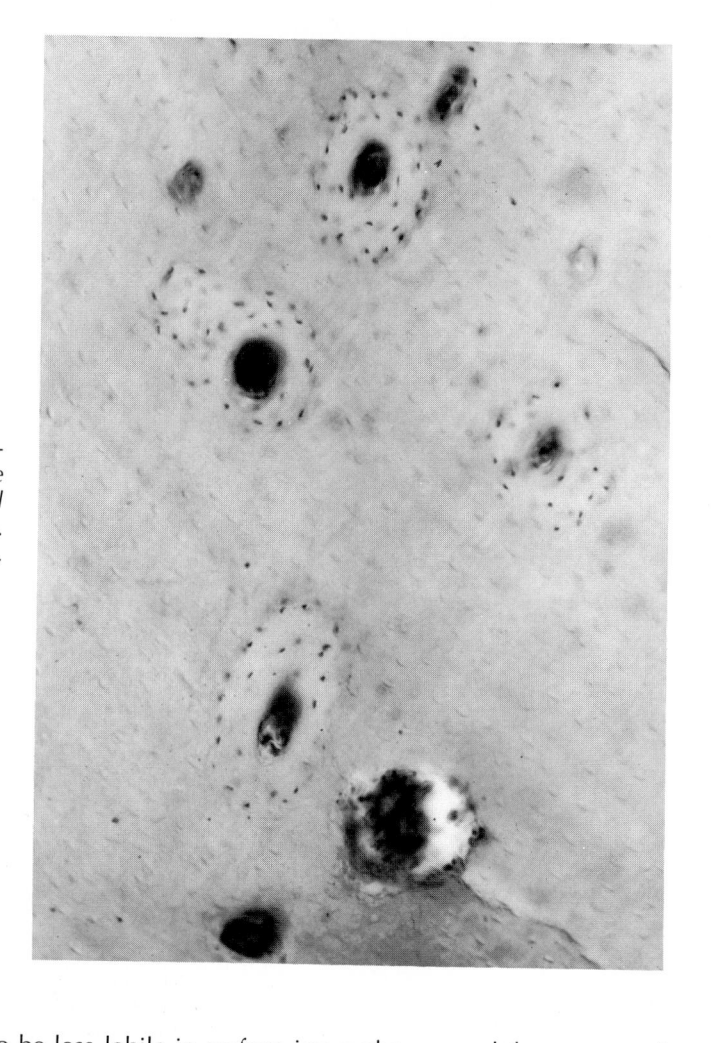

**FIGURE 11–45**

*Note the resorptive canal and the replacement secondary osteons that have developed within an area of cortical bone which is necrotic (empty lacunae). (From Enlow, D. H.: Am. J. Anat., 110:269, 1962.)*

ual. Aged bone tends to be less labile in surface ion exchange, and the process of secondary reconstruction provides calcium turnover.

Haversian systems are thus a feature of the **older** skeleton. They begin to appear during early maturity, and the distribution then accumulates over the years. A rat (as well as many other species) lacks haversian systems because (1) it does not live long enough, and (2) its cortical bone tends to be highly vascular in type or so thin that vessels in the periosteum and endosteum can effectively supply it.

The haversian system provides another key function. It is involved in the attachments and progressive reattachments of muscles on growing, remodeling surfaces of bones (Fig. 11–4). Even in the young child, haversian systems are formed specifically in some sites of muscle insertion. Here they function to secure the muscle and tendon on resorptive surfaces by anchoring the tendon into the bone deep to the resorptive surface (see p. 344). This has nothing to do with reconstruction of necrotic bone, which does not become a major process until much later in life. While other mechanisms are also operative in maintaining muscle attachments, the development of secondary osteons represents one process that does this (in larger mammals) wherever shifting tendons are inserted on a remodeling tubercle, tuberosity, or other localized resorptive surface (p. 342). This type of haversian bone is not nearly as widespread as the haversian system formation that takes place in adult bone.

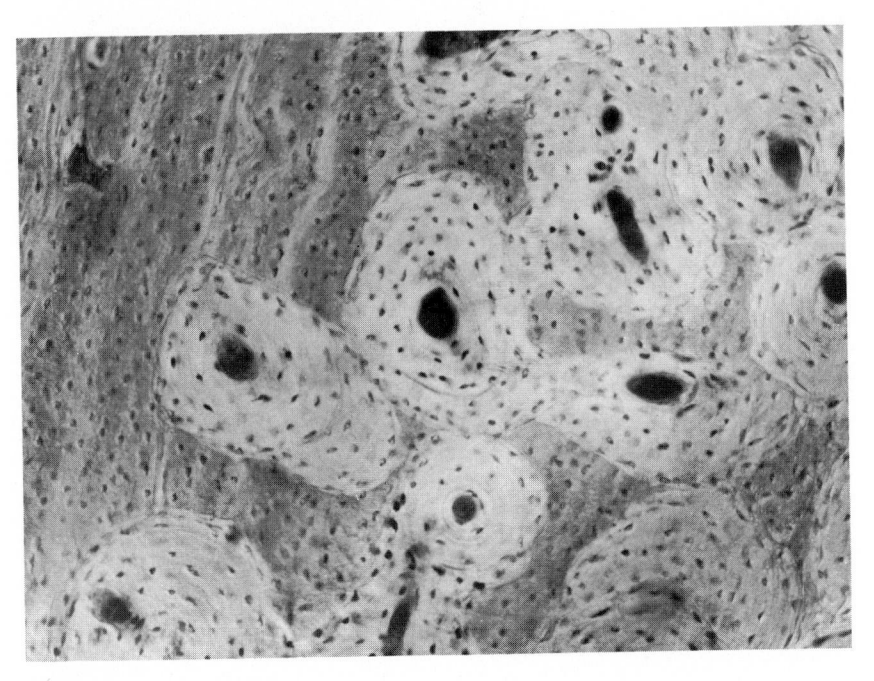

**FIGURE 11–46**

*Secondary osteons. Each haversian system is formed by a process of secondary reconstruction of existing primary or secondary bone. A resorption canal first develops (see Figs. 11–4 and 11–45). Concentric lamellae are then laid down within the enlarged tubular space until the resulting central canal has the diameter of the original primary vascular canal. Second* **generations** *of osteons have developed in this section (osteons superimposed over one another). Note remnants of the original primary lamellar bone between the haversian systems. Interestingly, haversian* **systems** *were not known to Clopton Havers, an English physician and anatomist. Osteons as such were not described until well into the following century. Havers, however, believed that he was the first to observe vascular* **canals** *in bone (during the 1680's), but Leeuwenhoek had actually published on their existence a decade earlier. Thus they should properly be called Leeuwenhoekian canals, but this is not likely. Since capillaries could not be seen at the time, Havers logically presumed that the smaller canals function to transport "medullary oil" to the joints for lubrication. (For an account of historical landmarks in bone biology, see Enlow, 1963.) (From Enlow, D. H.:* The Human Face. *New York, Harper & Row, 1968, p. 34.)*

"Primary" osteons are a feature of the young, fast-growing skeleton. They are formed by the deposition of new bone in cortical fine cancellous spaces (Fig. 11–47). The resultant structures resemble secondary osteons (haversian systems), but they are not formed by the process of secondary reconstruction involving the prior enlargement of canals by resorption. The bone between the primary osteons is always of the nonlamellar type, since it was fine cancellous in type (see above). Primary osteons are much smaller than haversian systems, and they are not surrounded by a "cementing line" (reversal line) because the formation of a resorption canal is not involved. Primary osteons often occur as narrow layers (Fig. 11–37) within the cortex, but large areas may also exist that are composed entirely of these structures. Primary osteons function to convert fine cancellous into compact bone.

**Compacted coarse cancellous bone is the most common of all bone types** (Figs. 11–48 and 11–49). It is found in almost all species, in almost all bones, and at all ages. It constitutes a major part of the compact, cortical bone in the growing child as well as in the adult. Incredibly, this major, ubiquitous type of bone tissue is not (at least at present) described in basic histology textbooks. Compacted coarse cancellous bone (which has no real formal name but is sometimes called, simply,

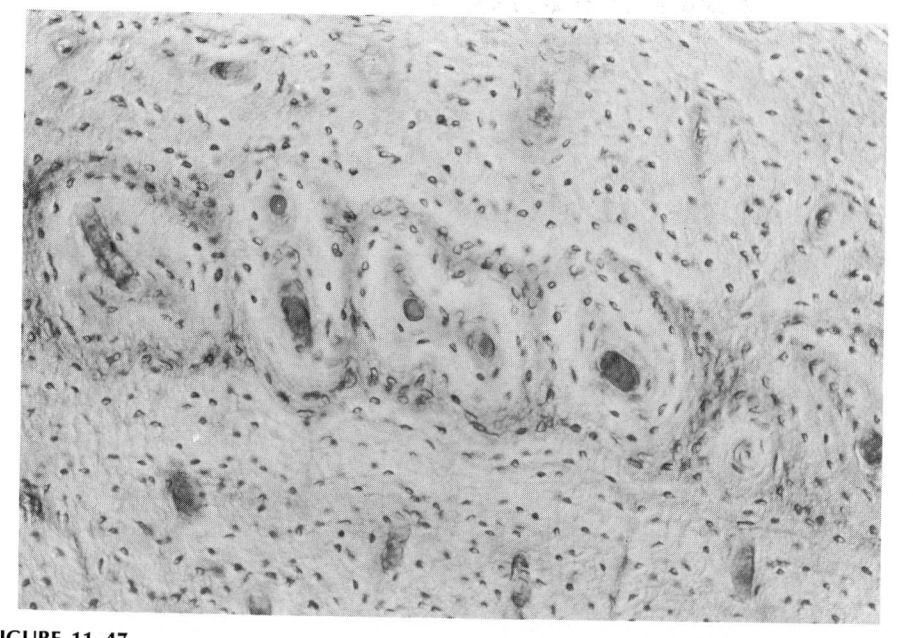

**FIGURE 11–47**

*A row of small primary osteons is seen in this section. Compare with the secondary osteons in Figure 11–46. (From Enlow, D. H.:* The Human Face. *New York, Harper & Row, 1968, p. 23.)*

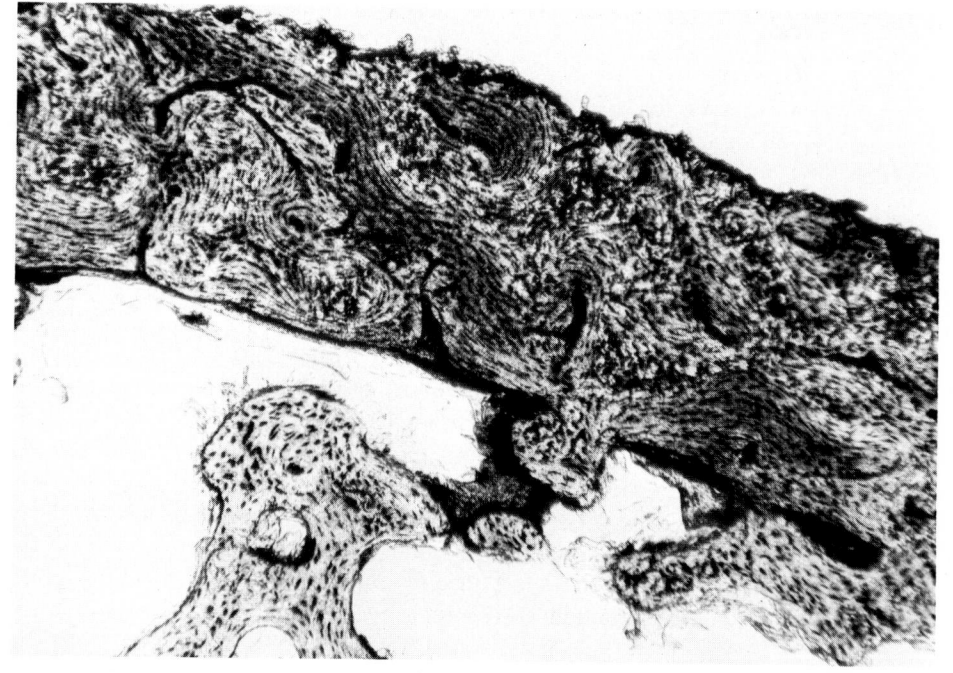

**FIGURE 11–48**

*This is endosteal cortical bone tissue that was formed by cancellous compaction. The periosteal surface is resorptive, and the endosteal side is depository. (From Enlow, Donald H.:* Principles of Bone Remodeling, *1963. Courtesy of Charles C Thomas, Publisher, Springfield, Illinois.)*

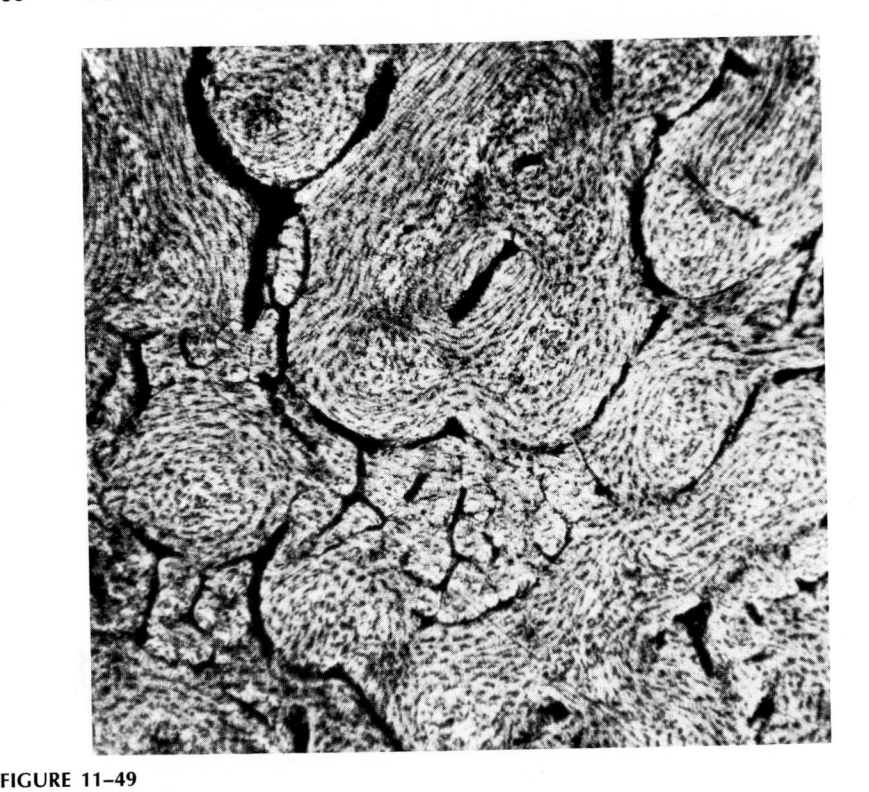

**FIGURE 11–49**

*Compacted cancellous bone. This was originally coarse cancellous bone in the medulla, but the endosteal mode of growth resulted in its conversion into cortical compact bone. (From Enlow, D. H.:* The Human Face. *New York, Harper & Row, 1968, p. 32.)*

"convoluted" bone tissue) is formed **only** by the endosteum during periods or at locations involving inward directions of cortical growth. This involves half to two-thirds of the skeletal formation process throughout the body (see p. 22). Convoluted bone forms by lamellar deposition within the large spaces between the convoluted trabeculae of cancellous bone. This is a process of compaction in which the **medullary cancellous** bone is converted into **compact cortical** bone. In Figure 11–48, the cortex is growing **inward**. It moves into the medullary area already occupied by cancellous bone. To change this medullary spongy bone into the compact bone of the inward-moving cortex, the cancellous spaces are filled in until each resulting lumen is reduced to the dimension of an ordinary cortical vascular canal. Because the cancellous trabeculae are the templates upon which the new bone is laid down, the microscopic appearance is one of irregular whorls and convolutions of compact bone, shown in Figure 11–50. When an outward reversal later takes place because this area becomes relocated to a new level, as seen in Figure 11–51, the endosteal layer then becomes covered by one of the periosteal varieties of bone (Fig. 11–52). In some medullary areas, cancellous bone may be absent, as in the mid-diaphysis of a long bone. Here uniform layers of **inner circumferential** lamellae are laid down rather than the convoluted type (Fig. 11–50). In other areas of the diaphysis, the endosteal surface becomes resorptive as the whole cortex grows outward, as seen in Figure 11–53. Whether the inner surface is depository or resorptive depends on the regional growth circumstances, as illustrated in Figure 11–38.

**Plexiform bone** is a type in which massive cortical deposits take place in rela-

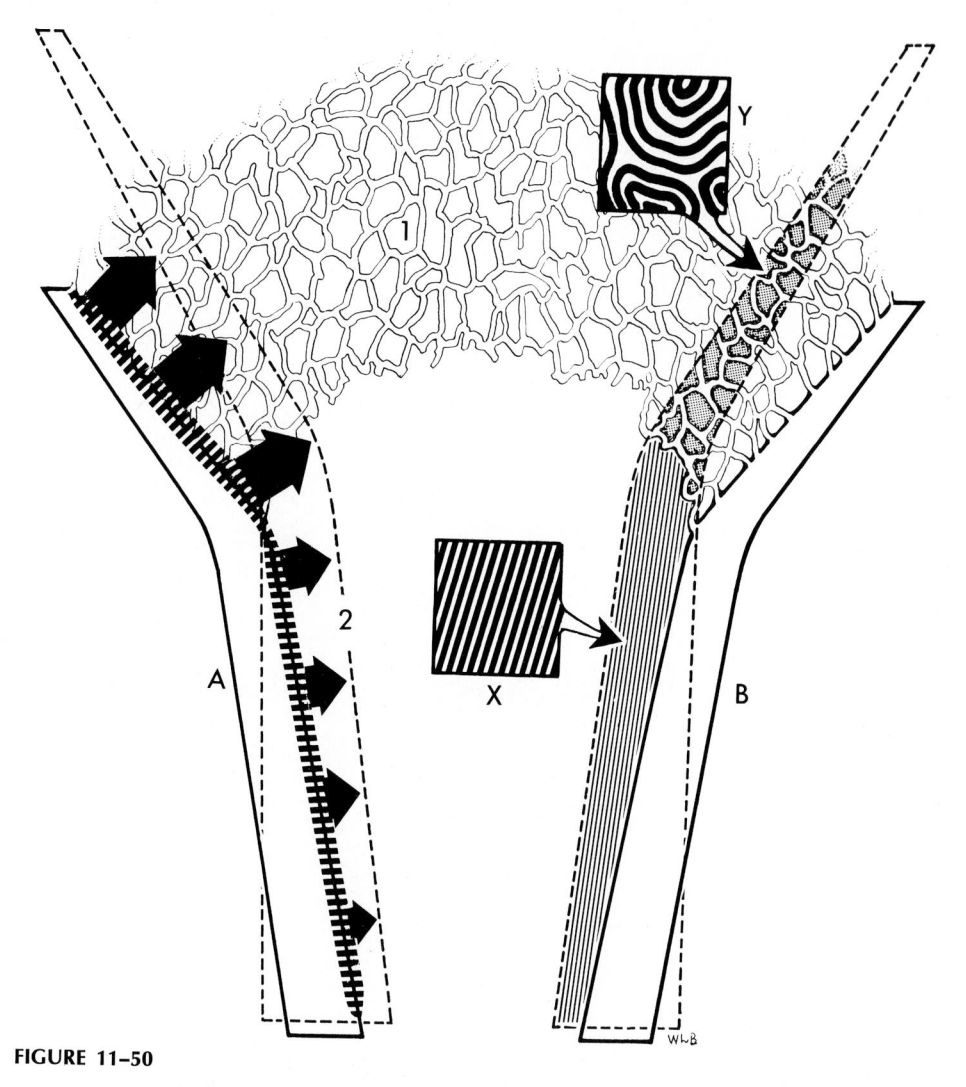

**FIGURE 11–50**

*In those areas of a bone that involve an* **endosteal** *direction of cortical growth (together with periosteal resorption), a process of direct conversion from medullary cancellous (1) to cortical compact bone (Y) takes place by filling in the cancellous spaces until they are "vascular canal" in diameter. In those places (A and B) where cancellous trabeculae are not present (2), inner circumferential lamellar or nonlamellar bone is laid down (X). See Figures 11–38 and 11–48. (From Enlow, Donald H.:* Principles of Bone Remodeling, *1963. Courtesy of Charles C Thomas, Publisher, Springfield, Illinois.)*

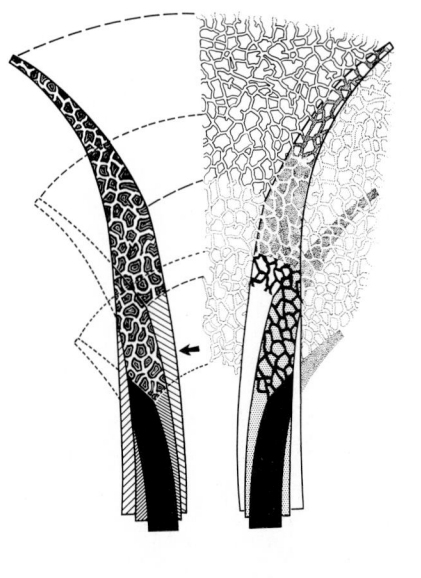

**FIGURE 11–51**

*The wide end of a bone grows in a longitudinal direction by deposition on the endosteal side and resorption from the periosteal side. This is because the* **inside** *surface actually faces toward the direction of growth. Medullary bone is converted into cortical bone by cancellous compaction. In areas where cancellous trabeculae are no longer present, however, inward growth involves deposition of inner circumferential bone (arrow). The inward mode of growth also serves to reduce the wide part into the more narrow part (diaphysis) as the whole bone lengthens. In the diaphysis, the direction of growth reverses, and periosteal bone is laid down. (From Enlow, Donald H.:* Principles of Bone Remodeling, *1963. Courtesy of Charles C Thomas, Publisher, Springfield, Illinois.)*

tively short periods of time (Fig. 11–54). It is most common in medium and large sized animals. In man, it is sometimes found in fast-forming areas such as the maxillary tuberosity. This bone type is composed of a symmetrical, three-dimensional plexus of primary vascular canals. The canals extend in longitudinal, radial, and circumferential directions, giving the cortex a brick wall–like appearance.

**Bundle bone** is so called because it contains massive bundles of parallel

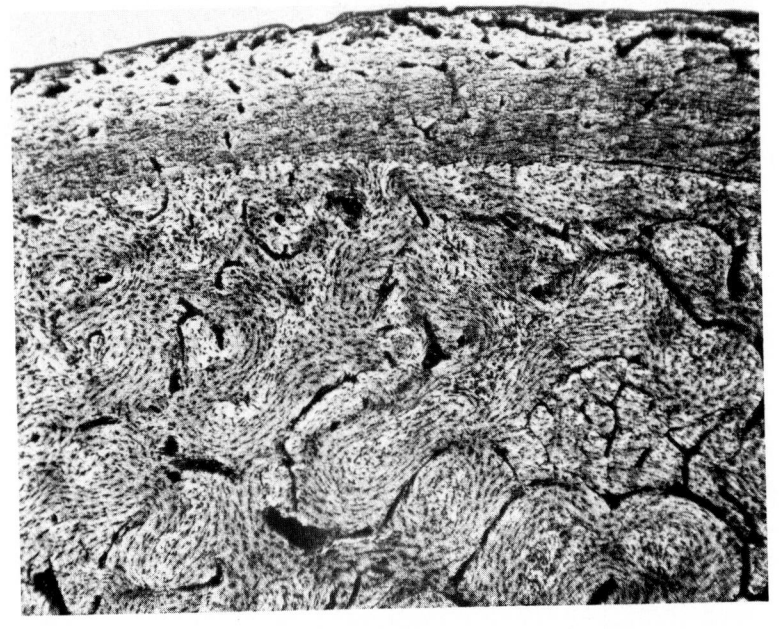

**FIGURE 11–52**

*The inner layer in this transverse section of cortical bone was produced during a period of inward (endosteal) growth and was formed by the process of cancellous compaction. Following outward reversal, a periosteal layer of primary vascular bone was subsequently laid down. Note the reversal front between these two zones. (From Enlow, D. H.: Am. J. Anat., 110:79, 1962.)*

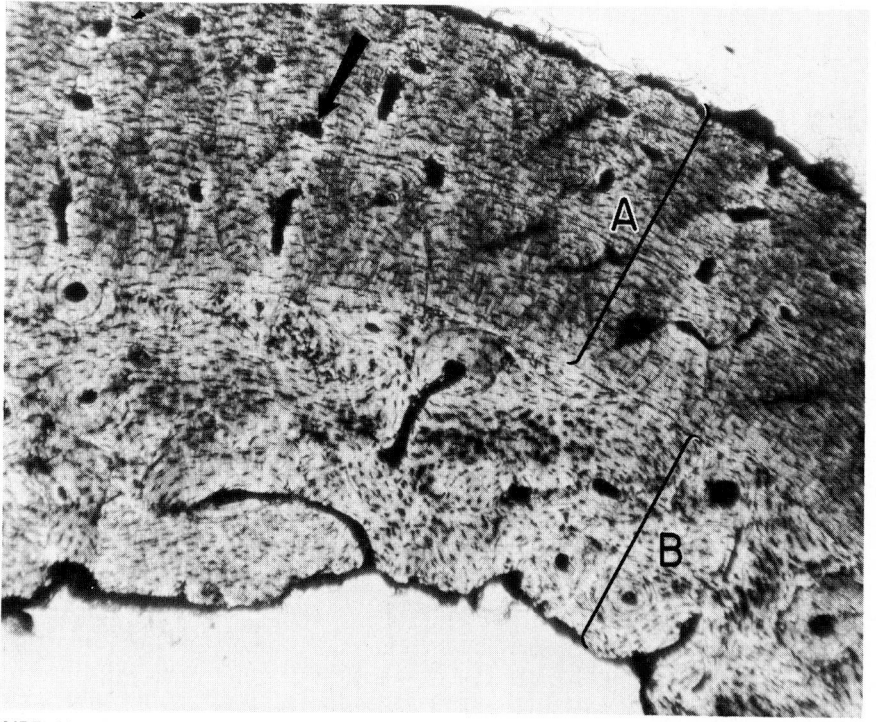

**FIGURE 11-53**

The cortex is often composed of two or more layers. Each layer is the result of past growth-remodeling changes, and the different bone types reflect the circumstances at the time of their formation. This cortical section shows an inner zone (B) that was produced during a former growth period involving an inward direction of growth and resultant endosteal bone formation. It is composed of compacted cancellous bone with superimposed secondary osteons that once functioned in muscle attachment on what was then a resorptive periosteal surface. A reversal then took place when this part of the bone became "relocated." The endosteal surface became resorptive, and a layer of periosteal bone was deposited (A). This outer layer is composed of primary vascular, lamellar bone tissue. ,After Enlow, Donald H.: Principles of Bone Remodeling, 1963. Courtesy of Charles C Thomas, Publisher, Springfield, Illinois.)

**FIGURE 11-54**

This is a section of plexiform bone tissue. It is characterized by a three-dimensional plexus of vascular canals, and the symmetrical arrangement of the bony laminae gives this bone type a characteristic "brick wall" appearance. Plexiform bone is a fast-growing type that provides large amounts of deposits for thick cortical areas. (From Enlow, D. H.: The Human Face. New York, Harper & Row, 1968, p. 26.)

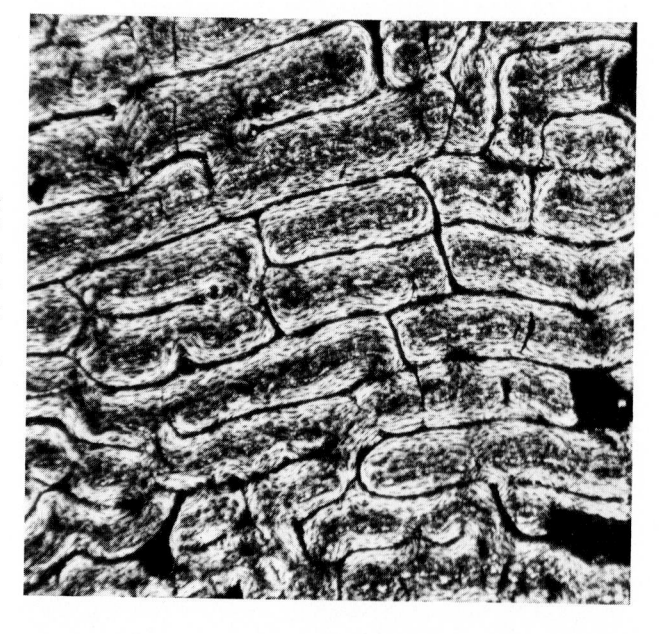

collagenous fibers attaching the periodontal membrane to the bony alveolar wall (Fig. 11–33). This type of bone is formed only on the depository ("tension") side of the socket. A reversal in drift direction may occur, however, and the surface of the bundle bone may then become resorptive on the new "pressure" side. Bundle bone is always laid down in a direction **toward** the moving tooth root, and as it forms, the outer periodontal fibers become enclosed within its matrix.

The bone on the pressure side of the socket is usually the "convoluted" type. This is because the bony alveolar plate drifts into the cancellous bone deep to it, and compaction of the spaces converts the cancellous trabeculae into the compact bone of the alveolar cortical lining. If the alveolar wall is very thin, however, with no cancellous bone separating adjacent sockets and with only a single bony plate between them, bundle bone is laid down in one socket and then removed by resorption on its other surface in the other socket as the teeth drift (Fig. 11–36).

**Chondroid bone** is found on the rapidly growing crests of alveolar ridges (Fig. 11–39). It forms as a cap on the apex of the crest. Chondroid bone may also develop elsewhere in the skeleton, such as the apex of a growing tubercle to which a muscle is attached. It has a distinctive, cartilage-like appearance with large, closely arranged cells and a nonlamellar, often basophilic, matrix. Chondroid bone undergoes direct metaplasia into ordinary nonlamellar bone, suggesting that this is one type that can actually undergo a kind of interstitial growth or remodeling. Little is known about the physical properties and functional nature of this bone type. It has not received the extensive study it needs (see Enlow, *Principles of Bone Remodeling,* 1963; *The Human Face,* 1968; and Chapter One in *International Workshop on Complete Denture Occlusion,* 1973; Gussen, 1968a and 1968b).

# Maturation of the Orofacial Neuromusculature*

## PART ONE

More attention has been given to the study of the growth of the craniofacial skeleton and the dentition than to the neuromusculature which activates the masticatory region. The methods of study of the neuromusculature are much more difficult; consequently, we know less about the facial and jaw muscles, and we are less certain of what we do know than we are about our knowledge of bones and teeth. Nonetheless, the basic rules of biology apply. There is just as much variability in the morphology and action of the muscles as there is in tooth anatomy or craniofacial profiles. Muscles grow, develop, and mature in a planned and scheduled way, even as teeth calcify and erupt and bones form and grow. Many malocclusions have their origins in abnormal neuromuscular behavior, and many an orthodontically treated malocclusion is not stable because occlusal stability, in the last analysis, could not be maintained by the muscles.

## CONCEPT 1: CLASSES OF NEUROMUSCULAR ACTIVITIES

**Unconditioned reflexes or responses** are those which are present at birth, having appeared as a normal part of the prenatal maturation of the neuromusculature. It is necessary for certain unconditioned congenital reflexes to be operative in the oropharyngeal region of the neonate in order for him to survive. **Conditioned reflexes** are of two types: those reflexes which appear with normal growth and development and those desirable or undesirable reflexes which have been learned as a singular part of the development of one child. Of course, no conditioned reflex is capable of being learned until all the necessary parts of the central nervous system and musculature have matured sufficiently to make that learning possible. In the orofacial region, the mature swallow and mastication are good examples of reflexes which normally appear with growth and development, while thumb-sucking is an example of an undesirable conditioned reflex. **Voluntary activities** are willful acts

---

*By Robert E. Moyers, D.D.S., Ph.D., Professor of Dentistry (Orthodontics), School of Dentistry, and Director, Center for Human Growth and Development, University of Michigan, Ann Arbor, Michigan.

under cortical control. Such volitional activities must, of course, be separated from the congenital unconditioned responses and the conditioned reflexes which are learned. Some things we do because we willfully choose to do them; other activities in the orofacial region we carry out because we have learned to do them that way, yet all mammals display instinctively primitive, unconditioned neuromuscular activities over which we have very little control.

## CONCEPT 2: PRENATAL MATURATION

During prenatal life, the orofacial region matures well ahead of the limb regions, since the mouth is so concerned with a number of vital functions that must be fully operative at the time of birth, such as respiration, nursing, and protection of the airway. Respiratory reflexes, jaw closure reflexes, gag reflex, suckling, and swallowing are all developed in a scheduled way between the fourteenth and thirty-second week of intrauterine life.

## CONCEPT 3: NEONATAL ORAL FUNCTIONS

The mouth at birth is a very active perceptual system. The infant uses his mouth and face for perceptual functions even more than he does his hands, and this then continues throughout life. The oral area presents in man the highest level of sensory-motor integrative functions.

**Infantile Suckling and Swallowing.** The infantile swallow is a part of the highly complicated suckling reflex. Both suckling and swallowing must be developed by birth so that the infant can feed himself. The infantile swallow is different from the mature swallow which appears later. The infantile swallow is characterized by (1) a positioning of the tongue between the gum pads holding the jaws apart as the swallow is completed, (2) a stabilization of the mandible by contractions of the **facial** muscles and the interposed tongue, and (3) a swallow which is initiated and to a great extent guided by the sensory interchange between the **lips** and the **tongue.** This infantile swallow normally is given up some time during the first year of life.

**Maintenance of the Airway.** The orofacial and jaw musculature is responsible for the vital positional relationships that maintain the airway. The physiologic maintenance of patency of the airway is of vital importance from the first day of extrauterine life. All **learned** jaw functions are built around and accommodated to the mandibular and tongue positions which make possible a clear airway.

## CONCEPT 4: EARLY POSTNATAL DEVELOPMENT OF ORAL NEUROMUSCULAR FUNCTIONS

**Mastication.** Mastication is a learned neuromuscular activity, but it cannot be learned until craniofacial growth has enlarged the intraoral volume, the teeth have erupted into occlusion, the musculature and the temporomandibular joint have matured, and the central nervous system's integrative and coordinative func-

tions are possible. Like the early stages of any new motor skill, the first chewing movements are irregular and poorly coordinated. Sensory guidance during this learning period is provided by receptors in the temporomandibular articulation, periodontal ligament, tongue, oral mucosa, and to some extent, the muscles. The individual's jaw movements during the chewing cycle are a developed, integrated pattern of many functional elements and are highly adaptive in the young child. Masticatory adaptive changes in later years are much more difficult.

**Facial Expression.**   Although many muscle patterns of facial expression are learned, largely through imitation, some facial responses are unlearned and are very similar to basic primitive reflexes seen in certain lower primates.

**Speech.**   While reflex infant cry is an unlearned activity, purposeful speech is much more complicated, for it must be performed on a background of stabilized and learned positions of the mandible, pharynx, and tongue. Speech requires complicated, sophisticated, varying sensory conditioning elements during learning. Infant cry is primitive and unlearned.

**Swallowing.**   The mature swallow usually appears in the latter half of the first postnatal year of life. The arrival of the erupted incisors cues the more precise opening and closing movements of the mandible, compels a more retracted tongue posture, and initiates the learning of mastication. The infantile swallow is related to suckling, the mature swallow to chewing. The transition from the infantile to the mature swallow takes place over several months, dependent upon the timing of the maturation of important developmental neuromuscular events, but most children achieve the mature swallow by 1 1/2 years. Several features characterize the mature swallow: (1) the teeth are together, (2) the mandible is stabilized by contractions of the mandibular elevators (rather than the facial muscles), (3) the tongue tip is held against the palate, above and behind the incisors, and (4) there are minimal contractions of the lips and facial muscles.

**Neural Regulation of Jaw Positions.**   Jaw position, like a number of other automatic somatic activities, normally is largely reflexly controlled, even though it can be altered voluntarily. The receptors of the temporomandibular capsular area are far more important in the control and guidance of jaw function and positions than has previously been thought and is ordinarily taught. Much of our knowledge of jaw position and its regulation has been derived from studies of adults, and it is hazardous to transfer casually to the growing child concepts which may be correct on older persons. Knowledge about the developmental aspects of jaw neurophysiology is most incomplete at this time.

## CONCEPT 5: OCCLUSAL HOMEOSTASIS

The goal of most occlusal treatment by the dentist is to achieve a self-stabilizing occlusal relationship. Modern clinicians have abandoned mechanistic models of occlusion and more practically view the occlusion as the relatively stabilized result of varied and discontinuous dynamic forces operating against the teeth. The sensory receptors of the temporomandibular joints, periodontal ligament, and other parts of the masticatory system provide a constant feedback mechanism for controlling forces against the teeth. Such factors as the growth of the facial bones, the force of muscle contractions during mastication, and the natural tendency of the teeth to drift are probably all far more important in maintaining occlusal homeostasis than the oft-mentioned factors of cuspal anatomy.

## CONCEPT 6: EFFECT OF NEUROMUSCULAR FUNCTION ON FACIAL GROWTH

The role of neuromuscular function in the growth of the craniofacial skeleton has only in recent years been brought into better perspective. Such factors as the growth of the muscles, their migration and attachment, variations in neuromuscular function, and abnormal function (e.g., mouth breathing) are now known to influence markedly some features of craniofacial growth and form.

## CONCEPT 7: EFFECTS OF ORTHODONTIC TREATMENT ON THE MUSCULATURE

The orthodontist's appliances and therapy are not directed solely towards improved positions of the teeth and altering skeletal relationships, for orthodontic treatment also influences the neuromusculature. It should (1) obliterate all neuromuscular reflexes adversely affecting the dentition and the craniofacial skeleton and (2) create an ideal occlusal relationship that is repeatedly stabilized reflexly by the unconscious swallow. The clinician removes disharmonious occlusal influences and utilizes the primitive reflex positions of the mandible to stabilize his clinical result. Obversely, since it has been found that severe malocclusion provokes changes in the temporomandibular articulation and the neuromusculature, proper orthodontic treatment usually results in a reduction in the range of mandibular positions and an improvement in the precise control of mandibular movements. Other adaptive muscular changes which may follow orthodontic treatment include altered lip posture, tongue posture, mandibular posture, chewing stroke, and method of breathing.

During prenatal life the neuromuscular system throughout the body does not mature evenly. It is not accidental that the orofacial region matures (in the neurophysiologic sense) ahead of limb regions, since the mouth is concerned with a number of vital functions which must be operative by birth—for example, respiration, nursing, and protection of the oropharyngeal airway. In the human fetus, by about the eighth week, generalized uniform reflex movements of the entire body can be elicited by tactile stimulation. A few spontaneous movements, in response to as yet unidentified stimuli, have been observed as early as 9 1/2 weeks. Localized specific and more peripheral responses can be produced before 11 weeks. At this time, stimulation of the nose-mouth region causes lateral body flexion. By 14 weeks, the movements have become much more individualized, and very delicate activities can be executed. When the mouth area is stimulated, general bodily movements no longer are seen; instead, facial and orbicular muscle responses are produced. Stimulating the lower lip, for example, causes the tongue to move. Stimulation of the upper lip causes the mouth to close and, often, deglutition to occur.

Respiratory movements of the chest and abdomen are first seen at about 16 weeks. The gag reflex has been demonstrated in the human fetus at about 18 1/2 weeks (menstrual age). By 25 weeks, respiration is shallow but may support life for a few hours if once established.

Stimulation of the mouth at 29 weeks has elicited suckling, although complete suckling and swallowing are not thought to be developed until at least 32 weeks.

Davenport Hooker and Tryphena Humphrey have shown us that there is an orderly, sequential staging of events in prenatal orofacial maturation—a staging seen throughout the body, but which is much more advanced in the oropharyngeal region. All this has to be established by the time of birth in order for the child to survive.

Probably the best single reference for further reading in this interesting area is that of Humphrey (1970).

## NEONATAL ORAL FUNCTIONS

At birth, the tactile acuity is much more highly developed in the lips and mouth than it is in the fingers. The infant carries objects to his mouth to aid in the perception of size and texture long before he inserts them into his mouth as a part of teething. The neonate slobbers, drools, chews his toe, sucks his thumb, and discovers that gurgling sounds can be made with his mouth.

Freudians consider all of this oral eroticism, as they do adult smoking; but in the infant surely it is also exploratory and exercises the most sensitive perceptual system in the body at that time. Oral functions in the neonate are guided primarily by local tactile stimuli, particularly those in the lips and the front part of the tongue.

The tongue at this age does not guide itself; rather, it **follows** superficial sensation. The posture of the neonate's tongue is between the gum pads, and it is often

far enough forward to rest between the lips where it can perform its role of sensory guidance more easily. The young infant, to a great extent, interprets the world with his mouth, and the integration of oral activities is therefore by sensory mechanisms.

If you touch a young child's lips or tongue and have him follow your finger, both his head and body turn. A little bit later he turns his head separately from his body, and still later he will move his mandible without moving his head. It is only last of all that he can follow with the tongue, while not moving his mandible. These stages appear in a natural sequence, just as teeth erupt on a kind of schedule.

The infant uses his mouth for many purposes. The perceptual functions of the mouth and face are combined with the sensory functions of taste, smell, and jaw position. The neonate's primary relationship to his environment is by means of his mouth, pharynx, and larynx. Here a high concentration of readily available receptors becomes stimulated and modulates the already matured brain stem coordinations which regulate respiration and nursing and determine head and neck positions during breathing and feeding.

The sensitivity of the tongue and lips is perhaps greater than that of any other body area. The sensory guidance for oral functioning, including jaw movements, is from a remarkably large area. These sensory inputs are compounded by many dual contacting surfaces, such as the tongue and lips, soft palate and posterior pharyngeal wall, and the compartments of the temporomandibular articulation. A great array of sensory signals is required for the integration, coordination, and interpretation of this complex system.

**Infantile Suckling and Swallowing.**   The effectiveness of these activities is a good indication of the neurologic maturation of premature infants. It has been found that a child will follow the same patterns in certain oral reflex movements years after initial learning. For example, a study was made of children whose records had been kept from infancy. As long as 5 years after weaning, if given a bottle from which to suckle, they produce the same suckling, swallowing, and respiratory rhythms as they had when infants. If they swallowed in a suckle-suckle-swallow type of pattern, which we call a two-for-one, this same rhythm appeared years later. It may be a three-for-one or even a four-for-one, but the pattern is maintained. Such primitive reflexes are difficult for us to change. How foolish it is for us, with our present ignorance about conditioning mechanisms, to try to alter some of these reflexes. We must spend more time with those problems that we have at least a theoretic chance to condition.

Rhythmic elevation and lowering of the jaw provide sequential changes in positions of the tongue in coordination with its suckling contractions. The activities of suckling are closely related temporally to the motor functions of positional maintenance of the airway.

Electromyographic studies in our own laboratory have confirmed visual observations reported in England by a number of people revealing that, while the mandibular movements are carried out by the muscles of mastication, the mandible is primarily stabilized during the actual act of infantile swallowing by concomitant contractions of the tongue and the facial (rather than masticatory) muscles. At the actual time of the infantile swallow, the tongue lies between the gum pads and in close approximation with the lingual surface of the lips. Thus, the infantile swallow is neuromuscularly a different mechanism from the mature swallow.

Characteristic features of the infantile swallow are that (1) the jaws are apart, with the tongue between the gum pads; (2) the mandible is stabilized primarily by contractions of the muscles of the seventh cranial nerve and the interposed tongue; and (3) the swallow is guided and to a great extent controlled by sensory interchange between the lips and tongue.

**Maintenance of the Airway.**    The oral-jaw musculature is responsible for the vital positional relationships which maintain the oral pharyngeal airway. While the infant is resting, a rather uniform diameter for the airway is provided by (1) maintaining the mandible anteroposteriorly and (2) stabilizing the tongue and posterior pharyngeal wall relationships.

The axial musculature around the vertebrae is also concerned. These primitive neonatal protective mechanisms provide the motor background upon which, with growth, all the postural mechanisms of the head and neck region are developed. Physiologic maintenance of the airway is of vital, continuing importance from the first day throughout life.

This little neonate who cannot focus his eyes, who cannot make a purposeful movement with any of his limbs, who cannot hold his head upright, who has absolutely no control of the lower end of his GI tract, has absolutely exquisite control of some functions in the orofacial regions. Why? He must in order to survive!

**Infant Cry.**    When the aroused baby is crying, the oral region is unresponsive to local stimulation. The mouth is held wide open, while the tongue is separated from the lower lip and from the palate. The steady stabilization of the size of the pharyngeal airway is given up during crying; and there are irregular, varying constrictions during expiration of the cry and large, reciprocal expansions during the alternating inspirations.

**Gagging.**    Gagging, the reflex refusal to swallow or accept foreign objects in the throat, is an exaggeration of the protective reflexes guarding the airway and alimentary tract. The gag reflex is present at birth, but it changes as the child grows older in order to accommodate visual, acoustic, olfactory, and psychic stimuli which are remembered and thus condition it.

## EARLY POSTNATAL DEVELOPMENT OF ORAL NEUROMUSCULAR FUNCTIONS

**Mastication.**    The interaction between the rapidly and differentially growing craniofacial skeleton and the maturing neuromuscular system brings about sequentially progressive modifications of the elementary oral functions seen in the neonate. Mandibular growth, downward and forward, is greater during this time than midfacial growth, giving rise to a greater separation of the thyroid bone and thyroid cartilage from the cranial base and mandible.

Maturation of the musculature and delineation of the temporomandibular joint help provide a more stable mandible. Although mandibular growth carries the tongue away from the palate and helps provide differential enlargement of the pharynx, patency of the airway is maintained—a most important point.

The soft palate and the tongue are commonly held in apposition, but as the tongue is no longer lowered by mandibular growth, its functional relationship with the lips is altered, an alteration aided by the vertical development of the alveolar process. So the morphologic relationship of the tongue and lips is strained. At rest now, the tongue is no longer in generalized apposition with the lips, buccal wall, and soft palate. The lips elongate and become more selectively mobile; the tongue develops discrete movements which are separate from lip and mandibular movements. The labial valve mechanism is constantly maintained during rest and feeding so that food is not lost.

The development of speech and mastication as well as facial expression requires a furthering of the independent mobility of the separate parts. In the neonate, however, the lips tightly surround a plungerlike tongue, moving in

synchrony with gross mandibular movements. Speech, facial expression, and mastication require the development of new motor patterns as well as greater autonomy of the motor elements. Not all the developmental aspects of these functions are known. But mastication certainly does not gradually develop from infantile nursing. Rather, it seems that the maturation of the central nervous system permits completely **new** functions to develop. These functions are triggered to an important extent by the eruption of the teeth.

One of the most important factors in the maturation of mastication is the sensory aspect of newly arriving teeth. The muscles controlling mandibular position are cued by the first occlusal contacts of the antagonistic incisors. Serial electromyographic studies at frequent intervals during the arrival of the incisors have demonstrated conclusively that the very instant the maxillary and mandibular incisors accidentally touch one another, the jaw musculature begins to learn to function in accommodation to the arrival of the teeth.

Thus, since the incisors arrive first, the closure pattern becomes more precise anteroposteriorly before it does mediolaterally. All occlusal functions are learned in stages. The central nervous system and the orofacial and jaw musculature mature concomitantly, and usually synchronously, with the development of the jaws and dentition.

The earliest chewing movements are irregular and poorly coordinated, like those during the early stages of learning of any motor skills. As the primary dentition is completed, the chewing cycle becomes more stabilized, using more efficiently the individual's pattern of occlusal intercuspation. In the very young child, sensory guidance for masticatory movement is provided by the receptors in the temporomandibular articulation, the periodontal ligament, the tongue, and the oral mucosa and muscles; of these, it seems by far that the most important are those of the temporomandibular articulations, and next those of the periodontal ligament. Cuspal height, cuspal angle, and incisal guidance (which is usually minimal in the primary dentition) play a role in establishment of chewing patterns in the infant. However, condylar guidance is not important, since the eminentia articularis is ill-defined and the temporal fossae are shallow.

Rather, it may be supposed that the bone of the eminentia articularis forms where temporomandibular function permits (or causes) it to develop. In a similar fashion, the plane of occlusion is established by the growth of the alveolar process, during eruption of the teeth, to heights permitted by the configuration and functioning of the neuromusculature.

The individual's movements during the chewing cycle are a developed, integrated pattern of many functional elements. In the young child, at the time of completion of the primary dentition, masticatory relationships are nearly ideal, since all three systems (bone, teeth, and muscle) still show the lability of development and are highly adaptive. Cusp height and overbite in the primary dentition are more shallow, bone growth more rapid and adaptive, and neuromuscular learning more easily obtained because pathways and patterns of activity are not yet well established. Adaptations to masticatory change are much more difficult in later years, as every dentist knows.

**Facial Expression.**    In a not dissimilar way, most subtle facial expressions are learned, largely by imitation, so we think, and begin about the time the primitive uses of the seventh nerve musculature for infantile swallowing are abandoned. Those of us who are parents imagine all sorts of facial expressions in the young neonate. Actually, observing the infant objectively, we must admit that the expression is often rather blank. The reason is that the facial muscles are busy being used

for the massive efforts of mandibular stabilization necessary during infantile swallowing. Eventually the seventh nerve muscles and the mandible become controlled and stabilized more by the muscles of mastication, particularly during unconscious reflex swallowing, and the delicate muscles of facial expression become free to be truly "muscles of facial expression."

Although many facial expressions are learned through imitation, some facial responses are unlearned and can be traced back to reflexes of earlier primates. Similar facial displays have evolved in the four lines of modern primates in which monkeylike forms have developed. Comparative studies have been made revealing similar reflex expressions of basic protective anger, for example, in various primates—the same primitive instinctive expressions you have seen on your best friend.

**Speech.**   Purposeful speech is different from reflex infant cry. Infant crying is associated with irregular tongue and mandibular positions related to sporadic inspirations and expirations during crying. Speech, on the other hand, is performed on a background of stabilized and learned positions of the mandible, pharynx, and tongue. Infant cry is usually a simple displacement of parts, accompanied by a single explosive emission, whereas speech can only be carried out by polyphasic and sequential motor activities synchronized closely with breathing. Speech is regular; infant cry is sporadic. Speech requires complicated, sophisticated, varying sensory conditioning elements during learning; infant cry is primitive and not learned.

Speech consists of four parts: (1) language—the knowledge of words used in communicating ideas; (2) voice—sound produced by air passing between the vibrating vocal cords of the larynx; (3) articulation—the movement of the speech organs used in producing a sound, that is, lips, tongue, teeth, mandible, palate, and so forth; (4) rhythm—variations of quality, length, timing, and stress of a sound, word, phrase, or sentence. If there is no impairment of hearing, sight, or oral sensation, the child will learn to speak from the speech he hears, reproducing as best he can what he has heard. Speech defects are a loss or disturbance of language, voice, articulation, and rhythm or combinations of such losses and disturbances.

**Mature Swallow.**   During the latter half of the first year of life, several maturational events occur that alter markedly the orofacial musculature's functioning. The arrival of the incisors cues the more precise opening and closing movements of the mandible, compels a more retracted tongue posture, and initiates the learning of mastication. As soon as bilateral posterior occlusion is established (usually with the eruption of the first primary molars), true chewing motions are seen to start, and the learning of the mature swallow begins. Gradually, the fifth cranial nerve muscles assume the role of muscular stabilization during the swallow, and the muscles of facial expression abandon the crude infantile function of suckling and the infantile swallow and then begin to learn the more delicate and complicated functions of speech and facial expressions. The transition from infantile to mature swallow takes place over several months, aided by maturation of neuromuscular elements, the appearance of upright head posture and, hence, a change in the direction of gravitational forces on the mandible, the instinctive desire to chew, the necessity to handle textured food, dentitional development, and so forth. Most children achieve most features of the mature swallow at 12 to 15 months. Characteristic features of the mature swallow are: (1) the teeth are together (although they may be apart with a liquid bolus); (2) the mandible is stabilized by contractions of the fifth cranial nerve muscles; (3) the tongue tip is held against the palate above and behind the incisors; and (4) minimal contractions of the lips are seen during the swallow.

**Neural Regulation of Jaw Positions.**    Jaw position, like a number of other au-
tomatic-somatic activities, normally is largely reflexly controlled, even though it
can be altered voluntarily. A surprising number of jaw functions are carried out at
the subconscious level, even though conscious control is possible and sometimes
necessary. Recent research has shown that the receptors in the temporomandibular
capsule area are far more important than previously thought. Many of the enigmas
of prosthetic dentistry have fallen into place logically as a result of the research efforts of Thilander in Sweden and Greenfield and Wyke in England.

Since most research on the neurophysiologic regulation of jaw position and
function has been done on the adult, there has been a tendency to transfer
prosthodontically oriented concepts, based on sound adult clinical practice, to
children. During development, before all of the system's parts have appeared and
while growth is dominant, it is hazardous to maintain the same clinical assumptions which are so useful in understanding the adult.

Our knowledge about the developmental aspects of orofacial and jaw neurophysiology is most incomplete at this time, though much research is under way.
We must remember that many of our attitudes are victims of our experience with
degenerating occlusions in adults, and the critical clinical factors which obtain
under those circumstances may not be present in the child or may have different
relative significance during development.

Unconditioned jaw positions and functions include mandibular posture for the
maintenance of the airway and the unconscious or reflex swallow. The neural
mechanisms which determine mandibular posture are important to the dentist,
since mandibular posture (sometimes in dentistry called the rest position) is a determinant of the vertical dimension of the face. In the opinion of many, the position of
the mandible during the unconscious swallow is an important factor in occlusal
homeostasis, since every time a person swallows unconsciously, he either stabilizes
his occlusal relationship or, because of tooth interferences, moves interfering teeth
until a stable occlusal relationship finally is obtained.

Conditioned jaw positions and functions include all those of mastication, the
mature swallow, facial expression, and speech.


## OCCLUSAL HOMEOSTASIS

Occlusal stability at any moment is the result of all the summated forces acting
against the teeth. Some of these forces have been measured in research study. Still,
it is not yet possible to describe precisely in summation all the forces and counter-
forces that produce occlusal homeostasis. Occlusal homeostasis is very dependent
upon elaborate and sophisticated sensory feedback mechanisms from the periodontal ligament, temporomandibular joint, and other parts of the masticatory system.
Such sensory feedback serves as a regulating mechanism helping to determine the
strength of muscle contractions. Each individual tooth is positioned between con-
tracting sets of muscles. It is also in contact with adjacent teeth and in occlusion
with the teeth of the opposite arch. A number of physiologic forces determine the
tooth's position occlusally, including eruption, the occlusal force during swallowing, the forces of mastication, occlusal wear of the crown of the tooth, etc.
Occlusal interferences in or near the unconscious swallowing position of the mandible tend to diminish reflexly the force of muscle contractions during the swallow.
Since reflex swallowing occurs so frequently, it acts as a dominant mechanism be-

tween the tooth's position and occlusal stability. Other factors involved in occlusal homeostasis include the natural mesial drifting tendencies of the teeth, the growth of the bones of the craniofacial complex, and alveolar bone growth and remodeling. It is now felt that the neuromuscular mechanisms and bone growth factors are far more important in the nature of occlusal relations than are the oft-mentioned factors of cuspal inclination, cusp height, condylar guidance, etc. The occlusal relationships are nowhere near as stable as depicted in some dental textbooks, if for no other reason than that occlusal adaptations must occur constantly to accommodate, in their way, changes in the neuromusculature and the craniofacial skeleton.

## EFFECT OF NEUROMUSCULAR FUNCTION ON FACIAL GROWTH

From the earliest periods of embryonic growth, an intimate functional relationship exists between muscles and the bones to which they are attached. Obviously, as the bones grow, the muscles must also change their size. Therefore, a relationship exists between the overall growth of any bone and the muscles attached to that bone, and adjustments between muscle and bone are a normal part of growth and development. During growth, muscles must migrate to occupy relatively different positions with time. As the skeleton grows, there is a constant adjustment of the attachment relationships between muscle and skeleton.

Functional use and disuse determine to some extent the thickness of the cortical plate of limb bones. However, the relationship of muscle function and bone growth in the craniofacial skeleton is much more difficult to assess. Certain parts of some of the facial bones are very dependent on function—for example, the alveolar process around the roots of the teeth and the coronoid process to which the temporal muscle is attached. In a more general way, the conformation of the bone and the craniofacial relationships are determined by such factors as mouth breathing, excessive masticatory function, etc. In the case of the cranium, cranial base, and nasomaxillary complex, functional features other than those of muscle apparently play an important role in development and growth—namely, the growth of the brain, the eyeballs, cartilage growth, etc. The mandible, with its important condylar cartilage, holds a special interest for dentists, particularly orthodontists. While there is general agreement that variations in muscle function affect markedly the areas of muscle attachment and that the development and use of the dentition affects the alveolar process, there is some dispute over whether or not muscle function can have a more general effect on the size and form of the mandible. The point is a very important one for orthodontists treating Class II malocclusions among children who are still growting. Although the evidence is still not complete, most workers now feel that function plays a more dominant role in the determination of mandibular size and conformation than was previously thought.

## EFFECTS OF ORTHODONTIC TREATMENT ON THE MUSCULATURE

It is known that severe malocclusion causes pathologic changes in the temporomandibular articulations, which in turn impair the sensory receptors within the joints, causing such orthodontic patients to have a less precise ability to determine

mandibular position than persons who have normal occlusion. After malocclusions have been treated orthodontically, there is a significant reduction in the range of mandibular position and an improvement in the precision of the determination of mandibular position. Occlusal equilibration on treated orthodontic patients has been shown to change a significant number of teeth-apart swallows to teeth-together swallows. Thus, orthodontic treatment, including occlusal equilibration, conditions swallowing reflexes, which in turn help stabilize the orthodontic occlusal result. Occlusal dysharmonies, at the end of orthodontic treatment, have been shown to be disruptive to the stability of treated orthodontic occlusions and thus a most important cause of relapse in treated malocclusions. Other adaptive muscular changes following orthodontic therapy may include altered lip posture, tongue posture, mandibular posture, chewing stroke, and method of breathing.

# The Edentulous Mandible

From a practical clinical point of view, alveolar bone is certainly an important tissue. It has been estimated that about 25 to 30 million people in the United States are edentulous in one or both jaws and that most wear dentures. Periodontal disease affects nearly 80 per cent of our adults, with over half of the population becoming partially or completely edentulous by the age of 60. The incidence is even higher in some other parts of the world. Basic causes of alveolar bone loss are poorly understood, and effective methods for controlling the integrity of alveolar bone before and after tooth loss are at present virtually unknown. The basic, underlying physiology of this important oral tissue is not really well understood, and such knowledge is thus not available for any rationally based clinical method for direct alveolar remodeling control. The overall situation can be regarded as a major and largely unsolved health problem, which can only be described as staggering in scope. New, really effective methods for both prevention and treatment of alveolar bone problems are needed. Many specialties within the dental, medical, biochemical, and biological fields are becoming involved, and this is essential. With regard to remodeling associated with edentulous alveolar bone, the physical, psychologic, and economic problems for millions of people all over the world represent a major, complex oral "disease" that is chronic, progressive, irreversible, and disabling (Atwood, 1971). The multifactorial origins are little understood, and seriously insufficient knowledge presently exists in the overall etiology of the problem.

Refer to page 348 for comments dealing with the basis of alveolar bone resorption when dentures are worn. A tooth in its alveolar socket is supported by the collagenous fibers of the periodontal membrane. This converts the **pressure** exerted by a tooth during mastication into **tension**. Alveolar bone and the periodontal membrane are tolerant of tension, but neither can withstand any undue level of compression. The denture, however, is placed on the bone so that pressure is applied directly to the bone's surface and its overlying soft tissues. This triggers resorption. A man-made denture thus violates the basic biological principle required for the maintenance of bone: the need for a tension type of relationship. The periodontal membrane provides for this, but ordinary denture design, at least at present, does not.

Edentulism does not just affect alveolar bone, however. The **entire** mandible becomes involved (and other bones as well). The distribution of resorptive and depository fields involved in the growth and remodeling of the child's mandible was described in Chapter 3. The pattern of these fields is shown in Figure 13–1 for comparison with the remodeling fields that occur in the **edentulous** mandible, as shown in Figure 13–2. As in the growing mandible of the child, areas of resorption and deposition are present on all the various bone surfaces, but the basic plan is quite

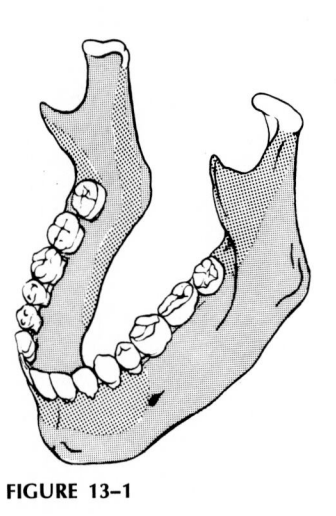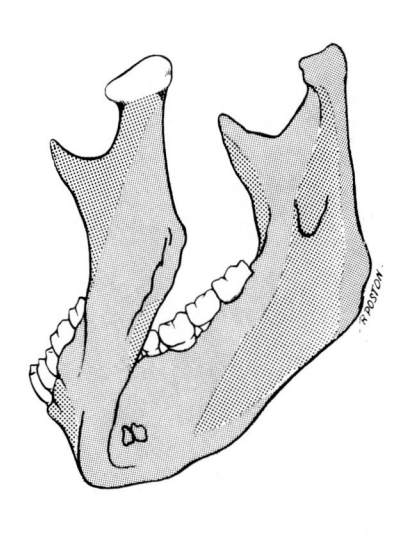

**FIGURE 13–1**

*Distribution of resorptive (darkly stippled) and depository (lightly stippled) fields of growth and remodeling in the mandible of the growing child.*

different. With loss of teeth, whether by periodontal disease, osteoporosis, or other underlying factors, the functional and structural relationships of the whole mandible become changed. The remodeling pattern illustrated in Figure 13–2 represents the means by which these structural changes are brought about. The occlusal relationships become altered, rotations of the entire mandible and also some of its parts occur, a redesign of the morphology of the corpus and ramus takes place, and decreased areas for some muscle attachments result.

In the edentulous (or nearly so) mandible, the overall horizontal dimension of the bony mandibular arch may actually increase (*1a* and *1b* in Fig. 13–3), because resorptive regression occurs along the vertical length of the anterior border of the ramus and temporal crest, and also because the mental protuberance has a depository type of remodeling field. This depository field on the chin's surface is left over

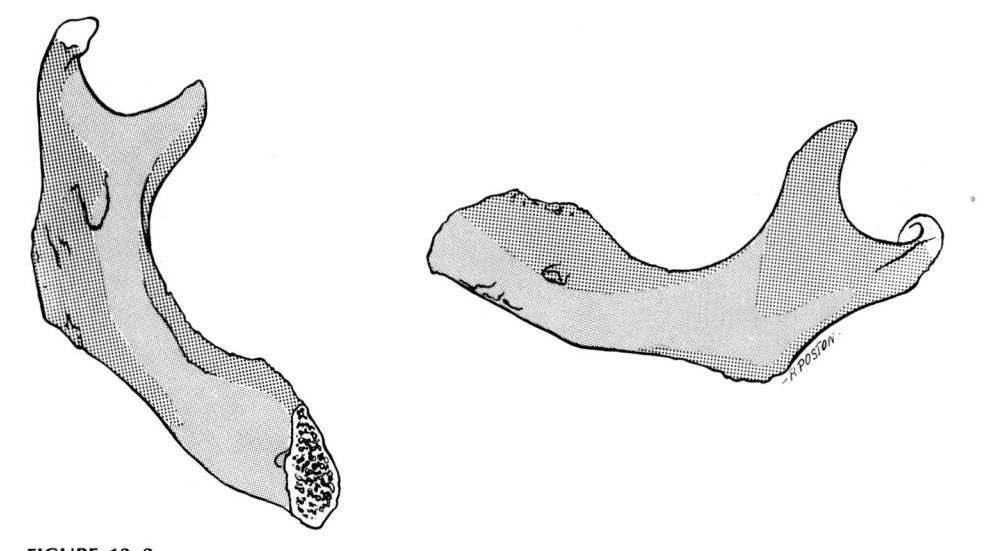

**FIGURE 13–2**

*Resorptive (darkly stippled) and depository (lightly stippled) fields in the edentulous mandible. Compare with Figure 13–1. See text for regional descriptions.*

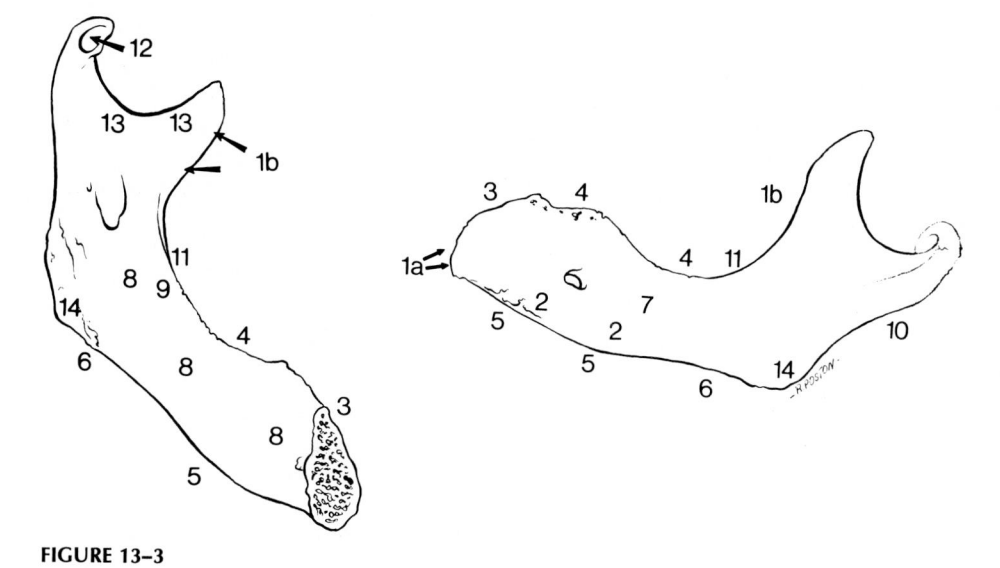

**FIGURE 13–3**

*See text for descriptions. Numbers correspond to text explanations of regional changes in mandibular structure related to edentulism.*

from the original growth field present during childhood mandibular development, and very little if any new bone is added following edentulism. The significance, however, is that this field does not undergo reversal to become actively resorptive in the edentulous mandible.

Because depository fields are present along the lateral (buccal) side of the corpus (2), the overall breadth of the bony **arch** can increase. It becomes more squared and less V-shaped. Each corpus itself also becomes transversely thicker, because deposition occurs on both the lingual and the buccal cortices. (As seen later, however, a common variation can take place in which both arch breadth and the cortex-to-cortex thickness become reduced.)

In the incisor region, an **inferior** and a **posterior** involution of the alveolar bone occurs (3). It is reduced progressively to a "residual ridge." This remodeling change does not directly affect the underlying basal bone, because, as seen above, the latter may actually grow even farther forward at the chin (although only slightly). In the edentulous **maxillary** arch, regression of alveolar bone takes place in a posterior direction back to the basal bone area, but this is located more posteriorly than in the mandible. This, combined with a forward rotation of the whole edentulous mandible (at the condylar pivot) and an increase in the downward angulation of the corpus at its junction with the ramus (which has the effect of lengthening the mandible), results in a marked protrusion of the mandible. Mandibular prognathism and maxillary regression are both characteristic features of the toothless face.

Alveolar involution is brought about by a "growth" movement of the inverted V-shaped ridge and underlying trabeculae toward the basal bone of the corpus (see p. 10). This is accomplished by periosteal resorption on the surface of the ridge and endosteal deposition within it*; the cortical bone thus moves gradually down-

---

*See the "V" principle, page 24.

ward (upward in the maxilla). It is interesting to note also that a **constructive** addi-tion of bone occurs in the empty alveolar **sockets**, while a **destructive** removal of bone occurs, at the same time, on the exterior of the alveolar ridges. Alveolar regression thus is not wholly a resorptive process. It should be pointed out that al-veolar resorption is undesirable when it **leads** to tooth loss. If the teeth are lost for other reasons, however, alveolar bone resorption is, indeed, actually desirable (among animals in general), because it removes the sharp edges that would otherwise damage enclosing soft tissues during mastication. Alveolar bone loss can be regarded as a "disease" only when it causes edentulism, not when it is the result of it (a statement that does not apply to man's use of dentures and the loss of resid-ual alveolar bone that can secondarily result, thereby loosening the denture; the bone loss, nonetheless, is a natural process in response to unusual physical forces).

Note that a difference in the locations of the **reversal lines** (between the resorptive alveolar and the depository basal bone areas) occurs on the lingual as compared with the labial sides in the anterior part of the arch. The line is much lower on the labial side. These reversal lines represent the inferior-most extent to which alveolar reduction is likely to progress unless overridden by heavy mechani-cal or unusual physiologic factors. Importantly, resorptive reduction is less marked and characteristically more stable on the **lingual** in comparison to the labial side in the incisor and canine region. Moreover, **this general area often remains elevated well above the bicuspid and molar regions** in the edentulous mandible in relation to the much higher placement of the reversal line in the anterior-lingual part of the jaw. (Whether this reversal placement is the cause or the effect of the higher incisor residual ridge, however, is not now known.) These factors also relate directly to the posterior directional tendency of alveolar reduction in the incisor region mentioned above, since the labial cortical plate moves in a more posterior direction, owing to its resorptive nature, than does the lingual plate. The labial cortex is also noticeably thin in contrast to the lingual side, which retains substantial thickness.

On the dorsum of the corpus, a downward mode of alveolar resorption takes place from the cuspid to the molar regions (4). The reversal line on the lingual side tends to be much lower than in the incisor region. It usually occurs along the oblique mylohyoid line. The residual ridge of the molar and premolar area is thus characterized by a pronounced dishing as it grades back from the more elevated in-cisor region.

Bone deposition takes place along the entire inferior border of the corpus (5), except in the antegonial region. The underside of the corpus in the antegonial area (6) is characteristically resorptive. These patterns relate to the remodeling changes that produce a more downward-inclined alignment of the whole corpus relative to the ramus, a change that favors somewhat better occlusal closure of the forward-ro-tated edentulous mandible. The effect of the forward and upward rotation of the en-tire mandible, due to loss of teeth, is thereby partially offset. A more pronounced antegonial notch results from this downward corpus alignment alteration, and this is augmented by deposition of bone along the remainder of the inferior border of the corpus and by resorption in the notch itself. A marked and characteristic curva-ture of the ventral profile of the corpus is produced as it grades up into the notch and the relatively high ventral margin of the gonial region (the latter is also resorp-tive, as will be seen). The opening of the angle between the corpus and ramus has a mandibular protrusive effect because the overall length of the mandible is in-creased. This contributes to the extent of edentulous prognathism. (Note: The "gonial angle" is not necessarily altered by these various remodeling changes. This

angle can remain relatively constant because other changes are also involved during ramus adjustments to the loss of teeth.)

Along the buccal side of the **basal bone** area of the corpus (7), bone depositon can occur. This adds to the squaring effect of the bony arch and produces a more pronounced lateral protrusion of the trihedral eminence. Generalized deposition of bone can also occur on the lingual side of the corpus below the mylohyoid line (8). Because this extends into the lingual fossa, the fossa tends to become leveled. A reversal is present along the crest of the mylohyoid line (9), and, superiorly, the alveolar surface is resorptive. In combination with the depository field in the lingual fossa just below the overlying, resorbing alveolar region, the dorsum of the corpus becomes **markedly flattened and widened.** The lingual tuberosity, a prominent feature of the tooth-bearing mandible, disappears.

Both the posterior (10) and anterior (1b) borders of the ramus (and the temporal crest) are resorptive. This narrows the **PA** breadth of the ramus (quite significantly in some individuals but **much** less so in others). The extent of backward remodeling on the anterior border approximates the amount of forward remodeling on the posterior border. The relative placement of the mandibular foramen thereby remains fairly constant (about midway) between the two margins.

As the whole edentulous mandible rotates forward and upward, the occurrence of resorption on the anterior border of the ramus serves to accommodate the mass of the temporalis muscle situated in a squeeze between the ramus and the maxillary tuberosity. Moreover, that part of the anterior ramus border which abuts the maxillary tuberosity during edentulous jaw closure shows a pronounced notching. This can give the coronoid process a flared and sometimes hooklike configuration. Resorption in the lower part of the anterior border of the ramus also functions to grade the surface contour of the ramus onto the resorbing alveolar surface of the dorsum of the corpus where it joins the ramus (11).

On the anterior side of the condylar neck, the subcondylar region has a characteristic resorptive field. This produces a localized dishing effect (12) to accommodate the abutment of the condylar neck with the articular tubercle of the glenoid fossa during edentulous jaw closure.

The lingual surface of the coronoid process and that part of the ramus just below the sigmoid (mandibular) notch are depository areas (13). On the contralateral buccal side, the surface is resorptive. Since the lingual side faces superiorly, this combination produces an elevation of the sigmoid notch, a vertical lengthening of the ramus superiorly, and an elongation of the coronoid process. The latter often rises well above the level of the condyle in many edentulous mandibles (a variation can exist, however, in which different remodeling patterns result in a shortening of the coronoid process, but this is less frequent).

A resorptive field occurs on the lingual side of the gonial region, and a depository field is often present on the opposite buccal side. This results in a tendency for a flaring and a buccal protrusion of the gonial region. A similar tendency, however, can also be found in many tooth-bearing adult mandibles.

The distribution of the resorptive and depository fields illustrated in Figure 13–2 is a composite of the most commonly encountered regional remodeling patterns. All of these specific remodeling features, however, are not usually found in every individual. One or more other remodeling variations, although less frequent in many cases, can exist. Outlined below are common, major variations that may be found in the different localized parts of the edentulous mandible.

A sizable field of resorption can occur on the lateral side of the corpus just an-

terior to the trihedral eminence. The bony arch is narrowed rather than widened by this remodeling change and becomes more V-shaped. In some mandibles, **both** the buccal and lingual sides of the corpus anterior to the trihedral eminence (the former premolar and cuspid area) can be resorptive. In addition to narrowing the arch, this reduces the transverse thickness of each corpus as well. The more stable ventral edge of the corpus, however, remains depository so that the vertical dimension of the corpus is reduced only by the resorptive reduction of the alveolar area.

In some edentulous mandibles, the area of the lingual fossa is resorptive rather than depository. The depth of the fossa is thereby increased in conjunction with the marked resorptive flattening of the overlying alveolar part previously housing the molars. This combination can cause a lingually overhanging bony ledge to form on the dorsum of the corpus.

Resorption can occur on the lingual side of the basal bone cortex in the former premolar and cuspid area, although it does so less frequently than surface deposition. Together with deposition on the contralateral buccal side, this increases arch width but does not materially change the transverse breadth (lingual to buccal cortex) of each side of the arch itself.

Variable areas of deposition can occur in the posterior part of the ramus on the lingual side, a region that is more often almost entirely resorptive in the edentulous mandible. This variation results in a lingual drift of this region and brings it more into line with the axis of the corpus (which lies well lateral to the arch in the tooth-bearing mandible).

In some edentulous mandibles, the entire inferior part of the ramus on the buccal side can be resorptive. Because the contralateral lingual side is also resorptive, each ramus is thereby thinned by reducing the volume of the medullary area between the lingual and buccal cortical plates. The endosteal surfaces of each cortex are depository.

On the lingual side of the ramus anterior to the mandibular foramen, a resorptive field may occur. The effect is a thinning of this region and a depression and flattening of the temporal crest. A decreased area for temporalis muscle insertion also results.

**Chapter Summary.**   In the edentulous mandible, the overall distribution of remodeling fields differs in many ways from that characterizing the growing mandible of the child. The surface of the basal bone on both the lingual and buccal sides of the **entire corpus** is most often of a depository nature, although specific resorptive types of variations can exist. This produces a tendency for an actual increase in arch width and a squaring of arch configuration. The overlying alveolar cortices, on both the buccal and lingual sides, are resorptive. In the anterior (incisor) part of the arch, the placement of the reversal line between the basal depository and alveolar resorptive regions, significantly, is much lower on the labial side (at the level of the mental foramen) than on the lingual side. This produces a posterior direction of alveolar regression. The reversal lines on **both** the lingual and labial sides between the alveolar resorptive and basal depository fields are markedly lower in the cuspid to molar region. These remodeling patterns relate to the characteristic, pronounced dishing of the contour of the residual ridge as it grades back and down from the resultant very high anterior to the resultant low posterior ends of the corpus. The lateral side of the ramus tends to be largely resorptive except for its inferior part, and the posterior half of the lingual side is also typically resorptive. Unlike the child's mandible, the posterior edge of the ramus is usually resorptive. The **PA** dimension of the ramus (**not** the whole mandible) is narrowed in conjunction with

resorption that also takes place on the anterior border. The amount removed from the anterior border is, in effect, added to the longitudinal dimension of the **corpus**. Unless condylar reduction is involved, bone removal from the posterior edge does not reduce the overall length of the mandible. Moreover, the corpus can become lengthened slightly because the mental protuberance is retained as a depository field. Overall mandibular length can become additionally increased by a more downward-inclined alignment of the corpus at its junction with the ramus (this augments the notching effect in the antegonial region, which is resorptive). Together with a forward rotation of the whole mandible because of tooth loss and the lingual direction of maxillary alveolar reduction, all of these various factors result in the multifactorial, composite basis for mandibular protrusion. Notching in the inferior part of the anterior ramus border and on the anterior surface of the condylar neck is associated with resorptive fields caused by pressure contacts with the maxilla and articular tubercle. Several common **variations** in the pattern of remodeling can occur. These affect the breadth of the arch and the thickness of each corpus, the flaring of the gonial region, the thickness of the ramus, and the placement of the corpus relative to the ramus.

# REFERENCES

Acheson, R. M., and M. Archer: Radiological studies of the growth of the pituitary fossa in man. J. Anat., 93:52, 1959.

Ackerman, J. L., Y. Tagaki, W. R. Proffit, and M. J. Baer: Craniofacial growth and development in cebocephalia. Oral Surg., 19:543, 1965.

Ackerman, J. L., J. Cohen, and M. I. Cohen: The effects of quantified pressures on bone. Am. J. Orthod., 52:34, 1966.

Adams, C. D., M. C. Meikle, K. W. Norwick, and D. L. Turpin: Dentofacial remodelling produced by intermaxillary forces in *Macaca mulatta*. Arch. Oral Biol., 17:1519, 1972.

Adams, D., and M. Harkness: Histological and radiographic study of the spheno-occipital synchondrosis in Cynomolgus monkeys, *Macaque irus*. Anat. Rec., 172:127, 1972.

Altemus, L. A.: A comparison of cephalofacial relationships. Angle Orthod., 30:223, 1960.

Amprino, R., and A. Bairati: Processi di ricostruzione e di riassorbimento nella sostanza compatta delle ossa dell'uomo. Z. Zellforsch., 24:439, 1936.

Amprino, R., and G. Godina: La struttura delle ossa nei vertebrati. Comment. Pont. Acad. Sci., 11:329, 1947.

Amprino, R., and G. Marotti: Topographic quantitative study of bone formation and reconstruction. In: *Proceedings of the First European Symposium on Bone and Tooth.* Ed. by H. H. J. Blackwood. New York, Macmillan, 1964.

Anderson, C. E.: The structure and function of cartilage. J. Bone Joint Surg., 44:777, 1962.

Anderson, C. E.: The mechanisms of cartilage growth and replacement in endochondral ossification. In: *Craniofacial Growth in Man.* Ed. by R. E. Moyers and W. M. Krogman. Oxford, Pergamon Press, 1971.

Anderson, J. H., L. Furstman, and S. Bernick: The postnatal development of the rat palate. J. Dent. Res., 46:366, 1967.

Angle, E. H.: Treatment of Malocclusion of the Teeth. Vol. 7. Philadelphia, White Dental Manufacturing Company, 1907, p. 132.

Angle, E. H.: Bone growing. Dent. Cosmos., 52:261, 1910.

Anson, B. J.: Development of the incus of the human ear. Illustrated in Atlas Series. Quart. Bull. Northwest, Univ. Med. Sch., 33:110, 1959a.

Anson, B. J.: Development of the stapes of the human ear. Illustrated in Atlas Series. Quart. Bull. Northwest. Univ. Med. Sch., 33:44, 1959b.

Anson, B. J., and T. H. Bast: The development of the auditory ossicles and associated structures in man. Ann. Otol., 55:467, 1946.

Anson, B. J., T. H. Bast, and S. F. Richany: The fetal and early postnatal development of the tympanic ring and related structures in man. Ann. Otol., 64:802, 1955a.

Anson, B. J., T. H. Bast, and S. F. Richany: The fetal development of the tympanic ring, and related structures in man. Quart. Bull. Northwest. Univ. Med. Sch., 29:21, 1955b.

Ashley-Montagu, M. F.: Form and dimensions of the palate in the newborn. Internatl. J. Orthod., 20:694, 1934.

Ashley-Montagu, M. F.: *Human Heredity.* New York, Mentor Books, 1960.

Ashton, E. H.: Age changes in the basicranial axis of primates. J. Anat., 91:601, 1957a.

Ashton, E. H.: Age changes in the axis of the anthropoidea. Proc. Zool. Soc. (Lond.), 129:61, 1957b.

Asling, C. W.: Congenital defects of the face and palate following maternal deficiency of pteroylglutamic acid. In: *Congenital Anomalies of the Face and Associated Structures.* Ed. by S. Pruzansky. Springfield, Ill., Charles C Thomas, Publisher, 1961.

Asling, C. W., and H. M. Evans: Anterior pituitary regulation of skeletal development. In: *The Biochemistry and Physiology of Bone.* Ed. by G. H. Bourne. New York, Academic Press, 1956.

Asling, C. W., and H. R. Frank: Roentgen cephalometric studies on skull development in rats. I. Normal and hypophysectomized females. Am. J. Phys. Anthrop., 21:527, 1963.

Atkinson, P. J.: Structural aspects of ageing bone. Gerontologia (Basel), 15:171, 1969.

Atkinson, P. J., and C. Woodhead: Changes in human mandibular structure with age. Arch. Oral Biol., 13:1453, 1968.

Atwood, D. A.: Reduction of residual ridges: A major oral disease entity. J. Prosthet. Dent., 26:266, 1971.

Avery, J. K.: Children with cleft lips and cleft palate; embryological basis for defects of the face and palate. Handicapped Children—Problems, Programs, Services in Michigan. University of Michigan Educational Series No. 93, 1961.

Avery, J. K., and R. K. Devine: The development of the ossification centers in the face and palate of normal and cleft palate human embryos. Cleft Palate Bull., 9:25, 1959.

Avis, V.: The relation of the temporal muscle to the form of the coronoid process. Am. J. Phys. Anthrop., 17:99, 1959.

Azuma, M.: Study of histologic changes of periodontal membrane incident to experimental tooth movement. Tokyo Med. Dent. Univ., 17:149, 1970.

Babineau, T. A., and J. H. Kronman: A cephalometric evaluation of the cranial base in microcephaly. Angle Orthod., 39:57, 1969.

Babula, W. J., G. R. Smiley, and A. D. Dixon: The role of the cartilaginous nasal septum in midfacial growth. Am. J. Orthod., 58:250, 1970.

Baer, M. J.: Patterns of growth of the skull as revealed by vital staining. Hum. Biol., 26:80, 1954.

Baer, M. J.: *Growth and Maturation: an Introduction to Physical Development.* Cambridge, Mass., Doyle Publishing Company, 1973.

Baer, M. J., and J. A. Gavan: Symposium on bone growth as revealed by *in vitro* markers. Am. J. Phys. Anthrop., 29:155, 1968.

Baer, M. J., and J. E. Harris: A commentary on the growth of the human brain and skull. Am. J. Phys. Anthrop., 30:39, 1969.

Bahreman, A. A., and J. E. Gilda: Differential cranial growth in rhesus monkeys revealed by several bone markers. Am. J. Orthod., 53:703, 1967.

Balbach, D. R.: The cephalometric relationship between the morphology of the mandible and its future occlusal position. Angle Orthod., 39:29, 1969.

Bang, S., and D. H. Enlow: Postnatal growth of the rabbit mandible. Arch. Oral Biol., 12:993, 1967.

Barber, C.: Effects of the physical consistency of diet on the condylar growth of the rat mandible. J. Dent. Res., 42:848, 1963.

Barber, T. K., and S. Pruzansky: An evaluation of the oblique cephalometric film. J. Dent. Child., 28:94, 1961.

Bassett, C. A. L.: Current concepts of bone formation. J. Bone Joint Surg., 44:1217, 1962.

Bassett, C. A. L.: Electrical effects in bone. Sci. Am., 213:18, 1965.

Bassett, C. A. L.: Electro-mechanical factors regulating bone architecture. In: *Proceedings of the Third European Symposium on Calcified Tissues.* Davos, Switzerland. New York, Springer-Verlag, 1966.

Bassett, C. A. L.: Biologic significance of piezoelectricity. Calcif. Tissue Res., 1:252, 1968.

Bassett, C. A. L.: A biological approach to craniofacial morphogenesis. Acta Morphol. Neerl. Scand., 10:71, 1972.

Baume, L. J.: Physiological tooth migration and its significance for the development of occlusion. II. The biogenesis of accessional dentition. J. Dent. Res., 29:331, 1950.

Baume, L. J.: A biologist looks at the sella point. Trans. Eur. Orthod. Soc., 1957:150, 1958.

Baume, L. J.: Principles of cephalofacial development revealed by experimental biology. Am. J. Orthod., 47:881, 1961a.

Baume, L. J.: The postnatal growth activity of the nasal cartilage septum. Helv. Odont. Acta, 5:9, 1961b.

Baume, L. J.: Ontogenesis of the human temporomandibular joint. I. Development of the condyles. J. Dent. Res., 41:1327, 1962.

Baume, L. J.: Patterns of cephalofacial growth and development. A comparative study of the basicranial growth centers in rat and man. Internatl. Dent. J., 18:489, 1968.

Baume, L. J.: Cephalofacial growth patterns and the functional adaptation of the temporomandibular joint structures. Eur. Orthod. Soc. Trans., 1969:79, 1970.

Baume, L. J., and H. Becks: The development of the dentition of the *Macaca mulatta*—its difference from the human pattern. Am. J. Orthod., 36:723, 1950.

Baume, L. J., and H. Derichsweiler: Is the condylar growth center responsive to orthodontic treatment? Oral Surg., 14:347, 1961a.

Baume, L. J., and H. Derichsweiler: Response of condylar growth cartilage to induced stresses. Science, 134:53, 1961b.

Baumhammers, A., and R. E. Stallard: $S^{35}$-sulphate utilization and turnover by the connective tissues of the periodontium. J. Periodont. Res., 3:187, 1968.

Baumhammers, A., R. E. Stallard, and H. A. Zander: Remodeling of alveolar bone. J. Periodontol., 36:439, 1965.

Baumrind, S.: Reconsideration of the propriety of the "pressure-tension" hypothesis. Am. J. Orthod., 55:12, 1969.

Baumrind, S.: Mapping the skull in 3-D. J. Calif. Dent. Assoc., 48:21, 1972.

Baumrind, S., and R. C. Frantz: The reliability of head film measurements (1). Am. J. Orthod., 60:111, 1971a.

Baumrind, S., and R. C. Frantz: The reliability of head film measurements (2). Am. J. Orthod., 60:505, 1971b.

Baumrind, S., and D. Miller: Computer-aided minimization of landmark location errors of head films. J. Dent. Res., 52:211, 1973, (Suppl.).

Becker, R. O.: The bioelectric factors in amphibian limb regeneration. J. Bone Joint Surg., 43:643, 1961.

Becker, R. O.: The direct current field: A primitive control and communication system related to growth processes. In: Proceedings of the SVI International Congress of Zoology. Vol. 3. Washington, D. C., 1963.

Becker, R. O., and D. G. Murray: A method for producing cellular dedifferentiation by means of very small electrical currents. Trans. N. Y. Acad. Sci., 29:606, 1967.

Becker, R. O., C. A. L. Bassett, and C. H. Bachman: The bioelectric factors controlling bone structure. In: Bone Biodynamics. Boston, Little, Brown and Company, 1964.

Beer, G. R., de: The Development of the Vertebrate Skull. Oxford, Clarendon Press, 1937.

Belanger, L. R.: Osteocytic osteolysis. Calcif. Tissue Res., 4:1, 1969.

Bergsma, D. (Ed.): Birth Defects Atlas and Compendium. Baltimore, Williams and Wilkins, 1973.

Bhaskar, S. N., J. P. Weinmann, and I. Schour: Role of Meckel's cartilage in the development and growth of the rat mandible. J. Dent. Res., 32:398, 1953.

Bimler, H. P.: Stomatopedics in theory and practice. Internatl. J. Orthod., 3:5, 1965.

Birch, R. H.: Foetal retrognathia and the cranial base. Angle Orthod., 38:231, 1968.

Birkby, W. H.: An evaluation of race and sex identification from cranial measurements. Am. J. Phys. Anthrop., 24:21, 1966.

Bjork, A.: The face in profile. Sven. Tandlak. Tidskr., 40:56, 1947.

Bjork, A.: Cranial base development. Am. J. Orthod., 41:198, 1955a.

Bjork, A.: Facial growth in man, studied with the aid of metallic implants. Acta Odont. Scand., 13:9, 1955b.

Bjork, A.: Variations in the growth pattern of the human mandible: longitudinal radiographic studies by the implant method. J. Dent. Res., 42:400, 1963.

Bjork, A.: Sutural growth of the upper face studied by the metallic implant method. Acta Odont. Scand., 24:109, 1966.

Bjork, A.: The use of metallic implants in the study of facial growth in children: Method and application. Am. J. Phys. Anthrop., 29:243, 1968.

Bjork, A.: Prediction of mandibular growth rotation. Am. J. Orthod., 55:535, 1969.

Bjork, A.: The role of genetic and local environmental factors in normal and abnormal morphogenesis. Acta Morphol. Neerl. Scand., 10:49, 1972.

Bjork, A., and T. Kudora: Congenital bilateral hypoplasia of the mandibular condyles. Am. J. Orthod., 54:584, 1968.

Bjork, A., and V. Skieller: Facial development and tooth eruption. Am. J. Orthod., 62:339, 1972.

Blackwood, H. J.: The double-headed mandibular condyle. Am. J. Phys. Anthrop., 15:1, 1957.

Blackwood, H. J.: Vascularization of the condylar cartilage. J. Dent. Res., 37:753, 1958.

Blackwood, H. J.: Vascularization of the condylar cartilage of the human mandible. J. Anat., 99:551, 1965a.

Blackwood, H. J.: Cell differentiation in the mandibular condyle of the rat and man. Calcif. Tissue Res., 1964:23, 1965b.

Blackwood, H. J.: Growth of the mandibular condyle of the rat studied with tritiated thymidine. Arch. Oral Biol., 11:493, 1966.

Bloore, J. A., L. Furstman, and S. Bernick: Postnatal development of the cat palate. Am. J. Orthod., 56:505, 1969.

Bohatirchuk, F.: Calciolysis as the initial stage of bone resorption; a stain historadiographic study. Am. J. Med., 41:836, 1966.

Bosma, J. F.: Maturation of function of the oral and pharyngeal region. Am. J. Orthod., 49:94, 1963.

Bosma, J. F.: Form and function in the infant's mouth and pharynx. In: Third Symposium on Oral Sensation and Perception. Ed. by James Bosma. Springfield, Ill., Charles C Thomas, Publisher, 1972.

Bowden, C. M., and M. W. Kohn: Mandibular deformity associated with unilateral absence of the condyle. J. Oral Surg., 31:469, 1973.

Boyne, P.: Autogenous cancellous bone and marrow transplants. Clin. Orthop., 73:199, 1970.

Brader, A. C.: Dental arch form related with intraoral forces: PR = C. Am. J. Orthod., 61:541, 1972.

Brash, J. C.: The Growth of the Jaws and Palate. London, Dental Board of the United Kingdom, 1924.

Brash, J. C.: The growth of the alveolar bone and its relation to the movements of the teeth, including eruption. Dent. Rec., 46:641, 1926.

Brash, J. C.: The growth of the alveolar bone and its relation to the movements of the teeth, including eruption. Dent. Rec., 47:1, 1927.

Brash, J. C.: The growth of the alveolar bone and its relation to the movements of the teeth, including eruption. Internatl. J. Orthod., 14:196, 1928.

Brash, J. C.: Some problems in the growth and developmental mechanics of bone. Edinb. Med. J., 41:305, 1934.

Brash, J. C., H. T. A. McKeag, and J. H. Scott: The aetiology of irregularity and malocclusion of the teeth. London, Dental Board of the United Kingdom, 1956.

Bremers, L. M. H.: De condylus mandibulae *in vitro*. Thesis. Katholieke Univ. te Nijmegen, 1973.

Broadbent, B. H.: A new x-ray technique and its application to orthodontia. Angle Orthod., 1:45, 1931.

Broadbent, B. H.: The face of the normal child. Angle Orthod., 7:183, 1937.

Brodie, A. G.: Present status of knowledge concerning movement of the tooth germ through the jaw. J.A.D.A., 21:1830, 1934.

Brodie, A. G.: Behavior of normal and abnormal facial growth patterns. Am. J. Orthod. & Oral Surg., 27:633, 1941a.

Brodie, A. G.: On the growth pattern of the human head. Am. J. Anat., 68:209, 1941b.

Brodie, A. G.: Facial patterns: a theme on variation. Angle Orthod., 16:75, 1946.

Brodie, A. G.: The growth of the jaws and the eruption of the teeth. Oral Surg., 1:334, 1948.

Brodie, A. G.: Cephalometric roentgenology: History, technics and uses. J. Oral Surg., 7:185, 1949.

Brodie, A. G.: Late growth changes in the human face. Angle Orthod., 23:146, 1953.

Brodie, A. G.: The behavior of the cranial base and its components as revealed by serial cephalometric roentgenograms. Angle Orthod., 25:148, 1955.

Brodie, A. G.: Craniometry and cephalometry as applied to the living child. In: *Pediatric Dentistry*. Ed. by M. M. Cohen. St. Louis, Mo., C. V. Mosby Company, 1961.

Brodie, A. G.: The apical base: zone of interaction between the intestinal and skeletal systems. Angle Orthod., 36:136, 1966.

Brookes, M.: Blood supply of developing bone and its possible bearing on malformation of the limb and face in congenital haemangiomatous disorders. Proc. R. Soc. Med., 65:597, 1972.

Bruce, R. A., and J. R. Hayward: Condylar hyperplasia and mandibular asymmetry. J. Oral Surg., 26:281, 1968.

Burdi, A. R.: Sagittal growth of the naso-maxillary complex during the second trimester of human prenatal development. J. Dent. Res., 44:112, 1965.

Burdi, A. R.: Catenary analysis of the maxillary dental arch during human embryogenesis. Anat. Rec., 154:13, 1966.

Burdi, A. R.: Morphogenesis of mandibular dental arch shape in human embryos. J. Dent. Res., 47:50, 1968.

Burdi, A. R.: Cephalometric growth analyses of the human upper face region during the last two trimesters of gestation. J. Anat., 125:113, 1969.

Burdi, A. R.: The premaxillary-vomerine junction: an anatomic viewpoint. Cleft Palate J., 8:364, 1971.

Burdi, A. R., and K. Faist: Morphogenesis of the palate in normal human embryos with special emphasis on the mechanisms involved. Am. J. Anat., 120:149, 1967.

Burdi, A. R., and R. G. Silvey: Sexual differences in closure of the human palatal shelves. Cleft Palate J., 6:1, 1969.

Burstone, C. J.: Integumental profile. Am. J. Orthod., 44:1, 1958.

Burstone, C. J.: Integumental contour and extension patterns. Angle Orthod., 29:93, 1959.

Burstone, C. J.: Biomechanics of tooth movement. In: *Vistas in Orthodontics*. Ed. by B. T. Kraus and R. A. Riedel. Philadelphia, Lea & Febiger, 1962.

Burstone, C. J.: Lip posture and its significance in treatment planning. Am. J. Orthod., 53:262, 1967.

Campo, R. D., and C. D. Tourtellotte: The composition of bone and cartilage. Biochem. Biophys. Acta, 141:614, 1967.

Cannon, J.: Craniofacial height and depth increments in normal children. Angle Orthod., 40:202, 1970.

Carlson, B. M.: Relationship between the tissue and epimorphic regeneration of muscles. Am. Zool., 10:175, 1970.

Castelli, W. A., P. C. Ramirez, and A. R. Burdi: Effect of experimental surgery on mandibular growth in Syrian hamsters. J. Dent. Res., 50:356, 1971.

Chaconas, S. J.: A statistical evaluation of nasal growth. Am. J. Orthod., 56:403, 1969.

Chalanset, C.: Utilisation de la methode des ellipses equiprobables dans l'étude de la variabilité de points cranio-faciaux, chez l'enfant de race blanche, au cours de la croissance. Orientation vestibulaire sur des projections sagittales. Thèse, Acad. Paris, Univ. Paris VI, 1973.

Charles-Severe, J.: Le coefficient (r') d'allongemenet elliptique. Son application a l'étude des nuages de points craniometriques d'une population de jeunes adultes. Thèse, Faculte libre de Medecine de Lille, 1972.

Charlier, J. P.: Les facteurs macaniques dans la croissance de l'arc basal mandibulaire a la lumiere de l'analyse des caracteres structuraux et des proprietes biologiques du cartilage condylien. L'Orthod. Franc., 38:1, 1967.

Charlier, J. P., and A. Petrovic: Recherches sur la mandibule de rat en culture d'organes: le cartilage condylien a-t-il un potentiel de croissance independant? Orthod. Fr., 38:165, 1967.

Charlier, J. P., A. Petrovic, and J. Herrmann: Déterminisme de la croissance mandibulaire: effets de l'hyperpulsion et de l'hormone somatotrope sur la croissance condylienne de jeunes rats. Orthod. Fr., 39:567, 1968.

Charlier, J. P., A. Petrovic, and J. Herrmann-Stutzmann: Effects of mandibular hyperpulsion on the prechondroblastic zone of young rat condyle. Am. J. Orthod., 55:71, 1969.

Charlier, J. P., A. Petrovic, and G. Linck: La fronde mentonniere et son action sur la croissance mandibulaire. Recherches experimentales chez la rat. Orthod. Fr., 40:99, 1969.

Chase, S. W.: The early development of the human premaxilla. J.A.D.A., 29:1991, 1942.

Christiansen, R. L., and C. J. Burstone: Centers of rotation within the periodontal space. Am. J. Orthod., 55:353, 1969.

Cicmanec, J. L., D. H. Enlow, and B. J. Cohen: Polyostotic osteophytosis in a rhesus monkey. Lab. Anim. Sci., 22:2, 1972.

Cleall, J. F.: Bone marking agents for the longitudinal study of growth in animals. Arch. Oral Biol., 9:627, 1964.

Cleall, J. F.: Normal craniofacial skeletal growth of the rat. Am. J. Phys. Anthrop., 29:225, 1968.

Cleall, J. F.: Growth of the craniofacial complex in the rat. Am. J. Orthod., 60:368, 1971.

Cleall, J. F.: Growth of the palate and maxillary dental arch. J. Dent. Res., 53:1226, 1974.

Cleall, J. F., G. W. Wilson, and D. S. Garnett: Normal craniofacial skeletal growth of the rat. Am. J. Phys. Anthrop., 29:225, 1968.

Clements, B. S.: Nasal imbalance in the orthodontic patient. Am. J. Orthod., 55:244, 1969.

Coben, S. E.: The integration of facial skeletal variants. Am. J. Orthod., 41:407, 1955.

Coben, S. E.: Growth concepts. Angle Orthod., 31:194, 1961.

Coben, S. E.: Growth and Class II treatment. Am. J. Orthod., 52:5, 1966.

Coben, S. E.: The biology of Class II treatment. Am. J. Orthod., 59:470, 1971.

Cochran, G. V. B., R. J. Pawluk, and C. A. L. Bassett: Stress generated electric potentials in the mandible and teeth. Arch. Oral Biol., 12:917, 1967.

Conklin, J. L., D. H. Enlow, and S. Bang: Methods for the demonstration of lipid as applied to compact bone. Stain Technology, 40:183, 1965.

Cousin, R. P., and R. Fenart: La rotation globale de la mandibule infantile envisagée dans sa variabilité. Etude en orientation vestibulaire. Orthod. Fr., 42:225, 1971.

Cousin, R. P., J. Dardenne, and R. Fenart: Apports de l'orientation vestibulaire a l'étude des correlations cranio-facilaes chez l'adulte. Trans. Eur. Orthod. Soc., Report of the 44th Congress of Monaco, 209, 1968.

Cox, N. H.: Morfogenese en vascularisatie van het secundaire palatum van de rat. Thesis, Katholieke Univ. te Nijmegen, 1973.

Craven, A. H.: Growth in width of the head of the Macaca Rhesus monkey as revealed by vital staining. Am. J. Orthod., 42:341, 1956.

Crelin, E. S., and W. E. Koch: An autoradiographic study of chondrocyte transformation into chondroclasts and osteocytes during bone formation in vitro. Anat. Rec., 158:473, 1967.

Dahl, E.: Craniofacial morphology in congenital clefts of the lip and palate. An x-ray cephalometric study of young adult males. Acta Odont. Scand., 28 (Suppl. 57):11, 1970.

Dahlberg, A. A.: Concepts of occlusion in physical anthropology and comparative anatomy. J.A.D.A., 46:530, 1953.

Dahlberg, A. A.: Evolutionary background of dental and facial growth. J. Dent. Res., 44(Suppl.):151, 1965.

Dale, J. G., A. M. Hunt, G. Pudy, and D. Wagner: Autoradiographic study of the developing temporomandibular joint. Can. Dent. Assoc. J., 29:27, 1963.

Dardenne, J.: Etude comparative des principaux parametres sagittaux de la face et du crane, chez l'Homme et les chimpanzes par la methode vestibulaire d'orientation. Thèse, Univ. de Lille, 1970.

Das, A., J. Meyer, and H. Sicher: X-ray and alizarin studies on the effect of bilateral condylectomy in the rat. Angle Orthod., 35:138, 1965.

Dass, R., and S. S. Makhni: Ossification of ear ossicles. The stapes. Arch. Otolaryngol., 84:88, 1966.

Davenport, C. B.: Postnatal development of the head. Proc. Am. Phil. Soc., 83:1, 1940.

Davidovitch, Z., J. L. Shanfeld, and P. J. Batastini: Increased production of cyclic AMP in mechanically stressed alveolar bone in cats. Eur. Orthod. Soc. Trans. p. 477, 1972.

De Angelis, V.: Autoradiographic investigation of calvarial growth in the rat. Am. J. Anat., 123:359, 1968.

De Angelis, V.: Observations on the response of alveolar bone to orthodontic force. Am. J. Orthod., 58:284, 1970.

DeBeer, G. R.: The Development of the Vertebrate Skull. London, Oxford, 1937.

DeCoster, L.: Une methode d'analyse des malformations maxillo-faciales. La Province Dentaire, 5:269, 1931.

DeCoster, L.: Une nouvelle ligne de reference pour l'analyse des tele-radiographics sagittales en orthodontie. Rev. Stomatol., 11:937, 1951.

Delaire, J.: La croissance des os de la voute du crane. Principes generaux. Rev. Stomatol., 62:518, 1961.

Delaire, J.: Malformations faciales et asymetrie de la base due crane. Rev. Stomatol., 66:379, 1965.

Delaire, J.: Considerations sur la croissance faciale (en particulier du maxillaire superieur). Deductions therapeutiques. Rev. Stomatol., 72:57, 1971.

Delaire, J., and J. Billet: Considerations sur la croissance de la region zygomato-malaire et ses anomalies morphologiques. Rev. Stomatol., 66:205, 1965.

Delattre, A., and R. Fenart: L'hominisation du crane. Editions du Centre National de la Recherche Scientifique, Paris, 1960.

Demoge, P. H.: La recherche en orthopedie dento-faciale. Orthod. Fr., 39:1, 1968.

Dempster, W. T.: Selected Dissections of the Facial Regions for Advanced Dental Students. Ann Arbor, Michigan, Overbeck Company, 1960.

Dempster, W. T.: Patterns of vascular channels in the cortex of the human mandible. Anat. Rec., 135:189, 1959b.

Dempster, W. T., and D. H. Enlow: Osteone organization and the demonstration of vascular canals in the compacta of the human mandible. Anat. Rec., 133:268, 1959a.

Dempster, W. T., W. J. Adams, and R. A. Duddles: Arrangement in the jaws of the roots of the teeth. J.A.D.A., 67:779, 1963.

Diamond, M.: Posterior growth of maxilla. Am. J. Orthod., 32:359, 1946.

Dixon, A. D.: The early development of the maxilla. Dent. Pract., 33:331, 1953.

Dixon, A. D.: The development of the jaws. Dent. Pract., 9:10, 1958.

Dixon, A. D., and D. A. N. Hoyte: A comparison of autoradiographic and alizarin techniques in the study of bone growth. Anat. Rec., 145:101, 1963.

Dolan, K.: Cranial suture closure in two species of South American monkeys. Am. J. Phys. Anthrop., 35:109, 1971.

Dorenbos, J.: Craniale Synchondroses. Doctoral thesis, Central Drukkerij n.v. Nijmegen, University of Nijmegen, 1971.

Dorenbos, J.: In vivo cerebral implantation of the anterior and posterior halves of the spheno-occipital synchondrosis in rats. Arch. Oral Biol., 17:1067, 1972.

Dorenbos, J.: Morphogenesis of the spheno-occipital and the presphenoidal synchondrosis in the cranial base of the fetal Wistar rat. Acta Morphol. Neerl. Scand., 11:63, 1973.

Doty, S. B., and B. H. Schofield: Electron microscope localization of hydrolytic enzymes in osteoclasts. Histochem. J., 4:245, 1972.

Doty, S. B., B. H. Schofield, and R. A. Robinson: The electron microscopic identification of acid phosphatase and adenosine triphosphatase in bone cells following parathyroid extract or thyrocalcitonin administration. In: *Parathyroid Hormone and Thyrocalcitonin (calcitonin).* Ed. by R. V. Talmage and L. F. Belanger. Amsterdam, Excerpta Medica, 1967.

Doty, S. B., R. Jones, and G. A. Finerman: Diphosphonate influence on bone cell structure and lysosomal activity. J. Bone Joint Surg., 54:1128, 1972.

Downs, W. B.: Variations in facial relations: Their significance in treatment and prognosis. Am. J. Orthod., 34:812, 1948.

Downs, W. B.: The role of cephalometrics in orthodontic case analysis and diagnosis. Am. J. Orthod., 38:162, 1952.

Downs, W. B.: Analysis of the dento-facial profile. Angle Orthod., 26:191, 1956.

Drachman, D. B. (Ed.): Trophic functions of the neuron. Ann. N.Y. Acad. Sci., 228:1, 1974.

Droel, R., and R. J. Isaacson: Some relationships between the glenoid fossa position and various skeletal discrepancies. Am. J. Orthod., 61:64, 1972.

Du Brul, E. L., and D. M. Laskin: Preadaptive potentiality of the mammalian skull. Anat. Rec., 138:345, 1960.

Du Brul, E. L., and D. M. Laskin: Preadaptive potentialities of the mammalian skull: An experiment in growth and form. Am. J. Anat., 109:117, 1961.

Du Brul, E. L., and H. Sicher: *The Adaptive Chin.* Springfield, Ill., Charles C Thomas, Publisher, 1954.

Dudas, M., and V. Sassouni: The hereditary components of mandibular growth: A longitudinal twin study. Angle Orthod., 43:314, 1973.

Dufresnoy, P.: Recherche des differences sexuelles du neurocrane sagittal par la methode des "droites frontieres." Thèse, Univ. de Nancy, 1973.

Dullemeijer, P.: Some methodology problems in a holistic approach to functional morphology. Acta Biotheor., 18:203, 1968.

Dullemeijer, P.: Comparative ontogeny and cranio-facial growth. In: *Cranio-facial Growth in Man.* Ed. by R. E. Moyers and W. M. Krogman, Oxford, Pergamon Press, 1971.

Durkin, J. F.: Secondary cartilage: a misnomer? Am. J. Orthod., 62:15, 1972.

Durkin, J. F., J. T. Irving, and J. D. Heeley: A comparison of the circulatory and calcification patterns in the mandibular condyle in the guinea pig with those found in the tibial epiphyseal and articular cartilages. Arch. Oral Biol., 14:1365, 1969.

Durkin, J. F., J. D. Heeley, and J. T. Irving: The cartilage of the mandibular condyle. Oral Sci. Rev., 2:29, 1973.

Duterloo, H. S.: In vivo implantation of the mandibular condyle of the rat. Doctoral Dissertation, U. of Nijmegen, 1967.

Duterloo, H. S., and D. H. Enlow: A comparative study of cranial growth in *Homo* and *Macaca.* Am. J. Anat., 127:357, 1970.

Duterloo, H. S., and H. W. B. Jansen: Chondrogenesis and osteogenesis in the mandibular condylar blastema. Eur. Orthod. Soc. Trans., 1969:109, 1970.

Duterloo, H. S., and J. M. Wolters: Experiments of the significance of articular function as a stimulating chondrogenic factor for the growth of secondary cartilages of the rat mandible. Eur. Orthod. Soc. Trans., 1971:103, 1972.

Eccles, J. D.: Studies on the development of the periodontal membrane: the principal fibers of the molar teeth. Dent. Pract., 10:31, 1959.

Edwards, L. F.: The edentulous mandible. J. Prosthet. Dent., 4:222, 1954.

Elgoyhen, J. C., R. E., Moyers, J. A. McNamara, Jr., and M. L. Riolo: Craniofacial adaptation of protrusive function in young rhesus monkeys. Am. J. Orthod., 62:469, 1972a.

Elgoyhen, J. C., M. L. Riolo, L. W. Graber, R. E. Moyers, and J. A. McNamara, Jr.: Craniofacial growth in juvenile *Macaca mulatta:* a cephalometric study. Am. J. Phys. Anthrop., 36:369, 1972b.

Ellison, M. L., and J. W. Lash: Environmental enhancement of *in vitro* chondrogenesis. Dev. Biol., 26:486, 1971.

Engel, M. B.: Lability of bone. Angle Orthod., 22:116, 1952.

Engel, M. B., and A. G. Brodie: Condylar growth and mandibular deformities. Surgery, 22:975, 1947.

Engel, M. B., J. B. Richmond, A. G. Brodie: Mandibular growth disturbance in rheumatoid arthritis of childhood. Am. J. Dis. Child., 78:728, 1949.

Enlow, D. H.: A plastic-seal method for mounting sections of ground bone. Stain Technol., 29:21, 1954.

Enlow, D. H.: Decalcification and staining of ground thin sections of bone. Stain Technol., 36:250, 1961.

Enlow, D. H.: Functions of the Haversian system. Am. J. Anat., 110:269, 1962a.

Enlow, D. H.: A study of the postnatal growth and remodeling of bone. Am. J. Anat., 110:79, 1962b.

Enlow, D. H.: *Principles of Bone Remodeling.* Springfield, Ill., Charles C Thomas, Publisher, 1963.

Enlow, D. H.: Direct medullary-to-periosteal transition and the occurrence of subperiosteal haematopoietic islands. Arch. Oral Biol., 10:545, 1965a.

Enlow, D. H.: Mesial drift as a function of growth. Symp. on Growth, Univ. West Indies, West Indian Med. J., 14:124, 1965b.

Enlow, D. H.: A comparative study of facial growth in *Homo* and *Macaca.* Am. J. Phys. Anthrop., 24:293, 1966a.

Enlow, D. H.: An evaluation of the use of bone histology in forensic medicine and anthropology. In: *Studies on the Anatomy and Function of Bone and Joints.* Ed. by F. G. Evans. New York, Springer-Verlag, 1966b.

Enlow, D. H.: A morphogenetic analysis of facial growth. Am. J. Orthod., 52:283, 1966c.

Enlow, D. H.: Osteocyte necrosis in normal bone. J. Dent. Res., 45:213, 1966d.

Enlow, D. H.: Morphogenic interpretation of cephalometric data. J. Dent. Res., 46:1209, 1967.

Enlow, D. H.: The bone of reptiles. In: *Biology of the Reptilia.* New York, Academic Press, 1968a.

Enlow, D. H.: *The Human Face: An Account of the Postnatal Growth and Development of the Craniofacial Skeleton.* New York, Hoeber Medical Division, Harper and Row, Publishers, 1968b.

Enlow, D. H.: Wolff's Law and the factor of architectonic circumstance. Am. J. Orthod., 54:803, 1968c.

Enlow, D. H.: Postnatal facial growth. In: *Cleft Lip and Palate.* Ed. by W. Grabb, S. W. Rosenstein, and K. R. Bzoch. Boston, Little, Brown and Company, 1971.

Enlow, D. H.: Facial growth and development. In: *Handbook of Orthodontics.* Ed. by R. E. Moyers, 1973a.

Enlow, D. H.: Alveolar bone. In: *International Workshop on Complete Denture Occlusion.* Univ. of Michigan, School of Dentistry, Report on the Workshop, 1973b.

Enlow, D. H.: Growth and the problem of the local control mechanism. Editorial, Am. J. Anat., 178:2, 1973c.

Enlow, D. H.: Croissance et architecture de la face. Pedod. Fr., 6:122, 1974a.

Enlow, D. H.: The PM boundary: a natural cephalometric plane. Anat. Rec., 178, 1974b.

Enlow, D. H.: Postnatal growth and development of the face and cranium. In: *Scientific Foundations of Dentistry.* Ed. by B. Cohen and I. R. H. Kramer. London, Heinemann Publishers, 1975.

Enlow, D. H., and M. Azuma: Functional growth boundaries in the human and mammalian face. In: *Morphogenesis and Malformations of the Face and Brain.* Langman, J. (Ed.). White Plains, N.Y., National Foundation, 1975.

Enlow, D. H., and S. Bang: Growth and remodeling of the human maxilla. Am. J. Orthod., 51:446, 1965.

Enlow, D. H., and S. O. Brown: A comparative histological study of fossil and recent bone tissues. Part I. Introduction, methods, fish and amphibian bone tissues. Tex. J. Sci., 7:405, 1956.

Enlow, D. H., and S. O. Brown: A comparative histological study of fossil and recent bone tissues. Part II. Reptilian and bird bone tissues. Tex. J. Sci., 9:186, 1957.

Enlow, D. H., and S. O. Brown: A comparative histological study of fossil and recent bone tissues. Part III. Mammalian bone tissues. General discussion. Tex. J. Sci., 10:187, 1958.

Enlow, D. H., and D. B. Harris: A study of the postnatal growth of the human mandible. Am. J. Orthod., 50:25, 1964.

Enlow, D. H., and W. S. Hunter: A differential analysis of sutural and remodeling growth in the human face. Am. J. Orthod., 52:823, 1966.

Enlow, D. H., and W. S. Hunter: Growth of the face in relation to the cranial base. Europ. Orthod. Soc., Report of the 44th Congress, 1968.

Enlow, D. H., and J. McNamara: The neurocranial basis for facial form and pattern. Angle Orthod., 1973a.

Enlow, D. H., and J. McNamara: Varieties of *in vivo* tooth movements. Angle Orthod., 1973b.

Enlow, D. H., and R. E. Moyers: Growth and architecture of the face. J.A.D.A., 82:763, 1971.

Enlow, D. H., J. L. Conklin, and S. Bang: Observations on the occurrence and the distribution of lipids in compact bone. Clin. Orthop., 38:157, 1965.

Enlow, D. H., R. E. Moyers, W. S. Hunter, and J. A. McNamara, Jr.: A procedure for the analysis of intrinsic facial form and growth. Am. J. Orthod., 56:6, 1969a.

Enlow, D. H., P. Williams, and K. Williams: An instrument for the analysis of facial growth. Angle Orthod., 39:316, 1969b.

Enlow, D. H., T. Kuroda, and A. B. Lewis: The morphological and morphogenetic basis for craniofacial form and pattern. Angle Orthod., 41:161, 1971a.

Enlow, D. H., T. Kuroda, and A. B. Lewis: Intrinsic craniofacial compensations. Angle Orthod., 41:271, 1971b.

Epker, B. N., and H. M. Frost: Correlation of bone resorption and formation with the physical behavior of loaded bone. J. Dent. Res., 44:33, 1965.

Epker, B. N., F. A. Henny, and H. M. Frost: Biomechanical control of bone modeling and architecture. J. Bone Joint Surg., 50:1261, 1968.

Evans, F. G.: *Stress and Strain in Bones*. Springfield, Ill., Charles C Thomas, Publisher, 1957.

Fastlicht, J.: Crowding of mandibular incisors. Am. J. Orthod., 58:156, 1970.

Fastlicht, J.: *The Universal Orthodontic Technique*. Philadelphia, W. B. Saunders Company, 1972.

Fawcett, E.: The development of the bones around the mouth. Five lectures on the growth of the jaws, normal and abnormal in health and disease. London, Dental Board of the United Kingdom, 1924.

Fell, H. B.: Chondrogenesis in cultures of endosteum. Proc. R. Soc. London, 112:417, 1933.

Felts, W. J. L.: Transplantation studies in skeletal organogenesis. I. The subcutaneously implanted immature long-bone of the rat and mouse. Am. J. Phys. Anthrop., 17:201, 1959.

Felts, W. J. L.: In vivo implantation as a technique in skeletal biology. Int. Rev. Cytol., 12:243, 1961.

Fenart, R.: Influence des modifications: experimentales et teratologiques de la station et de la locomotion, sur la morphologie cephalique des mammiferes quadrupedes. Etude par la methode vestibulaire. Arch. Anat. Histol. Embryol. (Strasb.), 69:5, 1966a.

Fenart, R.: Changements morphologiques de l'encephale, chez la rat ampute des membres anterieurs. J. Hirnforsch., 8:493, 1966b.

Fenart, R.: L'hominisation du crane. Bull. Acad. Dent. (Paris), 14:33, 1970.

Ferre, J.-C.: Contribution a l'etude du "syndrome asymetrique cranio-facial." Thèse, Univ. de Nantes, 1973.

Firschein, H. E.: Collagen and mineral dynamics in bone. Clin. Orthop., 66:212, 1969.

Ford, E. H.: Growth of the foetal skull. J. Anat., 90:63, 1956.

Ford, E. H.: Growth of the human cranial base. Am. J. Orthod., 44:498, 1958.

Forrester, D. J., N. K. Carstens, and D. B. Shurteff: Craniofacial configuration of hydrocephalic children. J.A.D.A., 72:1399, 1966.

Fraser, F. C.: Experimental teratogenesis in relation to congenital malformations in man. In: *Proceedings Second International Congress Congenital Malformations*. New York, International Medical Congress, 1964.

Fraser, F. C.: Gene-environment interactions in the production of cleft palate. In: *Methods for Teratological Studies in Experimental Animals and Man*. Ed. by H. Nishimura and J. R. Miller. Tokyo, Igaku Shoin, 1969.

Fraser, F. C.: The genetics of cleft lip and cleft palate. Am. J. Hum. Gen., 22:336, 1970.

Fraser, F. C.: Etiology of cleft lip and palate. In: *Cleft Lip and Palate*. Ed. by W. C. Grabb, S. W. Rosenstein, and K. R. Bzoch. Boston, Little, Brown and Company, 1971.

Fraser, F. C., and H. Pashayan: Relation of face shape to susceptibility to congenital cleft lip. J. Med. Genet., 7:112, 1970.

Fraser, F. C., B. E. Walker, and D. G. Trasler: Experimental production of congenital cleft palate; genetic and environmental factors. Pediatrics, 19:782, 1957.

Freeman, E., and A. R. Ten Cate: Development of the periodontium: An electron microscopic study. J. Periodont., 42:387, 1971.

Friedenberg, Z. B., R. H. Dyer, Jr., and C. T. Brighton: Electro-osteograms of long bones of immature rabbits. J. Dent. Res., 50:635, 1971.

Frommer, J.: Prenatal development of the mandibular joint in mice. Anat. Rec., 150:449, 1964.

Frommer, J., and M. R. Margolis: Contribution of Meckel's cartilage to ossification of the mandible in mice. J. Dent. Res., 50:1250, 1971.

Frommer, J., C. W. Monroe, J. R. Morehead, and W. D. Belt: Autoradiographic study of cellular proliferation during early development of the mandibular condyle in mice. J. Dent. Res., 47:816, 1968.

Frost, H. M.: In vivo osteocyte death. J. Bone Joint Surg., 42:138, 1960a.

Frost, H. M.: Micropetrosis. J. Bone Joint Surg., 42:144, 1960b.

Frost, H. M.: Tetracycline bone labeling in anatomy. Am. J. Phys. Anthrop., 29:183, 1968.

Fukada, E., and I. Yasuda: On the piezoelectric effect of bone. J. Physiol. Soc. Jap., 12:1158, 1957.

Furstman, L.: The early development of the human mandibular joint. Am. J. Orthod., 49:672, 1963.

Gans, B. J., and B. G. Sarnat: Sutural facial growth of the Macaca rhesus monkey: A gross and serial roentgenographic study by means of metallic implants. Am. J. Orthod., 37:827, 1951.

Garn, S. M.: Inheritance of symphyseal size during growth. Angle Orthod., 33:222, 1963.

Garn, S. M.: *The Earlier Gain and the Later Loss of Cortical Bone in Nutritional Perspective*. Springfield, Ill., Charles C Thomas, Publisher, 1970.

Garn, S. M., and B. Wagner: The adolescent growth of the skeletal mass and its implications to mineral requirements. In: *Adolescent Nutrition and Growth*. Ed. by F. P. Held. New York, Appleton-Century-Crofts, 1969.

Garn, S. M., C. G. Rohmann, and P. Nolan, Jr.: Developmental nature of bone changes during aging. In: *Relations of development and aging; a symposium presented before the Gerontological Society*. 15th Annual Meeting, Miami Beach, Florida. Ed. by J. E. Birren. Springfield, Ill., Charles C Thomas, Publisher, 1964.

Garn, S. M., A. B. Lewis, and R. M. Blizzard: Endocrine factors in dental development. J. Dent. Res., 44:243, 1965.

Garn, S. M., C. G. Rohmann, B. Wagner, and W. Ascoli: Continuing bone growth throughout life: A general phenomenon. Am. J. Phys. Anthrop., 26:313, 1967a.

Garn, S. M., C. G. Rohmann, and B. Wagner: Bone loss as a general phenomenon in man. Fed. Proc., 26:1729, 1967b.

Gasson, M. N., and M. A. Petrovic: Mecanismes et regulation de la croissance antero-posterieure du maxillaire superieur. Recherches experimentales, chez le jeune rat, sur le role de l'hormone somato-trope et du cartilage de la cloison nasale. Orthod. Fr., 43:255, 1972.

Gianelly, A. A.: Force-induced changes in the vascularity of the periodontal ligament. Am. J. Orthod., 55:5, 1969.

Gianelly, A. A., and H. M. Goldman: *Biologic Basis of Orthodontics.* Philadelphia, Lea & Febiger, 1971.

Gianelly, A. A., and C. F. A. Moorrees: Condylectomy in the rat. Arch. Oral Biol., 10:101, 1965.

Gillooly, C. J., Jr., R. T. Hosley, J. R. Mathews, and D. L. Jewett: Electric potentials recorded from mandibular alveolar bone as a result of forces applied to the tooth. Am. J. Orthod., 54:649, 1968.

Girgis, F. G., and J. J. Pritchard: Experimental production of cartilages during the repair of fractures of the skull vault in rats. J. Bone Joint Surg., 403:274, 1958.

Glasstone, S.: Differentiation of the mouse embryonic mandible and squamo-mandibular joint in organ culture. Arch. Oral Biol., 16:723, 1971.

Glimcher, M. J., E. P. Katz, and D. F. Travis: The solubilization and reconstruction of bone collagen. J. Ultrastruct. Res., 13:163, 1965.

Godard, H.: Les zones de croissance de la mandible. C. R. Assoc. Anat., 43:357, 1957.

Goland, P. P., and N. G. Grand: Chloro-s-triazines as markers and fixatives for the study of growth in teeth and bones. Am. J. Phys. Anthrop., 29:201, 1968.

Goldhaber, P.: The effect of hyperoxia on bone resorption in tissue culture. Arch. Pathol., 66:635, 1958.

Goldhaber, P.: Oxygen dependent bone resorption in tissue culture. In: *Parathyroids.* Ed. by R. O. Greep and R. V. Talmage. Springfield, Ill., Charles C Thomas, Publisher, 1961.

Goldhaber, P.: Some chemical factors influencing bone resorption in tissue culture. In: *Mechanisms of Hard Tissue Destruction; a Symposium.* Ed. by R. F. Sognnaes. A.A.A.S. Symposium, 1963.

Goldstein, M. S.: Changes in dimension and form of the face and head with age. Am. J. Phys. Anthrop., 22:37, 1936.

Goldstein, M. S.: Development of the head in the same individuals. Hum. Biol., 11:197, 1939.

Goodman, H. O.: Genetic parameters of dentofacial development. J. Dent. Res., 44:174, 1965.

Gordon, H. J.: Human cranial base development during the late embryonic and the fetal periods. Chicago, Univ. of Illinois M. S. (Orthodont.) Thesis, 1955.

Gorlin, R. J., and J. J. Pindborg: *Syndromes of the Head and Neck.* New York, McGraw-Hill Book Co., 1964.

Gorlin, R. J., J. Cervenka, and S. Pruzansky: Facial clefting and its syndromes. Birth Defects, 7:3, 1971.

Grabb, W. C., S. W. Rosenstein, and K. R. Bzoch: *Cleft Lip and Palate.* Boston, Little, Brown and Company, 1971.

Graber, T. M.: Implementation of the roentgenographic cephalometric technique. Am. J. Orthod., 44:906, 1958.

Graber, T. M.: Clinical cephalometric analysis. In: *Vistas of Orthodontics.* Ed. by B. S. Kraus and R. A. Reidel. Philadelphia, Lea & Febiger, 1962.

Graber, T. M.: A study of cranio-facial growth and development in the cleft palate child from birth to six years of age. In: *Early Treatment of Cleft Lip and Palate.* Ed. by R. Hotz. Berne, Switzerland, Hans Huber, 1964.

Graber, T. M.: *Orthodontics: Principles and Practice.* Philadelphia, W. B. Saunders Company, 1966.

Grant, D., and S. Bernick: Formation of the periodontal ligament. J. Periodont., 43:17, 1972.

Greenberg, A.: Life cycle of the human mandible. N.Y. Dent. J., 31:98, 1965.

Greenspan, J. S., and H. J. Blackwood: Histochemical studies of chondrocyte function in the cartilage of the mandibular condyle of the rat. J. Anat., 100:615, 1966.

Gregory, W. K.: *Our Face from Fish to Man.* New York, Putnam, 1929.

Gregory, W. K.: Certain critical stages in the evolution of the vertebrate jaws. Internatl. J. Orthod., 17:1138, 1931.

Greulich, W. W., and S. I. Pyle: *Radiographic Atlas of Skeletal Development of the Hand and Wrist.* Stanford, Stanford University Press, 1959.

Griffiths, D. L., L. Furstman, and S. Bernick: Postnatal development of the mouse palate. Am. J. Orthod., 53:757, 1967.

Grobstein, C., and G. Parker: *In vitro* induction of cartilage in mouse somite mesoderm by embryonic spinal cord. Proc. Soc. Exp. Biol. Med., 85:477, 1954.

Gugino, C. F.: An Orthodontic Philosophy. 6th Ed. RM/Communicators Division of Rocky Mountain/Associates International Inc., Denver, Colo., 1971.

Gussen, R.: The labyrinthine capsule: Normal structure and pathogenesis of otosclerosis. Acta Oto-laryngol. (Stockh.), Suppl 235, 1968a.

Gussen, R.: Articular and internal remodeling in the human otic capsule. Am. J. Anat., 122:397, 1968b.

Guth, L.: Regulation of metabolic and functional properties of muscle. In: *Regulation of Organ and Tissue Growth.* Ed. by R. J. Goss. New York, Academic Press, 1973.

Hall, B. K.: *In vitro* studies on the mechanical evocation of adventitious cartilage in the chick. J. Exp. Zool., 168:238, 1968.

Hall, B. K.: Differentiation of cartilage and bone from common germinal cells. J. Exp. Zool., 173:383, 1970.

Hall-Craggs, E. C. B.: Influence of epiphyses on the regulation of bone growth. Nature, 221:1245, 1969.

Hall-Craggs, E. C. B., and C. A. Lawrence: The effect of epiphysial stapling on growth in length of the rabbit's tibia and femur. J. Bone Joint Surg., 513:359, 1969.

Hanson, J. R., and B. J. Anson: Development of the malleus of the human ear. Illustrated in Atlas Series. Quart. Bull. Northwest. Univ. Med. Sch., 36:119, 1962.

Hanson, J. R., B. J. Anson, and T. H. Bast: The early embryology of the auditory ossicles in man. Illustrated in Atlas Series. Quart. Bull. Northwest. Univ. Med. Sch., 33:358, 1959.

Harris, J. E.: A cephalometric analysis of mandibular growth rate. Am. J. Orthod., 48:161, 1962.

Harris, J. E., C. J. Kowalski, and S. S. Watnick: Genetic factors in the shape of the cranio-facial complex. Angle Orthod., 43:107, 1973.

Hart, J. C., G. R. Smiley, and A. D. Dixon: Sagittal growth of the craniofacial complex in normal embryonic mice. Arch. Oral Biol., 14:995, 1969.

Harvold, E. P.: The asymmetries of the upper facial skeleton and their morphological significance. Eur. Orthod. Soc. Trans., 1951.

Harvold, E. P.: The role of function in the etiology and treatment of malocclusion. Am. J. Orthod., 54:883, 1968.

Harvold, E. P.: Skeletal and dental irregularities in relation to neuromuscular dysfunctions. In: *Patterns of Orofacial Growth and Development.* Report 6. Washington, D.C., Am. Speech and Hearing Assoc., 1971.

Harvold, E. P., and K. Vargervik: Morphogenic response to activator treatment. Am. J. Orthod., 60:478, 1970.

Harvold, E. P., G. Chierici, and K. Vargervik: Experiments on the development of dental malocclusions. Am. J. Orthod., 61:38, 1972.

Harvold, E. P., K. Vargervik, and G. Chierici: Primate experiments on oral sensation and dental malocclusion. Am. J. Orthod., 63:494, 1973.

Hasund, A., and R. Sivertsen: Dental arch space and facial type. Angle Orthod., 41:140, 1971.

Hellman, M.: A preliminary study in development as it affects the human face. Dent. Cosmos., 71:250, 1927a.

Hellman, M.: Changes in the human face brought about by development. Int. J. Orthod. Oral Surg., 13:475, 1927b.

Hellman, M.: An introduction to growth of the human face from infancy to adulthood. Int. J. Orthod., Oral Surg. Radiol., 18:777, 1932.

Hellman, M.: The face in its developmental career. Dent. Cosmos., 77:685, 1935.

Herovici, C.: A polychrome stain for differentiating precollagen from collagen. Stain Technol., 38:204, 1963.

Herring, S. W.: Sutures—A tool in functional cranial analysis. Acta Anat., 83:222, 1972.

Hinrichsen, G. J., and E. Storey: The effect of force on bone and bones. Angle Orthod., 38:155, 1963.

Hirschfeld, W. J., and R. E. Moyers: Prediction of craniofacial growth: The state of the art. Am. J. Orthod., 60:435, 1971.

Hirschfeld, W. J., R. E. Moyers, and D. H. Enlow: A method of deriving subgroups of a population: A study of craniofacial taxonomy. Am. J. Phys. Anthrop., 39:279, 1973.

Hixon, E. H.: Prediction of facial growth. Eur. Orthod. Soc. Rep. Congr., 44:127, 1968.

Hixon, E. H.: Cephalometrics: A perspective. Angle Orthod., 42:200, 1972.

Hixon, E. H., and S. L. Horowitz: *Nature of Orthodontic Diagnosis.* St. Louis, Mo., C. V. Mosby Company, 1966.

Holdaway, R. A.: Changes in relationship of points A and B during orthodontic treatment. Am. J. Orthod., 42:176, 1956.

Hooton, E. P.: *Up From the Ape.* 3rd Ed. New York, Macmillan, 1946.

Hopkin, G. B.: The cranial base as an aetiological factor in malocclusion. Angle Orthod., 38:250, 1968.

Horowitz, S. L.: Modifications of mandibular architecture following removal of the temporalis muscle in the rat. J. Dent. Res., 30:276, 1951.

Horowitz, S. L.: The role of genetic and local environmental factors in normal and abnormal morphogenesis. Acta Morphol. Neerl. Scand., 10:59, 1972.

Horowitz, S. L.: Variability of the maxillary complex in common dental malformations. Third Internatl. Orthod. Congr., London, 1975.

Horowitz, S. L., and R. Osborne: The genetic aspects of cranio-facial growth. In: *Craniofacial Growth in Man.* Ed. by R. E. Moyers and W. M. Krogman. Oxford, Pergamon Press, 1971.

Horowitz, S. L., and R. H. Thompson: Variations of the craniofacial skeleton in post-adolescent males and females with special references to the chin. Angle Orthod., 34:97, 1964.

Horowitz, S. L., R. H. Osborne, and F. V. DeGeorge: A cephalometric study of craniofacial variation in adult twins. Angle Orthod., 30:1, 1960.

Horowitz, S. L., L. J. Gerstman, and J. M. Converse: Craniofacial relationships in mandibular prognathion. Arch. Oral Biol., 14:121, 1969.

Houpt, M. I.: Growth of the craniofacial complex of the human fetus. Am. J. Orthod., 58:373, 1970.

Howells, W.: *Mankind So Far.* Garden City, New York, Doubleday, 1949.

Hoyte, D. A. N.: The relative contribution of sutural and ectocranial deposition of bone to cranial growth in rodents. J. Anat., 92:654, 1958.

Hoyte, D. A. N.: Resorption and skull expansion in rats. Anat. Rec., 139:307, 1961a.

Hoyte, D. A. N.: The postnatal growth of the ear capsule in the rabbit. Am. J. Anat., 108:1, 1961b.

Hoyte, D. A. N.: Facts and fallacies of alizarin staining: The exterior of the infant pig skull. J. Dent. Res., 43:814, 1964a.

Hoyte, D. A. N.: Facts and fallacies of alizarin staining: The interior of the infant pig skull. Anat. Rec., 148:292, 1964b.

Hoyte, D. A. N.: The role of the soft tissues in skull growth in rabbits. West Indian Med. J., 14:125, 1965a.

Hoyte, D. A. N.: The sphenoidal complex in the first three months of its postnatal growth in rabbits: An alizarin study. Anat. Rec., 151:364, 1965b.

Hoyte, D. A. N.: Experimental investigations of skull morphology and growth. Internatl. Rev. Gen. Exp. Zool., 2:345, 1966.

Hoyte, D. A. N.: Alizarin red in the study of the apposition and resorption of bone. Am. J. Phys. Anthrop., 29:157, 1968.

Hoyte, D. A. N.: Mechanisms of growth in the cranial vault and base. J. Dent. Res., 50:1447, 1971a.

Hoyte, D. A. N.: The modes of growth of the neurocranium: The growth of the sphenoid bone in animals. In: *Cranio-facial Growth in Man.* Ed. by R. E. Moyers and W. M. Krogman. Oxford, Pergamon Press, 1971b.

Hoyte, D. A. N.: Basicranial elongation. 2. Is there differential growth within a synchondrosis? Anat. Rec., 175:347, 1973a.

Hoyte, D. A. N.: Basicranial elongation. 3. Differential growth between synchondroses and basion. Proc. 3rd Europ. Anat. Congr., 231, 1973b.

Hoyte, D. A. N., and D. H. Enlow: Wolff's law and the problem of muscle attachment on resorptive surfaces of bone. Am. J. Phys. Anthrop., 24:205, 1966.

Hulanicka, B.: Anthroposcopic features as a measure of similarity. Nadbitka Z Nru 86, Materialow 1 Prac Antropologicznych Wroclaw, 115, 1973.

Humphry, G.: On the growth of the jaws. Cambridge, Trans. Cambridge Phil. Soc., 1864.

Humphrey, T.: Reflex activity in the oral and facial area of the human fetus. In: *Second Symposium on Oral Sensation and Perception.* Ed. by James Bosma. Springfield, Ill., Charles C Thomas, Publisher, 1970.

Humphrey, T.: The development of oral and facial motor mechanisms in human fetuses and their relation to craniofacial growth. J. Dent. Res., 50:1428, 1971a.

Humphrey, T.: Human prenatal activity sequences in the facial region and their relationship to postnatal development. In: *Patterns of Orofacial Growth and Development.* Report 6. Washington, D.C., Am. Speech and Hearing Assoc., 1971b.

Hunter, W. S.: A study of the inheritance of the craniofacial characteristics, as seen in lateral cephalograms of 72 like sexed twins. Eur. Orthod. Soc. Trans., 59:70, 1965.

Hunter, W. S., and S. Garn: Evidence for a secular trend in face size. Angle Orthod., 39:320, 1969.

Hunter, W. S., D. R. Balbach, and D. E. Lamphiear: The heritability of attained growth in the human face. Am. J. Orthod., 58:128, 1970.

Hurrell, D. J.: The vascularization of cartilage. J. Anat. (Lond.), 73:112, 1934.

Ingervall, B., and B. Thilander: The human spheno-occipital synchondrosis. I. The time of closure appraised macroscopically. Acta Odont. Scand., 30:349, 1972.

Inoue, N.: A study of the developmental changes in the dentofacial complex during fetal period by means of roentgenographic cephalometrics. Tokyo Med. Dent. Univ. Bull., 8:205, 1961.

Isaacson, J. R., R. J. Isaacson, T. M. Speidel, and F. W. Worms: Extreme variation in vertical facial growth and associated variation in skeletal and dental relations. Angle Orthod., 41:219, 1971.

Isotupa, K., K. Koski, and L. Makinen: Changing architecture of growing cranial bones at sutures as revealed by vital staining with alizarin red S in the rabbit. Am. J. Phys. Anthrop., 23:19, 1965.

Isreal, H.: Loss of bone and remodeling-redistribution in the craniofacial skeleton with age. Fed. Proc., 26:1723, 1967.

Isreal, H.: Continuing growth in the human cranial skeleton. Arch. Oral Biol., 13:133, 1968.

Isreal, H.: Age factor and the pattern of change in craniofacial structures. Am. J. Phys. Anthrop., 39:111, 1973a.

Isreal, H.: Recent knowledge concerning craniofacial aging. Angle Orthod., 43:176, 1973b.

Janzen, E. K., and J. A. Bluher: The cephalometric, anatomic and histologic changes in *Macaca mulatta* after application of a continuous action retraction force on the mandible. Am. J. Orthod., 51:823, 1965.

Jarabak, J. R., and J. R. Thompson: Cephalometric appraisal of the cranium and mandible of the rat following condylar resection. J. Dent. Res., 28:655, 1949.

Johnson, M., and P. W. Ramwell: Prostaglandin modification of membrane-bound enzyme activity: A possible mechanism of action? Prostaglandins, 3:703, 1973.

Johnston, L. E.: A statistical evaluation of cephalometric prediction. Master's Thesis, University of Michigan, Ann Arbor, 1964.

Johnston, M. C., et al.: An expanded role of the neural crest in oral and pharyngeal development. In:

*Oral Sensation and Perception: Development in the Fetus and Infant.* Ed. by J. Bosma. Washington, D.C., DHEW Pub. No. 73–546, 1973, pp. 37–52.

Joho, J. P.: Changes in form and size of the mandible in the orthopaedically treated *Macacus irus* (an experimental study). Eur. Orthod. Soc. Trans., 1968:161, 1969.

Jones, B. H., and H. V. Meredith: Vertical changes in osseous and odontic portions of human face height between the ages of 5 and 15 years. Am. J. Orthod., 52:902, 1966.

Joondeph, D. R., and L. E. Wragg: Facial growth during the secondary palate closure in the rat. Am. J. Orthod., 6:88, 1966.

Justus, R., and J. H. Luft: A mechanochemical hypothesis for bone remodeling induced by mechanical stress. Calcif. Tissue Res., 5:222, 1970.

Kallio, D. M., P. R. Garant, and C. Minkin: Ultrastructural effects of calcitonin on osteoclasts in tissue culture. J. Ultrastruct. Res., 39:205, 1972.

Kanouse, M. C., S. P. Ramfjord, and C. E. Nasjleti: Condylar growth in rhesus monkeys. J. Dent. Res., 48:1171, 1969.

Kawamura, Y.: *Physiology of Mastication.* Basel, S. Karger, 1974.

Keith, A.: Contribution to the mechanism of growth of the human face. Internatl. J. Orthod., 8:607, 1922.

Keith, B. S., and J. D. Decker: The prenatal inter-relationships of the maxilla and premaxilla in the facial development of man. Acta Anat., 40:278, 1960.

Kelsey, C. C.: Alveolar bone resorption under complete dentures. J. Prosthet. Dent., 25:152, 1971.

Kier, E. L.: The infantile sella turcica: New radiologic and anatomic concepts based on a developmental study of the sphenoid bone. Am. J. Roentgenol., 102:747, 1968.

Kisling, E.: *Cranial Morphology in Down's Syndrome.* Vol. 58. Copenhagen, Munksgaard, 1966, p. 106.

Klaauw, C. J. van der: Cerebral skull and facial skull. A contribution to the knowledge of skull structure. Arch. Neerl. Zool., 9:16, 1946.

Klaauw, C. J. van der: Size and position of the functional components of the skull. A contribution to the knowledge of the architecture of the skull, based on data in the literature. Arch. Neerl. Zool., 9:176, 1948.

Klaauw, C. J. van der: Size and position of the functional components of the skull (continuation). Arch. Neerl. Zool., 9:177, 1951.

Klaauw, C. J. van der: Size and position of the functional components of the skull (conclusion). Arch. Neerl. Zool., 9:369, 1952.

Klein, D. C., and L. G. Raisz: Role of adenosine 3', 5'-monophosphate in the hormonal regulation of bone resorption: Studies with cultured fetal bone. Endocrinology, 89:818, 1971.

Knott, V. B.: Ontogenetic change of four cranial base segments in girls. Growth, 33:123, 1969.

Knott, V. B.: Change in cranial base measures of human males and females from age 6 years to early adulthood. Growth, 35: 145, 1971.

Konjevich, N.: Origin and maturation of the spheno-occipital synchondrosis. University of Illinois M.S. (Orthodont.) Thesis, Chicago, 1963.

Koski, K.: Some aspects of the growth of the cranial base and the upper face. Odontol. Tidskr., 68:344, 1960.

Koski, K.: Cranial growth centers: Facts or fallacies? Am. J. Orthod., 54:566, 1968.

Koski, K.: Some characteristics of cranio-facial growth cartilages. In: *Cranio-facial Growth in Man.* Ed. by R. E. Moyers and W. M. Krogman. Oxford, Pergamon Press, 125, 1971.

Koski, K.: Variability of the cranio-facial skeleton. An exercise in roentgencephalometry. Am. J. Orthod., 64:188, 1973.

Koski, K.: The mandibular complex. Eur. Orthod. Soc. Trans., 1974 (in press).

Koski, K., and L. Makinen: Growth potential of transplanted components of the mandibular ramus of the rat. I. Suom. Hammaslaak. Toim., 59:296, 1963.

Koski, K., and K. E. Mason: Growth potential of transplanted components of the mandibular ramus of the rat. II. Suom. Hammaslaak. Toim., 60:209, 1964.

Koski, K., and O. Rönning: Growth potential of transplanted components of the mandibular ramus of the rat. III. Suom. Hammaslaak. Toim., 61:292, 1965.

Koski, K., and O. Rönning: Pitkan luun rustoisen paan siirrannaisen kasvupotentiaalista rotalla. Suom. Hammaslaak. Toim., 62:165, 1966.

Koski, K., and O. Rönning: Growth potential of subcutaneously transplanted cranial base synchondroses of the rat. Acta Odont. Scand., 27:343, 1969.

Koski, K., and Rönning: Growth potential of intracerebrally transplanted cranial base synchondroses in the rat. Arch. Oral Biol., 15:1107, 1970.

Koskinen, L., and K. Koski: Regeneration in transplanted epiphysectomized humeri of rats. Am. J. Phys. Anthrop., 27:33, 1967.

Kowalski, C. J., C. E. Nasjleti, and G. F. Walker: Differential diagnosis of adult male black and white populations. Angle Orthod., 44:346, 1974.

Kraus, B. S.: Prenatal growth and morphology of the human bony palate. J. Dent. Res., 39:1177, 1960a.

Kraus, B. S.: The prenatal inter-relationships of the maxilla and premaxilla in the facial development of man. Acta Anat., 40:278, 1960b.

Kraus, B. S.: *Human Dentition before Birth.* Philadelphia, Lea & Febiger, 1965.

Kraus, B. S., H. Kitamura, and R. A. Latham: *Atlas of Developmental Anatomy of the Face*. New York, Harper and Row, 1966.

Kraw, A. G., and D. H. Enlow: Continuous attachment of the periodontal membrane. Am. J. Anat., 120:133, 1967.

Kremenak, C. R., D. F. Hartshorn, and S. E. Demjen: The role of the cartilaginous nasal septum in maxillofacial growth: Experimental septum removal in beagle pups. J. Dent. Res., 48 (Abst. 32): 1969.

Krogman, W. M.: Studies in growth changes in the skull and face of anthropoids. Am. J. Anat., 46:315, 1930.

Krogman, W. M.: Principles of human growth. Ciba Found. Sympos., 5:1458, 1943.

Krogman, W. M.: The growth periods from birth to adulthood. Syllabus. Third Annual Midwestern Seminar of Dental Medicine, University of Illinois, 1950.

Krogman, W. M.: Craniometry and cephalometry as research tools in growth of head and face. Am. J. Orthod., 37:406, 1951a.

Krogman, W. M.: The problem of 'timing' in facial growth, with special reference to the period of the changing dentition. Am. J. Orthod., 37:253, 1951b.

Krogman, W. M.: *The Human Skeleton in Forensic Medicine*. Springfield, Ill., Charles C Thomas, Publisher, 1962.

Krogman, W. M.: Role of genetic factors in the human face, jaws and teeth: A review. Eugenics Rev., 59:165, 1967.

Krogman, W. M.: Growth of head, face, trunk and limbs in Philadelphia White and Negro children of elementary and high school age. Monogr. Soc. Res. Child Develop., Serial No. 136, 35:1, 1970.

Krogman, W. M., and D. D. B. Chung: The craniofacial skeleton at the age of one month. Angle Orthod., 25:305, 1965.

Krogman, W. M., and V. Sassouni: *Syllabus in Roentgenographic Cephalometry*. Philadelphia, Philadelphia Center for Research in Child Growth, 1957.

Krompecher, S., and L. Toth: Die Konzeption von Kompression, Hypoxie und konsekutiver Mucopolysaccharidbildung in der kausalen Analyse der Chondrogenese. Z. Anat. Entwicklungesch., 124:268, 1964.

Kuroda, T.: A longitudinal cephalometric study on the craniofacial development in Japanese children. Paper presented at the I.A.D.R., New York, 1970.

Kvinnsland, S.: Observation on the early ossification of the upper jaw. Acta Odont. Scand., 27:649, 1969.

Kvinnsland, S.: The sagittal growth of the upper face during foetal life. Acta Odont. Scand., 29:717, 1971a.

Kvinnsland, S.: The sagittal growth of the lower face during foetal life. Acta Odont. Scand., 29:733, 1971b.

Kvinnsland, S.: The sagittal growth of the foetal cranial base. Acta Odont. Scand., 29:699, 1971c.

Lacroix, P.: *The Organization of Bones*. London, J. & A. Churchill Ltd., 1952.

Langman, J.: The influence of teratogenic agents on serum proteins. In: *Congenital Anomalies of the Face and Associated Structures*. Ed. by S. Pruzansky. Springfield, Ill., Charles C Thomas, Publisher, 1961.

Langman, J.: *Medical Embryology*. Baltimore, Williams & Wilkins Company, 1969.

Latham, R. A.: Skull growth in the Rhesus monkey *(Macaca mulatta)*. J. Anat., 92:654, 1958.

Latham, R. A.: Observations on the growth of the cranial base in the human skull. J. Anat., 100:435, 1966.

Latham, R. A.: The sliding of cranial bones at sutural surfaces during growth. J. Anat. (Lond.), 102:593, 1968.

Latham, R. A.: The structure and development of the intermaxillary suture. J. Anat., 106:167, 1970a.

Latham, R. A.: Maxillary development and growth: The septopremaxillary ligament. J. Anat. (Lond.), 107:471, 1970b.

Latham, R. A.: Mechanism of maxillary growth in the human cyclops. J. Dent. Res., 50:929, 1971a.

Latham, R. A.: The development, structure, and growth pattern of the human mid-palatal suture. J. Anat., 108:1, 31–41, 1971b.

Latham, R. A.: The sella point and postnatal growth of the human cranial base. Am. J. Orthod., 61:156, 1972a.

Latham, R. A.: The different relationship of the sella point to growth sites of the cranial base in fetal life. J. Dent. Res., 51:1646, 1972b.

Latham, R. A., and W. R. Burston: The postnatal pattern of growth at the sutures of the human skull. Dent. Pract., 17:61, 1966.

Lavelle, C. L. B.: An analysis of foetal craniofacial growth. Ann. Hum. Biol., 1:3, 269–282, 1974.

Lebret, L.: Growth changes of the palate. J. Dent. Res., 41:1391, 1962.

Lecerf, J.-P.: Rapports ontogeniques et phylogeniques entre quelques plans manducatoires et le foramen magnum. These, Univ. de Lille, 1974.

Le Diascorn, H.: Anatomie et physiologie des sutures de la face. Thèse, Univ. de Nantes, 1971.

Levihn, W. C.: A cephalometric roentgenographic cross-sectional study of the craniofacial complex in fetuses from 12 weeks to birth. Am. J. Orthod., 53:822, 1967.

Lewis, A. B., and A. F. Roche: Elongation of the cranial base in girls during pubescence. Angle Orthod., 42:358, 1972.

Limborgh, J. van: The regulation of the embryonic development of the skull. Acta Morphol. Neerl. Scand., 7:101, 1968.

Limborgh, J. van: A new view on the control of the morphogenesis of the skull. Acta Morphol. Neerl. Scand., 8:143, 1970.

Limborgh, J. van: The role of genetic and local environmental factors in the control of postnatal craniofacial morphogenesis. Acta Morphol. Neerl. Scand., 10:37, 1972.

Limborgh, J. van, and H. L. Verwoerd-Verhoef: Effects of artifical unilateral facial clefts on growth of the skull in young rabbits. J. Dent. Res., 47:1013, 1968.

Liskova, M., and J. Hert: Reaction of bone to mechanical stimuli. Folia Morphol., (Warsz.), 19:301, 1971.

Long, R., R. C. Greulich, and B. G. Sarnat: Regional variations in chondrocyte proliferation in the cartilaginous nasal septum of the growing rabbit. J. Dent. Res., 47:137, 1968.

Longacre, J. J.: *Craniofacial Anomalies: Pathogenesis and Repair.* Philadelphia, J. B. Lippincott Company, 1968.

Lundstrom, A.: Importance of genetic and non-genetic factors in the facial skeleton studied in 100 pairs of twins. Eur. Orthod. Soc. Trans., 1954.

Lundstrom, A.: The clinical significance of profile x-ray analysis. Eur. Orthod. Soc. Trans., 31:190, 1955a.

Lundstrom, A.: The significance of genetic and non-genetic factors in the profile of the facial skeleton. Am. J. Orthod., 12:910, 1955b.

Maj, G., and C. Luzi: Analysis of mandibular growth on 28 normal children followed from 9 to 13 years of age. Eur. Orthod. Soc. Trans., 1962.

Manson, J. D.: *A Comparative Study of the Postnatal Growth of the Mandible.* London, Henry Kimpton, 1968.

Manson, J. D., and R. B. Lucas: A microradiographic study of age changes in the human mandible. Arch. Oral Biol., 7:761, 1962.

Margolis, H. I.: A basic facial pattern and its application in clinical orthodontics. Am. J. Orthod., 39:425, 1953.

Marshall, D.: Interpretation of the posteroanterior skull radiograph. Assembly of disarticulated bones. Dent. Radiogr. Photogr., 42:27, 1969.

Massler, M., and J. M. Frankel: Prevalence of malocclusion in children aged 14 to 18 years. Am. J. Orthod., 37:751, 1951.

Mauser, C.: A study of the prenatal growth of the human face and cranium. Thesis, West Virginia University, Morgantown, 1975.

Mauser, C., D. H. Enlow, D. O. Overman, and R. McCafferty: Growth and remodeling of the human fetal face and cranium. In: *Determinants of Mandibular Form and Growth.* Ed. by J. A. McNamara. Monograph 5. Craniofacial Growth Series. Center for Human Growth and Development, University of Michigan, Ann Arbor, 1975.

McKeown, M., and A. Richardson: The nature of cranial vault variation and its relation to facial height. Angle Orthod., 41:15, 1971.

McNamara, J. A., Jr.: *Neuromuscular and Skeletal Adaptations to Altered Orofacial Function.* Monograph No. 1. Craniofacial growth series. Center for Human Growth and Development, University of Michigan, Ann Arbor, 1972.

McNamara, J. A., Jr.: Neuromuscular and skeletal adaptation to altered function in the orofacial region. Am. J. Orthod., 64:578, 1973a.

McNamara, J. A., Jr.: Increasing vertical dimension in the growing face: An experimental study. Am. J. Orthod., 64:364, 1973b.

McNamara, J. A., Jr.: Procion dyes as vital markers in rhesus monkeys. J. Dent. Res., 52:634, 1973c.

McNamara, J. A., Jr., and L. Graber: Mandibular growth in *Macaca mulatta.* Am. J. Phys. Anthrop., 42:15, 1975.

McNamara, J. A., Jr., M. L. Riolo, and D. H. Enlow: Growth of the maxillary complex in the rhesus monkey *(Macaca mulatta).* Am. J. Phys. Anthrop., 1975, (in press).

Mednick, L. W., and S. L. Washburn: The role of the sutures in the growth of the braincase of the infant pig. Am. J. Phys. Anthrop., 14:175, 1956.

Meikle, M. C.: The role of the condyle in the postnatal growth of the mandible. Am. J. Orthod., 64:50, 1973.

Meikle, M. C.: In vivo transplantation of the mandibular joint of the rat—An autoradiographic investigation into cellular changes at the condyle. Arch. Oral Biol., 18:1011, 1973.

Melcher, A. M.: Behaviour of cells and condylar cartilage of foetal mouse mandible maintained *in vitro.* Arch Oral Biol., 16:1379, 1971.

Melsen, B.: Time of closure of the spheno-occipital synchondrosis determined on dry skulls: A radiographic craniometric study. Acta Odont. Scand., 27:73, 1969.

Melsen, B.: Computerized comparison of histological methods for the evaluation of craniofacial growth. Acta Odont. Scand., 29:295, 1971a.

Melsen, B.: The postnatal growth of the cranial base in *Macaca rhesus* analyzed by the implant method. Tandlaegebladet, 75:1320, 1971b.

Melsen, B.: Time and mode of closure of the spheno-occipital synchondrosis determined on human autopsy material. Acta Anat., 83:112, 1972.

Melsen, B.: The cranial base. Acta Odont. Scand., 32: (Suppl. 62), 1974.

Meredith, H. V.: Serial study of change in a mandibular dimension during childhood and adolescence. Growth, 25:229, 1961.

Meredith, H. V.: Childhood interrelations of anatomic growth rates. Growth, 26:23, 1962.

Merow, W. W.: A cephalometric statistical appraisal of dento-facial growth. Angle Orthod., 32:205, 1962.

Mestre, J. C.: A cephalometric appraisal of cranial and facial relationships at various stages of human fetal development. Am. J. Orthod., 45:473, 1959.

Michejda, M.: Ontogenic changes of the cranial base in *Macaca mulatta:* Histologic study. Proc. 3rd Internatl. Congr. Primat., Zurich, 1:215, 1970.

Michejda, M.: The role of the basicranial synchondroses in flexure processes and ontogenetic development of the skull base. Am. J. Phys. Anthrop., 37:143, 1972a.

Michejda, M.: Significance of basiocranial synchondroses in nonhuman primates and man. Medical Primatology: Proc. 3rd Conf. Exp. Med. Surg. Primates, Lyon. Vol. I. Basel, S. Karger, 1972b, p. 372.

Michejda, M., and D. Lamey: Flexion and metric age changes of the cranial base in the *Macaca mulatta.* I. Infant and juveniles. Folia Primatol. (Basel), 14:84, 1971.

Midy, J.: Trajet et inclinaison des germes dentaires et des dents temporaires et permanentes dans les axes vestibulaires d'orientation au cours de l'ontogenese humaine. Thèse, Academie de Paris, Univ. Paris VI, 1973.

Miura, F., N. Inoue, and S. Kazuo: The standards of Steiner's analysis for Japanese. Bull. Tokyo Med. Dent. Univ., 10:387, 1963.

Miura, F., N. Inoue, M. Azuma, and G. Ito: Development and organization of periodontal membrane and physiologic tooth movements. Bull. Tokyo Med. Dent. Univ., 17:123, 1970.

Moffett, B. C., Jr.: The prenatal development of the human temporomandibular joint. Contrib. Embryol. Carneg. Instn., 36:19, 1957.

Moffett, B. C., Jr.: A research perspective on craniofacial morphogenesis. Acta Morphol. Neerl. Scand., 10:99, 1972.

Moffett, B. C., Jr., L. C. Johnson, J. B. McCabe, and H. C. Askew: Articular remodeling in the adult human temporomandibular joint. Am. J. Anat., 115:119, 1964.

Moore, A. W.: Head growth of the Macaque monkey as revealed by vital staining, embedding, and undecalcified sectioning. Am. J. Orthod., 35:654, 1949.

Moore, A. W.: Observations on facial growth and its clinical significance. Am. J. Orthod., 45:399, 1959.

Moore, A. W.: A critique of orthodontic dogma. Angle Orthod., 39:69, 1969.

Moore, A. W.: Cephalometrics as a diagnostic tool. J.A.D.A., 82:775, 1971.

Moore, K. L.: *Before We Are Born.* Philadelphia, W. B. Saunders Company, 1974.

Moore, W. J.: Masticatory function and skull growth. J. Zool., 146:123, 1965.

Moore, W. J.: Skull growth in the albino rat *(Rattus norvegicus).* J. Zool. (Lond.), 149:137, 1966.

Moore, W. J., and T. F. Spence: Age changes in the cranial base of the rabbit *(Cryctolagus cuniculus).* Anat. Rec., 165:355, 1969.

Moorrees, C. F. A.: Normal variation and its bearing on the use of cephalometric radiographs in orthodontic diagnosis. Am. J. Orthod., 39:942, 1953.

Moorrees, C. F. A.: Natural head position, a basic consideration in the interpretation of cephalometric radiographs. Am. J. Phys. Anthrop., 16:213, 1958.

Moorrees, C. F. A.: *Dentition of the Growing Child, a Longitudinal Study of Dental Development Between 3 and 18 Years of Age.* Cambridge, Harvard University Press, 1959.

Moorrees, C. F. A.: Register of longitudinal studies of facial and dental development. International Society of Craniofacial Biology, 1967.

Moss, M. L.: Postnatal growth of the human skull base. Angle Orthod., 25:77, 1955a.

Moss, M. L.: Correlation of cranial base angulation with cephalic malformations and growth disharmonies of dental interest. N.Y. Dent. J., 21:452, 1955b.

Moss, M. L.: Rotations of the cranial components in the growing rat and their experimental alteration. Acta Anat., 32:65, 1958a.

Moss, M. L.: Fusion of the frontal suture in the rat. Am. J. Anat., 102:141, 1958b.

Moss, M. L.: Embryology, growth and malformations of the temporomandibular joint. In: *Disorders of the Temporomandibular Joint.* Ed. by L. Schwartz. Philadelphia, W. B. Saunders Company, 1959.

Moss, M. L.: Functional analysis of human mandibular growth. J. Prosthet. Dent., 10:1149, 1960a.

Moss, M. L.: A functional approach to craniology. Am. J. Phys. Anthrop., 18:281, 1960b.

Moss, M. L.: The functional matrix. In: *Vistas of Orthodontics.* Ed. by B. S. Kraus and R. A. Riedel. Philadelphia, Lea & Febiger, 1962.

Moss, M. L.: Functional cranial analysis of mammalian mandibular ramal morphology. Acta Anat., 71:423, 1968a.

Moss, M. L.: The primacy of functional matrices in orofacial growth. Dent. Pract., 19:65, 1968b.

Moss, M. L.: The primary role of functional matrices in facial growth. Am. J. Orthod., 55:566, 1969.

Moss, M. L.: Functional cranial analysis and the functional matrix. Am. Speech Hear., Assoc. Rep., 6:5, 1971a.

Moss, M. L.: Neurotropic processes in orofacial growth. J. Dent. Res., 50:1492, 1961b.

Moss, M. L.: Ontogenic aspects of cranio-facial growth. In: *Cranio-facial Growth in Man.* Ed. by R. E. Moyers and W. M. Krogman. Oxford, Pergamon Press, 1971c, p. 109.

Moss, M. L.: The regulation of skeletal growth. In: *Regulation of Organ and Tissue Growth.* Ed. by R. J. Goss. New York, Academic Press, 1972a.

Moss, M. L.: An introduction to the neurobiology of orofacial growth. Acta. Biotheor., 22:236, 1972b.

Moss, M. L.: Twenty years of functional cranial analysis. Am. J. Orthod., 61:479, 1972c.

Moss, M. L.: Neurotrophic regulation of craniofacial growth. In: *Determinants of Craniofacial Form and Growth.* Craniofacial Growth Series, Center for Human Growth and Development, University of Michigan, 1975.

Moss, M. L., and S. N. Greenberg: Functional cranial analysis of the human maxillary bone. Angle Orthod., 37:151, 1967.

Moss, M. L., and M. A. Meehan: Functional cranial analysis of the coronoid process in the rat. Acta Anat., 77:11, 1970.

Moss, M. L., and R. M. Rankow: The role of the functional matrix in mandibular growth. Angle Orthod., 38:95, 1968.

Moss, M. L., and L. Salentijn: The primary role of functional matrices in facial growth. Am. J. Orthod., 55:566, 1969a.

Moss, M. L., and L. Salentijn: The capsular matrix. Am. J. Orthod., 56:474, 1969b.

Moss, M. L., and L. Salentijn: Differences between the functional matrices in anterior open and deep overbite. Am. J. Orthod., 60:264, 1971a.

Moss, M. L., and L. Salentijn: The logarithmic growth of the human mandible. Acta Anat., 77:341, 1970.

Moss, M. L., and L. Salentijn: The unitary logarithmic curve descriptive of human mandibular growth. Acta Anat., 78:532, 1971.

Moss, M. L., C. R. Noback, and G. G. Robertson: Growth of certain human fetal cranial bones. Am. J. Anat., 98:191, 1956.

Moss, M. L., B. E. Bromberg, I. C. Song, and G. Eisenman: The passive role of nasal septal cartilage in mid-facial growth. Plast. Reconstr. Surg., 41:536, 1968.

Moss, M. L., M. A. Meehan, and L. Salentijn: Transformative and translative growth processes in neurocranial development of the rat. Acta Anat., 81:161, 1972.

Moyers, R. E.: Temporomandibular muscle contraction patterns in Angle Class II, Division 1 malocclusions: An electromyographic analysis. Am. J. Orthod., 35:837, 1949.

Moyers, R. E.: Periodontal membrane in orthodontics. J.A.D.A., 40:22, 1950.

Moyers, R. E.: Some recent electromyographic findings in the oro-facial muscles. Eur. Orthod. Soc. Trans., 32:225, 1956.

Moyers, R. E.: Role of musculature in malocclusion. Eur. Orthod. Soc. Trans., 37:40, 1961.

Moyers, R. E.: Development of occlusion. Dent. Clin. North Am., 13:523, 1969.

Moyers, R. E.: Some comments about the nature of orthodontic relapse. Orthodoncia, 54:215, 1970.

Moyers, R. E.: Postnatal development of the orofacial musculature. In: *Patterns of Orofacial Growth and Development.* Report 6. Washington, D.C., Am. Speech and Hearing Assoc., 1971.

Moyers, R. E.: *A Handbook of Orthodontics.* 3rd ed. Chicago, Year Book Medical Publishers, 1972.

Moyers, R. E., and F. Muira: The use of serial cephalograms to study racial differences in development. I. and II. Trans. VIII Congress of Anthrop. and Ethnol. Sci., Tokyo, 284, 1968.

Moyers, R. E., J. Elgoyhen, M. Riolo, J. McNamara, and T. Kuroda: Experimental production of Class III in rhesus monkeys. Eur. Orthod. Soc. Trans., 46:61, 1970.

Mugnier, A., and M. Schouker-Jolly: Physio-pathologic des malocclusions dento-maxillaires moyens prophylactiques et therapeutiques precoces. Pedod. Fr., 5:101, 1973.

Murad, F., H. B. Brewer, Jr., and M. Vaughan: Effect of thyrocalcitonin on cAMP formation by rat kidney and bone. Proc. Natl. Acad. Sci. U.S.A., 65:446, 1970.

Muzj, E.: *Oro-facial Anthropometrics.* Hempstead, Index Publishers, Inc., 1970.

Nahoum, H. I., S. L. Horowitz, and E. A. Benedicto: Varieties of anterior open bite. Am. J. Orthod., 61:486, 1972.

Nanda, R. S.: Rates of growth of several facial components measured from serial cephalometric roentgenograms. Am. J. Orthod., 41:658, 1955.

Nanda, R. S., and R. C. Taneja: Growth of the face during the transitional period. Angle Orthod., 42:165, 1972.

Nemeth, R. B., and R. J. Isaacson: Vertical anterior relapse. Am. J. Orthod., 65:565, 1974.

Noback, C. R.: Developmental anatomy of the human osseous skeleton during embryonic fetal and circumnatal periods. Anat. Rec., 88:91, 1944.

Noback, C. R., and G. G. Robertson: Sequences of appearance of ossification centers in the human skeleton during the first five prenatal months. Am. J. Anat., 89:1, 1951.

Oberg, T.: Morphology, growth and matrix formation in the mandibular joint of the guinea pig. Trans. Roy. Schools of Dent., Stockholm & Umea, No. 10, 1964.

Odegaard, J.: Mandibular rotation studied with the aid of metal implants. Am. J. Orthod., 58:448, 1970.

Odegaard, J., and A. G. Brodie: On the growth of the human head from birth to the third month of life. Anat. Rec., 103:311, 1949.

Patten, B. M.: *Human Embryology.* Third edition. New York, McGraw-Hill, 1968.

Patten, B. M.: Embryology of the palate and maxillofacial region. In: *Cleft Lip and Palate,* Ed. by W. C. Grabb, S. W. Rosenstein, and K. R. Bzoch. Boston, Little, Brown and Company, 1971.

Persson, M.: Mandibular asymmetry of hereditary origin. Am. J. Orthod., 63:1, 1973a.

Persson, M.: Structure and growth of facial sutures. Odontol. Revy, 24:26, 1973b.

Perry, H. T.: The temporomandibular joint. Am. J. Orthod., 52:399, 1966.

Perry, H. T.: Relation of occlusion to temporomandibular joint dysfunction: The orthodontic viewpoint. J.A.D.A., 79:137, 1969.

Petit-Maire, N.: Morphogenese du crane de primates. L'Anthropologie, 75:85, 1971.

Petrovic, A.: Recherches sur les mechanismes histophysiologiques de la croissance osseuse cranio-faciale. Ann. Biol., 9:63, 1970.

Petrovic, A.: Mechanisms and regulation of condylar growth. Acta Morphol. Neerl. Scand., 10:25, 1972.

Petrovic, A., and J. P. Charlier: La synchondrose spheno-occipitale de jeune rat en culture d'organes: mise en evidence d'un potential de croissance independent. C. R. Acad. Sci. (D), (Paris) 265:1511, 1967.

Petrovic, A., and J. Stutzmann: Le muscle pterygoidien extarne et la croissance du condyle man-dibulaire. Recherches experimentales chez le jeune rat. Orthod. Fr., 43:271, 1972.

Petrovic, A., J. P. Charlier, and J. Herrman: Les mechanismes de croissance du crane. Recherches ser le cartilage de la cloison nasale et sur les sutures craniennes et faciales de jeunes rats en culture d'organes. Bull. Assoc. Anat. (Nancy), 143:1376, 1968.

Petrovic, A., C. Oudet, and N. Gasson: Effects des appareils de propulsion et de retropulsion man-dibulaire sur le nombre des sarcomeres en serie du muscle pterygoidien externe et sur la croissance du cartilage condylien de jeune rat. Orthod. Fr., 44:191, 1973a.

Petrovic, A., J. Stutzmann, and C. Oudet: Effects de l'hormone somatotrope sur la croissance du cartilage condylien mandibulaire et de la synchondrose spheno-occipitale de jeune rat, en culture organotypique. C. R. Acad. Sci. (D) (Paris) 276:3053, 1973b.

Pimenidis, M. Z., and A. A. Gianelly: The effect of early postnatal condylectomy on the growth of the mandible. Am. J. Orthod., 62:42, 1972.

Poswillo, D. E.: The late effects of mandibular condylectomy. Oral Surg., 33:500, 1972.

Poswillo, D. E.: Orofacial malformations. Proc. R. Soc. Med., 67:343, 1974.

Poulton, D. R.: Influence of extraoral traction. Am. J. Orthod., 53:8, 1967.

Powell, T. W., and A. G. Brodie: Laminagraphic study of the spheno-occipital synchondrosis. Anat. Rec., 147:15, 1963.

Prescott, G. H., D. F. Mitchell, and H. Fahmy: Procion dyes as matrix markers in growing bone and teeth. Am. J. Phys. Anthrop., 29:219, 1968.

Pritchard, J. J.: The control or trigger mechanism induced by mechanical forces which cause responses of mesenchymal cells in general and bone application and resorption in particular. Acta. Morphol. Neerl. Scand., 10:63, 1972.

Pritchard, J. J., J. H. Scott, and F. G. Girgis: The structure and development of cranial and facial sutures. J. Anat. (Lond.), 90:73, 1956.

Proffit, W. R., and J. L. Ackerman: Rating the characteristics of malocclusion: A systematic approach for planning treatment. Am. J. Orthod., 64:258, 1973.

Pruzansky, S.: Congenital anomalies of the face and associated structures. Springfield, Ill., Charles C Thomas, Publisher, 1961.

Pruzansky, S., and E. F. Lis: Cephalometric roentgenography of infants: Sedation, instrumentation, and research. Am. J. Orthod., 44:159, 1958.

Pruzansky, S., and J. B. Richmond: Growth of the mandible in infants with micrognathia. Am. J. Dis. Child., 88:29, 1954.

Ramfjord, S. P., and R. D. Enlow: Anterior displacement of the mandible in adult rhesus monkeys: Long-term observations. J. Prosthet. Dent., 26:517, 1971.

Rees, L. A.: The structure and function of the temporomandibular joint. Br. Dent. J., 96:125, 1954.

Reitan, K.: Tissue behavior during orthodontic tooth movement. Am. J. Orthod., 46:881, 1960.

Reitan, K.: Bone formation and resorption during reversed tooth movement. In: *Vistas in Orthodontics.* Ed. by B. T. Kraus and R. A. Riedel. Philadelphia, Lea & Febiger, 1962.

Reitan, K.: Biomechanical principles and reactions. Chapter 2 In: *Current Orthodontic Concepts and Techniques.* Ed. by T. M. Graber. Philadelphia, W. B. Saunders Company, 1969.

Richardson, A. S.: Dental development during the first two years of life. J. Can. Dent. Assoc., 33:418, 1967.

Ricketts, R. M.: Planning treatment on the basis of the facial pattern and an estimate of its growth. Angle Orthod., 27:14, 1957.

Ricketts, R. M.: Cephalometric synthesis. Am. J. Orthod., 46:647, 1960a.

Ricketts, R. M.: The influence of orthodontic treatment on facial growth and development. Angle Orthod., 30:103, 1960b.

Ricketts, R. M.: The evolution of diagnosis to computerized cephalometrics. Am. J. Orthod., 55:795, 1969.

Ricketts, R. M.: A principle of arcial growth of the mandible. Angle. Orthod., 42:368, 1972a.

Ricketts, R. M.: The value of cephalometrics and computerized technology. Angle Orthod., 42:368, 1972b.

Ricketts, R. M., R. W. Bench, J. J. Hilgers, and R. Schulhof: An overview of computerized cephalome-trics. Am. J. Orthod., 61:1, 1972.

Riedel, R.: The relation of maxillary structures to cranium in malocclusion and in normal occlusion. Angle. Orthod., 22:142, 1952.

Riedel, R.: An analysis of dentofacial relationships. Am. J. Orthod., 43:103, 1957.

Riedel, R.: A review of the retention problem. Angle Orthod., 30:179, 1960.

Riolo, M. L.: Growth and remodeling of the cranial floor: A multiple microfluoroscopic analysis with serial cephalometrics. M. S. Thesis, Georgetown University, Washington, D.C., 1970.

Riolo, M. L., and J. A. McNamara, Jr.: Cranial base growth in the rhesus monkey from infancy to adulthood. J. Dent. Res., 52:249, 1973.

Riolo, M. L., R. E. Moyers, J. A. McNamara, and W. S. Hunter: *An Atlas of Craniofacial Growth: Cephalometric Standards from the University School Growth Study, The University of Michigan.* Monograph 2. Craniofacial Growth Series. Center for Human Growth and Development, University of Michigan Series. Ann Arbor, 1974.

Robinson, I. B., and B. G. Sarnat: Growth pattern of the pig mandible. A serial roentgenographic study using metallic implants. Am. J. Anat., 96:37, 1955.

Roche, A. F.: Increase in cranial thickness during growth. Hum. Biol., 25:81, 1953.

Roche, A. F., K. Manuel, and F. S. Seward: Unusual patterns of growth in the frontal and parietal bones. Anat. Rec., 152:459, 1965.

Roger-Thooris, M. O.: Relations entre: croissance et variabilite sur le profil cranien. Thèse, Univ. de Nancy, 1973.

Rönning, O.: Observations on the intracerebral transplantation of the mandibular condyle. Acta Odont. Scand., 24:443, 1966.

Rönning, O., and K. Koski: The effect of the articular disc on the growth of condylar cartilage transplants. Eur. Orthod. Soc. Trans., 1969:99, 1970.

Rönning, O., and K. Koski: The effect of periostomy on the growth of the condylar process in the rat. Proc. Finn. Dent. Soc., 70:28, 1974.

Rönning, O., K. Paunio, and K. Koski: Observations on the histology, histochemistry and biochemistry of growth cartilages in young rats. Suom. Hammaslaak. Toim., 63:187, 1967.

Rosenstein, S. W.: Pathological and congenital disturbances: The orthodontic viewpoint. J.A.D.A., 82:763, 1971.

Ruff, R.: Orthodontic treatment in the mixed dentition. Am. J. Orthod., 52:502, 1970a.

Ruff, R.: Orthodontic treatment in the permanent dentition. Am. J. Orthod., 58:597, 1970b.

Salentijn, L., and M. L. Moss: Morphological attributes of the logarithmic growth of the human face: gnomic growth. Acta Anat., 78:185, 1971.

Salzmann, J. A.: The research workshop on cephalometrics. Am. J. Orthod., 46:834, 1960a.

Salzmann, J. A.: *Roentgenographic Cephalometrics. Proceedings of the second research workshop.* Philadelphia, J. B. Lippincott Company, 1960b.

Salzmann, J. A.: *Practice of Orthodontics.* Philadelphia, J. B. Lippincott Company, 1966.

Sarnat, B. G.: Facial and neurocranial growth after removal of the mandibular condyle in the Macaca rhesus monkey. Am. J. Surg., 94:19, 1957.

Sarnat, B. G.: Postnatal growth of the upper face: Some experimental considerations. Angle Orthod., 33:139, 1963.

Sarnat, B. G.: *The Temporomandibular Joint.* 2nd Ed. Springfield, Ill., Charles C Thomas, Publisher, 1964.

Sarnat, B. G.: The face and jaws after surgical experimentation with the septovomeral region in growing and adult rabbits. Acta Otolaryngol., Suppl., 268, 1970.

Sarnat, B. G.: Clinical and experimental considerations in facial bone biology: Growth, remodeling, and repair. J.A.D.A., 82:876, 1971a.

Sarnat, B. G.: Surgical experimentation and gross postnatal growth of the face and jaws. J. Dent. Res., 50:1462, 1971b.

Sarnat, B. G., and M. B. Engel: A serial study of mandibular growth after the removal of the condyle of the Rhesus monkey. Plast. Reconstr. Surg., 7:364, 1951.

Sarnat, B. G., and H. Muchnic: Facial skeletal changes after mandibular condylectomy in the adult monkey. J. Anat., 108:338, 1971a.

Sarnat, B. G., and H. Muchnic: Facial skeletal changes after mandibular condylectomy in growing and adult monkeys. Am. J. Orthod., 60:33, 1971b.

Sarnat, B. G., and P. D. Shanedling: Postnatal growth of the orbit and upper face in rabbits. Arch. Ophthalmol., 73:829, 1965.

Sarnat, B. G., and M. R. Wexler: Growth of the face and jaws after resection of the septal cartilage in the rabbit. Am. J. Anat., 118:755, 1966.

Sassouni, V.: Roentgenographic cephalometric analysis of cephalo-facio-dental relationships. Am. J. Orthod., 41:734, 1955.

Sassouni, V.: Diagnosis and treatment planning via roentgenographic cephalometry. Am. J. Orthod., 44:433, 1958.

Sassouni, V.: *The Face in Five Dimensions.* Philadelphia, Philadelphia Center for Research in Child Growth, 1960.

Sassouni, V.: *Heredity and Growth of the Human Face.* Pittsburgh, University of Pittsburgh, 1965.

Sassouni, V., and E. J. Forrest: *Orthodontics in Dental Practice.* St. Louis, Mo., C. V. Mosby Company, 1971.

Savara, B. S., and I. J. Singh: Norms of size and annual increments of seven anatomical measures of maxillae in boys from three to sixteen years of age. Angle Orthod., 38:104, 1968.

Sawin, P. B., M. Ranlett, and D. D. Crary: Morphogenetic studies of the rabbit. XXV. The spheno-occipital synchondrosis of the dachs (chondrodystrophy) rabbit. Am. J. Anat., 105:257, 1959.

Scammon, R. E., J. A. Harris, C. M. Jackson, and D. G. Patterson: *The Measurement of Man*. Minneapolis, University of Minnesota Press, 1930.

Scheiman-Tagger, E., and A. G. Brodie: Lead acetate as a marker of growing calcified tissues. Anat. Rec., 150:435, 1964.

Schouker-Jolly, M.: Utilisation d'appareillages extra-oraux recents dans le prognathisme mandibulaire, associe a une hypoplasie maxillaire. Méd. Infant., 6:479, 1972.

Schudy, F. F.: Cant of the occlusal plane and axial inclinations of teeth. Angle Orthod., 33:69, 1963.

Schudy, F. F.: Vertical growth vs. anteroposterior growth as related to function and treatment. Angle Orthod., 34:75, 1964.

Schudy, F. F.: The rotation of the mandible resulting from growth: Its implications in orthodontic treatment. Angle Orthod., 35:36, 1965.

Schudy, F. F.: The control of vertical overbite in clinical orthodontics. Angle Orthod., 38:19, 1968.

Scott, J. H.: The cartilage of the nasal septum. Br. Dent. J., 95:37, 1953.

Scott, J. H.: Growth and function of the muscles of mastication in relation to the development of the facial skeleton and of the dentition. Am. J. Orthod., 40:429, 1954a.

Scott, J. H.: The growth of the human face. Proc. R. Soc. Med., 47:91, 1954b.

Scott, J. H.: Craniofacial regions: Contribution to the study of facial growth. Dent. Pract., 5:208, 1955.

Scott, J. H.: Growth at facial sutures. Am. J. Orthod., 42:381, 1956.

Scott, J. H.: The growth in width of the facial skeleton. Am. J. Orthod., 43:366, 1957.

Scott, J. H.: The cranial base. Am. J. Phys. Anthrop., 16:319, 1958a.

Scott, J. H.: The analysis of facial growth. Part I. The anteroposterior and vertical dimensions. Am. J. Orthod., 44:507, 1958b.

Scott, J. H.: The analysis of facial growth. Part II. The horizontal and vertical dimensions. Am. J. Orthod., 44:585, 1958c.

Scott, J. H.: The face in foetal life. Eur. Orthod. Soc. Trans., 37:168, 1961.

Scott, J. H., and A. D. Dixon: *Anatomy for Students of Dentistry*. 3rd Ed. Edinburgh and London, Churchill & Livingstone, 1972.

Sekiguchi, T., and B. Savara: Variability of cephalometric landmarks used for face growth studies. Am. J. Orthod., 61:603, 1972.

Servoss, J. M.: An in vivo and in vitro autoradiographic investigation of growth in synchondrosal cartilages. Am. J. Anat., 136:479, 1973.

Sherman, M. S.: The nerves of bone. J. Bone Joint Surg., 45:522, 1963.

Sicher, H.: The growth of the mandible. Am. J. Orthod., 33:30, 1947.

Sicher, H.: Some aspects of the anatomy and pathology of the temporomandibular articulation. Bur., 48:14, 1948.

Sicher, H., and E. L. DuBrul: *Oral Anatomy*. 5th Ed. St. Louis, Mo., C. V. Mosby Company, 1970.

Sicher, H., and J. P. Weinmann: Bone growth and physiologic tooth movement. Am. J. Orthod. Oral Surg., 30:109, 1944.

Siegel, M. I.: The facial and dental consequences of nasal septum resections in baboons. Med. Primat., 1972:204, 1972.

Silbermann, M., and J. Frommer: Further evidence for the vitality of chondrocytes in the mandibular condyle as revealed by $^{35}$S-sulfate autoradiography. Anat. Rec., 174:503, 1972a.

Silbermann, M., and J. Frommer: Vitality of chondrocytes in the mandibular condyle as revealed by collagen formation. An autoradiographic study with $^{3}$H-proline. Am. J. Anat., 135:359, 1972b.

Silbermann, M., and J. Frommer: The nature of endochondral ossification in the mandibular condyle of the mouse. Anat. Rec., 172:659, 1972c.

Sillman, J. H.: Relationship of maxillary and mandibular gum pads in the newborn infant. Am. J. Orthod., 24:409, 1938.

Simon, M., and M. L. Moss: Dynamics of sigmoid notch configuration and its relationship to total normal morphology. Acta Anat., 85:133, 1973.

Sognnaes, R. F.: Calcification process. J.A.D.A., 62:516, 1961.

Solow, B.: The pattern of craniofacial associations. A morphological and methodological correlation and factor analysis study on young male adults. Acta Odontol. Scand., 24 (Suppl. 46): 1966.

Solow, B.: Automatic processing of growth data. Angle Orthod., 39:186, 1969.

Solow, B.: Factor analysis of cranio-facial variables. In: *Craniofacial Growth in Man*. Ed. by R. E. Moyers and W. M. Krogman. Oxford, Pergamon Press, 1971.

Speidel, T. M., R. J. Isaacson, and F. W. Worms: Tongue-thrust therapy and anterior dental open bite. Am. J. Orthod., 62:287, 1972.

Stanley, R. B., and R. A. Latham: The regression pattern of Meckel's cartilage in normal mandibular development. IADR Abstr., 1973:216, 1973.

Steiner, C. C.: Cephalometrics for you and me. Am. J. Orthod., 39:729, 1953.

Steiner, C. C.: Cephalometrics in clinical practice. Angle Orthod., 29:8, 1959.

Stenstrom, S. J., and B. L. Thilander: Effects of nasal septal cartilage resections on young guinea-pigs. Plast. Reconstr. Surg., 45:160, 1970.

Steuer, L.: The cranial base for superimposition of lateral cephalometric radiographs. Am. J. Orthod., 16:493, 1972.

Stockli, P. W., and H. G. Willert: Tissue reactions in the temporomandibular joint resulting from anterior displacement of the mandible in the monkey. Am. J. Orthod., 60:142, 1971.

Storey, A. T.: Physiology of a changing vertical dimension. J. Prosthet. Dent., 1:912, 1962.

Storey, E.: Growth and remodeling of bone and bones. Am. J. Orthod., 62:142, 1972.

Stramrud, L.: External and internal cranial base: A cross-sectional study of growth and of association in form. Acta Odont. Scand., 17:239, 1959.

Strong, R. M.: The order, time and rate of ossification of the albino rat skeleton. Am. J. Anat., 36:313, 1925.

Stutzmann, J., and A. Petrovic: Particularites de croissance de la suture palatine sagittale de jeune rat. Bull. Assoc. Anat. (Nancy), 148:552, 1970.

Subtelny, J. D.: Longitudinal study of soft tissue facial structures and their profile characteristics, defined in relation to underlying skeletal structures. Am. J. Orthod., 45:481, 1959.

Subtelny, J. D.: The soft tissue profile, growth and treatment changes. Angle Orthod., 31:105, 1961.

Subtelny, J. D., and M. Sakuda: Muscle function, oral malformation and growth changes. Am. J. Orthod., 52:495, 1966.

Subtelny, J. D., and J. Subtelny: Oral habits—Studies in form, function, and therapy. Angle Orthod., 43:347, 1973.

Swindler, D. R., J. E., Sirianni, and L. H. Tarrant: A longitudinal study of cephalofacial growth in *Papio cynocephalus* and *Macaca nemestrina* from three months to three years. Symp. IVth Int. Cong. Primat. Vol. 3. Craniofacial Biology of Primates. Basel, S. Karger, 1973.

Symons, N. B. B.: Studies on the growth and form of the mandible. Dent. Rec., 71:41, 1951.

Symons, N. B. B.: The development of the human mandibular joint. J. Anat., 86:326, 1952.

Symons, N. B. B.: A histochemical study of the secondary cartilage of the mandibular condyle in the rat. Arch. Oral Biol., 10:579, 1965.

Szego, C. M.: The role of cyclic AMP in lysosome mobilization and their nucleotropic translocation in steroid hormonal target cells. Adv. Cyclic Nucleotide Res., 1:541, 1972.

Takagi, Y.: Human postnatal growth of vomer in relation to base of cranium. Ann. Otol., 73:238, 1964.

Tanner, J. M.: *Growth at Adolescence*. 2nd Ed. Oxford, Blackwell, 1962.

Ten Cate, A. R.: Development of the periodontium. In: *Biology of the Periodontium*. Ed. by A. H. Melcher and W. H. Bowen. New York, Academic Press, 1969.

Ten Cate, A. R., C. Mills, and G. Solomon: The development of the periodontium. A transplantation and autoradiographic study. Anat. Rec., 170:365, 1971.

Terk, B.: Modifications apportées chez le rat, à l'évolution des dents, par l'hydrocéphalie expérimentale étude en orientation vestibulaire. Thèse, Acad. Paris, Univ. Paris VI., 1973.

Tessier, P. J.: Ostéotomies totales de la face. Syndrome de Crouzon. Syndrome d'Apert. Oxycéphalies. Scaphocéphalies. Turriecephalies. Ann. Chir. Plast., 12:273, 1967.

Tessier, P., J. Delaire, J. Billet, and H. Landais: Considerations sur le development de l'orbite; ses incidences sur la croissance faciale. Rev. Stomatol., 63:1–2, 27–39, 1964.

Theunissen, J. J. W.: Het fibreuze periosteum. Thesis, Katholieke Univ. te Nymegen, 1973.

Thilander, B.: Innervation of the temporomandibular joint capsule in man. Trans. Roy. Schools Dent. Stockholm, 7:9, 1961.

Thilander, B.: The structure of the collagen of the temporomandibular joint disc in man. Acta. Odont. Scand., 22:135, 1964.

Thompson, D. W.: *On Growth and Form*. 2nd Ed. London, Cambridge University Press, 1952.

Thurow, R. C.: Cephalometric methods in research and private practice. Angle Orthod., 21:104, 1951.

Thurow, R. C.: *Atlas of Orthodontic Principles*. St. Louis, Mo., C. V. Mosby Company, 1970.

Todd, T. W.: Prognathism; a study in development of the face. J.A.D.A., 19:2172, 1932.

Tonna, E. A., and E. P. Cronkite: Cellular response to fracture studied with tritiated thymidine. J. Bone Joint Surg., 43:352, 1961.

Tonna, E. A., and L. Pentel: Chondrogenic cell formation via osteogenic cell progeny transformation. Lab. Invest., 27:418, 1972.

Trelstad, R. L.: The developmental biology of vertebrate collagens. J. Histochem. Cytochem., 21:521, 1973.

Tracy, W. E., B. S. Savara, and J. W. A. Brant: Relation of height, width and depth of the mandible. Angle Orthod., 35:269, 1965.

Turpin, D. L.: Growth and remodeling of the mandible in the *Macaca mulatta* monkey. Am. J. Orthod., 54:251, 1968.

Tweed, C. H.: The frankfort-mandibular plane angle in orthodontic diagnosis, classification, treatment planning, and prognosis. Am. J. Orthod. Oral Surg., 32:175, 1946.

Tweed, C. H.: The frankfort-mandibular incisor angle (FMIA) in orthodontic diagnosis, treatment planning and prognosis. Angle Orthod., 24:121, 1954.

Urist, M. R., and B. S. Strates: Bone morphogenetic protein. J. Dent. Res., 50:1392, 1971.

Urist, M. R., B. Silverman, K. Buring, F. Dubuc, and J. Rosenberg: The bone induction principle. Clin. Orthop., 53:243, 1967.

Utley, R. K.: The activity of alveolar bone incident to orthodontic tooth movement as studied by oxytetracyline-induced fluorescence. Am. J. Orthod., 54:167, 1968.

Vaes, G.: On the mechanism of bone resorption. The action of parathyroid hormone on the excretion

and synthesis of lysosomal enzymes and on the extracellular release of acid by bone cells. J. Cell Biol., 39:676, 1968.

Vaes, G.: Inhibitory actions of calcitonin on resorbing bone explants in culture and on their release of lysosomal hydrolases. J. Dent. Res., 51:362, 1972.

van der Linden, F. P. G. M.: Interrelated factors in the morphogenesis of teeth, the development of the dentition and craniofacial growth. Schweiz. Monatsschr. Zahnheilkd., 80:518, 1970.

van der Linden, F. P. G. M.: A study of roentgenocephalometric bony landmarks. Am. J. Orthod., 59:111, 1971.

van der Linden, F. P. G. M., and D. H. Enlow: A study of the anterior cranial base. Angle Orthod., 41:119, 1971.

Vermeij-Keers, C.: Transformation in the facial region of the human embryo. Advances in Anatomy, Embryology, and Cell Biology, 46:5, 1972.

Verwoerd-Verhoef, H. L.: Schedelgroei onder invloed van aangezichtsspleten. Academisch proefschrift, Universiteit van Amsterdam, 1974.

Vidic, B.: The morphogenesis of the lateral nasal wall in the early prenatal life of man. Am. J. Anat., 130:121, 1971.

Vilmann, H.: The growth of the cranial base in the albino rat revealed by roentgencephalometry. J. Zool. (Lond.), 159:283, 1969.

Vilmann, H.: The growth of the cranial base in the Wistar albino rat studied by vital staining with alizarin red S. Acta Odont. Scand., 29 (Suppl. 59). 1971.

Vilmann, H.: Osteogenesis in the basioccipital bone of the Wistar albino rat. Scand. J. Dent. Res., 80:410, 1972.

Voorhies, J. W., and J. W. Adams: Polygonic interpretations of cephalometric findings. Angle Orthod., 21:194, 1951.

Walker, G.: The composite vectorgram: A key to the analysis and synthesis of craniofacial growth. J. Dent. Res., 50:1508, 1971.

Walker, G.: A new approach to the analysis of craniofacial morphology and growth. Am. J. Orthod., 61:221, 1972.

Walker, G., and C. J. Kowalski: A two-dimensional coordinate model for the quantification, description, analysis, prediction and simulation of craniofacial growth. Growth, 35:119, 1971.

Walker, G., and C. J. Kowalski: On the growth of the mandible. Am. J. Phys. Anthrop., 36:111, 1972a.

Walker, G., and C. J. Kowalski: Use of angular measurements in cephalometric analysis. J. Dent. Res., 51:1015, 1972b.

Washburn, S. L.: The effect of facial paralysis on the growth of the skull of rat and rabbit. Anat. Rec., 94:163, 1946a.

Washburn, S. L.: The effect of removal of the zygomatic arch in the rat. J. Mammal., 27:169, 1946b.

Washburn, S. L.: The relation of the temporal muscle to the form of the skull. Anat. Rec., 99:239, 1947.

Watson, E. H., and G. H. Lowery: *Growth and Development of Children.* Chicago, Year Book Medical Publishers, 1967.

Weidenreich, F.: The special form of the human skull in adaptation to the upright gait. Z. Morphol. Anthropol., 24:157, 1924.

Weidenreich, F.: The brain and its role in the phylogenetic transformation of the human skull. Trans. Am. Phil. Soc., 31:321, 1941.

Weidenreich, F.: *Apes, Giants, and Man.* Chicago, University of Chicago Press, 1946a.

Weidenreich, F.: Generic, specific, and subspecific characters in human evolution. Am. J. Phys. Anthrop., 31:413, 1946b.

Weinmann, J. P.: Adaptation of the periodontal membrane to physiologic and pathologic changes. Oral Surg., 8:977, 1955.

Weinmann, J. P., and H. Sicher: *Bone and Bones.* 2nd Ed. St. Louis, Mo., C. V. Mosby Company, 1955.

Weinstein, S., D. C. Haack, L. Y. Morris, B. B. Snyder, and H. E. Attaway: On an equilibrium theory of tooth position. Angle Orthod., 33:1, 1963.

West, E. E.: Facial patterns in malocclusion. J. Dent. Res., 31:464, 1952.

West, E. E.: Analysis of early Class I, Division 1 treatment. Am. J. Orthod., 43:769, 1955.

Wexler, M. R., and B. G. Sarnat: Rabbit snout growth after dislocation of nasal septum. Arch. Otolaryngol., 81:68, 1965.

Woo, J. K.: On the asymmetry of the human skull. Biometrika, 22:324, 1931.

Woo, J. K.: Ossification and growth of the human maxilla, premaxilla and palate bones. Anat. Rec., 105:737, 1949.

Wood, N. D., L. E. Wragg, O. G. Stuteville, and R. G. Oglesby: Osteogenesis of the human upper jaw. Proof of the nonexistence of a separate premaxillary centre. Arch. Oral Biol., 14:1331, 1969.

Woodside, D. G.: Distance, velocity and relative growth rate standards for mandibular growth for Canadian males and females age three to twenty years. Toronto, Canada, American Board of Orthodontics Thesis, 1969.

Worms, F. W., L. H. Meskin, and R. J. Isaacson: Open-bite. Am. J. Orthod., 59:589, 1971.

Wright, D. M., and B. C. Moffett: The postnatal development of the human temporomandibular joint. Am. J. Anat., 141:235, 1974.

Wright, H. V., I. Kjaer, and C. W. Asling: Roentgen cephalometric studies on skull development in rats. II. Normal and hypophysectomized males; sex differences. Am. J. Phys. Anthrop., 25:103, 1966.

Wylie, W. L.: Assessment of antero-posterior dysplasia. Angle Orthod., 17:97, 1947.

Wylie, W. L., and E. L. Johnson: Rapid evaluation of facial dysplasia in the vertical plane. Angle Orthod., 22:164, 1952.

Young, R. W.: The influence of cranial contents on postnatal growth of the skull in the rat. Am. J. Anat., 105:383, 1959.

Young, W. F.: The influence of the growth of the teeth and nasal septum on growth of the face. In: *Report of 36th Congress.* Ed. by G. E. M. Hallett. Eur. Orthod. Soc., London, 1959, p. 385.

Youssef, E. H.: The development of the skull in a 34 mm human embryo. Acta Anat., 57:72, 1964.

Youssef, E. H.: The chondrocranium in the albino rat. Acta Anat., 64:586, 1966.

Youssef, E. H.: Development of the membrane bones and ossification of the chondrocranium in the albino rat. Acta Anat., 72:603, 1969.

Yuodelis, R. A.: The morphogenesis of the human temporomandibular joint and its associated structures. J. Dent. Res., 45:182, 1966.

Zengo, A. N., C. A. L. Bassett, R. J. Pawluk, and G. Prountzos: In vivo bioelectric potentials in the dentoalveolar complex. Am. J. Orthod., 66:130, 1974.

Zuckerman, S.: Age changes in the basicranial axis of the human skull. Am. J. Phys. Anthrop., 13:521, 1955.

Zwarych, P. D., and M. B. Quigley: The intermediate plexus of the periodontal ligament: history and further investigations. J. Dent. Res., 44:383, 1965.

# Index